Advances in Radiation Protection and Dosimetry in Medicine

ETTORE MAJORANA INTERNATIONAL SCIENCE SERIES
Series Editor:
Antonino Zichichi
European Physical Society
Geneva, Switzerland

Advances in Radiation Protection and Dosimetry in Medicine

Edited by

Ralph H. Thomas

Lawrence Berkeley Laboratory
University of California at Berkeley
Berkeley, California

and

Victor Perez-Mendez

University of California at San Francisco
San Francisco, California

Springer Science+Business Media, LLC

Library of Congress Cataloging in Publication Data

International School of Radiation Damage and Protection, 3d, Erice, Italy, 1979.
  Advances in radiation protection and dosimetry.

  (Ettore Majorana international science series: Life sciences; v. 2)
Proceedings of the third course of the International School of Radiation Damage and
Protection, held Sept. 16–29, 1979, in Erice, Italy.
  Includes index.
  1. Radiology, Medical—Congresses. 2. Radiation dosimetry—Congresses. 3. Particle
beams—Diagnostic use—Congresses. 4. Particle beams—Therapeutic use—Congresses.
5. Radiation—Physiological effect—Congresses. I. Thomas, Ralph H. II. Perez-
Mendez, Victor. III. Title. IV. Series: [DNLM: 1. Radiation protection—Congresses. 2. Radiation dosage—Congresses. 3. Radiometry—Congresses. WL ET712M
v. 2/WN650 I6555 1979a]
R895.A2I54 1979   616.07'572   80-20218

ISBN 978-1-4757-1717-4   ISBN 978-1-4757-1715-0 (eBook)
DOI 10.1007/978-1-4757-1715-0

Proceedings of the Symposium on Advances in Radiation Protection Dosimetry in
Medicine, held in Italy, September 16–25, 1979.

ACKNOWLEDGMENTS

The Editors of these proceedings (and co-directors of the course) wish to acknowledge the assistance they have received from Drs. G. T. Chen, S. B. Curtis and W. Schimmerling in organizing the course held at Erice in September 1979. They played an important role in arrangements made from Berkeley.

Dr. Rindi, Director of the International School of Radiation Damage and Protection, was invaluable in making early arrangements within Western Europe, while Dr. A. Gabriele, Scientific Secretary of the "Ettore Majorana" Centre, made the many detailed arrangements in Erice that were necessary for a successful course.

We are indebted also to Italian Ministry of Public Education, The Italian Ministry of Scientific and Technological Research, The Italian National Research Council, The Sicilian Regional Government, and the Italian Radiation Protection Association for sponsoring this course.

Great thanks are also due to Professor A. Zichichi, who had the great energy and imagination to create the "Ettore Majorana" Centre for Scientific Culture in the beautiful town of Erice-- whose citizens so warmly welcome participants.

Finally no course would be possible without participants-- students and lecturers. Thank you for having attended!

Ralph H. Thomas
Victor Perez-Mendez

CONTENTS

CONTENTS

OPENING OF THE COURSE

Alessandro Rindi

Director of the School INFN-LNF
Frascati (Italy)

When Ralph Thomas and I started this School with the first
course in 1975, I was a little stirred and shy because it seemed
to me a rather large responsibility to introduce Health Physics
to the "E. Majorana" Centre. This authoritative and very influen-
tial centre welcomes eminent scientific authorities - several
Nobel laureates just a few days ago sat in these very chairs - to
discuss problems and theories at the frontier of scientific
research. Health Physics has often been considered the Cinderella
of science and there has been a tendency to see health physicists
as mere "surveyors" who limit themselves to practical applications
of knowledge and methods developed in other branches of science.

On the contrary, I firmly believe that Health Physics is a
field of theoretical and applied research in its own right that,
like all branches of science, takes advantage of the results in
other fields and, at the same time, provides original results that
are used by other scientific disciplines.

The first course, directed by Ralph Thomas and me, both at
that time from Lawrence Berkeley Laboratory (USA), was held from
1 to 10 October 1975 and was devoted to "High Energy Radiation
Dosimetry and Protection". Thirty-eight scientist from twelve
countries participated to the course. The lectures and seminars
were presented by J. V. Bailey from NASA (USA), S. B. Curtis from
LBL (USA), E. Freytag from DESY (Germany), P. J. Gollon from FNAL
(USA), M. Ladu from the University of Cagliari (Italy), W. R.
Nelson from SLAC (USA), M. Pelliccioni from LNF (Italy),
V. Perez-Mendez from LBL (USA), S. Pszona from B. Jadrowych
Instîtute (Poland), H. H. Rossi from Columbia University (USA),
J. T. Routti from University of Helsinki (Finland), and G. R.

Stevenson from CERN (Switzerland). The lectures covered the following fields of interest in dosimetry and protection around
accelerators and in space missions: the physical specification
of radiation fields and their measurement; the precise definition
of dose equivalent and its derivation from physical measurements;
the design of instrumentation to implement adequate measurement;
an understanding of the interaction of the various components of
a radiation field with matter and the related calculation and
design of shielding for radiation in the human body; the evaluation of risk from radiation of high LET such as those encountered
around accelerators and in space. The proceedings of the course
were published in IEEE Transactions on Nuclear Science (Vol. NS-23
No. 4, 1976).

After the first course, a large number of possible topics
for other courses were proposed by several scientists interested
in Health Physics.

The second course was held from October 26 to November 3,
1978. It was directed by W. R. Nelson of SLAC (USA) and devoted
to "The Use of Computers in Health Physics". Forty-two people
from fourteen countries participated. Twenty-seven lectures and
four seminars were presented by T. W. Armstrong from Science
Application Inc. (USA), H. Dinter from DESY (West Germany),
W. W. Engle from ORNL (USA), T. A. Gabriel from ORNL (USA),
T. Nakamura from the University of Tokyo (Japan), K. O'Brien from
the Health and Safety Laboratory of the Dept. of Energy (USA),
C. Ponti from Euratom (Italy), J. Ranft from K. Marx University
(East Germany), J. T. Routti from University of Helsinki (Finland),
G. R. Stevenson from CERN (Switzerland) and A. Van Ginneken from
FNAL (USA). The lectures covered the following fields: review
of the basic mathematical techniques used in radiation transport
like analytic, Monte-Carlo, discrete ordinates etc; presentation
of the main codes from electromagnetic shower and hadronic cascade calculations; basic techniques and codes used in activation
spectrum analysis; use of these techniques for solving problems
in radiation dosimetry, shielding, spectrum analysis, radiation
therapy, detector design, cosmic ray physics etc. The proceedings of the course have been published by Plenum Publishing Corp.
(E. Majorana International Science Series, Vol. 3, New York,1979).

The scientific success of these two courses assured Health Physics
a dignified place in the "E. Majorana" cultural "family" and we
were encouraged to go on.

So, I am very glad to open this course of the School.

The "new" particles, at present widely used in medicine, are
not new to the nuclear physicist. Indeed, electron and proton
beams of energy and intensity adequate for application to

radiotherapy have been available for more than thirty years.

As early as 1946, R. R. Wilson suggested (<u>Radiology</u>, 47: 487) that the energy-loss mechanisms of protons made them capable of delivering (distributions of) absorbed-dose in tissue which are far superior to those produced by ^{60}Co photons. Considerable experience in the use of protons and helium ions in radiotherapy has been obtained at some cyclotron laboratories, e.g. Berkeley (USA), Dubna (USSR), Harvard (USA) and Uppsala (Sweden).

The clinical use of protons has developed only slowly. However, since the 70's interest has increased and at present the University of Uppsala, Harvard University, the Fermilab and other institutes are using protons in the energy range up to 100-200 MeV in the treatment of some cancers.

Electron accelerators were used in the fifties primarily as sources of energetic bremsstrahlung but there has been increasing interest in the direct use of electrons to irradiate tumors. The possibility of obtaining very intense and short pulses may prove to be important.

The first experience using neutrons in the thirties and forties was not encouraging, but evidence is now available of their therapeutic value from studies made in Holland and in England. Neutron factories for medical use are under study in the USA.

Following the suggestion by P. H. Fowler and D. H. Perkins in 1961 (<u>Nature</u> 189:524), the application of negative pions in radiotherapy has been investigated experimentally and clinical trials are now underway at Los Alamos (USA), Villigen (Switzerland) and Vancouver (Canada).

In 1971 radiological studies were initiated at the Lawrence Berkeley Laboratory (USA) using heavy-ion beams and since that time measurements have been made with several different species of ions up to argon. In mid-1977, clinical trials were initiated using carbon ions.

Trials for using "new" radiations to replace or to complement X and gamma rays in clinical diagnosis are underway.

The application of these "new" particles to medicine gives a stimulus to the study of beam design and dosimetry. In addition, the use of computers in patient treatment-planning and dosimetry becomes more and more indispensable. Problems new to the health physicist in the design of shielding and beam collimators and in the evaluation of the risk from exposure to high LET radiation of increasing numbers of people will have to be faced.

A course on these topics is, I think, very timely.

I then make haste to open the third course of our School on "Advances in Radiation Protection and Dosimetry in Medicine".

On behalf also of Professor A. Zichichi, Director of the Centre, I welcome all of you to Erice and thank you for your participation.

THE NEW PARTICLES AND THEIR APPLICATION IN MEDICINE

Stanley B. Curtis

Biology and Medicine Division
Lawrence Berkeley Laboratory
Berkeley, CA 94720

INTRODUCTION

In recent years there has been increasing interest in the use of "new" particle beams for various medical applications. The particles being used, fast neutrons, protons, helium ions, negative pions, and heavy ions, have in fact been known and studied by nuclear and high-energy physicists for over forty years. They are new only in the sense that, until recently, the medical community has had little experience with them in the clinic. One of the particles, the neutron, is not really new even to radiotherapy. In the late thirties Stone et al. treated 226 patients using cyclotron-produced neutrons (1940), but they discontinued therapy five years later because of unexpected severe late reactions (Stone, 1948). It was not until the sixties that it became generally recognized that neutron irradiation is relatively more effective compared to x-rays at the lower doses per fraction used in therapy than at the higher single doses commonly used in animal and cell experiments. This difference in effectiveness becomes even more important at the lower doses encountered in radiation protection, and it is interesting that it played an inhibiting role in the early days of particle radiotherapy (Brennan and Phillips, 1971). This paper will review the applications of these "new" particles to clinical medicine, primarily in the areas of tumor radiotherapy and diagnostic radiography.

APPLICATIONS IN TUMOR RADIOTHERAPY

Physical Dose Distributions of Charged Particle Beams

A basic problem facing the radiotherapist is how to deliver a
tumoricidal dose while limiting the damage to normal tissues so
that the patient can recover and continue to live in relative
comfort. Several ways are being employed to try to accomplish
this. Among these are multiport and rotational therapy, combined
chemo- and radiotherapy, and the simple and standard procedure of
fractionating the total dose into small daily doses over perhaps
a five or six week period. The latter procedure is advantageous
presumably because the normal tissue can recover between doses to
a greater extent than the tumor tissue, thus allowing a larger
overall dose to be given to the tumor before net normal tissue
damage becomes intolerable.

The ideal beam for radiotherapy would deposit the entire dose
within the tumor volume and none to the normal tissue. No beam
or particle type has been found that will do this, but charged
particle beams approach this ideal much more closely than the
standard beams used in the clinic today (e.g., ^{60}Co gamma rays,
ortho- and megavoltage x rays, and megavoltage electrons). The
reason for this is a direct result of the difference in the
physical processes by which heavy charged particles slow down in
matter as compared with those processes which slow down electrons
and cause the absorption of electromagnetic radiation. For all
charged particles heavier than electrons, the dominating energy-
loss process is the ionization (and excitation) of the atoms of
the medium into which the charged particles penetrate. The rate
of energy loss by this process increases as the particles slow
down. This causes an increase in energy deposition, hence
absorbed dose, near the end of the particle's range. Examples of
the dose vs. depth characteristics of helium, carbon, neon and
argon ion beams are shown in Fig. 1 for two different ranges
(Lyman and Howard, 1977). Note the sharp peak at a depth in
water corresponding to the range of the particle. The small
residual dose beyond the peaks of the carbon, neon, and argon
beams is from the secondary fragments created by a small fraction
of the primary particles undergoing nuclear interactions and
fragmenting into lighter and thus more penetrating ions. Note
also the flatness of dose in the so-called plateau or shallow
region of the curve near the surface.

Because the peaks of these beams are so sharp and narrow,
they are not useful for treating the extended and often irregular
tumors encountered in the clinic. Thus, these unmodified beams
are generally modified for clinical application by the introduction of a variable thickness absorber sometimes called a
"ridge" filter to spread the peak region. These absorbers move

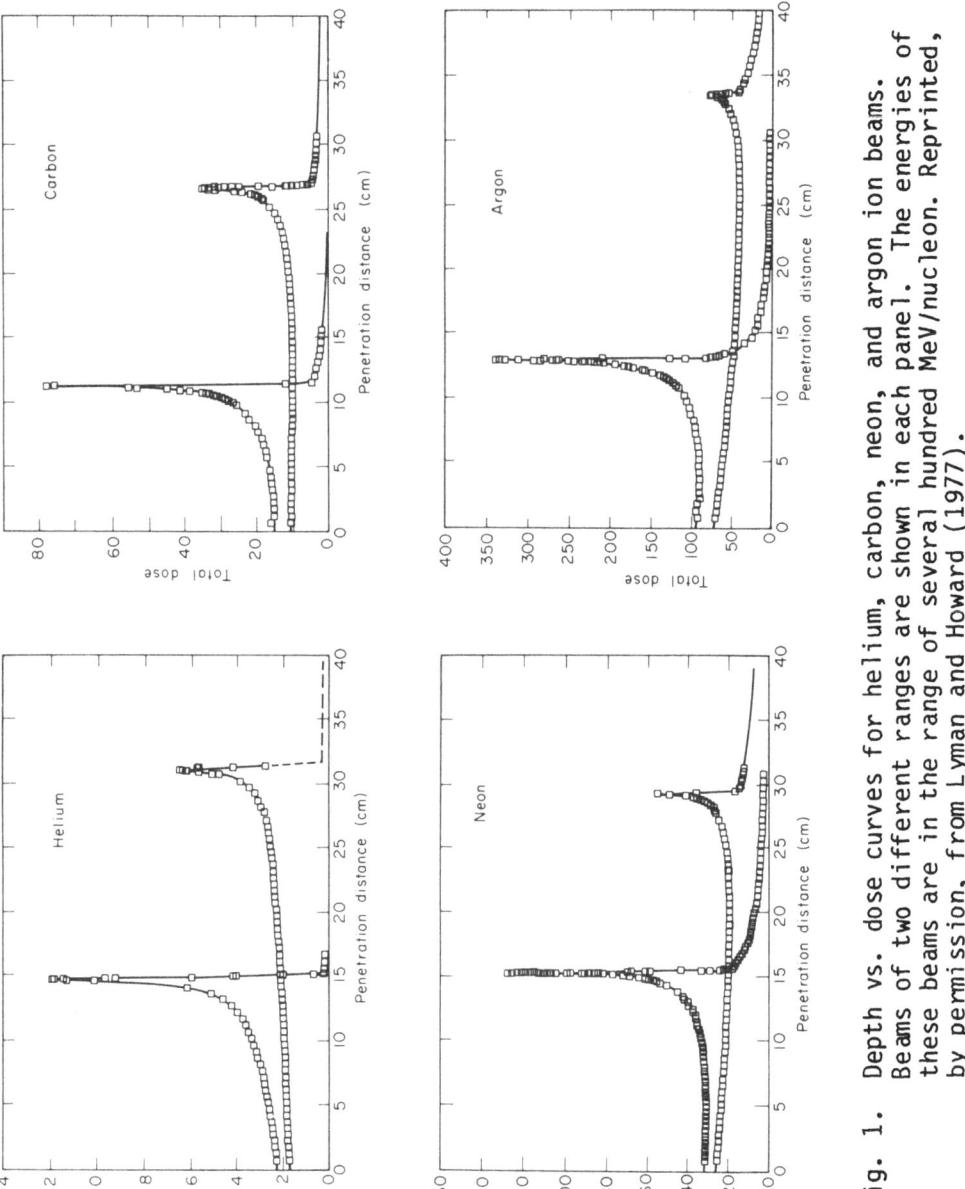

Fig. 1. Depth vs. dose curves for helium, carbon, neon, and argon ion beams. Beams of two different ranges are shown in each panel. The energies of these beams are in the range of several hundred MeV/nucleon. Reprinted, by permission, from Lyman and Howard (1977).

in the beam to provide a varying thickness to any point as a
function of time, thus effectively providing that point with
doses from different positions in the dose vs. depth curve. The
peak is thus spread through a broad region which is determined
solely by the design of the filter. Figure 2 shows an example of
a peak spread to 4 cm (Lyman and Howard, 1977). The peak-to-
plateau dose ratio decreases from over 4:1 to less than 1.5:1.
The subject of treatment planning with particle beams will be
covered in another chapter. Figure 3 is a treatment plan for a
pancreatic tumor designed with a Clinac 18 (a) compared to one
designed with a carbon beam (b) (G. T. Y. Chen, private communication). The Clinac 18 plan is a four-port plan and is considered an optimum plan for this site using conventional beams.
The carbon plan is a two-port plan using wedge compensation. It
is clear that the spinal cord and right kidney (to the lower left

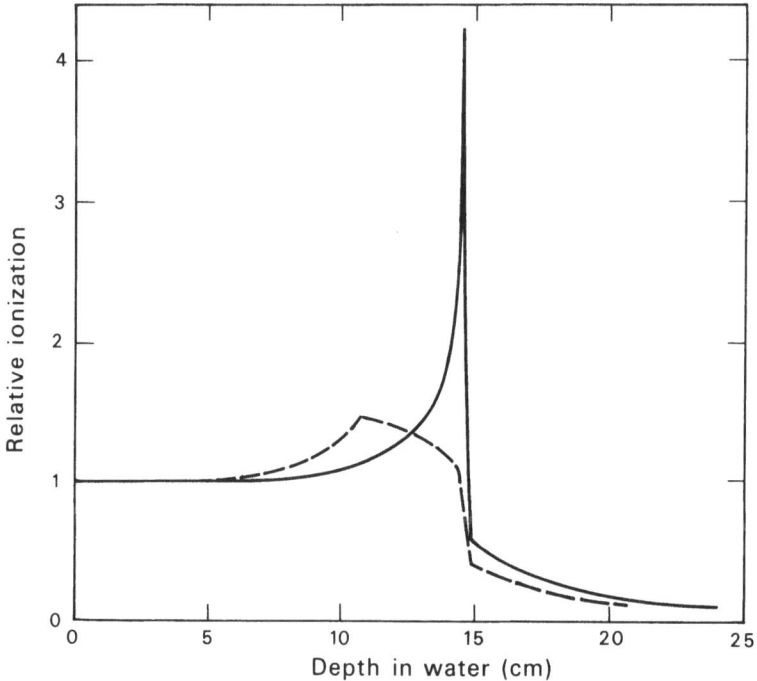

Fig. 2.   The effect on the depth vs. dose curve of the
          introduction of a variable absorber or "ridge
          filter" into an unmodified beam. The sharp
          peak is broadened (in this case to 4 cm) and
          the peak-to-plateau dose ratio is significantly
          decreased. Reprinted, by permission, from
          Lyman and Howard (1977).

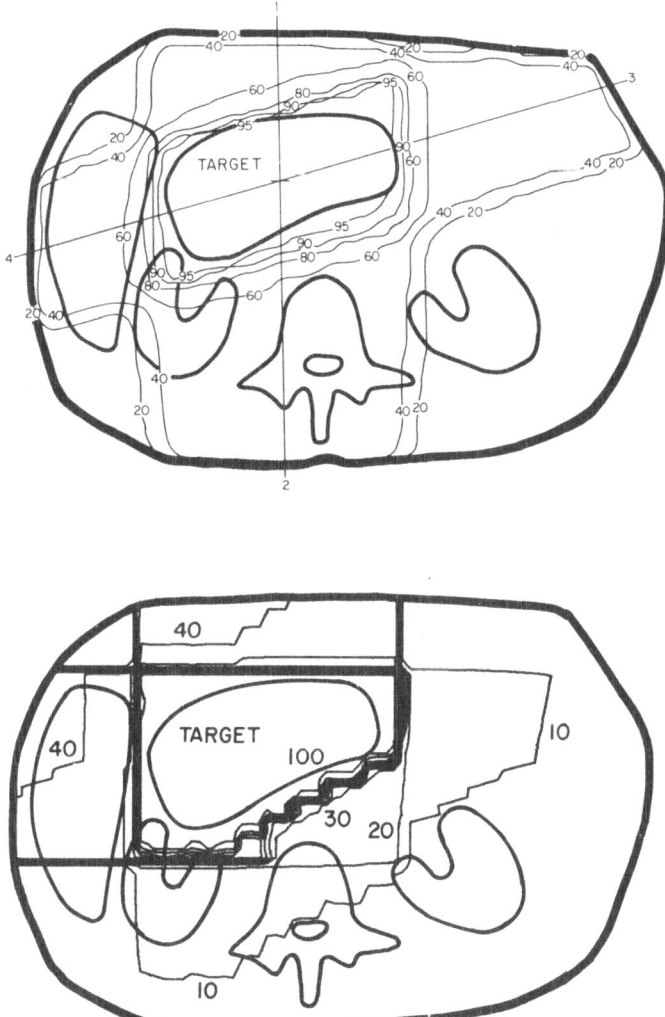

Fig. 3. A comparison of treatment plans for a
pancreatic tumor developed with photons
from a Clinac 18 (a) and with carbon
ions (b). Inspection of the exposure
levels outside the target volume shows
that considerably less dose is deposited
in the normal tissue for the carbon plan
than for the photon plan. Reprinted by
permission from G. T. Y. Chen.

of the target volume) receive considerably less dose in the carbon plan.

Heavy ions have been used as an example of charged particle beams in the above discussion, but the qualities of proton, helium-ion, and pion beams differ only slightly from heavy-ion beams. Proton and helium-ion beams have negligible secondary radiation, and potentially will provide the sharpest delineation of dose within a volume. Thus, inhomogeneities within the tissue itself become a major consideration when planning treatments, and will probably be the factor limiting the accuracy with which a given amount of radiation can be deposited within a specified target volume (Goitein, 1977).

The energy deposition from a negative pion beam is complicated by the inevitable nuclear interactions that the pions undergo as they are captured by the nuclei of the atoms in the region where they come to rest. Pion beams are secondary beams; that is, they must be created by the bombardment of a target material by a primary beam of other particles, such as protons or electrons. The pions are "radioactive" in the sense that they have a well-defined lifetime (τ = 2.6 x 10^{-8} sec) and may decay into a muon and a neutrino. If they survive until they come to rest, negative pions will always be captured in a Bohr orbit, creating a pi-mesic atom. The total rest energy of the pion (140 MeV) is available to break apart the nucleus and give the various nuclear fragments kinetic energy. These fragments have a short range and the heavier ones, including alpha particles, have quite high rates of energy loss and contribute significantly to the absorbed dose. Negative pion beams are contaminated by muons, which result from the direct decay of the pions in flight, and electrons from gamma rays, which result mainly from neutral pion decay in the target. The various components in a negative pion beam are shown schematically in Fig. 4 (Curtis and Raju, 1968). The "star" contribution is caused by the nuclear disintegrations mentioned above. The peak in ionization is not as high as the heavy-ion peaks because typical momentum spreads for pion beams, including the one chosen for this calculation, are much larger than for the heavy-ion beams. Thus, in a sense, this pion beam is already slightly spread. Note the small muon peak beyond the pion peak. The electrons form a more or less uniform background. Design of the beam itself can alter the relative proportion of pions, muons and electrons.

A comparison of experimental data of the dose vs. depth curves for all the "new" particles is shown in Fig. 5 (Raju et al., 1978). All the curves were normalized to the center of the modified peak region. All charged particle beams were modified to give peak regions 10 cm in depth. The reason the carbon beam shows more dose beyond the peak region is because it

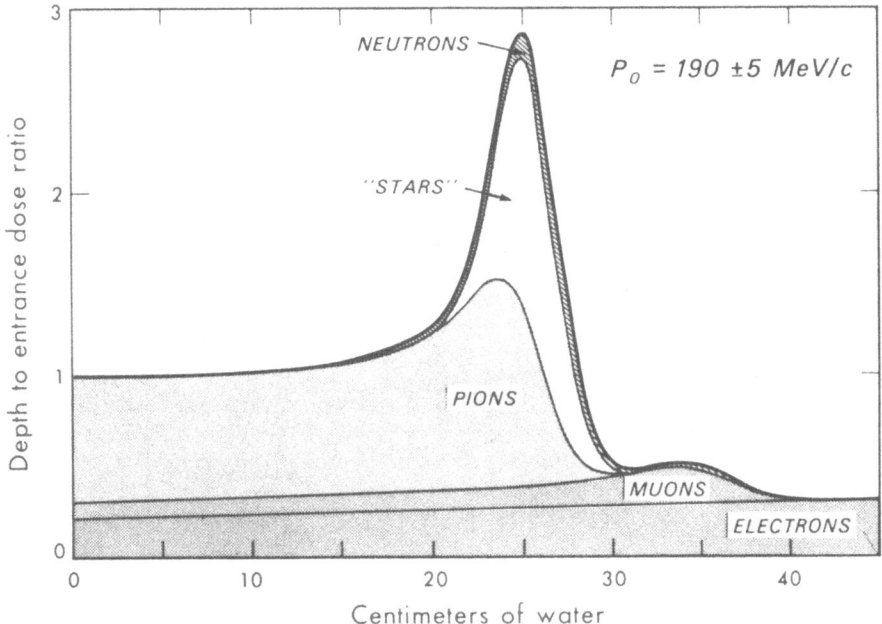

Fig. 4. A calculated depth vs. dose curve for a negative pion
         beam showing representative values for the contribu-
         tions to the total dose of pions, muons, electrons,
         and the "stars" or nuclear fragments from the capture
         interactions as the pions come to rest. Reprinted, by
         permission, from Curtis and Raju (1968).

was a higher energy beam (i.e., more penetrating) than the neon
beam, and so more fragmentation occurred.

High LET, Relativistic Biological Effectiveness and the Oxygen Effect

The highly ionizing nature of the pion and heavier ion beams
in the peak region produce not only an increased dose but also an
enhanced biological effect. The quantity that is a measure of
the average rate at which a particle loses energy is its dE/dx
value or LET (linear energy transfer). In general, except for
very high LET, biological effects increase with increasing LET.

It is important now to define another term, the relative
biological effectiveness or RBE. It is used to compare the
biological effectiveness of a given beam or radiation to that of

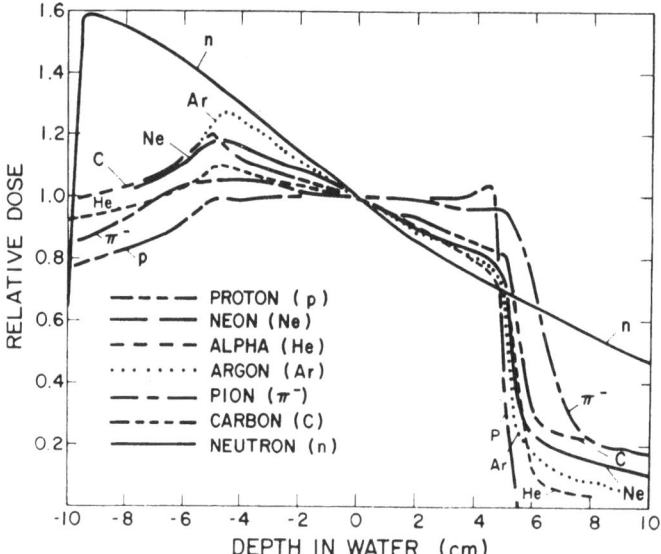

Fig. 5. A comparison of experimentally obtained depth vs. dose
curves for proton, neutron, pion, helium, carbon, neon,
and argon beams, normalized to unity at the center of
the spread peak dose region. Reprinted, by permission,
from Raju et al. (1978).

a standard radiation, usually 250 kV x rays. It is the ratio of
absorbed dose of the standard radiation to the absorbed dose of
the radiation in question necessary to cause equal biological
effect. That is,

$$RBE = \frac{Dose_{standard}}{Dose_{high\ LET}} \quad ,$$

where both doses produce the same response. Thus, the RBE
depends not only on the particle type and type of end point
selected, but also on the magnitude of the effect selected.

A typical dependence of RBE on LET is shown in Fig. 6 (Hall, 1978). Here we see three regions: one where the ionizations are random at low LET, one where the RBE is maximum and the ionizations are optimal for the particular effect, and one at very high LET where the RBE decreases with LET, in the so-called saturation or overkill region.

Now we turn to another problem facing the radiotherapist, that of cells within tumors that do not respond well to conventional radiation. Hypoxic cells are more radioresistant than well-oxygenated cells (Barendsen et al., 1966), and some tumors contain pockets or regions away from blood vessels where there are hypoxic cells. It has been conjectured that it is the surviving hypoxic cells that in some cases cause the failure of local tumor control (Gray, 1961). The sizes of tumors that have a 90 percent chance of cure if given a dose of ^{60}Co gamma rays are shown in Fig. 7 for three different fractions of hypoxic cells, 0, 0.1 and 1.0 (Fowler et al., 1963). Clearly, a considerably larger dose is necessary to "cure" a tumor with hypoxic cells than one of the same size with no hypoxic cells.

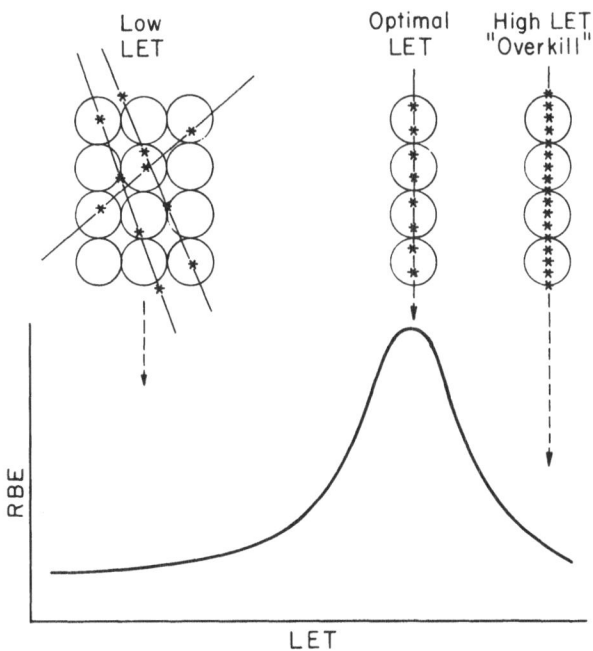

Fig. 6. A typical dependence of RBE on LET: an initial increase is followed by a maximum and then a decrease at even higher LET. Reprinted, by permission, from Hall (1978).

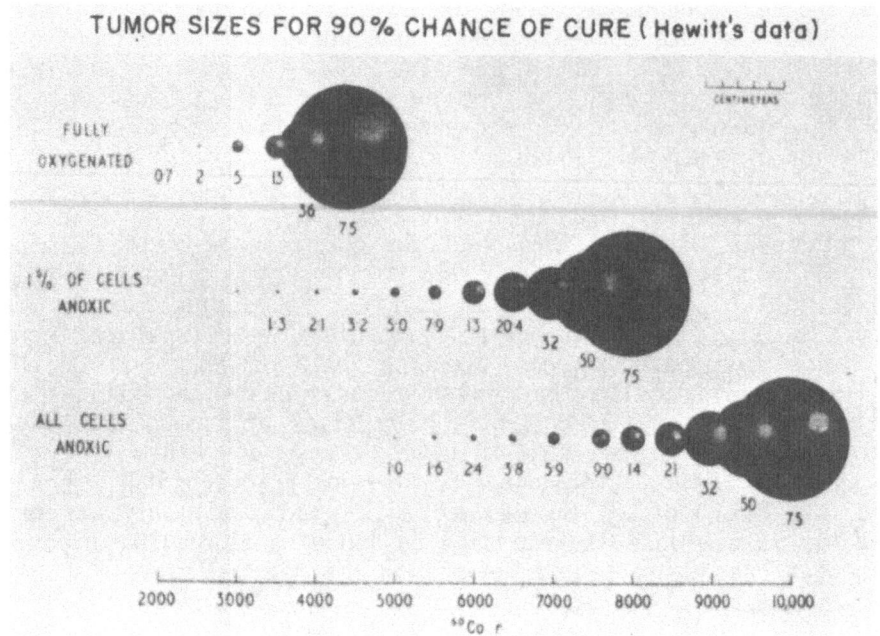

Fig. 7.   Calculated doses required to reduce the number of cells
          surviving in each tumor volume to 0.1, to give a
          90 percent chance of "cure." Upper row of tumors with
          cells fully oxygenated, middle row with 1 percent
          hypoxic cells, and the bottom row with all cells
          hypoxic. Reprinted, by permission, from Fowler et al.,
          (1963).

Since high LET particles have an increased capacity to kill
cells, independent of their oxygen status (Barendsen et al.,
1966), high LET beams of neutrons (Fowler, 1967; Fowler et al.,
1963), pions (Fowler, 1965; Richman et al., 1966), or heavy ions
(Lawrence and Tobias, 1967; Tobias and Todd, 1967) have been
suggested to treat tumors with large fractions of hypoxic cells.
Unfortunately, the hypoxic fraction of cells in human tumors
cannot be accurately determined at present.

Experimental data, however, have been accumulating on animal
tumors. Some of the more recent data will be reviewed in later
lectures. Barendsen and Broerse (1969) performed a classic
experiment with a rhabdomyosarcoma in a rat. They found evidence
of the presence of hypoxic cells in the tumor, and then studied
the extent to which a 14 MeV neutron beam could cut down the
oxygen effect. The results are shown in Fig. 8. The survival of
cells after irradiation <u>in situ</u> and subsequent cell dispersion,

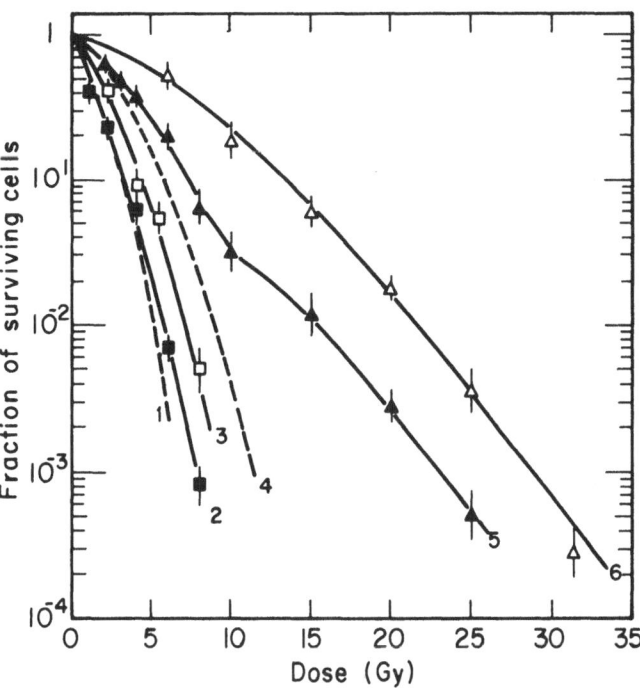

Fig. 8. Survival of R-1 rhabdomyosarcoma tumor cells irradiated
         in rats with neutrons (curves 1, 2, 3) and x-rays
         (curves 4, 5, 6). The dashed lines, curves 1 and 4,
         show results of cells irradiated in vitro. Curves 2
         and 5 show results of tumors irradiated in vivo with
         subsequent excision, cell dispersion, and plating, and
         curves 3 and 6 show results obtained for hypoxic popu-
         lations. The break in curve 5 at about 10 gray is
         evidence of an hypoxic fraction of about 15 percent.
         Reprinted, by permission, from Barendsen and Broerse
         (1969).

trypsinization and plating is shown for x-irradiated tumors in
both air-breathing (curve 5) and asphyxiated animals (curve 6).
The latter curve is representative of an hypoxic cell

population. Note the distinct break in curve 5, clearly
indicating the existance of two populations of cells with
different radiosensitivities. Curves 2 and 3 are for neutron
irradiated tumors. Their position on the graph shows not only a
rather large RBE (~3.0) for this end point relative to 300 keV
x rays, but also a significant decrease in the oxygen effect for
the neutron beam.

A quantity used to measure the extent to which the oxygen
effect is decreased is the oxygen enhancement ratio or OER. This
is defined as the ratio of absorbed dose necessary to produce a
given effect in hypoxic cells to that necessary to produce the
same effect in oxygenated cells. Thus, the OER is large (i.e.,
between 2.5 and 3.0) for x rays and decreases with increasing
LET, becoming close to one for high LET radiations (Barendsen
et al., 1966).

Another quantity used to evaluate the relative radio-
resistance of hypoxic cells is the oxygen gain factor or OGF,
which is defined as the ratio $OER_{x\ ray}/OER_{high\ LET}$. It is a
convenient quantity for comparing results from different cell
lines and in different laboratories and beams. Large OGF values
denote a large reduction of the oxygen effect. A comparison of
OGFs for the "new" particles is shown in Fig. 9 (Raju, 1979).

EVALUATING THE PARTICLES FOR THERAPY

This is a very difficult subject because criteria for
evaluating beams vary widely. Two separate evaluations of the
particles will be presented here. The first, Fig. 10, plots the
"physical" or dose vs. depth advantage along the ordinate and the
"biological" or possible low OER advantage along the abscissa
(Raju, 1978). The various particle beams are placed rather sub-
jectively in positions relative to each other. The number of
dollar signs is a rough estimation of the cost involved in implementing any of the modalities for patient treatment. The second
evaluation, Fig. 11, presents the relative biologically effective
dose between peak and plateau vs. OER (C. A. Tobias, private
communication). Here large values of the abscissa and small
values of the ordinate are advantageous. It appears that heavy
ions compare quite favorably with the other modalities.

APPLICATION TO DIAGNOSTIC RADIOLOGY

Charged particle beams are being used to image structures in
the body that are difficult to image with more conventional
techniques. The sharp, well-defined range of charged particle
beams suggested that they might be used to image structures with
slightly different densities within the body. As we have seen,
the energy deposited by these beams is maximum in the stopping

region.   Imaging with proton beams has been accomplished and shows promise for diagnosing tumors of the breast and brain (Steward and Koehler, 1973, 1974). Helium ions have also been used to obtain axial tomographs of the human brain (Crowe et al., 1975).   The heavier ions are being used to image the breast,

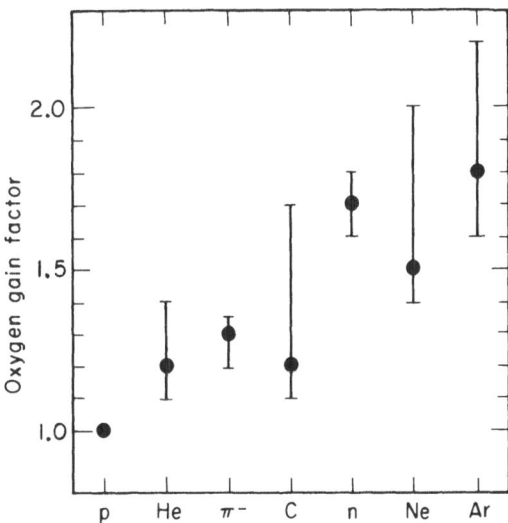

Fig. 9.   A comparison of oxygen gain factors experimentally determined for the various particles throughout the peak region. The spread of values reflect the fact that beam quality (and there-fore the effect of oxygen on cell survival) varies throughout the spread region. Reprinted, by permission, from M. R. Raju (1979).

abdominal region, and extremities (Sommer et al., 1978). The stopping points of heavy ions such as carbon and neon ions can be accurately recorded on stacks of plastic sheets placed behind the specimen, because the high LET nature of the particles renders the plastic preferentially etchable near the positions where the

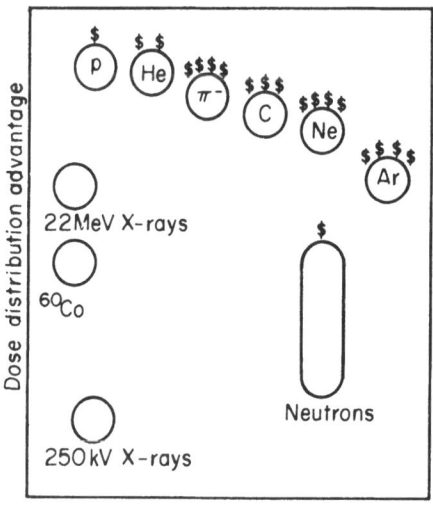

High LET advantage?

Fig. 10. An attempt to compare the thera-
        peutic advantages of the various
        particles. The advantage due to
        a favorable dose distribution is
        plotted against a possible advan-
        tage due to an increased LET (e.g.,
        high RBE and low OER). Particles
        in the upper right, are assumed to
        be the most advantageous for therapy.
        The number of dollar signs give a
        rough indication of the relative
        expense of implementing the various
        modalities in the hospital environ-
        ment. Reprinted, by permission,
        from M. R. Raju (1978).

particles come to rest. The large mass of the ions keeps
scattering and straggling very low. Thus, the position of the
plastic sheet in which the particle stops is a sensitive measure

of the range, and therefore the average density through which the particle passed. If an object of slightly higher density is traversed, the particles will stop in a different plastic sheet, as shown in Fig. 12. Thus, two images of the object will be seen

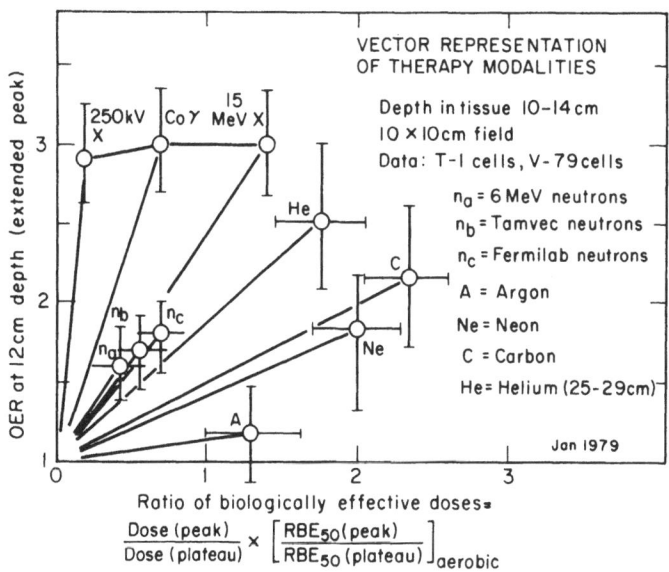

Fig. 11. Another attempt to graph the relative merits of the various particles. Here the OER is plotted against the ratio of "biologically effective" dose at the peak to that at the plateau. Here the more therapeutically advantageous particles lie in the lower right-hand region of the figure. Reprinted, by permission, from C. A. Tobias.

on different sheets, one the negative of the other. Figure 13 compares an x-ray of a human foot and a computer reconstructed neon-ion radiograph of that foot made from a composite of all the exposed plastic sheets. Considerably more soft tissue detail is seen in the radiograph, including such structures as the Achilles tendon. The dose to the patient from the neon ions was estimated to be between 100 and 200 millirad.

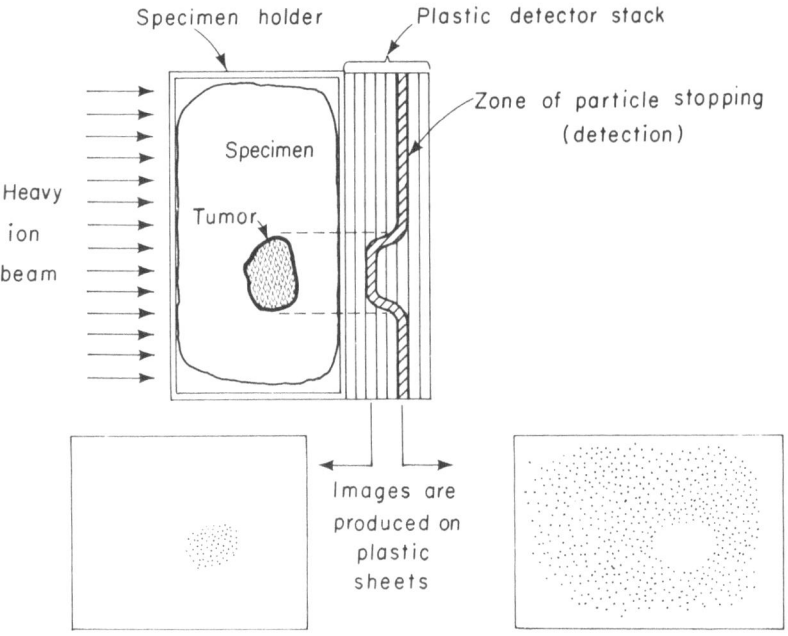

Fig. 12.   The concept of heavy-ion radiography. Heavy particles
           penetrating a region of higher density will stop in
           different plastic detector sheets than the particles
           on either side. The plastic sheets in which the
           particles stop show etchable tracks. Reprinted, by
           permission, from Sommer et al. (1978).

APPLICATION TO NUCLEAR MEDICINE

    All the new particle beams can be used to make radioisotopes,
which can then be used in most of the standard techniques of
nuclear medicine. However, there is another potential use of at
least the heavy ion beams. A portion of the heavy-ion beam
particles undergo nuclear fragmentation, and some of these high
energy fragments are radioactive, for instance, ^{11}C or ^{19}Ne.
These fragments can be focussed into a second beam and allowed to
penetrate the body, coming to rest, for example, inside a blood
vessel. The subsequent positron decay produces annihilation
gamma radiation that can be detected in an appropriately designed
counter array. The rate of decay of radioactivity after the beam
has been turned off measures not only the usual decay of the
isotope but also the flow rate of the blood carrying the isotope

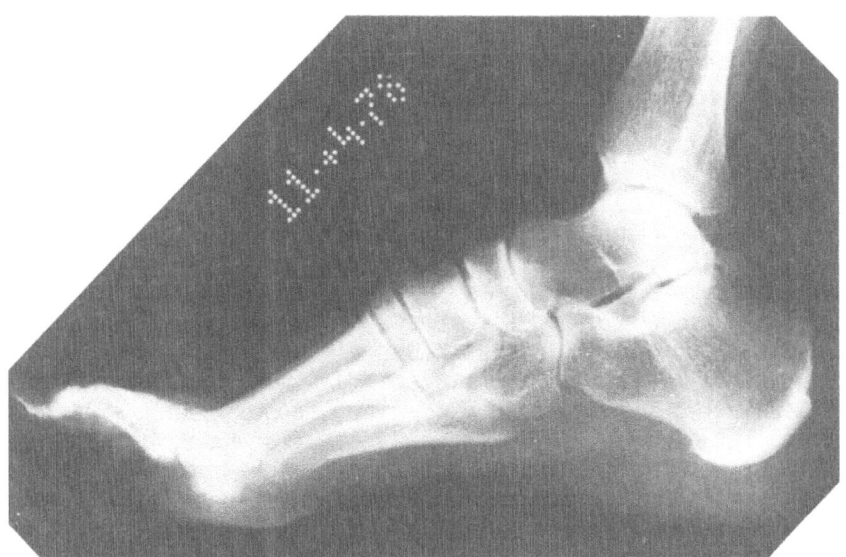

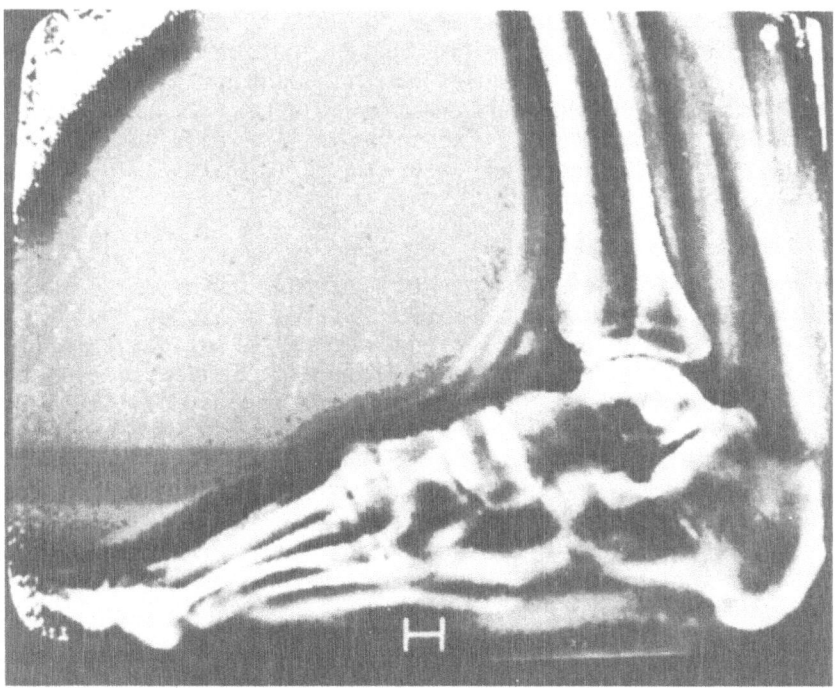

Fig. 13. A comparison of a conventional x-ray and a computer reconstructed heavy-ion radiograph of a human foot. Soft tissue details are more evident in the heavy ion radiograph than in the x ray. Reprinted by permission from C. A. Tobias.

away from the stopping site. Studies of such noninvasive techniques for measuring blood flow are presently underway (Chatterjee and Tobias, 1977).

CURRENT MEDICALLY ORIENTED PROJECTS INVOLVING PARTICLE BEAMS

The new particles are presently being used in the therapeutic and diagnostic applications discussed above. Neutron therapy machines can be found in nine different countries: Belgium, East Germany, England, Japan, The Netherlands, Poland, Scotland, the United States and West Germany; several of these countries have both cyclotrons and d-t generators. Clinical trials are underway with varying degress of success (Dutreix and Tubiana, 1979). The most encouraging results have been with tumors of the head and neck, salivary glands, cervical lymph nodes, and soft tissue sarcomas (Catterall, 1979). Good results have been obtained treating gynecologic tumors with mixed high and low LET beams (two fractions of neutrons and three fractions of high energy x rays per week) (Peters et al., 1979). The results on brain tumors, on the other hand, have been rather discouraging.

Proton beams are being used in Sweden, the United States, and the Soviet Union. In the United States, patients are given boost therapy (i.e., proton irradiation in addition to conventional therapy) for prostatic carcinoma and other sites where dose localization is important. Proton beams are also being used to treat patients with chorodial melanoma (Gragoudas, 1978), and the results are quite encouraging.

A randomized trial of carcinoma of the pancreas is underway with a helium ion beam in the United States (Castro, 1979). This was initiated after an earlier pilot group of 41 patients showed an encouraging response. Of 34 evaluable patients, 10 either are now without cancer or at least showed local control of the tumor.

Pion programs are underway in three countries: Canada, Switzerland, and the United States. A clinical program is in progress in the United States (Kligerman, 1979) and will be started in the other two countries in the near future. The results so far are encouraging.

Initial patient studies are in progress using carbon, neon, and argon beams in the United States (Castro, 1979), but it is too early to assess the efficacy of these beams.

Proton, carbon, and neon beams are being used for radiography. Results have indicated that there are definite details on the particle radiographs that cannot be seen on the best CT scans or xerographs. There have been at least two cases where a malignant region in a breast was identified on a particle radiograph and not seen by conventional means (C. A. Tobias, private communication).

CONCLUDING REMARKS

In conclusion, we see that the age of new particles in clinical medicine has arrived. There is tremendous activity within these programs all over the world. Much of the progress in tumor localization, dose delivery, computerization of treatment planning, and patient immobilization has been to a considerable extent motivated by an attempt to solve the many problems involved in particle therapy.

Finally, along with the potential these high LET radiation modalities hold for increased tumor cure rates and decreased failures in local tumor control, there are also the responsibilities of protecting the patient and personnel from the increased biological effect of any residual high LET radiation. It is then with the excitement of knowing that we are reviewing here a rapidly growing field important to radiation protection, tumor diagnosis, and therapy that I open the discussion of these subjects, and hope the papers to be presented here will stimulate further interest in the field.

ACKNOWLEDGEMENTS

It is a pleasure to thank G. T. Y. Chen, M. R. Raju, and C. A. Tobias for allowing use of their unpublished data. The skilled editorial assistance of M. C. Pirrucello is gratefully acknowledged. Many of the studies reported here were supported by the National Cancer Institute, and the U. S. Department of Energy under contract No. W-7405-ENG-48.

REFERENCES

Barendsen, G. W. and Broerse, J. J., 1969, Experimental radiotherapy of a rat rhabdomyosarcoma with 15 MeV neutrons and 30 kV x rays. I. Effects of single exposures, Europ. J. Cancer, 5, 373.

Barendsen, G. W., Koot, C. J., van Kersen, G. R., Bewley, D. K., Field, S. B., and Parnell, C. J., 1966, The effect of oxygen on impairment of the proliferative capacity of human cells in culture by ionizing radiations of different LET, Int. J. Radiat. Biol. 10, 317.

Brennan, J. T., and Phillips, T. L., 1971, Evaluation of past experience with fast neutron teletherapy and its implications for future applications, Europ. J. Cancer 7, 219.

Castro, J. R., 1979, Progress report on heavy particle clinical radiotherapy trial at Lawrence Berkeley Laboratory, July 1975–July 1979, Lawrence Berkeley Laboratory Report No. LBL-9738.

Catterall, M., 1979, Observations on the reactions of normal and malignant tissues to a standard dose of neutrons, Proceedings, Third Meeting on "Fundamental and Practical Aspects of the Application of Fast Neutrons and Other High-LET Particles in Clinical Radiotherapy," The Hague, The Netherlands, September 1978, Europ. J. Cancer, in press.

Chatterjee, A., and Tobias, C. A., 1977, Radioactive beams, in: "Biological and Medical Research with Accelerated Heavy Ions at the BEVALAC 1974–1977," Lawrence Berkeley Laboratory Report LBL-5610.

Crowe, K. M., Budinger, T. F., Cahoon, J. L., et al., 1975, Axial scanning with 900 MeV alpha particles, IEEE Trans. Nucl. Sci., NS-22, 1952.

Curtis, S. B., and Raju, M. R., 1968, A calculation of the physical characteristics of negative pion beams—energy-loss distribution and Bragg curves, Radiat. Res. 34, 239.

Dutreix, J., and Tubiana, M., 1979, Evaluation of clinical experience concerning tumor response to high LET radiation, Proceedings, Third Meeting on "Fundamental and Practical Aspects of the Application of Fast Neutrons and Other High-LET Particles in Clinical Radiotherapy," The Hague, The Netherlands, September 1978, Europ. J. Cancer, in press.

Fowler, P. H., 1965, Pi mesons versus cancer, Proc. Phys. Soc. 85, 1051.

Fowler, J. F., 1967, Fast neutron therapy—physical and biological considerations, in: "Modern Trends in Radiotherapy," Vol. 1, T. J. Deeley and C. A. P. Wood, eds., Butterworth, London.

Fowler, J. F., Morgan, R. L., and Wood, C. A. P., 1963, Pre-therapeutic experiments with the fast neutron beam from the medical research council cyclotron I. The biological and physical advantages and problems of neutron therapy, Brit. J. Radiol. 36, 77.

Goitein, M., 1977, The measurement of tissue heterodensity to guide charged particle radiotherapy, Int. J. Radiat. Oncol., Biol. Phys. 3, 27.

Gragoudas, E. S., Goitein, M., Koehler, A., Constable, L. J., Wagner, M. S., Verhey, L., Tepper, J., Suit, H. D., Broskhurst, R. J. Schneider, R. J., and Johnson, K. N., 1978, Proton irradiation of choroidal melanomas, Arch. Ophthalmol., 96, 1583.

Gray, L. H., 1961, Radiobiologic basis of oxygen as a modifying factor in radiation therapy, Am. J. Roentgenol., 85, 803.

Hall, E. J., 1978, "Radiobiology for the Radiologist," 2nd ed., Harper and Row, Hagerstown, Maryland.

Kligerman, M., 1979, Results of clinical applications of negative
    pions at Los Alamos, Proceedings, Third Meeting on "Funda-
    mental and Practical Aspects of the Application of Fast
    Neutrons and Other High–LET Particles in Clinical Radio-
    therapy," The Hague, The Netherlands, September 1978, Europ.
    J. Cancer, in press.
Lawrence, J. H. and Tobias, C. A., 1967, Heavy particles in
    therapy, in: "Modern Trends in Radiotherapy Vol. 1," T. J.
    Deeley and C. A. P. Wood, eds., Butterworth, London.
Lyman, J. T. and Howard, J., 1977, Dosimetry and instrumentation
    for helium and heavy ions, Int. J. Radiat. Oncol. Biol.
    Phys. 3, 81.
Peters, L. J., Hussey, D. H., Fletcher, G. H., and Warton,
    J. J., 1979, Second preliminary report of the M. D. Anderson
    study of neutron therapy for locally advanced gynecological
    tumors, Proceedings, Third Meeting on "Fundamental and
    Practical Aspects of the Application of Fast Neutrons and
    Other High–LET Particles in Clinical Radiotherapy," The
    Hague, The Netherlands, September 1978, Europ. J. Cancer, in
    press.
Raju, M. R., 1979, "Heavy Particle Radiotherapy," Academic Press,
    New York, in press.
Raju, M. R., 1978, Continued studies on the potential for heavy
    particle radiation therapy, Los Alamos Scientific Laboratory
    Report Preprint LA–UR–78–3107.
Raju, M. R., Amols, H. I., DiCello, J. F., Howard, J., Lyman, J.
    T., Koehler, A. M., Graves, R., and Smathers, J. B., 1978, A
    heavy particle comparative study. Part I: depth–dose
    distributions, Brit. J. Radiol. 51, 699.
Richman, C., Aceto, H., Raju, M. R. and Schwartz, B., 1966, The
    therapeutic possibilities of negative pions: preliminary
    physical experiments, Am. J. Roentgen., 96, 777.
Sommer, F. G. Capp, M. P., Tobias, C. A., Benton, E. V.,
    Woodruff, K. H. Henke, R. P., Holley, W., and Genant, H. K.,
    1978, Heavy–ion radiography: density resolution and specimen
    radiography, Invest. Radiol., 13, 163.
Steward, V. W., and Koehler, A. M., 1973, Proton beam
    radiography in tumor detection, Science, 179, 913.
Steward, V. W., and Koehler, A. M., 1974, Proton radiography in
    the diagnosis of breast carcinoma, Radiology, 110, 217.
Stone, R. S., 1948, Neutron therapy and specific ionization,
    Am. J. Roentgenol., 59, 771.
Stone, R. S., Lawrence, J. H., and Aebersold, P. C., 1940, A
    preliminary report on the use of fast neutrons in the treat-
    ment of malignant disease, Radiology, 35, 322.
Tobias, C. A., and Todd, P. W., 1967, Heavy charged particles in
    cancer therapy, in: "Radiobiology and Radiotherapy," U. S.
    National Cancer Monograph 24, National Cancer Institute,
    Bethesda, MD.

INTRODUCTION AND FUNDAMENTALS

Ralph H. Thomas

Lawrence Berkeley Laboratory
University of California
Berkeley, California 94720 USA

"Conception, my boy, *fundamental brain work*, is what
makes the difference in all art."[1]
                        Dante Gabriel Rossetti

INTRODUCTION

   Some may think that a lecture on "fundamentals" is perhaps
inconsistent with the title of this course which deals with
"advances" in the areas of radiation protection and dosimetry in
medicine. The Oxford English Dictionary[2] defines fundamental
as "Pertaining to the foundations or groundwork, going to the
root of the matter." The English language is so abused these
days that it may be a shock to some to learn that an equivalent
to "fundamental" is "radical." Dante Gabriel Rossetti was able
to very clearly perceive that "fundamentals" were, in fact, the
basis for all fine creativity--if I may be allowed the presump-
tion that he would have agreed that the distinction between
artistic and scientific creativity is a fine one, if indeed it
exists at all. This course will discuss advances in the
fundamental sciences which underlie that applied science which
is variously called health physics, radiological protection,
or radiation protection.

   It is my hope in this lecture to lay before you, so to speak,
the menu. To continue to metaphore I will, as a good waiter
should, indicate the special virtues of each course and mention
the specialities of the restaurant. In order to do this

effectively I will first discuss some of the underlying
assumptions of radiation protection and, later, describe some of
the developments in the radiological sciences—both fundamental
and administrative acts by the ICRP and ICRU—which have taken
place since the first course was given under the auspices of the
International School of Radiation Damage and Protection in
1975.[3] These foundations and developments will then be related
to the topics to be discussed in this course.

Measuring the Risks of Ionizing Radiations

It is often claimed that more is known about the harmful
effects of ionizing radiation than about any other toxic agent, a
claim that can probably be supported by the great volume of
published literature on the subject. As Morgan has written elsewhere, "Perhaps there has never before been an enterprise that
was planned so carefully for its safety and never before a risk
that has been so thoroughly studied and guarded against as has
been the case with the nuclear energy industry and its concern to
avoid unnecessary exposure to ionizing radiation."[4] If this is
indeed so, we are presented with the conundrum that of all the
toxic agents, the public has the greatest fear of ionizing radiations. Perhaps we have evidence here of yet another variation of
that great universal principle known as Parkinson's Law.[5] In
this variation it takes the form "Public Apprehension increases
in direct proportion to the frequency with which a potential
hazard is discussed."

To the great mass of the general public the degree of
authoritive standing of the discussion does not seem to carry
much weight—it is important only that the potential hazard be
discussed. Such is the power of the producers of television news
programs that people in California requested whole-body counts
for fear that they had been contaminated by the radioactive
releases at Three-Mile Island, some 2,500 miles away.

It is perhaps necessary from time to time to stand back and
take a broad view of toxicology and be somewhat selfcongratu-
latory in realizing that the radiation protection is quite
soundly based. Any other branch of toxicology that could produce
as authoritative a volume as "Sources and Effects of Ionizing
Radiations" would consider itself fortunate.[6]

Lawrence[7] suggests that there are four basic lines of
investigation to assess the risk of any hazard: "Measurements
are made to:

1. Define the conditions of exposure

2. Identify adverse effects

3. Relate exposure with effect

4. Estimate the overall risk."

It is of some interest for us to see how well our assessment of risk measures up to these requirements.

Conditions of Exposure

It is important to remark at the onset that one of the extremely valuable properties of ionizing radiations is that they are readily detected. This property has led to extensive documentation of the sources of exposure to ionizing radiation. Personal dosimetry records are extensive and—compared with estimates of exposure to other toxic agents—relatively accurate. In most cases internal contamination is easily detected by whole body counting or bioassay-techniques. By the same token, the measurements of any releases of radioactivity to the environment may be carried out with great sensitivity.

This ease of detection of ionizing radiation means, in the United States at least, that we have an increasing body of data, which is essentially public information, concerning exposures of both radiation workers and the general public to radiation. It is, for example, the policy of the United States Department of Energy (DOE) to publish annual summaries of personal monitoring and environmental monitoring data of all its programs. We may conclude, therefore, that the conditions and magnitude of exposure are quite well known.[6]

Figure 1 shows a summary of the major sources of human radiation exposure at the present time, as determined by the U. S. Environmental Protection Agency (USEPA), and the figure also gives predictions until the year 2,000 A. D. Ionizing radiation is seen to be an unusual "pollutant" in that the major exposure to man is of natural origin. If low levels of ionizing radiations were dominant in their influence among environmental toxic agents it might therefore be expected that one would be able to observe significant correlations between the incidence of cancer and external radiation levels.

Frigerio has in fact studied the correlation between leukemia rates in the United States and natural radiation background. Perhaps surprisingly, he finds a weakly negative but significant correlation.[9] This is not, of course, to say that ionizing

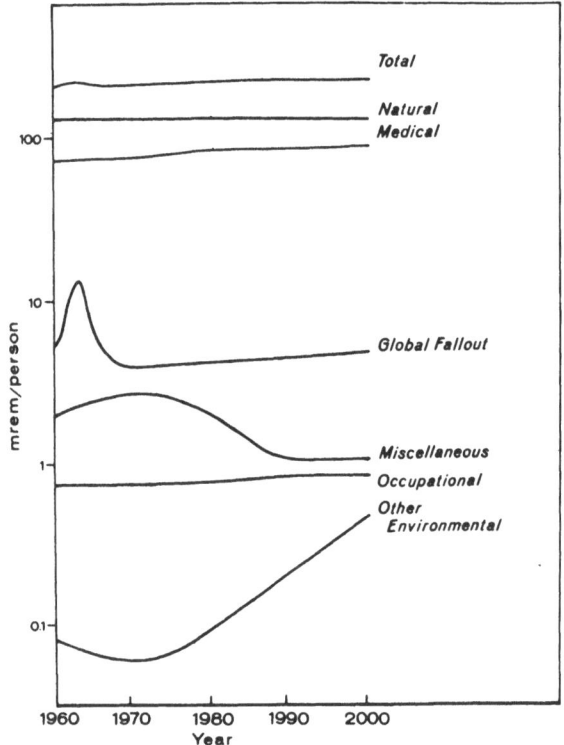

Figure 1. A summary of the major sources
of human radiation exposure
until 2000 A.D., prediction by
the Environmental Protection Agency.

radiation has been demonstrated to be beneficial to man; but
perhaps it does indicate that, even if ionizing radiations are
carcinogenic in man at dose rates of ~0.1 rem/yr there are
environmental agents which are of much greater importance.
Nevertheless, this has not prevented many who should know better
from making claims that we now know that a significant fraction
of cancer is caused by the natural background of ionizing radiation. Thus, for example, Rowe of the USEPA has made the claim
that natural background causes 13,000 "health effects" (a
euphemism for cancers) per year in the United States.[10]

Figure 1 shows the very important role of diagnostic
radiology as a source of radiation exposure in man. It is, and
no doubt will continue to be, the dominant source of radiation
exposure. However, Chamberlain[11] has made the point that:
"Not only is this true, . . . but it would be surprising if it

were not true. Medical radiology is the only legitimate situation in which radiation exposure is purposefully given to human beings for their own benefit. It would be more remarkable, if not alarming, if any other man—made source exceeded it in human exposure."

Thomas and Busick[12] have suggested that, at present exposure levels in the United States, diagnostic radiology is at least ten thousand times more beneficial than harmful. Nevertheless in the absence of a satisfactory resolution to the low dose problem it is prudent to assume that all radiation exposures are harmful—no matter how small the dose. Considerations such as these have led to suggestions, sometimes stated in a rather shrill fashion, that there is an urgent need to reduce exposure to ionizing radiation from medical sources.[13,14]

Morgan supports such a view by citing five arguments:

1. Medical exposures in the United States are higher by a factor of 2 to 10 than in most advanced countries in the world.

2. There is no threshold to the deleterious effects of radiation.

3. Linear extrapolations of biological effects observed at high radiation doses may underestimate radiation effects at low doses.

4. Despite current concern as to the somatic effects of radiation, the genetic risks may still be the limiting form of radiation damage.

5. Low level exposures such as those in medical diagnosis may be harmful.

It is certainly true, as Morgan says, that medical exposures in the United States are higher than in any advanced country in the world. However, the mere fact of exposure does not of itself establish any detrimental health effects. There is much to consider at a meeting such as this in what Morgan says, and I look forward to discussion of these topics particularly in the talks by Silini[15] and Paretzke.[16]

The International Commission on Radiological Protection (ICRP) has published two reports which are great value in reducing patient exposure.[17,18] Both reports are currently being revised by the ICRP, which also plans to issue a third report dealing with the protection of patients during radiotherapy.

There are many other pressures in the United States which tend to reduce radiation exposures from medical sources. These include consumer organizations, trade unions, professional societies, the news media and federal agencies.[19] Recently, for example, the Environmental Protection Agency and Bureau of Radiological Health[20] jointly published in the Federal Register guidelines for the use of X-rays in medicine by agencies of the U. S. Government. These guidelines discourage the routine radiological examinations without symptomatic or other medical reasons.[21]

These and other pressures have led to increasing efforts to reduce radiation exposures to levels as low as readily achievable[22] and to the widespread use of the concept of population dose equivalent as an index of harm resulting from the exposure of populations to low levels of ionizing radiations.[23] While the former effort is of undoubted value, provided it is achieved by some means of risk-benefit analysis, the widespread use of the quantity population dose equivalent has been questioned by some and will no doubt be discussed during this course.

ADVERSE EFFECTS OF IONIZING RADIATION

There are basically three sources of information on the adverse effects of ionizing radiation on humans. These are:

1. Fundamental radiobiology

2. Animal experiments

3. Epidemiological studies.

We will hear more on these separate topics in lectures by Silini, Paretzke, Rossi, Larsson, Broerse, and Curtis in this course[15,16,24-29] and in some of the panel discussions. Nevertheless, I would like to make some comments from the empirical point of view of the health physicist, recognizing full well my lack of qualifications to make such comments.

It is necessary that radiation protection standards at the present time are properly based upon the data obtained from epidemiological studies of humans exposed to ionizing radiation. This is in no way intended to derogate fundamental radiobiology or animal studies. However, even the extrapolation from "mice to men" is fraught with difficulties. Evans has often alluded to the large differences between rats, beagles, and man in radiosensitivity to radium, and to the need to choose the correct radiobiological end-point.[30]

With such difficulties evident in animal studies, it is apparent that extrapolations from cellular radiology, or from studies of particular tumor systems are fraught with uncertainty. Rossi and his colleagues have pointed to the complexity of tumor induction processes when hormones are involved, [31], [32] when discussing the incidence of mammary neoplasms in Sprague-Dawley rates Rossi and Kellerer[31] conclude: "With the complexity of the tumor induction process established, there remains little justification for linear extrapolations, and this conclusion, in turn, removes apparent inconsistencies between the dose-effect relation postulates and histological evidence to the effect that carcinogenesis requires the transformation of several contiguous cells. There is, however, at present insufficient evidence for numerical estimation of tumor incidence based on linear or other extrapolations."

If we are to rely upon epidemiological studies it is most important to understand their limitations. The late Sir Ronald Fisher always began his series of undergraduate lectures to his students at Cambridge by demonstrating that there was a highly significant correlation between the population density of storks and the birth rate. He earnestly implored his students not to draw the conclusion that a causal relationship had been established!

Jablon[33] has stated the difficulty with great clarity: "as it happens, it is almost always possible to find flaws in surveys (not excluding the Oxford survey*). By the very nature of the thing we do not have experimental control in a survey. We cannot guarantee that two groups of people are wholly comparable save with respect to the one factor that we wish to study. We cannot regard any single survey, no matter how carefully done, as the equivalent of the large, well-controlled experiment that, ideally, we would prefer. We therefore invoke what is, I suppose, a new principle: If many surveys, none perfect but conducted in diverse contexts, all point to the same conclusion, then that conclusion is likely to be right. So it has been with the issue of smoking and lung cancer; and similarly, after the first report from the Oxford Survey,* a number of investigators, working in differing settings and using varying materials have studied the question, but here the results have not been unanimous."[33]

---

*The "Oxford Survey" refers to studies of the incidence of leukemia in children irradiated *in utero* made by A. M. Stewart and her colleagues. [34], [35]

Patterson and Thomas[36] have shown, using a simple model, that the number of man years, M, required to form the basis of a study to reveal radiation induced decrease in populations chronically exposed to a radiation level D rem/y greater than the controls given by:

$$M \simeq \frac{4}{rD} \left[1 + \frac{2f}{rD} \right] \qquad (1)$$

where r = risk of cancer per rad
f = "natural" probability of cancer per year.

Substitution into equation (1) gives the results shown in Table 1.

Inspection of Table 1 shows that, taking the risk of radiation induced cancer induction as 10^{-4} rem^{-1} (Ref. 37), we see that a population of radiation workers exposed to 1 rem year of 40,000 would have to be studied for 30 years for an effective epidemiological study! One might therefore be forgiven for viewing with scepticism some recent studies that claim to show significant findings with studies of populations of a few thousand or less and with no reliable dose estimates. The ICRP has, in its most recent recommendations, published a brief summary of the risks of ionizing radiation and recommends risk estimates to be used in radiation protection.[38]

Despite the rather wide-spread unanimity of these risk estimates upon which radiation safety standards presently rest, there has been over the past decade a steady barrage of criticism from a rather small, but vocal, rump group. This group has objected that presently accepted risk estimates are too low by at least an order of magnitude. The literature contains many such

Table 1. Number of Man Years Needed for an Epidemiological Study of Radiation Induced Cancer.

Dose Rate (rem/yr)	Radiation Risk (cancers/rem)	Man Years M
0.1	10^{-2}	1.6×10^{4}
	10^{-3}	1.2×10^{6}
	10^{-4}	1.2×10^{8}
0.1	10^{-2}	5.2×10^{2}
	10^{-3}	1.6×10^{4}
	10^{-4}	1.2×10^{6}

Table 2. Risk Factors for Radiation Protection.

Tissue	Risk Factor
Red bone marrow	2×10^{-3} Sv^{-1}
Bone	5×10^{-4} Sv^{-1}
Lung	2×10^{-3} Sv^{-1}
Thyroid	5×10^{-4} Sv^{-1}
Breast	2.5×10^{-3} Sv^{-1}
All other tissues	5×10^{-3} Sv^{-1}

claims, but it is possible to mention only a few. For example, Sternglass[39] invoked the influences of radioactive fallout from nuclear weapons tests to explain a diminution in the rate of decrease in infant mortality in the early 1960's. This suggestion was ably dismissed by Lindop and Rotblat[40] but, nevertheless, there was great prominence given to the orginal suggestion of Sternglass. Recent studies do not provide support for the view that infant mortality in the 1960's was significantly affected by radioactive fallout.[41]

Sternglass[42] also claimed to be able to show correlations between nuclear power reactor operation and the incidence of leukemia in populations close by. These claims were refuted by Tompkins.[43]

Of more interest are the studies by Stewart[34,35] and other authors[44,46] of the incidence of leukemia and other cancers in children who were irradiated in utero incidentally to diagnostic procedures performed on the mother. For example, the Oxford study of Stewart and Kneale[35] found an increase of ~50 percent per rad in the exposed group when compared with the non-exposed group used for control. They also suggest their study is consistent with a linear dose-effect relationship.

These suggestions, if confirmed, would be of great importance in public health because they would provide evidence of deleterious effects to humans at dose levels of a few tenths of a rad. However, the conclusions drawn from these studies are not generally agreed upon. Stewart herself has drawn attention to the fact that several prospective studies[34] do not show such an effect. Jablon and Kato[47] have indicated inconsistencies between the Stewart data and those of the Atomic Bomb Casualty Commission (ABCC) studies of the Hiroshima and Nagasaki bomb victims, where no elevation in leukemia or cancer in children irradiated in utero is found. Jablon,[48] in a later paper, speculated that "some of the relationships that have been reported reflect, in reality, selection of gravida for x-ray." More recently Oppenhiem et al.[49] in a prospective study of 1000 children irradiated in the course of routine pelvimetry found no conclusive radiation effect. However, they found several possible sources of error in such studies—particularly in the improper choice of control groups. They suggest that the reported radiation effects on man "have been due to bases introduced in the selection process, rather than to the radiation itself." In addition, Shore et al.[50] have disputed Stewart and Kneale's claim that their data are consistent with a linear dose relationship. When the Stewart data are divided into five year periods, the average case/control ratio fluctuates. For the most recent period (1960–1965) the case/control ratio was −0.05 ± 0.26, indicating no significant elevation of cancer incidence.

There is, then, no general agreement as to whether irradiation of the human foetus to ~1 rad does produce an elevation in observed cancer or leukemia rates. This is due to the intrinsic difficulty in drawing conclusions from epidemiological data.[33] Nevertheless, in the best interests of public health, until this issue is resolved both the ICRP and NCRP have recommended limiting radiation exposure to pregnant women of reproductive capacity.[51,52]

In some cases incorrect conclusions are drawn which suggest risk factor much higher than is likely. For example Morgan[53] has interpreted the data of Modan[54] and his colleagues to suggest that exposures of a few rads to the thyroid will significantly increase the risk of thyroid cancer. Modan et al. noted an elevation of thyroid cancer in individuals whose heads were irradiated in the treatment of tinea capititis. The thyroid gland was in the radiation field and received absorbed doses of a few rads. The pituitary gland was also in the radiation field and the absorbed dose to the pituitary was an order of magnitude higher than the absorbed dose to the thyroid. The most probable cause of the elevation of thyroid cancers observed by Modan et al. is therefore the irradiation of the pituitary and not the thyroid.

Many other examples may be found in the literature and they continue to appear. Recent examples are the Mancuso study,[55] the Tri-State studies[56] and the Portsmouth shipyard workers study.[57] Each must be studied on its merits. If the effects claimed are confirmed, there will be important ramifications on our radiation protection standards.

DOSE EFFECT RELATIONSHIPS

Much of the present controversy concerning radiation effects lies in the fact that we are forced to extrapolate risk estimates obtained from epidemiological studies of persons exposed to tens or even hundreds of rads (and usually at very high dose rates) down to the absorbed doses and absorbed doses rates of concern in radiation protection (a few rads per year, or less). This extrapolation has led to a great deal of discussion, and it will be discussed during this course in lectures by Partezke,[15] Silini,[16] and Rossi.[24] Suffice it to say at the present time that two of the pieces of data used as a basis for present radiation protection standards show non-linear dose-effect relationships.

Our protection standards for internal radiation are, in part, founded upon the studies of the radium-dial painters. These studies represent one of the few pieces of chronic human exposure data we have. Figure 2 shows the observed tumor cumulative incidence in epidemiologically suitable cases as a function of average skeletal cumulative rad dose as reported by Evans et al.[58,59] Evans, in fact, hypothesizes that his data may show evidence of an effective threshold. Figure 3 shows a plot of tumor appearance time versus average skeletal cumulative rads. There is a suggestion that the tumor appearance time increases as the dose decreases. When the tumor appearance is longer than the life expectancy for an individual an "effective threshold" may be said to exist.

The concept of an "effective threshold" is of great interest and has been invoked in other areas of toxicology. Thus, in the analysis of data on dose of a carcinogen versus the period between exposure and tumor appearance, Jones and Grendon[60] find that, for all carcinogens examined including radiation and a number of chemicals and in a variety of mammalian species including man, the relationship between latent period, L, and dose, D, can be expressed as: $L = CD^x$, where x is generally close to 1/3. A biological model has been developed to account for this similarity of behavior among such divergent carcinogenic agents. If the generalization proves valid, it will justify the use of "practical thresholds" in estimating the effects of radiation and other carcinogens at low levels. Dinman[61] has persuasively argued that stochastic determinants impose a lower limit on the dose-response relationship between cells and chemicals, and one would assume similar arguments applied to cellular-radiation effects.

It is of great interest whether effective thresholds of absorbed dose or absorbed dose rate exists for all radiation effects in man. It does seem clear from the data presented in

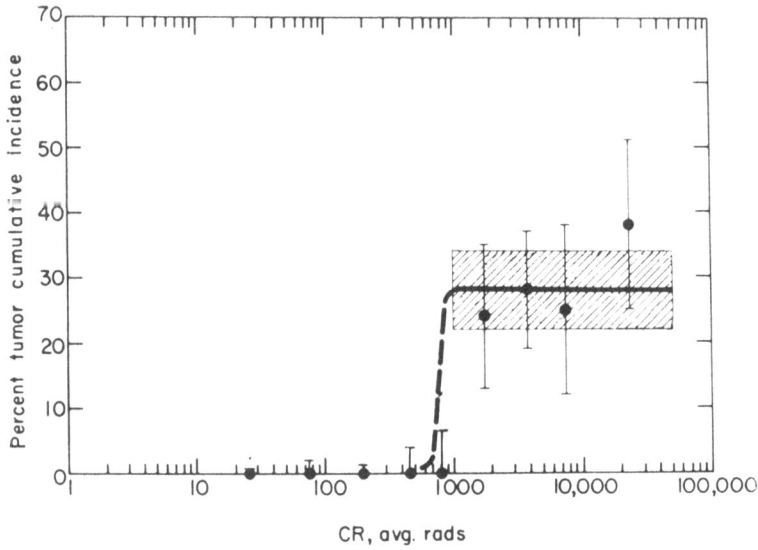

Figure 2. The observed tumor cumulative incidence or
           occurrence in an epidemiologically suitable
           group of radium dial painters (Evans et al., 1972).

Fig. 2 that the dose-effect relationship is non-linear. Gofman
and Tamplin[62] have suggested that these data are consistant with
linearity but the chance that this is so is 10^8 to 1 against.[59]

External radiations protection standards are, in part, based
upon data on the incidence of leukemia and other cancers in the
survivors of the nuclear weapons explosions at Hiroshima and
Nagasaki. Analysis of the Japanese survivor data is difficult,
and this difficulty is compounded by the difference in radiation
fields produced by the two weapons. At Hiroshima the influence of
a substantial absorbed dose caused by neutrons has made dose
estimates and interpretation difficult. At Nagasaki, however, the
situation is somewhat simpler in that the neutron contribution to
absorbed dose was smaller and "wide ranges in the assumed neutron
potency factor cause only a small variation (±10 percent) in the
calculated risk rate coefficient for Nagasaki."

Figure 4 shows the estimated annual leukemia deaths per 10^6
persons for the exposed population at Nagasaki during the period
1950-1970. Mays et al.[63] conclude:

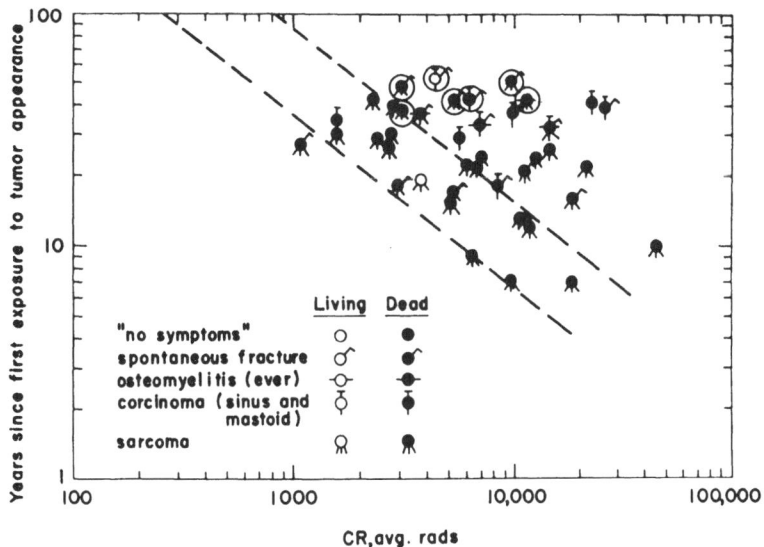

Figure 3. Log—log presentation of tumor appearance time versus average skeletal cumulative rads (CR) for radiogenic tumors (Evans et al., 1972).

"The non-linear appearance of the plotted dose-response curve for Nagasaki raises reasonable doubt on whether the dose-response is really linear (Fig. 4). Among the 4931 persons exposed at Nagasaki to 10-99 rads, 7.2 total cases of leukemia are predicted (4.7 natural plus 2.5 induced according to the 'preferred' linear estimate), whereas only 2 leukemia cases were actually observed. A linear relationship predicting 7.2 cases when only 2 were observed is rejected significantly (P = 0.03). An excellent fit to the Nagasaki incidence rate is made by the fitted dose squared relationship of 0.003 induced leuk per year/10^6 person rem^2 starting at a natural incidence rate of 52 leuk per year/10^6 persons, and assuming an average neutron potency factor of 9. This dose squared relationship will be used to provide alternative estimates of risk.

"Now, the lifetime risks will be estimated for leukemia induced by total body x-ray irradiation at high dose-rates (10-1000 rem/min) such as received

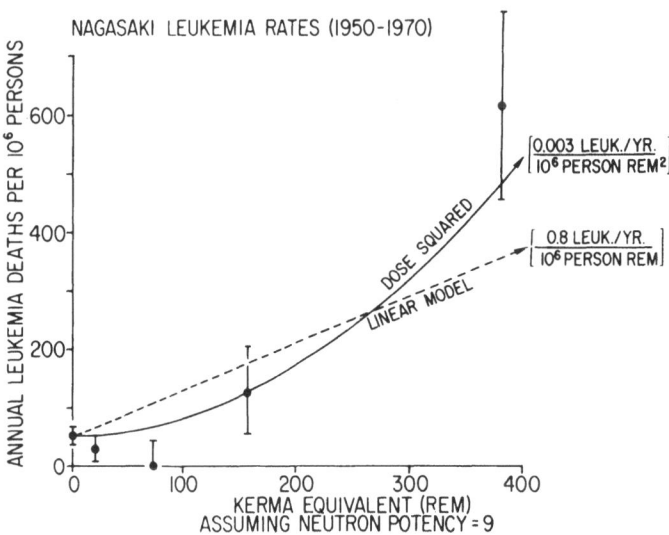

Figure 4.   The leukemia deaths (expressed as deaths
             per million people) observed among the
             Nagasaki survivors, plotted as a function
             kerma equivalent (rem) (A neutron RBE of 9
             has been assumed). The error bars show one
             standard deviation; the darkened line, a dose
             squared model (Mays et al., 1973).

received by the A-bomb survivors. Assuming the average
death rate from induced leukemia was the same in the
unobserved interval 0 to 5 years after irradiation as
in the observed 5 to 25 year interval, the total incidence during the first 25 years following irradiation
based on the preferred linear model would be (25 yr)
(0.8 leuk per yr/10^6 person rem) = 20 leuk/10^6
person rem. Based on present trends about 80 percent
of the lifetime leukemia risk should be expressed at
25 years. Therefore, the preferred linear estimate for
the lifetime risk from leukemia is (20 leuk/10^6
person rem)/0.8 = 25 leuk/10^6 person rem. The higher
and lower linear estimates and the dose squared
estimate were calculated similarly, and are shown in
Table 3."[63]

Table 3. Lifetime Risk from Leukemia at High Dose-Rates
          (from a total body γ-ray dose "D").

Higher linear estimate	= (40 leuk/10^6 person rem) D
Preferred linear estimate	= (25 Leuk/10^6 person rem) D
Lower linear estimate	= (14 leuk/10^6 person rem) D
Dose squared estimate	= (0.1 leuk/10^6 person rem) D^2

Thus, in some of the better epidemiological data available, there is strong evidence that a linear extrapolation from high dose data to the radiation protection dose region will give conservative risk factors. I would like to go further to stimulate discussion and suggest that, at the present time, we have no generally agreed data upon that demonstrate that exposure to x or γ-radiation below ten rads are harmful to humans. Furthermore, even if such data are eventually forthcoming the effect of dose-rate must be carefully studied.

On the other hand an entirely different situation may exist with high LET radiation and I look forward to the lecture by Silini on the effects of radiation quality upon biological effects.[25]

THE BASIC ASSUMPTIONS OF CURRENT RADIATION PROTECTION STANDARDS

Despite the great increase in our understanding in the biological effects of ionizing radiation, our present protection standards are based upon simplying assumptions that originate from the early 1950's. It is assumed that any radiation exposure may carry with it some detriment.

"A linear non-threshold model was specifically chosen on a basis of mathematical simplicity and prudence to represent the upper limit of risk in the low-dose domain, for somatic radiobiological effects which had been observed only in a higher-dose domain. The linear nonthreshold model was not based on radio-biological data for somatic effects in the low-dose domain.

"As originally introduced, care was always taken in protection committee reports to point out that the true risk in the low-dose domain would be expected

to lie between zero and the upper limit given by the
linear non-threshold approximation."[58]

For the purposes of radiation protection it is assumed that a
quantity dose equivalent, H, is related to the presumed radiation
risk.

"In radiation protection it has been found convenient to introduce a quantity that correlates better
with the more important deleterious effects of exposure
to radiation, more particularly with the delayed
stochastic effects. This quantity, called dose equivalent, is the absorbed Dose, D, weighted by the
modifying factors Q and N.

"The dose equivalent, H, at a point in tissue is
given by the equation: H = DQN, where D is the
absorbed dose, Q is the quality factor and N is the
product of all other modifying factors specified by the
Commission. Such factors might take account, for
example, of absorbed dose rate and fractionation."[64]

The importance of dose equivalent for radiation protection is
also indicated by International Commission of Radiation Units and
Measurements. In ICRU Report 19A, Dose Equivalent is stated as
being "related to the presumed radiation risk."[65]

"The quality factor, Q, is intended to allow for
the effect on the detriment of the microscopic distribution of absorbed energy. It is defined as a
function of the collision stopping power (L_∞) in water
at the point of interest. Interpolated values of Q as
a function of L_∞ can be obtained from the figure,
which is based upon the values shown in the table
[Table 4]."[66]

THIS COURSE AND RADIATION PROTECTION

We have seen in the previous sections that the present basis
of radiological protection includes many assumptions. The
fundamental assumption is that the deleterious effects to man of
ionizing radiation are directly proportional to the absorbed dose
in the irradiated tissue. This assumption may well be shown to
be false or at least an approximation under very special circumstances. Thus, Mole finds that a dose-response of the form
$aD^2e^{-\lambda D}$ fits the data for the incidence of both thyroid
cancer and breast cancer in the survivors of Hiroshima.[67]

Table 4. L_∞-Q Relationship

L_∞ in water (KeV/μm)	Q
3.5 (and less)	1
7	2
23	5
53	10
175 (and above)	20

Kellerer and Rossi[68] have proposed a theory of dual radiation action which explains radiation effects observed at the cellular level. In general neither the dose-response relationships proposed by Mole or by Kellerer and Rossi are linear. We may expect to hear more of these topics in the lectures by Paretzke[69] and Rossi.[24]

A second assumption used in radiation protection is that there is no diminution in the biological effect produced by a given absorbed dose as the absorbed dose rate is reduced. The recent statement by the ICRP that the risk estimates contained in Publications 26 (Ref. 70) represent best estimates (rather than upper limits) is not universally accepted. Paretzke, in his lecture on "Late Effects of Low Doses and Dose Rates"[15] and Silini on "The Estimation of Radiation Risk in Man"[16] will address some of these issues.

The third basic assumption of radiation protection which may be challenged during this course is L_∞-Q relationship. This relationship is now some twenty years old,[71] and some redifinition is perhaps overdue. There have been recent challenges of the basic concept of this relationship. Mole has suggested that "a question for the future is whether a single value of Q for a given quality of radiation can continue to be used in radiological protection, regardless of the particular tissues exposed and the nature of the biological effects which is to be minimized."[72] Rossi has in fact proposed substantial increase in the Quality Factor for neutrons,[73] based upon studies of the incidence of leukemia in the survivors of Hiroshima and Nagasaki[74] and other considerations. Furthermore, Rossi has suggested that radiation quality may be better described by a microscopic quantity which he terms the lineal energy, rather than by L_∞.[75] Lectures by Paretzke,[69] Rossi,[76] and Silini[25] will discuss some of these issues. Because their is no unanimity evident in the literature we may expect some lively discussions.[32,77–81]

One of the facts that will make these discussions informative is that we are now beginning to accumulate a great deal of information on high LET radiation effects. Following the first acceleration of heavy ions to high energies at Princeton[82] and a few months later at Berkeley,[83] high LET radiations have been available which permit not only studies of cellular radiobiology but also whole body irradiation of small mammals. Schimmerling[84] will discuss the new sources of radiation that have become available in the past decade, while Curtis will discuss their application to medicine.[85]

Perhaps we can hope that, for this course at least, the views of Francis Bacon will prevail: "If a man begin with certainties, he shall end in doubts; but if he will be content to begin with doubts, he shall end in certainties."[86]

REFERENCES

1. D. G. Rossetti, 1882, Letter to Hall Caine in "Recollections of Rossetti" by H. Caine.
2. Oxford English Dictionary, Compact Edition, 1971, Oxford University Press, New York.
3. R. H. Thomas and A. Rindi, Eds., 1976, "Proceedings of the First Course on High Energy Radiation Dosimetry and Protection" Erice, Sicily, Italy. IEEE Trans. Nuc. Sci. NS 2, 1315–1427.
4. K. Z. Morgan, 1975, Suggested Reduction of Permissible Exposure to Plutonium and other Transuranium Elements. Am. Ind. Hyg. Assoc. J., 36, 567.
5. C. Parkinson, Northcote, 1957, "Parkinson's Law." Houghton Mifflin Co., Boston.
6. United National Scientific Committee on the Effects of Atomic Radiation Sources and Effects of Ionizing Radiations, "UNSCEAR 1977 Report to the General Assembly," United Nations, 1977.
7. W. W. Lawrence, 1976, "Of Acceptable Risk: – Science and the Determination of Safety" William Kauffmann, Inc., Los Altos, California.
8. op. cit. ref. (7) p. 18.
9. N. A. Frigerio, and Stowe, R. S., 1975. "Carcinogenic and Genetic Hazard from Background Radiation," IAEA SM 202/805 Symp. On Biological Effects of Low Level Radiation.

10. W. D. Rowe, 1974, "EPA Approach to Environmental Analysis and Control of Radiation," Course on Environmental Analysis and Environmental Monitoring for Nuclear Power Generation, University of California, Berkeley.

11. R. H. Chamberlain, 1973, Facts, Fables and Follies in Medical Radiology in Proc. Third International Congress of IRPA. Sept. 9-14, Washington, D. C. USAEC Report CONF - 73-0097. (2 Vols.) p. 791.

12. R. H. Thomas and D. D. Busick, 1976, "Reducing patient exposure to Ionizing Radiation - is it really necessary?" Am. Ind. Hy., Assoc. Journal. 37, 567.

13. K. Z. Morgan, 1975, Reducing Medical Exposure to Ionizing Radiation - Landauer Memorial Lecture. Am. Ind. Hyg. Assoc. J. 36, 358.

14. C. Nader, 1975, The Dispute Over Safe Uses of X-Rays in Medical Practice, Health Phys. 29, 181.

15. H. G. Paretzke, 1979, Late Effects of Low Doses and Dose Rates, Lecture No. 6 in "Advances in Radiation Protection and Dosimetry in Medicine" Course Proceedings, International School of Radiation Damage and Protection, Ettore Majorana Centre for Scientific Culture, Erice, Italy.

16. G. Silini, 1979. The Estimation of Radiation Risk in Man, Lecture No. 7, in "Advances in Radiation Protection and Dosimetry Medicine" Course Proceedings, International School of Radiation Damage and Protection, Ettore Majorana Centre for Scientific Culture, Erice, Italy.

17. ICRP Publication 16, 1970, "Protection of the Patient in X-ray Diagnosis," Pergamon Press, Oxford.

18. ICRP Publication 17, 1971, "Protection of the Patient in Radionuclide Investigation," Pergamon Press, Oxford.

19. V. Cohn, 1979, "Anybody Can Administer X-Rays, Critics Complain" Washington Post (July 25, 1979. p. A2).

20. U. S. Dept. of Health Education and Welfare, 1979, "Califano Calls for Accelerated Efforts to Reduce Medical X-Ray Exposure," Bureau of Radiological Health Bulletin XIII No. 5.

21. Radiation Protection Guidance to Federal Agencies for Diagnostic X-Rays. Federal Register 43, No. 22 (Wednesday, February 1, 1978, Part 5, Title 3. p. 4377).

22. ICRP Publication 22, 1973, "Implications of Commission Recommendations that Doses be kept as Low as Readily Achievable," Pergamon Press, Oxford.

23. ICRP Publication 27, 1977, "Problems of Commission Recommendations that Doses be kept as Low as Readily Achievable," Pergamon Press, Oxford.

23. ICRP Publication 27, 1977, "Problems Involved in Developing An Index of Harm," Pergamon Press, Oxford.

24. H. H. Rossi, 1979, The role of the Theory of Dual Radiation Action in Radiation Protection, Lecture No. 8, in "Advances in Radiation Protection and Dosimetry in Medicine" Course Proceedings, International School of Radiation Damage and

Protection, Ettore Majorana Centre for Scientific Culture, Erice, Italy.

25. G. Silini, 1979, Biological Effects of Radiation and their variation with Radiation Quality, Lecture No. 5, "Advances in Radiation Protection and Dosimetry in Medicine" Course Proceedings, International School of Radiation Damage and Protection, Ettore Majorana Centre for Scientific Culture, Erice, Italy.

26. B. Larsson, 1979, Dosimetry and Radiobiological Effects of Protons. Lecture No. 16, Advances in Radiation Protection and Dosimetry in Medicine" Course Proceedings, International School of Radiation Damage and Protection, Ettore Majorana Centre for Scientific Culture, Erice, Italy.

27. T. J. Broerse, 1979, Radiobiological Effects of Neutrons, Lecture No. 18, "Advances in Radiation Protection and Dosimetry in Medicine" Course Proceedings, International School of Radiation Damage and Protection, Ettore Majorana Centre for Scientific Culture, Erice, Italy.

28. J. DiCello, 1979, Radiobiological Effects of Pions, "Advances in Radiation Protection and Dosimetry in Medicine" Course Proceedings, International School of Radiation Damage and Protection, Ettore Majorana Centre for Scientific Culture, Erice, Italy.

29. S. B. Curtis, 1979, Radiobiological Effects of Heavy Ions, Lecture No. 22, in "Advances in Radiation Protection and Dosimetry in Medicine" Course Proceedings, International School of Radiation Damage and Protection, Ettore Majorana Centre for Scientific Culture, Erice, Italy.

30. R. D. Evans, Private Communication.

31. H. H. Rossi and A. M. Kellerer, 1972, Radiation Carcinogenisis at Low Doses, Science 175, 200.

32. H. H. Rossi and W. E. Mays, 1979, Reply to G. W. Beebe and C. W. Lande, Health Physics 36: 4666.

33. S. Jablon, 1973, Letter to the Editor, Health Physics 24: 223.

34. A. M. Stewart, 1973, The Carcinogenic Effects of Low Level Radiation, A re-appraisal of Epidemiologist's Methods and Observations, Health Physics 24: 223.

35. A. M. Stewart and G. W. Kneale, 1968, Lancet 1: 104.

36. H. W. Patterson and R. H. Thomas, Radiation and Risk - the Source Data, in "Proc. Sixth Berkeley Symposium on Mathematical Statistics and Probability," University of California Press, Berkeley, p. 313.

37. ICRP Publication 26, 1977, "Recommendations of the International Commission on Radiological Protection," Pergamon, Oxford.

38. op. cit. ref. 37, "Tissues at risk," pp. 9-13.

39. E. J. Sternglass, 1969, Bull, Atomic Scientists XXV (No. 4): 18.

40. P. J. Lindop and J. Rotblat, 1969, Nature, Dec. 27, 1969: 224.

41. V. R. Fuchs, Private Communication.

42. E. J. Sternglass, "Environmental Radiation and Human Health," op. cit. ref. 36, p. 145.
43. E. Thompkins, "Infant Mortality Around Three Nuclear Plants," op. cit. ref. 36, p. 279.
44. T. L. T. Lewis, 1960, Brit. Med. J. ii: 1551.
45. J. Wells and C. M. Steer, 1961, Am. J. Obstet. Gynocol. 81: 1059.
46. W. M. Court-Brown, R. Doll, and A. B. Hill, 1960, Brit. Med. J. ii: 1539.
47. S. Jablon and H. Kato, 1970, Lancet ii: 1000.
48. S. Jablon, 1974, "The Late Effects of Acute External Exposure to Ionizing Radiation in Man," paper read at the Fifth International Congress of Radiation Research, July 14-20, 1974, Seattle.
49. B. E. Oppenheim, M. L. Griem, and P. Meier, 1975. "The Effects of Prenatal Exposure to Diagnostic X-rays; Are they real ", paper read at the twentieth Annual Meeting of the Health Physics Society, Buffalo.
50. F. J. Shore, et al., 1973, Childhood Cancer Following Obstetric Radiography, Health Physics. 24: 258.
51. op. cit. ref. 37 Paragraphs. 115 and 116.
52. NCRP Rep. 53, 1971, Review of NCRP Radiation Dose Limit for Embryo and Fetus in Occupationally Exposed Women.
53. K. Z. Morgan, 1976, "Yes is the answer to the question of R. H. Thomas and D. D. Busick, "Is it really necessary to reduce patient exposure" Am. Ind. Hyg. Assoc. J. 37, 655.
54. B. Modan, H. Mart, D. Baidatz, R. Steinitz, and S. G. Levin, 1974, Radiation-induced Head and Neck Tumors, Lancet 1:277.
55. T. F. Mancuso, A. Stewart, and G. Kneale, 1979, Radiation Exposures of Hanford Workers Dying from Cancer and other Causes, Health Physics 33: 369.
56. I. D. Bross, and N. S. Natarajan, 1977, Genetic Damage from Diagnostic Radiation, JAMA 237: 2399.
57. T. Najarian, and T. Colton, 1978, Mortality from Leukemia and Cancer in Shipyard Nuclear Workers, Lancet May 13, 1978: 1018.
58. R. D. Evans, A. T. Keane, and M. M. Shanahan, 1972, Radiogenic Effects in Man of Long-Term Skeletal Alpha Irradiation, in Radiobiology of Plutonium, Eds., B. J. Stove and W. S. S. Jee, J. W. Press, Salt Lake City, p. 431.
59. R. D. Evans, 1974, Radium in Man, Health Physics, 27: 497.
60. H. B. Jones and A. Grendon, 1975, "The Effect of Dose on Latent Period in Carcinogenesis," paper read at the Twentieth Annual Meeting of the Health Physics Society, Buffalo, July, 1975.
61. B. D. Dinman, 1972, "Non-Concept of No Threshold Chemicals in the Environment," Science 175: 495.
62. J. Gofman, and A. Tamplin, 1971, The Question of Safe Radiation Thresholds for Alpha Emitting Bone Seekers in Man, Health Physics 21:47.

63. C. W. Mays, P. Lloyd, and J. H. Marshall, 1973, "Malignancy Risks to Humans from Total Body γ-Irradiation," Proc. Third Int. Cong. of the International Radiation Protection Association, September 1973, Washington D. C., USAEC Report, CONF 730907-VI, p. 47.
64. op. cit. ref. 37, paragraphs 17 and 18.
65. International Commission on Radiation Units and Measurements, 1973, "Supplement to ICRU Report 19 - Dose Equivalent," ICRU, Washington, D. C.
66. op. cit. ref. 37, paragraph 19.
67. R. H. Mole, 1972, British Journal of Radiology, 48: 1571975.
68. A. M. Kellerer and H. H. Rossi, 1978, A Generalized Formulation of Dual Radiation Action, Radiation Res. 75: 2.
69. H. G. Paretzke, 1979, Advances in Energy Deposition Theory, Lecture No. 3, in "Advances in Radiation Protection and Dosimetry in Medicine" Course Proceedings, International School of Radiation Damage and Protection, Ettore Majorana Centre for Scientific Culture, Erice, Italy.
70. ICRP, 1978, ICRP Statement from the 1978 Stockholm Meeting of the ICRP, Annuals of the ICRP 2, No. 1.
71. RBE Committee, 1963, Report of the RBE Committee to the International Commissions on Radiological Protection and on Radiological Units and Measurements, Health Physics 9: 357.
72. R. H. Mole, 1979, RBE for Carcinogenis by Fission Neutrons, Health Physics 36: 463.
73. H. H. Rossi, 1977, A Proposal for the Revision of Quality Factor, R. Envir. Biophys. 14: 275.
74. H. H. Rossi, and C. W. Mays, 1978, Leukemia Risk from Neutrons, Health Physics 34: 353.
75. J. A. Dennis, 1978, The Sixth Symposium on Microdosimetry, Brussels, May 1978 (A Review), Radiological Protection Bulletin. No. 25: 36.
76. H. H. Rossi, 1979. Microdosimetry and its Application to Biology and Medicine. Lecture No. 4, in "Advances in Radiation Protection and Dosimetry in Medicine" Course Proceedings, International School of Radiation Damage and Protection, Ettore Majorana Centre for Scientific Culture, Erice, Italy.
77. G. W. Beebe, and C. E. Land, 1979, Comments on "Leukemia Risk from Neutrons," Health Physics 36: 465.
78. H. H. Rossi, and W. E. Mays, 1979, Reply to Dr. Mole, Health Physics 36: 464.
79. V. P. Bond, Effects of Quantitative Risk on Modifying Factors and Dose Equivalent (submitted to Health Physics).
80. V. P. Bond, 1978, The Risk from Fast Neutron Exposure, paper presented at the Health Physics Society Meeting, Minneapolis, June 18-23, 1978 (submitted as a note to Health Physics).
81. C. E. Land, G. W. Beebe, and S. Jablon, 1978, Role of Neutrons in Rate Effects of Radiation Among A-Bomb Survivors,

paper presented at the Health Physics Society Meeting, Minneapolis, June 18-23, 1978.

82. M. G. White, 1971, The acceleration of heavy ions to very high energies and their scientific significance in Proc. Particle Accelerator Conference, Chicago, p. 1115.

83. H. A. Grunder, et al., 1971, Acceleration of Heavy Ions at the Bevatron, Science 174: 1128.

84. W. S. Schimmerling, 1979, New Sources of Radiation, Lecture No. 11, in "Advances in Radiation Protection and Dosimetry in Medicine." Course Proceedings, International School of Radiation Damage and Protection, Ettore Majorana Centre for Scientific Culture, Erice, Italy.

85. Curtis, S. B., The New Particles and Their Application to Medicine, Lecture in "Advances in Radiation Protection and Dosimetry in Medicine." Course Proceedings, International School of Radiation Damage and Protection, Ettore Majorana Centre for Scientific Culture, Erice, Italy.

86. F. Bacon, Advancement of Learning. Book I, V 8 (1605).

ACKNOWLEDGEMENT

This work was supported by the U.S. Department of Energy under contract No. WO7405-ENG-48.

ADVANCES IN ENERGY DEPOSITION THEORY

Herwig G. Paretzke

Institut für Strahlenschutz
Gesellschaft für Strahlen- und Umweltforschung
D-8042 Neuherberg, Fed. Rep. Germany

1 INTRODUCTION

The subject "energy deposition" of radiation has several theoretical aspects, which could be classified e.g. according to the scale or object of interest as
- macroscopic (e.g. dose distribution in extended bodies or an organ),
- microscopic (e.g. in a mammalian cell or a makromolecule),
- radiation field related (e.g. radiation shielding, differential energy loss), and
- target related (e.g. yields of chemically reactive species, induction of mutations).
It is clear, that it will not be possible to give you even only a rough overview over the whole field within the time allotted to this lecture. Since we are here mainly concerned with radiation protection and dosimetric problems in medicine, I thought you might be interested in recent advances in the area of microscopic target related studies. Only a few remarks will be made as to the manifold possibilities we have at hand today to perform energy deposition calculations on a macroscopic scale e.g. to compute organ doses in heterogeneous human phantoms. However, even this restriction in the subject does not allow to discuss in full detail all the interesting ideas and results which could be obtained in recent years in this field. Thus a selection of a few highlights had to be made and I can only hope that my necessarily subjective choice of the menu fits your taste. To this purpose I will also retain from bothering you with too many formulae or particulars

51

which are only of interest to the specialists, but will
rather try to give you some impressions of our present
knowledge and remaining problems in energy deposition
theory and of what can or cannot be calculated today.

2 IMPORTANCE OF ENERGY DEPOSITION AND CHARGED PARTICLE TRACK STRUCTURE

Already decades ago it has been realized that the
knowledge of dose alone is not sufficient to specify
the amount of changes produced by interaction of radiation with matter. An example for this is given in fig. 1
showing that in thin scintillators (here Na I doped with
Tl) the yield of luminescence photons per absorbed energy
for fast ions is not a constant and even is not uniquely
related to the rate of energy loss LET of the particles
(after Meyer and Murray (1962)).

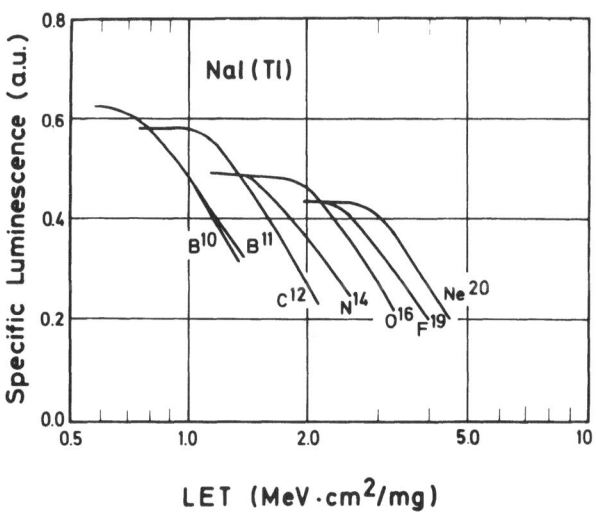

Fig. 1 Specific luminescence of various fast ions in a
       Na I (Tl) scintillator as a function of LET (after
       Meyer and Murray (1962)).

Detailed information on the microscopic pattern of new chemically active species has been found to be an indispensable prerequisite for any deeper understanding of the fundamental mechanisms of most radiation effects. Such knowledge is also necessary to predict radiation effects of a certain radiation field from knowledge on the effects of other radiations.

This importance of the so-called physical phase of energy deposition essentially stems from the fact that it represents the initial condition for all subsequent chemical and biological processes initiated by the radiation interaction (Table 1). That means that, under exactly fixed conditions for the residual parameters, different final effects produced by same doses of two radiation fields can only be due to differences in their respective pattern of primary events.

Table 1. Schematic list of elements of the causality chain and typical time scales of biological radiation effects

	Elements	Time Scale
1	Indirectly Ionizing Radiation (X- or γ-rays, neutrons)	
		$10^{-24} - 10^{-14}$ sec
2	Directly Ionizing Radiation (electrons, protons, alpha particles, recoil atoms, etc.)	
		$10^{-16} - 10^{-14}$ sec
3	Energy Loss and Deposition Stage (Primary Physical Track Structure)	$< 10^{-13}$ sec
4	Physico-Chemical Stage	$10^{-12} - 10^{-8}$ sec
5	Chemical Process in Radiation Thermal Equilibrium	10^{-7} sec - hrs
6	Subcellular and Cellular Biochemical Reactions	msec - hrs
		hrs - days
7	Final Early Cellular Effects	
		years - centuries
8	Final Late Somatic or Genetic Effects	

Because of this basic role in radiation research,
many experimental and theoretical attempts have been made
during the past decades to obtain such information on the
microtopography of molecular changes. By the way, it is
surprising (and sometimes somewhat frustrating to us),
that some of the results obtained more than 30 years ago
without our present knowledge of cross sections, reaction
mechanisms, etc. (e.g. by Lea (1944)) and without the
help of computers only from estimates and guesses, now
turn out to be rather accurate and useful. The approaches to characterize the energy deposition pattern can
roughly be devided into two classes, namely
- the micro-dosimetric approach putting emphasis on the
  experimental or theoretical determination of the dose
  (i.e. absorbed energy per mass element) in microscopic
  volumes and on the statistical variation of this quantity, and
- the charged particle track structure approach putting
  emphasis on the determination of the spatial distribution of localized primary events and their correlations,
  i.e. the determination of the joint probabilities p
  $(A_1, \vec{X}_1, A_2, \vec{X}_2, \ldots A_n, \vec{X}_n)$ to find new active spezies
of reasonably long lifetime (e.g. 10-11 sec) of types A_i
at locations $\vec{x}_i$ in a charged particle track.

The essential difference between both strategies is
that microdosimetry is dose-oriented and implicitely uses
the so-called optical approximation (see e.g. Platzman
(1967)) which leads to a constant mean yield of new spezies per absorbed energy. This assumption is based on the
Bethe-Born theory and is applicable to structureless particles of high velocities and for low atomic activation
energies (i.e. up to, say, several ten eV). The track
structure theory, on the other hand - as mentioned by
Fano (1970) - tries directly to describe and predict the
spatial distribution of localized events with a minimum
of detailed analysis and of assumptions concerning preliminary physical processes. That is, specification of
the nature of the activations and of their spatial correlations are the main tasks in track structure theory.

It is evident that from knowledge of such spatial
event distributions microdosimetric quantities can readily be and have already been calculated. Also I believe
that the physical concepts of microdosimetry represent
simplifications with rather severe limitations. For this
reason and because the following lecture will deal in detail with the concepts and application of microdosimetry,
I will here put more weight on recent advances in our
knowledge about energy deposition as they result from
track structure studies.

3 ROLE OF ELECTRONS IN ENERGY DEPOSITION

    Whenever photons interact with matter, they create
secondary electrons. They do so, either by photo- or
Comptoneffect or by pair production, whereby the relative
importance shifts with increasing energy from the first
to the last mechanism. Thus, the energy deposition of
X- or γ-ray fields on a microscopic scale is determined
almost exclusively by energy losses of their secondary
electrons.

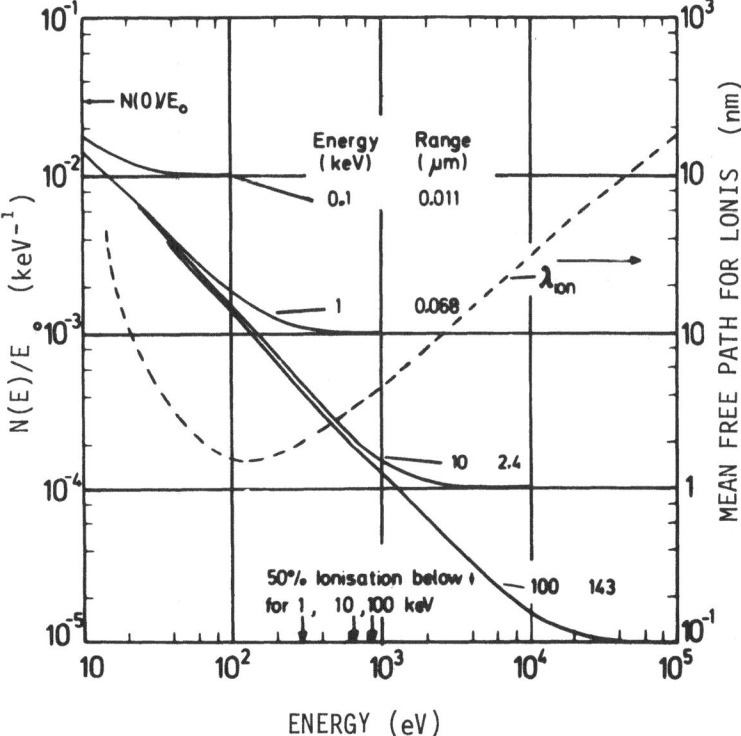

Fig. 2 Relative multiplication factors, ranges
        and mean free paths for ionizations in water
        vapour for 0.1, 1, 10, 100 keV electrons. Indi-
        cated is also the mean electron energy as regards
        the ion pair production and the W-value = $N(0)/E_0$

It is almost superfluous to state that also the
energy deposition pattern of fast electrons essentially
is determined by their secondary electrons. However, it
should be pointed out that even for MeV electrons most
of the excitations and ionizations are due to their low
energy higher order electrons. In fig. 2, we find some
useful information on the energy degradation of 0.1, 1,
10, 100 keV primary electrons in water vapour, which is
considered as tissue equivalent in this context, namely
- their ranges,
- the mean free paths between successive (primary) ionizations,
- the total number of electrons divided by the primary
  energy that can be found with starting energies larger
  than the abscissa value in matter, when one primary
  electron slows completely down,
- the energy value, where 50 % of all ion pairs are produced by electrons while having an actual energy above
  or below that value
- the high energy W-value which is identical to the number of all secondary and higher order electrons starting with energies greater than 0 eV divided by the
  primary electron energy.

We find that even for a 100 keV electron 50 % of the
yield of ion pairs is created by secondary electrons with
an actual energy below 1 keV, and that about 100 secondary electrons are set in motion with start energies above
100 eV and about 3300 electrons at all by a single fast
primary electron. The mean distance between two successive
ionizations is about 0.2 µm at 100 keV but only 15 Å at
100 ev. Evidently such a difference in ionization density
can also lead to differences in the local relative biological effectiveness of such electrons (Harder (1964),
Paretzke (1975) and (1979 a.)).

When neutrons penetrate matter, their interactions
produce recoil atoms and for MeV neutrons typically about
90 % of the neutron dose is due to energy deposition by
their recoil protons. It can be shown by simple integration over the energy transfer cross sections that about
75 % of the energy loss of fast ions (which may be either
recoil particles in neutron fields or accelerated beam
particles) is used to produce secondary electrons. Thus
also for neutron or heavy ion irradiations electrons play
an important role. Their lineal frequency of emission
along ion paths is given in fig. 3 where a typical molecular dimension of 10 Å has been used as unit of lenght.
This figure shows that protons of 0.1 MeV on an average

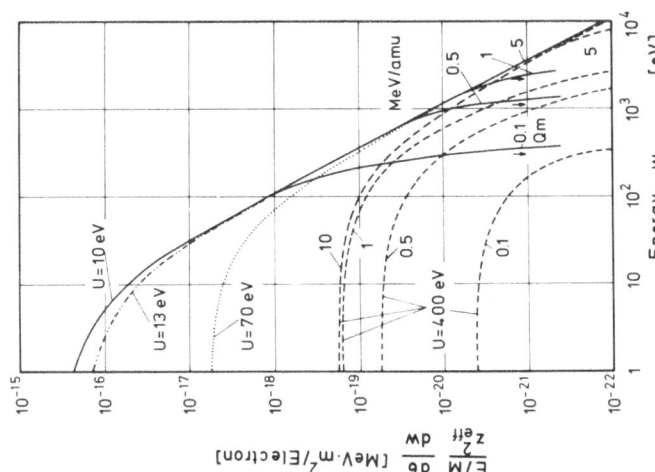

Fig. 4 Differential secondary electron emission cross section per target electron bound with potential energy U for ions with specific energy E/M and charge z eff. Parameters are the specific ion energy and the potential (binary encounter approximation)

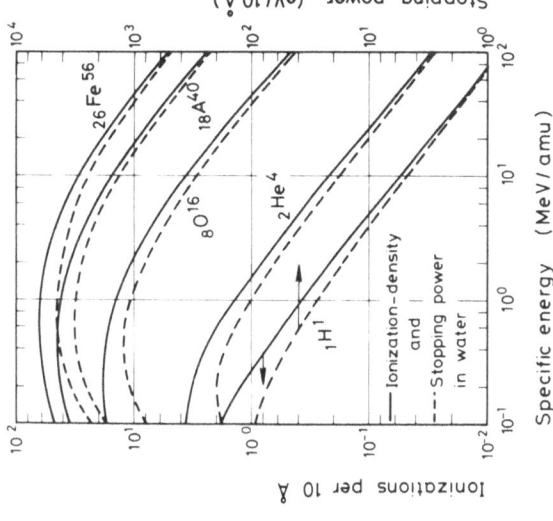

Fig. 3 Ionization density and stopping power of various ions in water

ionize such a molecule even twice during penetration whereas a 10 MeV p ionizes only every 200 Å. An alpha particle produces roughly a four times higher lineal ionization density leading also to a higher relative efficiency for biological damage. Fig. 4 gives approximative starting energy spectra of the secondary electrons for ions of specific energy E/M and effective charge Z_{eff} and for a target ionization potential of U ev. Here I only want to draw your attention to the steep (approximately quadratic) decrease of the cross section with increasing secondary electron energy. Therefore most of the secondary electrons have an excess energy of only a few eV, and electrons with starting energies above a few hundred eV are very seldom emitted along an ion path as will be seen later.

I hope that it became clear from these few examples which central role especially slow electrons (say, below 10 keV) play for the microscopic energy deposition of X- and γ-rays, of neutrons and of charged particles. As regards the energy deposition on a makroscopic, i.e. on a centimeter scale, e.g. in phantoms, however, secondary electrons may safely be neglected because of their small, i.e. micrometer ranges (except for a few exceptions, like surface of interface effects).

4 TRACK STRUCTURES OF ELECTRONS

Since the importance of the energy deposition by electrons is common to all radiation fields, we will start with a discussion of their track structure. Most detailed theoretical information on the microtopography of events in a charged particle track can be obtained from Monte Carlo-type calculations simulating all essential interactions of the primary and all secondary and higher order particles in as much detail as necessary and/or possible. To this purpose one first has to analyse all interaction processes and possible molecular changes as regards their importance to the problem under consideration. Then data for the relevant processes have to be extracted from literature and evaluated. This is by far the most difficult task of such calculations and the derivation of one reliable and consistent set of relevant cross sections for electron energy degradation in one target gas takes easily one or two man-years. I would also like to point out that close collaboration and friendly personal relations to other theoretical or experimental groups are an essential necessity for any progress in this complex and interdisciplinary area, and I appreciate very much the great help I obtained from numerous colleagues over the years.

Such cross section evaluations can be found e.g.
in papers of Green and his coworkers (e.g. Olivero,
Stagat and Green, 1972; Kutcher and Green, 1976), Kim
and Inokuti (e.g. Kim, 1975; Kim and Inokuti, 1973) and
of Paretzke and Berger (1979). Fig. 5 gives you an example
of an intercomparison of the total inelastic (after the
article mentioned last),the total ionization and the total
excitation cross sections of two sets which turned out to
be rather satisfactory, although sometimes quite different assumptions were made. In our simulation code
for water vapour e.g. we presently take into account 14
types of excitations, ionization from 5 molecular shells
and elastic scattering. Thus rather detailed information
is available on types and locations of events in electron
tracks.

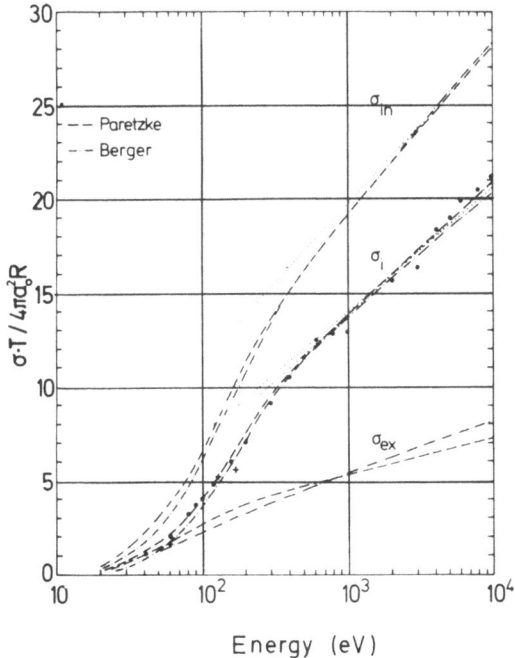

Fig. 5 Fano plots of total inelastic, total ionization,
and total excitation cross sections for electron
impact on water vapour as a function of electron
energy (intercomparison of two cross section sets)

Other electron transport codes of Berger (1974),
Terrisol and Patau (1974) and Hamm et al. (1975) are
equally complex. They all lead to essentially the same
results as regards the quantities and the accuracy required in radiation biology, although they are based on
partly different approximations. This fact and the agreement that has been found in comparisons with experimental
results (Paretzke, 1974; Combecher et al., 1979; Wilson
et al., 1979; Paretzke, 1979 b) gives some confidence in
the theoretical results obtained.

In this lecture, I do not want to confuse you with
too many quantitative results which can now be derived
from Monte Carlo track structure codes. One example shall
be representative for many. Often I am asked to calculate
"cluster size" distributions as they are observed in cloud
chamber pictures. However, in practise it turns out to be
rather difficult to define a "cluster" as regards the distance up to which you include ions as belonging to a certain cluster. Fig. 6 gives you the probabilities to find
more than N ion pairs within a sphere of a radius given on
the abscissa in a complete track of a 0.2, 0.5, or 5 keV
electron. From such data you may derive your own distributions according to your personally preferred cut-off
radius!

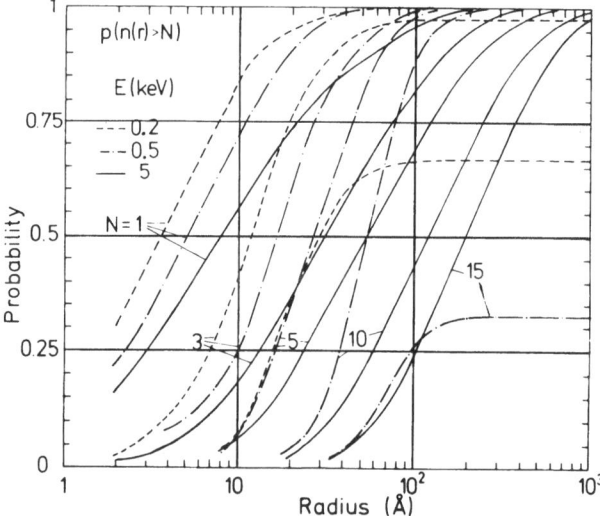

Fig. 6 Probability to find more than N ionizations within a sphere as a function of its radius for complete tracks of 0.2, 0.5, and 5 keV electrons in
water vapour.

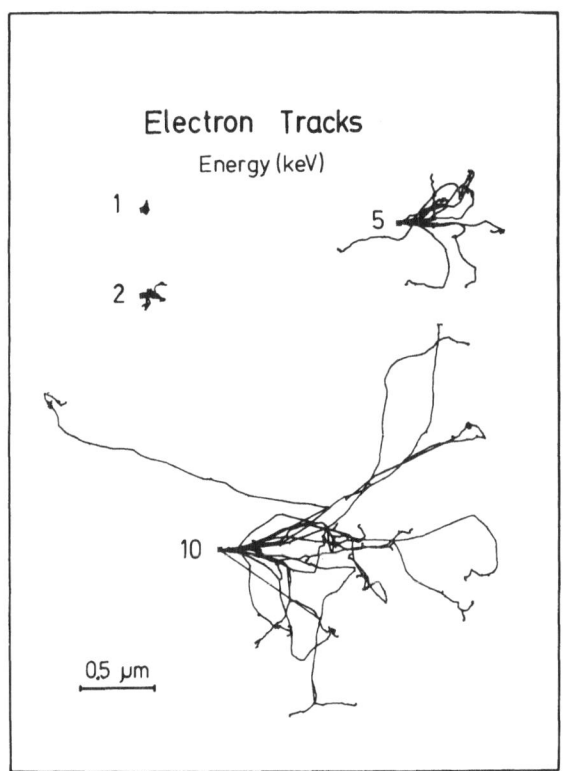

Fig. 7 Ten tracks each for 1,2,5 and 10 keV electrons
in water vapour (unit density)

You might be interested to see how simulated electron tracks look like. In figs. 7,8 and 9 you find some
electron tracks, which were not selected according to

some criterion but taken as they were created and drawn
on the X-Z-plane of the co-ordinate system by the computer with the given starting number for the random number
generator. Fig. 7 shows 10 tracks of 1,2,5 and 10 keV
electrons without symbols at the locations of events.
From this qualitative picture one gets the impression
that ten 1 keV electrons because of their higher event
density could well show a larger biological effect as
one 10 keV electron. Fig. 8 shows three absorption
events of Al-K$_\alpha$- photons (1.5 keV) as they were used in

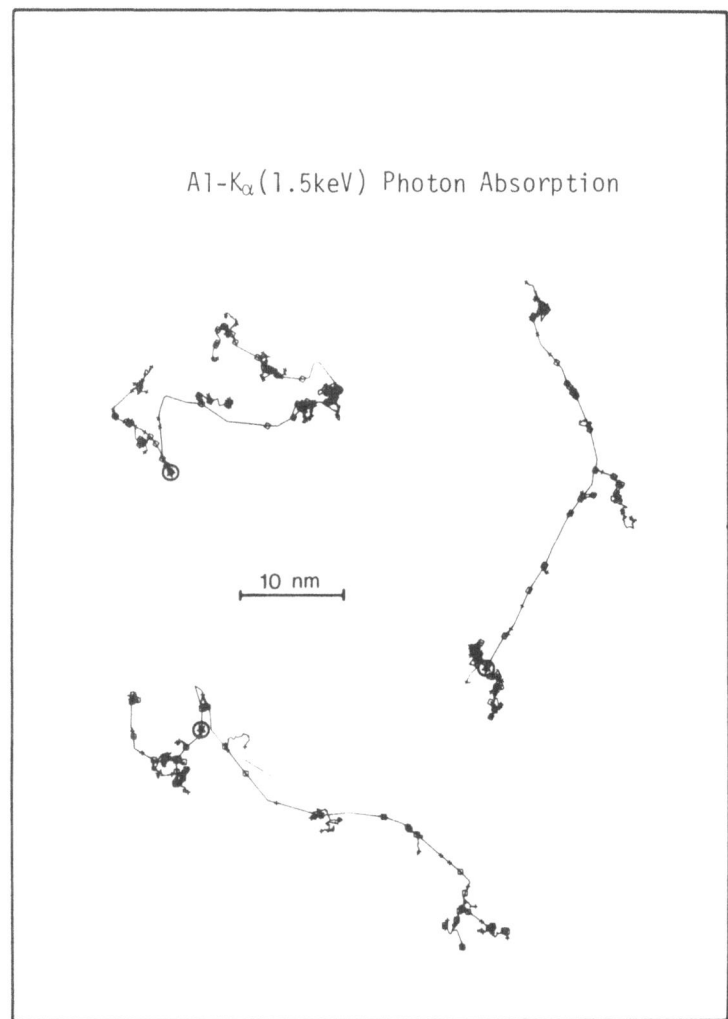

Al-K$_\alpha$(1.5keV) Photon Absorption

10 nm

Fig. 8 Three photon absorption events in water at the
       indicated locations leading typically to a 1 keV
       photo electron and a 0.5 keV Auger electron.

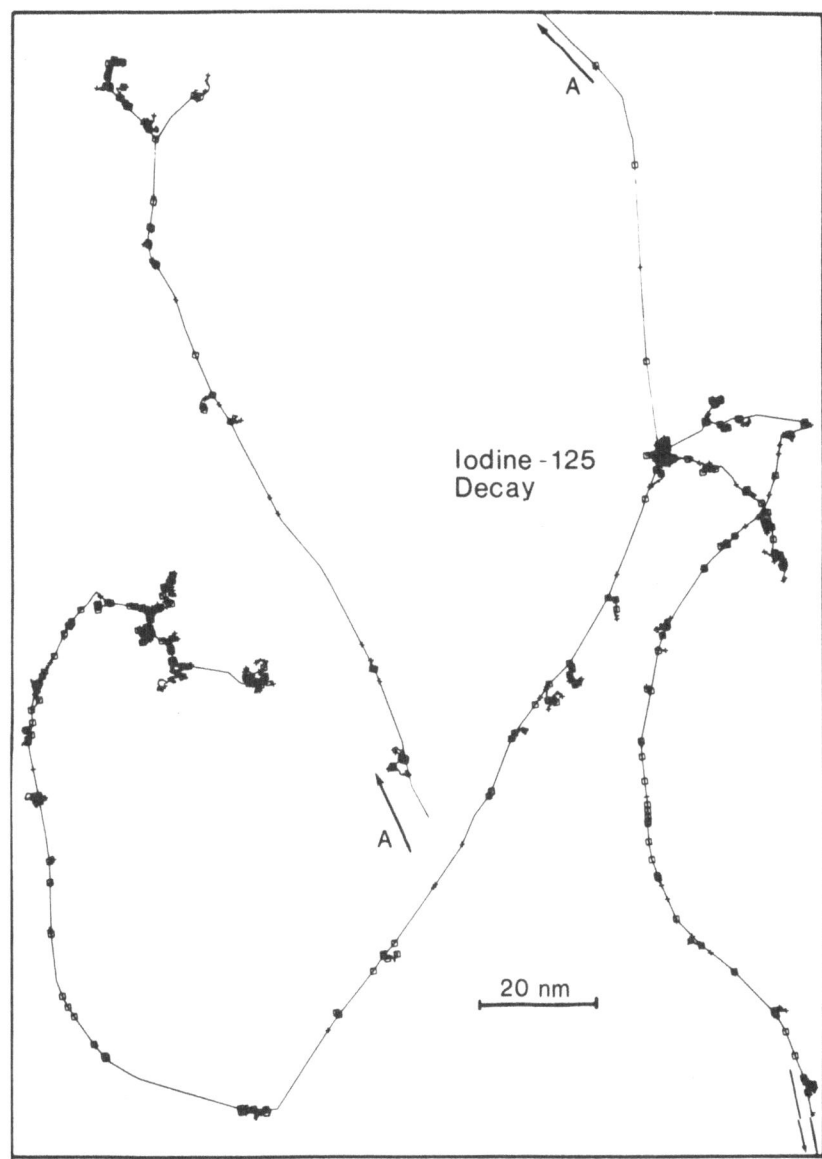

Fig. 9 One decay of Iodine-125 leading to 20 electrons
       (decay spectrum from J. Booz, KFA Jülich). Note
       the break and shift at point A for graphical
       reasons.

survival experiments. Locations of excitations are indicated with a "+ ", those of ionizations with a "▫ ". In
the computer, however, the actual type of ionization or
excitation is stored.

   In fact, we have available 13 data per event in our
book-keeping system, among which are the type of event,
the three cartesian co-ordinates, the three direction
cosines and the actual energy of the electron before the
impact, etc. As a rule of fist we have about 1 inelastic
event per 10 eV energy loss, i.e. the complete slowing
down of one 10 keV electron and all its secondaries down
to 10 eV (where we use to stop the histories) leads to
1000 inelastic and at least twice as many elastic events.
One history takes about 1 sec / keV, i.e. 10 sec in this
case . Finally, fig. 9 shows one decay of the important
internal emitter Iodine-125. This burst of 20 electrons
leads to high event concentrations around the nuclide
with a few higher energy electrons sticking out and de-
positing their energy with lower concentration one or a
few hundred nanometer away. The decay energy spectrum
was calculated by J. Booz, KFA Jülich, with whom together we presently study the energy deposition characte-
ristica of that nuclide.

   I hope that those few track structure pictures give
you some impressions which amount of basic information
has recently become available on electron energy deposition and is waiting to be used for interpretations
of radiation effects.

4 TRACK STRUCTURES OF HEAVY CHARGED PARTICLES

   As mentioned above, a large part of the energy loss
of heavy charged particles is used to emit secondary
electrons along the ion path, whereby about 20 % of the
LET are consumed to overcome the ion potential and about
50 - 60 % is converted to excess kinetic energy of the
ejected electrons. The slowing down history of these
predominantly slow electrons can be treated independently
from the further ion history, and thus the resulting contributions to the ion track structure can be calculated
straight forward as discussed in the chapter before. About
20 % of the stopping power is due to primary excitations.

   Unfortunately, as regards the availability of theoretical or experimental basic interaction cross sections
the situation is much worse for fast ions than for electrons even for gaseous targets.

Apparently there are only rather few experimental groups (mainly L.H. Toburen and W.E. Wilson, Battelle Richland; M.E. Rudd et al., University of Nebraska; and N. Stolterfoht et al., Universität Berlin) interested in measuring absolute differential ionization cross sections for fast ions over a wide enough secondary electron energy range and with acceptable error margins. For protons and structureless fast charged particles (without lightly bound electrons in the projectile) we have already some confidence in the evaluated experimental data. For projectiles with own electrons bound with a few ten eV only, however, we are not yet in a position to make reasonably accurate assumptions on the secondary electron spectrum differential in energy and angle. W.E. Wilson has developed from their experimental data an algorithm for computation of such spectra for 0.3 to 3 MeV protons in water vapour. His subroutine is now used in all our ion track structure programs and substitutes in the low energy range a modified binary encounter approximation which appears at higher ion and secondary electron energies rather acceptable.

There is only very scarce experimental information on total excitation cross sections for fast ions. Although excitations contribute only with 15 - 20 % to their total stopping power, better data are badly needed especially below a few MeV/amu where scaling from equal velocity electron cross sections might lead to rather large errors.

Nevertheless, besides for this problem we are rather confident that our calculated track structures of protons above 0.3 MeV simulate the reality to an acceptable degree (fig. 10). You clearly see the expected behaviour that with increasing energy the density of primary events decreases. The fact that a 3 MeV proton can emit higher energy electrons compared to a 0.3 MeV proton cannot likely be seen on an arbitrary picture of a short track segment because these are rather rare events.

The density of primary events increases roughly with the second power of the ion charge and decreases inversely proportional to the specific ion energy. This effect can clearly be seen on fig. 11, 12 and 13 showing track segments of Helium, Carbon, and Neon ions of 1 and 3 MeV/amu. Fig.11 displays two pictures for each specific energy to give you a slight impression of the range of variation of actual track structures.

In the pictures for the higher charge ions you clearly can distinguish a very dense central zone and an area

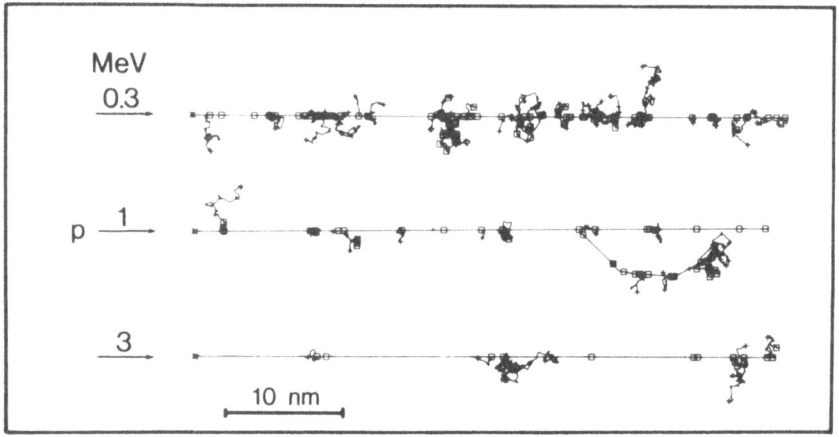

Fig.10   Tracks of 0.3, 1, and 3 MeV protons in water

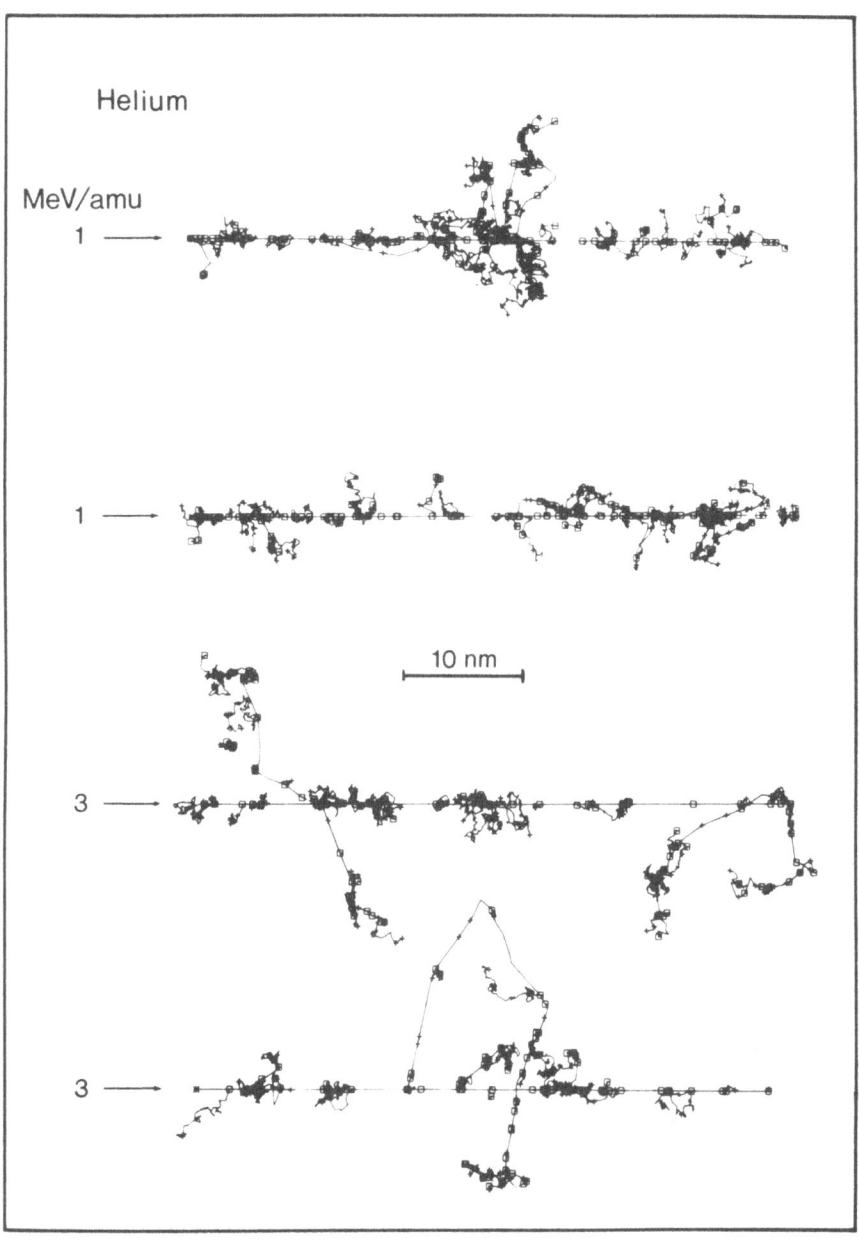

Fig. 11 Two tracks each of 1 and 3 MeV/amu He-ions
in water

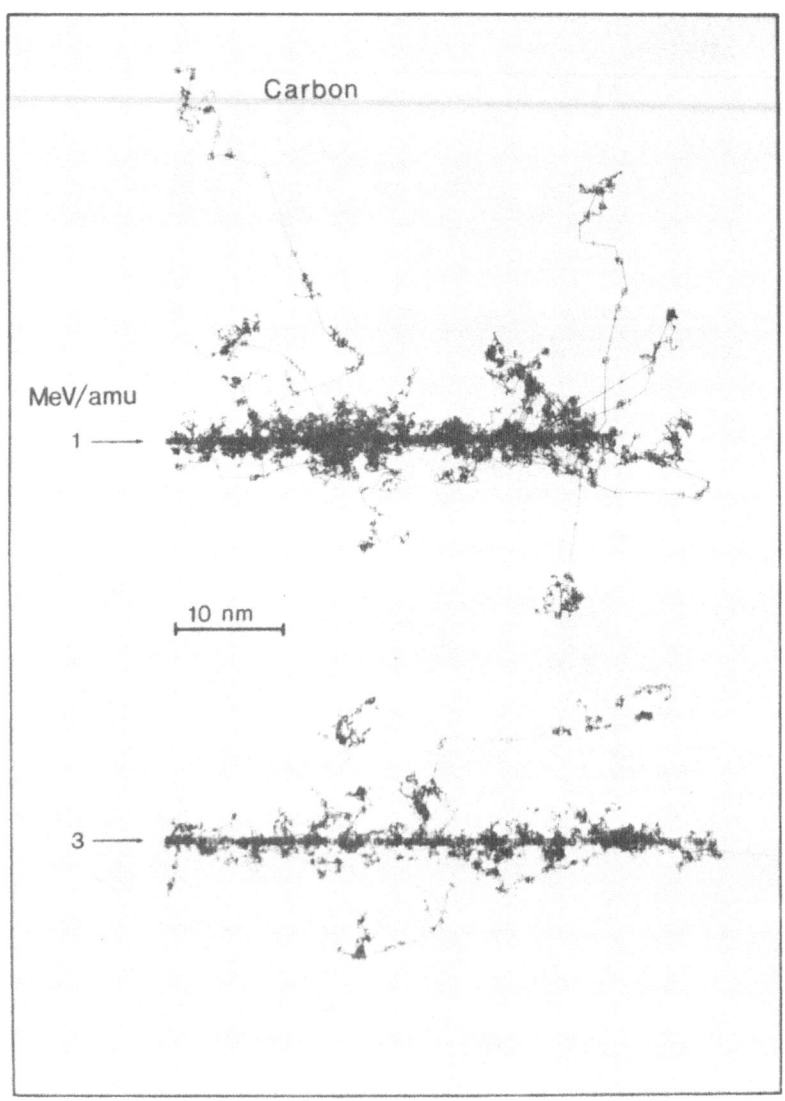

Fig. 12 Tracks of 1 and 3 MeV/amu C-ions in water

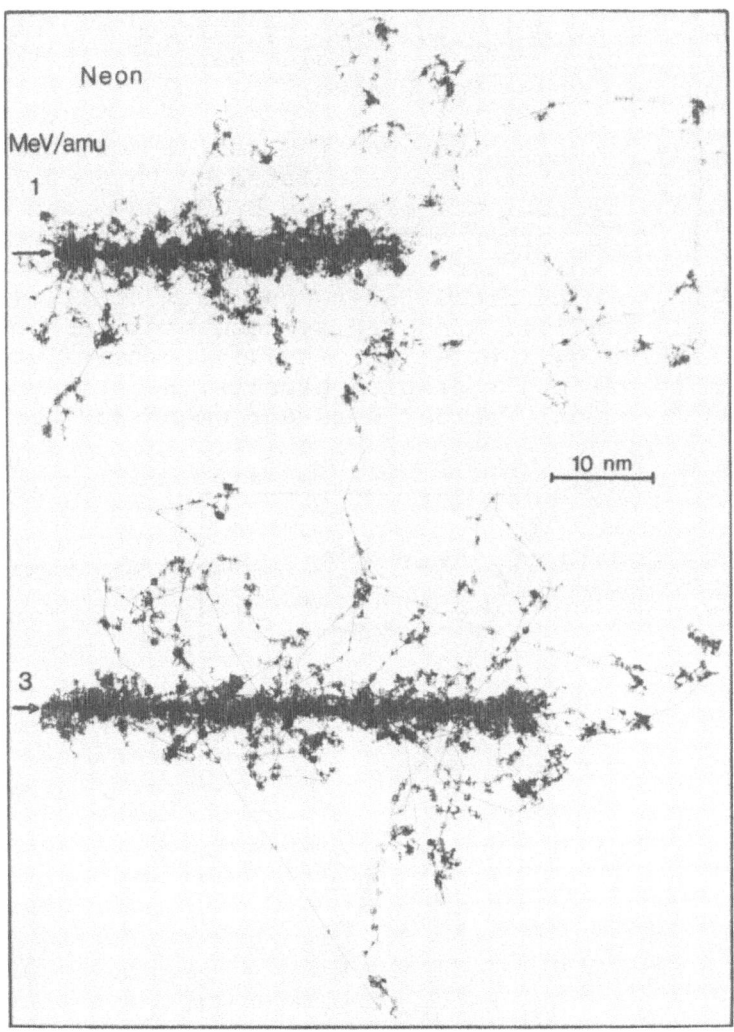

Fig. 13 Tracks of 1 and 3 MeV/amu Ne-ions in water

of low event density which are sometimes called track
core and penumbra or intratrack and infratrack. It might
well be possible that we will have to take into account
non-linear effects in such intense damaged track cores
as they are created e.g. by fission fragments and by
HZE-particles. Since such particles are not too important for radiation protection, we shall not go into further detail as to which interesting effects and mechanisms might play a role in damaging permanently the matter surrounding these ion tracks.

Fig. 14 finally gives you the radial dose profiles
in a 1 MeV proton track as calculated by three different
computational methods and measured by Wingate and Baum
(1967). You find rather good agreement between theory
and experiment which has been observed also for other
ions and up to much higher energies (Baum et al. (1973);
Varma et al. (1975)). Such radial dose profiles have been
used by Katz (e.g. Katz et al. (1971)) to correlate track
structure to various radiation effects. He succeeded in
obtaining frequently rather good agreement and his theory

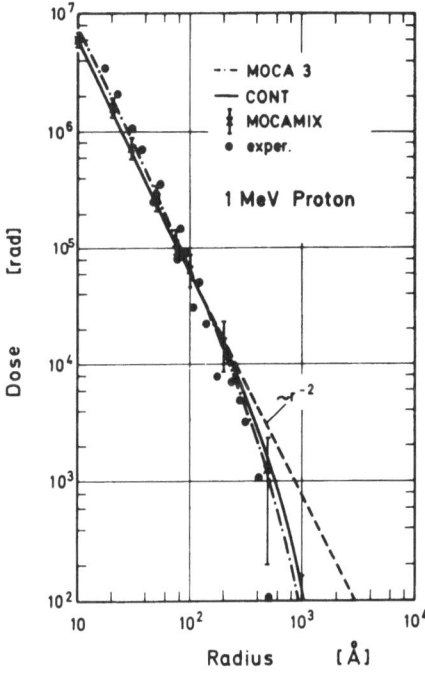

Fig. 14 Dose as a function or radius in 1 MeV proton
         tracks as calculated by three methods and as
         measured.

can be used to empirically predict the relative effects
of heavy ions for many effects. However, from a comparison of fig. 10 with fig. 14 serious doubts could arise
about the validity and usefulness of a concept with ave-
rages the energy deposition at fixed radii along the direction of the ion path.

5 CONCLUDING REMARKS

        In the past ten years some remarkable progress could
be made as regards our knowledge on the energy deposition
of ionizing radiation. This progress essentially is due
to two facts, namely the availability of necessary cross
sections and of sufficiently large and fast computers.
Today we find in the literature computer codes for the
computation of makroscopic energy deposition quantities
as depth and lateral dose in homogeneous phantoms and as
organ dose in structured phantoms. Such calculations can
be made for x- and γ-rays, for neutrons, for fast electrons and ions, and even for π-mesons. In this lecture
I tried to give you some ideas on the recent advances made
in our knowledge on microscopic energy deposition only.

        I will not end without pointing out at least two
problems remaining open in the latter field. We do not
see major problems with track structure calculations in
homogeneous gases and even not for structured targets.
However, problems connected with gas mixtures have not
even been touched yet. In addition, the conversion of
transferred energy into new chemically active species
might be rather different in condensed matter than in
gases especially for energy transfers around the ionization potential. Although such problems are common to
most existent approaches which are alternatives to track
structure theory, this fact does not overcome the enormous difficulties we encounter with yield calculations
for condensed matter. I do not see too many open problems
with energy deposition calculations because these are not
too sensitive to the physical phase state and the exact
chemical nature of the target. However, I am convinced
that the field of track structure theory, i.e. where the
nature and location of primary events play an essential
role, will continue to be an interesting scientific chal-
lenge for some more decades. We even do not yet have good
concepts as to how to describe charged particle track
structures in general by some data reduction method
without losing essential information.

REFERENCES

Baum, J.W., Varma, M.N., Wingate, C.L., Paretzke, H.G.,
    and Kuehner, A.V., 1973, Nanometer Dosimetry of
    heavy ion tracks, BNL-Report 18219
Berger, M.J., 1974, Some new transport calculations on
    the deposition of energy in biological materials
    by low-energy electrons, in : "Fourth Symposium
    on Microdosimetry", Booz, J., Ebert, H.G., Eickel,
    R., and Waker, A., eds., CEC, Brussels, EUR 5122
Combecher, D., Kollerbaur, J., Leuthold, H., Paretzke,
    H.G., and Burger, G., 1974, Energy spectra of de-
    graded electrons in water vapour and in carbon,
    loc. cit. Paretzke (1979 a)
Fano, U., 1979 , The formulation of track structure the-
    ory, in : "Charged Particle Tracks in Solids and
    Liquids", Adams, G.E., Bewley, D.K., and Boag, J.W.,
    eds., The Institute of Physics, London.
Hamm, R.N., Wright, H.A., Ritchie, R.H., Turner, J.E.,
    and Turner, T.P., Monte Carlo calculation of trans-
    port of electrons through liquid water, loc. cit.
    Paretzke (1975).
Harder, D., 1964, Physikalische Grundlagen zur relativen
    biologischen Wirksamkeit verschiedener Strahlenar-
    ten, Biophysik, 1:225.
Katz, R., Ackerson, B., Homayoonfar, M., and Sharma, S.,
    1971, Inactivation of cells by heavy ion bombard-
    ment, Radiat. Res. 47:402.
Kim, Y.-K., 1975, Energy distribution of Secondary Elec-
    trons, Radiat. Res. 64: 96
Kim, Y.-K., and Inokuti, M., 1973, Slow electrons ejec-
    ted from He by fast charged particles, Phys. Rev.
    A 7:1257.
Kutcher, G.J. and Green, A.E.S., 1976, Energy deposition
    in liquid water , Radiat. Res.67: 408
Lea, D.E., 1944, "Actions of Radiation on Living Cells",
    University Press, Cambridge.
Meyer, A., and Murray, R.B., 1962. Effect of Energetic
    Secondary Electrons on the Scintillation Process
    in Alkali Halide Crystals, Phys. Rev., 128:98.
Olivero, J.J., Stagat, R.W., and Green, A.E.S., 1972,
    Electron deposition in water Vapor with atmosphe-
    ric Applications, J. Geophys. Res. 77:4797
Paretzke, H.G., 1974, Comparison of track structure cal-
    culations with experimental results, loc. cit.
    Berger (1974)
Paretzke, H.G., 1975, An Appraisal of the Relative Im-
    portance for Radiobiology of Effects of Slow Elec-
    trons, in: "Fifth Symposium on Microdosimetry",
    Booz, J., Ebert, H.G., and Smith, B.G.R., eds.,

Comm. Europ. Communities, Brussels, EUR 5452.

Paretzke, H.G., 1979 a, On Limitations of Classical Microdosimetry and Advantages of Track Structure Analysis for Radiation Biology, in: "Sixth Symposium on Microdosimetry", Booz, J., and Ebert, H.G., eds., Harwood Academic Publishers Ltd., Brussels.

Paretzke, H.G., 1979 b, Track Structure calculations and their accuracy, 6th Int. Congr. on Radiat. Res., Tokyo, in press.

Platzman, R.L., 1967, Energy Spectrum of Activations in the Action of Ionizing Radiation, in : "Radiation Research, Proceed. of the 3rd Intern. Congress", Silini, G., ed., Cortina d Ampezzo

Terrisol, M., and Patau, J.P., 1974, Simulation du transport d électrons d énergie inférieure à un keV par une méthode de Monte-Carlo, loc. cit. Berger (1974).

Varma, M.N., Paretzke, H.G., Baum, J.W., Lyman, J.T., and Howard, J., 1975, Dose as a function of radial distance from a 930 MeV He-4 ion beam, BNL-Report 20476 R

Wilson, W.E., Toburen, L.H., and Paretzke, H.G., Calculation of energy deposition spectra in small gaseous sites and its applicability to condensed phase, loc. cit. Paretzke (1979 a).

Wingate, C.L., and Baum, J.W., 1967, Micro-radial distribution of dose and LET for alpha and proton beams, BNL-Report 14767.

MICRODOSIMETRY AND ITS APPLICATION TO BIOLOGY

Harald H. Rossi

Radiological Research Laboratory, Department of
Radiology, Cancer Center/Institute of Cancer Research
Columbia University College of Physicians and Surgeons
New York, NY   10032

Among the various subspecialities of biology, radiobiology has
attracted perhaps the most extensive interest of physicists. Some
of the most fundamental contributions to radiobiology have been made
by physicists and even engineers who became completely engrossed in
the subject and its biological ramifications to such an extent that
it is often not realized that they started their career in nuclear
physics or electrical engineering.

The probable explanation for this association is that ionizing
radiation is not only one of the most important physical agents in
biology, but also that the very process of energy absorption by
irradiated tissues implies certain characteristics of the resulting
biophysical process and especially its kinetics. It is of fundamental importance that at the cellular level energy absorption is
quantal in that a few events and sometimes even only one event (the
passage of a charged particle) can cause cell injury or death. The
absorbed dose, D, which is the most important physical quantity in
radiobiology does, however, not directly relate to this mechanism.
The absorbed dose which is energy absorbed in an irradiated volume
divided by the mass in that volume, is an average value of energy
concentration that can be determined only when energy is deposited
by many particles. It follows that at doses that are of interest
to radiobiology, energy concentration at the cellular level fluctuates greatly and may be quite different from the absorbed dose.

This is illustrated in Figure 1 which is a schematic representation of a monolayer of some 150 cells that have received a dose
of 10 milligray (1 rad) of gamma rays or neutrons. In either case,
the radiation energy is of the order of 1 MeV (e.g., one may be

75

CELL DIAMETER $\simeq$ 5 μm

a   b

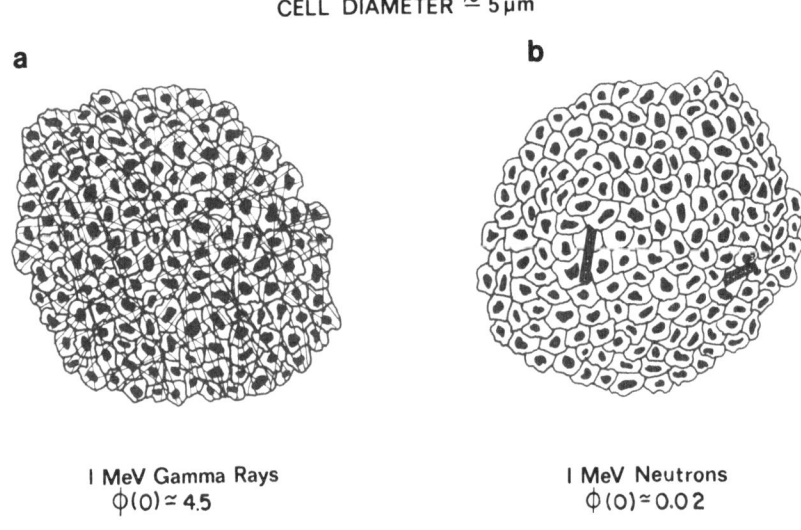

I MeV Gamma Rays
$\phi(0) \approx 4.5$

I MeV Neutrons
$\phi(0) \approx 0.02$

$\phi(0)$ is the Event Frequency per Cell

Fig. 1.   Schematic representation of track patterns produced in
about 150 cells by 10 mGy of gamma radiation (a) and
10 mGy of neutrons (b).

dealing with ^{60}Co gamma radiation and fission neutrons). It will
be seen that in the case of gamma radiation each cell is traversed
by some five secondary electrons whilst in the case of neutrons only
about 1 cell in 50 receives any energy at all. This is an immediate
consequence of a more than hundred-fold difference in mean energy
loss in the tracks of protons and electrons in this energy range.
Figure 1 suggests the need for a quantity that is similar to the
absorbed dose but takes into account such extreme fluctuations.
This quantity is the specific energy, z, which is the energy
actually absorbed in a (small) specified volume divided by the mass
in this volume. For a given D, z must always have a range of possible values which depends on the number of events taking place in the
volumes of interest and on the trajectory of the particles causing
these events. However, there is always a finite probability that
z is zero. This is quite evident in Figure 1-b. However, in
Figure 1-a the specific energy in the cells irradiated by gamma
radiation is usually not too different from the absorbed dose and

one is dealing with a probability distribution of specific energies
that is a reasonably narrow gaussian distribution about the absorbed
dose. However, this is evidently not true if instead of considering
a specific energy in the cell as a whole, one considers the specific
energy in some substantially smaller subcellular structure.

Increasing variability of specific energy is also observed when
with a given radiation and a given volume the absorbed dose is
decreased. As shown in Figure 2, $f(z)$, the probability distribution
of z, is a narrow gaussian distribution centered about high absorbed
doses. As D is decreased, the distribution broadens, but when sufficiently small values of D are reached, the distribution acquires
a more or less fixed shape and merely decreases in amplitude. This
is due to the fact that at sufficiently small doses it is unlikely

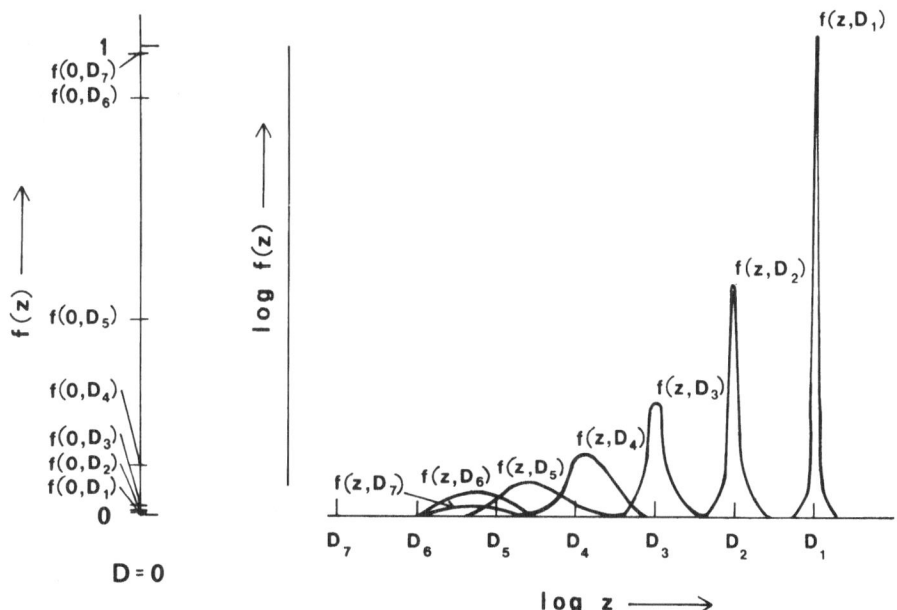

Fig. 2. Schematic representation of $f(z)$ the probability of z in
a fixed volume as a function of decreasing absorbed dose
(D_1, D_2, etc.). At large values of D, $f(z)$ is a narrow
gaussian centered at D. As D decreases the width of the
distribution increases. However, at low values of D the
distribution approaches a fixed shape and merely decreases
in amplitude. This is the spectrum of energy loss of
single particles. As D decreases the probability of
$z = 0$ increases from 0 to 1 (left side of graph).

that more than one particle traverses the cell and as the dose is
reduced, the energy distribution due to single traversals remains
the same and there is merely an increased probability that no event
occurs. This situation obtains at much larger neutron doses.
However, in all cases the mean value of specific energy, $\bar{z}$, is
equal to D.

It follows that D would be an adequate quantity for cellular
radiobiology if the biological effectiveness of radiation were
proportional to specific energy. This would require that biological
effectiveness is independent of radiation quality which certainly
is not the case. Indeed, in the example shown in Figure 1 neutrons
are far more effective biologically than gamma rays. At a dose of
10 milligray, the RBE (relative biological effectiveness) of neutrons
relative to gamma rays is of the order of 50 for many biological
systems. It follows that in plots showing biological rather than
physical equivalence, the total length of electron track should not
be 200, but rather about 10,000 times as long as the total amount of
proton track. It is virtually impossible to provide a graphical
illustration of this, but it is evident that in this case the
specific energy due to gamma radiation would be very nearly equal
to the absorbed dose in volumes that are a great deal smaller than
that of the cells.

The non-linear dependence of biological action on energy concentration becomes also apparent when one investigates the dependence
of RBE on dose. Evidently in the example given in Figure 1-b reduction of the neutron dose by a factor of two must halve the number
of cells killed or injured in any way. However, in order to obtain
the same reduction in effect, the gamma dose must usually be divided
by a factor that is only about 1.4.

It can be shown (Rossi, 1970; Kellerer and Rossi, 1972) that
this behavior indicates that the response of the cell to ionizing
radiation depends basically on the square of the energy that is
deposited. It would thus appear that the fundamental action of
radiation on the cell is one involving second order kinetics. This
observation forms the basis of the theory of dual radiation action
which postulates that cellular impairment by radiation is due to the
formation of lesions which in turn arise from the combination of
pairs of sublesions. It has also been deduced that the mean initial
separation of combining sublesions is about 1 micrometer and that
consequently the physical quantity of basic importance is the square
of the specific energy in sites that have a diameter of the order
of one micron.

It can be shown that the mean value of z^2 is given by

$$\overline{z^2} = \zeta D + D^2 \tag{1}$$

where ζ is the average of the square of the specific energy produced in single events.

It follows that $\varepsilon(D)$, the yield of lesions, should be given by

$$\varepsilon(D) = k(\zeta D + D^2) \tag{2}$$

where k is a constant.

In Eq. 2 the left-hand side is plainly a biological quantity which in fact is frequently considered to be the negative logarithm of the surviving fraction in an irradiated population. D and ζ are obviously purely physical quantities. Consequently k must be a mixed quantity. It can be considered to be proportional to the yield of lesions divided by z^2. A more basic interpretation would be that the $\sqrt{k}$ is proportional to the number of sublesions produced per unit of z.

Although the very simple Eq. 2 has been found to be useful in a variety of circumstances it must be realized that it is an approximation. Most of its limitations were identified when the theory was first presented (Kellerer and Rossi, 1972) and it is hardly surprising that the theory has been found not to apply under conditions where the basic simplifying assumptions were no longer justified.

Perhaps the most important of the limitations rests in the concept of the site. There is no cytological or other physical evidence for the existence for regions within the cell nucleus that have diameters of the order of 1 or 2 microns and where one might expect that the molecular alterations which constitute the sublesions combine with essentially constant probability. It is far more likely that the site diameter is simply an effective mean separation of combining sublesions and that combinations can occur with varying probability throughout the nucleus. This more general formulation known as the distance model has been considered in some detail (Kellerer and Rossi, 1978).

The distance model has been supported by the results of the molecular ion experiment (Rossi, et al., 1978). This involves the irradiation of cells by pairs of charged particles (usually deuterons) which traverse cells at mean separations that range from a tenth of a micrometer to about a micrometer. An analysis of the findings shows not only that the probability of sublesion combination depends on separation, but also that it follows a highly skewed bimodal distribution. This has been taken to indicate that radiosensitive material is distributed in the cell nucleus in an inhomogeneous form which might be approximated by separated floccules of DNA. It would appear that lesions can be produced by combinations of sublesions that are produced in the same floccules or alternatively by sublesions that were produced in different floccules. Since either

type can result from single track action, it would seem to follow
that even the linear component of the survival curve is due to two
types of damage that might differ with respect to recovery rate,
sensitivity to radiation modifiers, etc. While some of these conclusions may still be speculative, there can be little doubt that
the site model has been shown to be only an approximation and that
it must be expected that a given amount of energy absorbed in a
volume having a dimension of the order of a micrometer is biologically more effective if it is deposited in a short track rather than
in a longer one which traverses the volume.

There is probably another effect which influences dual radiation action at an even smaller scale of distance and that is varying
efficiency in the production of sublesions (Rossi, 1972). These
molecular alterations (possibly double strand breaks of DNA) may
well depend on energy concentration at the nanometer level and they
would be reflected in a dependence of k on LET. There is increasing
experimental evidence for the existence of this effect and although
it may not be very large, it may cause changes of k by perhaps a
factor of 3.

A further effect which exists beyond doubt is that of saturation. Plainly there must be a limit to the increase of biological
effectiveness that accompanies an increase in local energy concentration and the point must be reached where there is saturation with
consequent waste of energy at still higher concentrations. A correction for this effect has been provided (Kellerer and Rossi,
1972), but it is at this point purely empirical.

Finally, it must be recognized that at least in some instances
radiant energy may not directly cause DNA damage, but that it may
produce relatively simple highly reactive products (such as active
radicals produced in water) that may diffuse some distance from the
point of origin before interacting with DNA. This process blurs
the original pattern of energy deposition (Kellerer and Chmelevsky).

It is a remarkable and rather fortunate fact that none of these
complicating features alter the dependence of lesion formation on
only two terms, one of which is linear in dose with the other being
quadratic in dose. However, the physical interpretation of the
coefficients in these two terms are altered in ways which can be
complicated and which are in part only poorly known. However, even
without detailed knowledge one can rather definitely identify the
kinds of biological and biomedical data for which interpretation
in terms of the theory should be sensitive to these complications
and those that are more likely to be interpreted in terms of Eq. 2.

Turning first to experiments in which the effects of different
radiations are compared (i.e., RBE experiments) it is evident that
all four of the factors given above may cause complications because

every one of them may differ with radiation type. Nevertheless,
application of Eq. 2 has resulted in at least a moderately good
agreement between theory and experiment when the RBE as a function
of neutron energy was investigated after correction for the saturation effect was made (Hall, et al., 1975). The remaining discrepancy
is a somewhat high predicted RBE at high neutron energies. This is
very likely due to the far greater range of the protons and a lower
k than would be expected at low LET. Unpublished experiments
designed to determine RBE of heavy ions relative to x and gamma rays
do in fact disclose a variation of k with LET (Bird, et al). However,
one would expect that regardless of these complications the relation
between RBE and dose of a high LET radiation should contain a wide
range in which the RBE is inversely proportional to the square root
of the neutron dose. This seems to be almost universally observed.

In experiments involving only one type of radiation, one is
not dealing with any differences in k but saturation could not only
change the value of ζ at low doses, but also conceivably produce a
distortion of the dose-effect curve at high doses. If the theory is
applied to survival curves for irradiated cells, it is usually
assumed that the inactivation is proportional to the number of
lesions produced. One therefore obtains the relation

$$S = e^{-(\alpha D + \beta D^2)} \tag{3}$$

To the extent that the various complications can be disregarded,
$\alpha = \kappa\zeta$ and $\beta = \kappa$. Survival curves for all types of radiations are
now frequently analyzed according to Eq. 3, although it must be
realized that in the usual case of asynchronous cells, α and β are
mean values over the cell cycle and that if there are substantial
variations in sensitivity through the cycle, Eq. 3 may not strictly
apply.

Perhaps the most straightforward application of the theory can
be made in various experiments where the dose is fractionated or
the dose rate is varied. Such experiments are particularly important for low LET radiations where the quadratic component predominates. For instance, if a low LET dose is given in separate fractions and sublesions produced in each fraction can be partially or
completely repaired between fractions, they can no longer combine
with sublesions produced in later fractions. Thus, in the simplest
case in which there is complete recovery between two equal fractions,
one obtains

$$\varepsilon(2 \times D/2) = 2(k[\zeta D/2 + (D/2)^2] = k(\zeta D + D^2/2) \tag{4}$$

In the extreme case where D is much larger than ζ, the recovery
reduces the yield of lesions by a factor of 2. Various applications
of Eq. 4 and similar equations have shown good agreement between
predicted and observed fractionation effects (Rossi and Kellerer,1973).

Recent, thus far unpublished work, has also shown that it is possible to account quantitatively for the effects observed when two different types of radiation are administered to biological systems with varying time delay. A more general formulation of Eq. 4 which applies to various doses of two different radiations and variable recovery is given by

$$\varepsilon(D_1 + D_2) = k_1\zeta_1 D_1 + k_2\zeta_2 D_2 + \sqrt{k_1 k_2} \ (D_1{}^2 + D_2{}^2 + 2qD_1 D_2) \quad (5)$$

where the subscripts apply to each radiation type. q ranges from one to zero depending on the degree of recovery between the two irradiations.

Most of the pronouncements of the theory of dual radiation action apply to single, autonomous (non-interacting) cells. Application of the theory to radiotherapy must, therefore, have the same limitations as the applications of other aspects of cellular radiobiology. They can often provide useful and revealing guidance, but it must always be remembered that the complexities of organ and tissue responses may strongly modify cellular effects.

ACKNOWLEDGEMENTS

This investigation was supported by Contract EP-78-S-02-4733 from the Department of Energy and by Grant Nos. CA 12536, CA 15307 to the Radiological Research Laboratory/Department of Radiology, and by Grant No. 13696 to the Cancer Center/Institute of Cancer Research, awarded by the National Cancer Institute, DHEW.

REFERENCES

R.P. Bird   (In preparation).
E.J. Hall, J.K. Novak, A.M. Kellerer, H.H. Rossi, S. Marino and L.J. Goodman, RBE as a function of neutron energy, I. Experimental observations, Radiat. Res. 64: 245-255(1975).
A.M. Kellerer and D. Chmelevsky (In preparation).
A.M. Kellerer and H.H. Rossi, A Generalized Formulation of Dual Radiation Action, Radiat. Res. 75: 471-488 (1978).
A.M. Kellerer and H.H. Rossi, The Theory of Dual Radiation Action, Current Topics in Radiat. Res. Quart. 8: 85-158 (1972).
H.H. Rossi, Neutron radiation quality, Proc. First Symp. on Neutron Dosimetry in Biol and Med., Munich EUR 4896 d-f-e, p. 57-65 (1972).
H.H. Rossi, The effects of small doses of ionizing radiation, Phys. Med. Biol. 15: 255-262 (1970).
H.H. Rossi, R.P. Bird, R.D. Colvett, A.M. Kellerer, N. Rohrig and Y.M.P. Lam, The Molecular Ion Experiment, Proc. Sixth Int. Symp. on Microdosimetry, Euratom, J. Booz and H. Ebert (eds) p. 937-947 (1978).
H.H. Rossi and A.M. Kellerer, Biological implications of microdosimetry, I. Temporal aspects, Proc. 4th Symp. on Microdosimetry Verbania, Italy, p. 315-330 (1973).

BIOLOGICAL EFFECTS OF RADIATION AND THEIR

VARIATION WITH RADIATION QUALITY

Giovanni Silini

Division of Radiation Protection
C.S.N. Casaccia
Rome, Italy

INTRODUCTION

Many lecturers during the present course will be dealing with the effects of ionizing radiations of various types and energies and each of them will provide the details of technique and the findings relevant to any specific quality of radiation. As an introduction to these lectures dealing with protons, neutrons, pions and heavy ions, I have been asked to lead the way with a rather more general outline of the radiobiology of high-LET radiations in relation to the complexity of the biological organization. I will give you therefore a prospective view of the problems applying to this type of studies and I should warn you right from the start that in my lecture you will not find the latest details but generalities, because I will deliberately insist on the formulation of problems rather than pointing to their possible solutions in the different instances.

A RETROSPECTIVE VIEW

One of the oldest and most general observation in radiation biology is that the degree of biological change that follows the absorption of a given amount of energy is different when such energy is released to the medium by radiations of different type. To say it in other words, equal doses of radiations of different types usually produce effects which are quantitatively different. It follows that this difference of effects should be attributed to changes in the <u>way</u> the radiation energy is deposited. Since the spatial distribution of the energy released to the medium on a submicroscopic scale differs for different radiations, it is natural that we should attribute these differences in effect to the characteristics of the structure of the track.

A precise description of the primary physical events occurring along the track of an ionizing particle is a very difficult exercise and one for which much effort has been spent. I would not presume to enter here a field that others will cover with much greater authority, but let me just mention the various phases in the development of concepts. From the very initial formulations in which the mean linear ion density over the total length of the track was the reference quantity, more refined models were derived where the energy released per unit length of the track and its variation along the track were used to describe these phenomena. The most recent microdosimetric approaches attempt a description in terms of the probability distribution along the track of events in which a specified amount of energy is released.

All these are alternative ways to disentangle a very complex situation with many interplaying variables. To mention just a few: the energy transfers due to the primary ionizing particles; the ionizations and excitations from the secondaries; the heterogeneity of the physical mechanisms through which different amounts of energy are eventually delivered to the medium and their dependence on the kinetic energy of the particle; the role of energy transfer mechanisms within or between molecules by physical actions or by physico-chemical intermediaries.

In parallel with the evolution of the physical concepts, great advances in the understanding of the molecular structure of the living systems were also taking place, so that one might have expected that cross-fertilization between the two disciplines could result in an enhanced knowledge of radiobiological problems. But the results from biology were analogous with those from physics: the improved understanding of the basic mechanisms was insufficient to explain all the intermediate steps up to the final effects at high levels of the biological organization.

It was precisely to overcome this difficult reality that the fathers of radiation biology, in the impossibility to identify molecular mechanisms, developed abstract semiquantitative models of the radiation action under the forms of the so-called "theories" of radiation effects. In these the heterogeneity of the primary biophysical reactions was glossed over by the use of the term "hit" and the absence of knowledge about the biological interaction sites was covered by the concept of "target".

We must acknowledge today, after more than fifty years during which these concepts have been with us, that the complexity of the initial goal has defeated all attempts for more analytical descriptions. To the eyes of the biologist the essence of the problem remains that of explaining how under circumstances which would tend naturally to average out any specificity of the initial events due to radiations of different types one might still resolve differences in the degree of the final effects. And this question is particularly relevant when one deals with irradiated biological objects where the physical and physico-chemical phases of radiation action are only the beginning of a further chain of events at the chemical and biological level determining the expression of the final damage at times far removed from the initial energy deposition. This is why we must still use today (and I shall be using during this lecture) the semi-empirical language of models to justify phenomena whose intimate essence is far from being fully understood.

Having thus dealt with this retrospective view of the field, I would propose to develop the main subject of my lecture systematically with a review of the relationships between RBE and LET at various levels of the biological scale with some remarks about their possible interpretation.

A REVIEW OF THE LET-RBE RELATIONSHIPS

It should first be realized that a detailed discussion of this point would be equivalent to reviewing an impressive amount of radiobiological literature which has led by now to widely accepted generalizations. Therefore, rather than referring to the original contributions, it would be sufficient for our purpose to outline such generalities.

LET-RBE relationships are conventionally studied by graphs in which some form of average LET is plotted in the abscissa on a log scale and the efficacy of radiation to produce a given biological effect is given on the ordinate. This latter quantity if often expressed as the ratio of slopes of dose-effect relationships obtained with a test radiation and with a beam of reference low-LET radiation. One difficulty with such plots is the choice of the LET values to be used, since under most circumstances there is no single average LET value that may properly describe the irradiation conditions under study. The distribution or spectrum of LET values may in fact change substantially throughout the irradiated biological objects.

There is probably no single field of radiation biology where theoretical and practical aspects are so closely tied as this one: actually, the application of LET-RBE relationships spans from the fundamental studies of the observed relationships in order to test models of radiation action for various biological end-points, to very practical applications in estimating the Quality Factors for radiation protection purposes.

Elementary Biological Units

When dealing with comparatively simple biological objects such as small protein molecules, enzymes, molecules of nucleic acids, virus particles, the inactivation of a specific function by radiation is often the biological end-point investigated. The distinctive character of these elementary biological units is the absence of a biochemical machinery through which the energy deposited by radiation may develop into a complex biological effect. The simple disruption of the structural integrity of the molecule is therefore to be considered itself the final effect, because it cannot be further modified by energy transfers or by the action of repair mechanisms. This is particularly true when molecules are irradiated in the dry state and in the presence of oxygen to fix the damage.

If one further considers that at least for the smallest molecules the whole molecule acts as a target, it is easy to understand why energy deposition anywhere within the unit should directly result in the loss of the molecule's specific function. Qualitatively speaking, the target for this effect must be fairly simple and easy to inactivate by a single event of energy deposition. With such simple biological objects the relationship between inactivation and absorbed dose is a negative exponential of the form

$$S_D = e^{-D/D_0} \qquad\qquad\qquad\qquad\qquad (1)$$

where S_D is the surviving fraction of biological units at dose D and D_0 is the mean inactivation absorbed dose.

This form of the relationship, taken together with other biological and physical evidence, suggests that inactivation depends upon the random deposition of a dis-crete amount of energy (a "hit") within a defined volume of the biological object (the "target") which may or may not coincide with the physical volume of the object. The efficiency with which a given dose of a given type of radiation might produce inactivation is bound to depend of course upon the volume (and shape) of the target, even though the exponential trend of the dose-relationship cannot as such be taken as sufficient evidence that inactivation does result from single-hit kinetics.

It follows from these postulates that the deposition within the target of energy in amounts exceeding that strictly necessary to produce the loss of function does not result in any additional biological effect, although it is measured as an increase in the dose absorbed. Such a situation is observed most typically when the spacing between ionizations is small compared with the target size, so that more than one single event of energy deposition may occur within the target volume. Since one deals here with a "all-or-none" type of effect and a given biological unit cannot be inactivated twice, there will be in this case a "waste" of energy, by comparison with that sufficient to inactivate the function. With in-creasing LET the ratio of the efficiency of the given dose of the test radiation to that of the reference rad-iation will thus decrease, giving rise to a negative trend of the RBE _versus_ LET.

It should be added that if the inactivation kinetics is purely exponential for all radiation types and en-ergies, then the RBE of different radiations is unchanged at different levels of dose and effect.

Cells

The situation changes when biological structures of more complex sort are irradiated, such as most typically in the case of pro- and eu-karyotic cells. Although fair-ly elementary in terms of biological organization, cell-ular structures are biophysically very complicated indeed. To start with, the organization of the presumed target molecule, the DNA-protein complex, is of a higher order because its size is much larger, because there is redundancy of genetic information and there is more than one set of genes. Biochemically, there are various types of intracellular repair mechanisms which may modify the expression of the damage by introducing alternative pathways that the cell may follow in order to overcome the effect of a primary damaged site. Functionally, there is a great number of effects that one could think of testing in relation to radiation dose, each of them specific with respect to the testing technique but non-specific with respect to target site and induction mechanisms.

In practice, for a number of very good reasons that we may not go into here, inactivation of the capacity for the cell's indefinite reproduction is the end-point most frequently studied and the loss of this function is defined as the failure of the inactivated cells to give rise to a line of daughter cells developing into a visible colony in vitro or in vivo.

With these biological objects the larger volume of the targets increases the inactivation efficacy of a given radiation type, because for the same number of ionizations produced per unit volume of the irradiated object (i.e., for a given dose) the lesion of the target is more likely. On the other hand, one single event of energy deposition is insufficient to produce inactivation of the function because repair is operating and more energy must be released to the target within a given time in order to produce inactivation.

The form of the dose-inactivation curve for these objects is frequently of the so-called sigmoid type with radiations of low-LET and is characterized by a fairly shallow increase of the effect for low doses, followed by a gradual transition into a nearly-exponential trend at high doses. Such curves are usually and most conveniently approximated at high dose values by relationships of the general form

$$S_D = n \cdot e^{-D/D_0} \tag{2}$$

where n is formally equivalent to the number of targets to be inactivated each by a single "hit" (in a "multi-target single-hit" model) in order to produce the loss of function. In other words, these curves are described by a D_0 value and by a value for the "shoulder" at low doses, expressed either by an extrapolation number, n, or by a quasi-threshold dose, D_q, these quantities being related by the following expression:

$$\log_e n = D_q / D_0 \qquad\qquad (3)$$

If we discard for the moment the effect of the shoulder and only consider D_0 as the quantity representing the intrinsic radiosensitivity of the cells, we may qualitatively define the RBE as the ratio between the D_0 of the standard radiation's inactivation curve and the D_0 of the test radiation. If we thus plot the different RBE's obtained by irradiating cells with radiations of increasing LET, we usually see that the RBE values increase at increasing LET up to values of about 100 keV/µm. Then the curve goes through a maximum and declines for still higher LET's.

What might be the general significance of such a finding? Following the same kind of qualitative picture described earlier, we may interpret the curve to mean that the energy requirements to produce inactivation of the cells are more easily satisfied when more energy is deposited simultaneously within the target volume, until an optimum LET value corresponding to the observed maximum of the RBE. For still higher densities of the ionization, the energy deposited is in excess of that required for inactivation under the specific experimental conditions and the situation described earlier for wasted ionization begins to apply.

Let us now consider the form of a sigmoid dose-inactivation curve and ask ourselves what the presence of the shoulder may mean. Generally speaking, we might interpret it by assuming that summation of discrete events of energy deposition is required in order to produce inactivation. As an example, imagine that each cell has ten targets and that, on the average, nine of them receive a hit each. At this point the probability that the tenth target is inactivated will follow a random distribution and the inactivation will increase exponentially. It is easy to visualize that the above conditions will be reached more rapidly with high than with low-LET radiations. For this very reason radiations of progressively higher LET will produce shoulders of a decreased width, until the shoulder will be completely abolished for radiations of very high LET.

The result of such a condition is that the form itself of the dose-effect curve changes from low-LET to high-LET radiations and, as a consequence, the RBE values

versus LET depend on the different levels of the effect
and increase for progressively lower doses, by comparison
with the values that would be obtained according to the
simple ratio of the final exponential slopes. There is a
very important practical corollary to this statement,
since the instances when lower doses are delivered are
those of relevance for radiation protection. It is necessary therefore that, for a given radiation, the Quality Factor (the practical counterpart of the RBE concept
used in radiation protection) be set high enough to be
applicable to very small doses.

Cell Populations and Tissues

We may now go to the next higher level of organization and consider populations of cells all of the same
kind or tissues where different types of stem, differentiating and functional cells are represented. At this
level, the concept of indefinite cell reproduction is
only applicable to the stem cells which are a very small
minority of the whole population. Most cells are instead
differentiating and have thus by definition a limited
mitotic history.

In addition to the functional organization described
for single cells, we have here other complexities
introduced by the fact that cells divide with mitotic and
doubling times which are, on the average, characteristic
of any given cell line under any given condition of
growth, although there is variability from cell to cell.
And when various lines of cells are present, often
interdependent because of their hierarchical arrangement
or by feed-back regulation mechanisms as in self-renewing
tissues, maturation and regulation phenomena
characteristic of each differentiating line of cells increase the structural complexity of the tissue.

In parallel with this more variable and rich
expression of the function, the effects of radiation are
also more complex. The magnification effect introduced by
the division of radiation-surviving cells transforms the
simple expression of a "yes-or-no" inactivation of cells
into a graduated tissue response where different levels
of reaction correspond to different doses absorbed. If
at the level of a single cell it was only the survival of
that cell that mattered, for tissue damage it is the loss
of capacity of that cell to effectuate itself into a
small cohort of functional cells that gives rise to
losses of large numbers of functional elements and there-

fore to failure of the specific function of that cell line. The rapidity and the extent of such failure is mainly governed by the kinetics of each tissue. For each differentiative line the relative size of the stem, maturative and functional compartments, the relative transit time of cells from one compartment to the next, the permanence time and the birth rate of the component cells into each compartment are - together with the loss of division capacity of the stem elements - the parameters of importance. The difficulties in defining all these effects quantitatively is determined, among other things, by the fact that these very mechanisms may be altered or disrupted by irradiation.

With increased organizational complexity, the tissues also acquire further capabilities to repair the damage inflicted upon them and thus repopulation by cell division of the depopulated compartments completes the action of intracellular repair mechanisms to restore the functional integrity of the irradiated biological system. To sum up, I would say that involvement of many functional cells, the modulation of the response, the cell and tissue kinectic parameters, the repopulation of the surviving cells are the main features at this level of organization.

The question is therefore how the ionization density of various radiations may influence the final expression of tissue damage. We may answer by saying, as a general proposition, that the scale at which all these mechanisms operate is by far larger with respect to the submicroscopic scale distribution of the energy deposition events to be of real significance in modifying the LET dependencies of the effects. These dependencies are therefore of the same type as in the case of single cells with an increase, a maximum and a subsequent decrease of RBE as a function of increasing ionization density, as though all the functional steps previously outlined would simply project on a magnified scale phenomena that have been fully expressed within the single dividing cell, the death of which is at the origin of the tissue effects.

The above statement does not imply, of course, that there may not be differences between tissues in respect to the position of the maximum along the abscissa or to its height on the ordinate scale. Actually, each tissue recognizes some distinctive features superimposed on the general relationship. But the patterns of the phenomena, as reflected by the RBE versus LET curve remains essentially the same as for single cells.

Whole Animal

    Moving now to the whole-body level, we are confront-
ed with at least two classes of effects. The first class
includes those that are termed "non-stochastic" in
conventional radiobiological jargon and may simply be
viewed as the projection of effects already analyzed for
tissues. The second class of effects called "stochastic"
is of a completely different nature.

    Let us start with a few words on the characteristics
of the system, in this case a living animal. Integration
of functions into a coherent set of tissue components is
here the distinctive character. New functions are becoming expressed like the hormonal, the immunological and
the nervous activities, each of them geared for an optimal adaptation of the species to the environment, to
individual interrelationships, to survival of the single
animal and maintenance of the species. We could elaborate
a lot on these concepts, almost to the borderline of
philosophy, but it would be more profitable to remain
firmly rooted on our ground.

    Non-Stochastic Effects. I would propose that surviv-
al of the irradiated animal is the effect of relevance
here. According to the dose, mode and time of irradiation, lesion of one of the self-renewing cell lines leading to one of the main acute radiation syndromes (nervous, intestinal or hemopoietic) is the end-point for
discussion. For each given line of cells there is at any
given time from irradiation a minimum number of functional cells compatible with the survival of the irradiated individual. If that minimum is exceeded death will
ensue, always as the result of a very profound alteration
involving the failure of a great number of cells. There
is nothing peculiar with respect to the tissue phenomena
previously described, except for the fact that tissues
are now part of a living organism and the depression of
their function is so extreme to be incompatible with the
life of that organism.

    If this is so, the relationships of RBE versus LET
for the animal's death should be of the same kind as for
the relevant tissues, except for small departures resulting from the interplay of effects on other tissues or
from the superimposition of the regulatory functions
already alluded to before. And in fact, if we take for
example the $LD_{50/30}$ to be the expression of the lesion of
bone marrow stem cells and as a typical non-stochastic
effect of irradiation at the whole-body level, we find a

relationship of the kind we have described before for single cells and for isolated tissues.

I would comment in addition that the dose levels at which these phenomena become manifest are comparatively high, so that their study is only of academic interest for a general interpretation and systematization of radiobiological effects. In practice, they concern only the fairly rare cases of accidental exposure, but the existence of a practical threshold in the dose-induction relationships renders these effects of no relevance whatsoever at the low doses of interest for radiation protection.

<u>Stochastic Effects</u>. These effects appear pathologically as leukemias and tumors in an exposed population. While a causal relationship may be empirically established for each case of non-stochastic damage between the dose absorbed and the conditions of exposure, on the one hand, and the occurrence of the clinical conditions, on the other, it is impossible to demonstrate such relationship in each single case for stochastic effects.

The existence of a correlation between exposure and effect may be derived by the occurrence of an extra increase of these diseases in an irradiated population (by comparison with a non-exposed one) over and above the level at which any animal population expresses these diseases for apparently "natural" causes. The study of the radiation-induced stochastic effects is rendered very difficult by their comparative rarity (which requires analysis of large animal populations) and by the fact that "natural" and supposedly "radiation-induced" diseases are absolutely identical: only that there are more of these diseases in the exposed than in the non-exposed populations, following some function of the absorbed dose.

The most striking observation that can be made at the present state of our knowledge with respect to tumor induction is the great variability in relation to radiation dose of the tumor systems tested. This is why radiation protectionists assume for practical purposes that there is a linear relationship between dose and probability of induction: this assumption is taken to be valid in all cases and to overestimate the risk of induction at low doses and dose-rates.

The second observation relates to the extremely limited amount of generalized information that we have been able to gain so far with radiations of different LET. There are indeed reasons for this, due partly to the intrinsic difficulties and to the length of the experiments required and partly to the lack of systematic data collection. In spite of this limited experience, we do know that increasing the LET of the radiation does lead to an increased yield of tumors for the same dose, although we find it very hard to establish a value of the RBE which might be reasonably applicable to any tumor model system in the experimental animal and even less to the human situation.

Yet, for practical reasons we are in great need for this information. Such a gap in our knowledge would be frustrating enough at relatively high doses causing substantial and measurable tumor induction in vivo . But our need for precise estimates of the RBE for tumor induction is most vital at the very low doses where experiments are made impossible by the extremely rare occurrence of radiation-induced neoplastic conditions which are masked by the very high background of "naturally" occurring ones.

This does not allow to establish with precision the trend of the tumor induction curves at points near the origin and their possible changes as a function of LET. It is easy however to visualize that if the linear component of the curves would prevail over the higher order component for high-LET radiations, very high values of the RBE might apply at the very low values of absorbed doses with obvious important ripercussions over the practical recommendations for radiation protection. We must recognize that still today this is one of the main gaps in our radiobiological knowledge that badly needs some answer.

SUMMARY

To conclude in just a few words, any type of radiobiological effect at any level of the biological scale depends on the amount of energy released to the sensitive site for the effect under study, following relationships expressed by the so-called dose-effect curves. But biological effects depend also on the way the energy is released, that is, on the spacing between the events of energy deposition along the tracks of the ionizing particles: these relationships are expressed by the LET-RBE curves.

Increasing the ionization density of radiation may lead to essentially different types of LET-RBE relationships:

<u>a</u>) an increase of the relative effectiveness of the radiation treatment, when the energy needs for the given effect are more easily met by the simultaneous deposition of more energy within the sensitive site; or

<u>b</u>) a decrease of the relative effectiveness, when the extra energy deposited by densely-ionizing radiations is in excess of that necessary for the production of the effect. There is also (and it is indeed most common) a combination of the two above phenomena into a complex relationship,

<u>c</u>) where an increase, a maximum and an ensuing decrease of the effectiveness are seen for progressively higher values of the LET.

The decreasing pattern <u>b</u>) is most typically observed for simple molecules or elementary biological units where one or very few energy deposition events are sufficient to cause the biological reaction. The pattern <u>a</u>) is seen for cells at relatively low values of LET, where the biological complexity of the object and the presence of repair require more energy to be deposited for inactivation. However pattern <u>a</u>) is almost invariably accompanied by a decrease in a pattern of type <u>c</u>), if the LET range is sufficiently extended at values beyond 100 keV/μm.

Since the scale at which the energy deposition phenomena apply is that of the μm or lower, there is no trend of the LET-RBE relationship characteristic of the higher levels of biological complexity (cell populations, tissues, whole animals): they show or are expected to show essentially the same relationships as the component cells, whose death or survival determines the response of the integrated systems.

LATE EFFECTS OF LOW DOSES AND DOSE RATES

Herwig G. Paretzke

GSF - Institut für Strahlenschutz
D-8042 Neuherberg, Fed.Rep.Germany

1 INTRODUCTION

Studies on the quantification of health hazards to man due to exposure by small doses of ionizing radiation delivered at low dose rates belong to the most difficult problems in radiation research. Sometimes this field is even associated with "parascience" because of the difficulties encountered in searching for any significant effect in this region. Whereas the lectures 5 and 7 by Silini put more emphazis on a description of the types of biological radiation effects and on the risk estimates actually used in practical radiological protection work, this lecture will put more weight on an outline of the spectrum of problems and approaches used in work on the derivation of quantitative prognoses of late effects in man of low doses and dose rates. Because of the wide scope of such risk analytic investigations, in one lecture it will only be possible to give you some impression of the complexity of this topic and of concepts to cope with it, rather than a thorough discussion of our present knowledge. At this point, however, I would like to stress that most of these problems are of basic scientific nature, and that it is much easier to make sound recommendations for practical applications in radiological protection with the duty not to under-estimate possible effects for an average person.

Here, we will deal with
- the origins of some principal problems encountered in radiation risks assessments,

- some definitions and explanations of useful quantities,
- methods of deriving risk factors from biological data,
- methods of deriving risk factors from epidemiological data,
- concepts of risk evaluation and problems of acceptance.

2 SOME PRINCIPAL PROBLEMS ENCOUNTERED IN RADIATION RISK ASSESSMENT

There are several facts which cause severe problems in any investigation on the assessment of health hazards due to ionizing radiations. Among these facts we should mention as first order and insoluble problems:
- small doses of ionizing radiation can only lead to such types of health defects which occur also naturally,
- these types of health defects in addition occur naturally in human populations rather frequently compared to expected additional radiation induced frequencies,
- small doses (say below 10 rem), fortunately, lead with such small a probability only to serious health effects, that these radiation induced frequencies are small compared even to the natural variation of normal cases,
- we do not have methods at hand to discriminate a naturally occurring case, e.g. of leukemia or cancer, against a radiogenic one.

The second order problems are due to the pluridisciplinary nature of the field. To be able to deal properly with the task of analytic and prognostic radiation risk assessment, you have to try to be up-to-date in quite a number of scientific fields, namely e.g.
- radiation physics
- radiation chemistry
- radiation instruments and technologies
- biology
- medicine
- statistics and epidemiology,
- radioecology, systems analysis and operations research,
- agriculture, soil science, and forestry,
- dietics and food distribution pathways,
- aeronomy, hydrology and marine sciences.

With some simplification, one might say that knowledge in the first fields is needed to assess the probability of health effects from a given radiation exposure, whereas knowledge in the second group must be used to estimate in a prognostic way details of the radiation exposure of individuals and / or the whole population from information on the releases of radionuclides from technical facilities.

Because of the pluridisciplinary nature of radiation risk analysis work, most groups active in this field are composed of scientists from various disciplines. In addition, they use to collaborate with other groups much more frequently than average for the same reason.

There is at least one further order of problems to be taken into account in this field especially when dealing with radiological risks assessments for nuclear facilities or for medical procedures:
- sociology and ecology,
- politics and economy,
- journalism and public non-scientific discussions.

It is particularly in the latter region where scientists often encounter rather severe difficulties because of their lacking experience and training. In this lecture, however, this topic will be neglected and more attention will be given to problems encountered in the derivation of risk factors and their relative weights.

3 WHAT ARE LATE EFFECTS OF LOW DOSES AND DOSE RATES?

What are we talking about when we deal with "late effects of low doses and dose rates"?

In this context, "late" means for somatic effects, say, ten to fourty years after irradiation or begin of chronic exposure. Mean latency periods of 20-25 years until radiation induced malignant neoplasms are detected and 10-15 years for leukemia have often been reported. The reasons for this long time passing between the primary radiation insult and the appearance of somatic effects in the low dose range are still not yet fully understood. For genetic defects due to radiation induced irreversible changes of the genetic material, late means one or more generations later.

What types of serious health "effects" have to be expected after exposure to low doses? Although this topic has been extensively dealt with already in the last lecture, we should recapitulate here that we have to expect
- somatic effects, i.e. effects in the irradiated individuum itself, like leukemia and cancers e.g. of the lung, thyroid, breast, GI-tract, bone, etc.
- genetic effects, i.e. effects in later generations of an irradiated individuum like hereditary diseases, anomalies of the skeleton, mental defects, change of eye colour, etc.

To our present knowledge, somatic effects appear
to be more important than genetic effects (by a factor
of, say, 3-6). Both are of stochastic nature, i.e. only
the probability of their appearance is determined by dose whereas the severity of the effects is almost completely independent of dose.

How small are "low doses"? An reasonable estimate
of the average annual dose to man from natural radiation
sources is about 150 mrem. Hereby, as a rule of thumb,
about one half of the exposure has its origin in external irradiation by cosmic and terrestrial (mainly γ-and
n -) radiation, and the other half stems from internal
irradiation after ingestion and inhalation of natural
radionuclides (mainly α- or β-particle emitters). Integrated over a life span of 70 years, the total natural
radiation dose equivalent to man amounts to about 10 rem.
The dose burden from diagnostic radiation applications in
medicine, though not too well-known, might typically also lead to additional 150-250 mrem/a. Thus, the sum of
both, i.e. about 20 rem, or at the most 50 rem (if we
take in addition average occupational doses into account)
might be a reasonable proposal for the upper limit of
the typical "low dose" region.

You will encounter severe difficulties in defining
in a general way a "low dose rate". Such a definition
depends more on the actual topic under consideration
than usually anticipated. Let us come back to the two
major sources of human radiation exposure for demonstration of these difficulties. The irradiation period during
which an diagnostic x-ray picture is taken by a radiologist can be as small as fractions of a second. Thus the
relation between dose rate and time is made up from one
or more high spikes during the year and zero medical dose rate in between. On the other hand, natural radiation
sources have a rather continuous character and lead to
an almost constant, low irradiation rate over the year.
Therefore, the maximum dose rates for an individual from
these two major sources differ by about 8 orders of
magnitude! However, in considerations of the reactions
of single biological cell nuclei from which this individual is made up, this enormous differences in dose rates
can completely be neglected in certain cases! Often the
interaction between adjacent cells may be disregarded.
Then is the only difference among irradiated cell nuclei
at sufficiently low total doses, whether or not a cell
has been affected by an energy deposition event. Such an
event takes place in less than 10^{-12} sec anyway and this
is independent of dose rate! In the very low dose region
the probability of hitting the same nucleus twice by two

independent interaction processes is a rather rare event.
Therefore the value of the dose can be considered as a
measure for the number of cells affected by irradiation.
Dose rate is then a measure for the time interval after
which the same cell might be affected by a second event.
If this time is long compared to the time a cell needs
for recovery and repair of the first damage because of
a very low dose rate or because of a sufficiently small
total dose, we are in the so called "low dose - low dose
rate" region which is of high importance to radiation
protection.

4 EXPLANATION OF USEFUL QUANTITIES

It might be adequate to introduce here shortly a
few terms which you will frequently encounter in literature on risk assessment for late somatic radiation effects. First, we should discriminate between "incidence"
and "mortality". The former gives the number of individuals who developed a certain disease during a certain
period of time, whereas the latter takes into account
only lethal histories of this disease. There are various
ways to specify these quantities either as integral or
differential quantities with respect to time or dose.
The quantity "differential cancer rate r(t)" e.g. is the
dose, age, time, sex, radiation quality, etc. dependent
probability per unit time for an individual to die from
a somatic late effect due to radiation exposure (Fig.1):

$$\frac{1}{n} \frac{dn}{dt} = \sum_i a_i(t) + r(t)$$

where $a_i(t)$ are other, radiation independent risks of
death.

Evidently this is one of the most basic quantities
in this context and we would like to know it accurately
for all the types of late effects and for all relevant
parameters. The related quantity "integral cancer rate
R(t)" gives the time integral of r(t):

$$R(t) = \int_0^t r(t') \, dt'$$

If there are only a few cases of cancer in a population,
R(t) might be the only statistically useful quantity
which can be derived. In experiments animals sometimes
develop more than one cancer. Then one must discriminate
in addition between "number of cases with at least one
tumor" and "number of tumors". Clearly the number of ca-

ses with at least one cancer is given by
$$N(t) = 1 - e^{-R(t)}.$$

In epidemiological studies often the number of excess cases divided by the "person-years-at risk (PYR)" is reported as a function of the mean dose in a cohort:

$$R(D) = \frac{\text{Observed-expected cases}}{\text{PYR}}$$

For this purpose, the irradiated population and a suitable control population must be matched for age and sex distribution, and the numbers are usually given for a nominal population size of one million persons.

Sometimes the excess cases are only divided by the number of irradiated individuals (normalized to 10^6 persons) without consideration of the actual time span of observation or/of persons lost from the irradiated population due to other, competing risks. Under the usual assumption of a linear dose-effect relationship, the proportionality factor f giving the excess cases per dose unit per 10^6 persons plays an important practical role:

$$f = \frac{\text{Excess cases}}{\text{rem} \cdot 10^6 \text{ persons}}$$

These risk faktors f for mortality averaged over a whole population are given e.g. by the ICRP report 26 (Tab.1).

Finally, one should understand the difference between an absolute and a relative risk model. The absolute risk hypothesis assumes that radiation exposure adds an absolute risk (e.g. $R_r = f \cdot D$) independently from any pre-existing risk R_0 to the total risk R_t of an individual e.g. to die from lung cancer:

$$R_t = R_0 + f \cdot D .$$

The relative risk hypothesis is used e.g. by the BEIR-committee, and it assumes that radiation exposure enhances the pre-existing risk e.g. proportional to dose:

$$R_t = R_0 (1 + f' \cdot D).$$

This is a very important principal difference. At the present state, however, we do not yet know in general which hypothesis describes the real situation more correctly although some evidence speaks in favour of the absolute risk model. The main reason for this lacking knowledge is that in most cases either the observation periods are too short or the sample sizes too small.

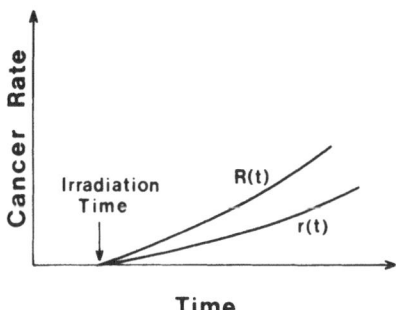

Fig.1   Differential and integral cancer rates as
        a function of time after single exposure

Table 1   Expectation values of population averaged
         absolute risk factors for mortality due to
         radiation exposure of organs and tissues
         recommended by ICRP Report 26 and by
         Jacobi and Paretzke (1979)

Tissue	Mortality Risk / Organ dose (cases per 10^6 persons per rem)	
	ICRP 26	Jacobi and Paretzke
Red bone marrow	20	15 - 40
Breast	25	15 - 40
Lung	20	10 - 30
Thyroid	5	5 - 10
Bone surface	5	$\leq$ 5
Total GI-tract	50	20 - 50
All other tissues		10 - 30
Total somatic risk after whole body irradiation	125	80 - 200

5 DERIVATION OF RISK FACTORS FROM BIOLOGICAL AND MEDICAL DATA

A detailed discussion of the available data sets from which radiation risk factors can be derived is given e.g. in the comprehensive UNSCEAR-Report "Sources and Effects of Ionizing Radiation" (1977). This 700 page report has been prepared for the UNO and it is available from all scientific bookstores or the UN sales sections in New York and Geneva (price 28 US $). We do not have sufficient time to repeat in this lecture such a detailed discussion nor to give an overview on all these interesting data. But it might be useful for those not daily involved in risk analysis work to learn here something about the practical problems encountered in the derivation of as accurate as possible dose-effect relationships from biological data.

Generally we can discriminate at least three types of data sources for quantitative radiation risk analysis, namely data from groups of persons exposed to radiation for various reasons, experimental results for animals and experimental results for cells, etc. As regards animal experiments, the incidence and mortality rates from cancer in animals vary widely among species and even among different inbred strains of the same species. Even the shapes of dose-response curves cannot safely be generalized or even extrapolated from animal data to man. Therefore, at present somatic radiation risk factors for men are only derived from experience with irradiated human groups. However, lacking any safe conclusions at low doses from human data, animal and cellular or subcellular data are nevertheless used to obtain some indication e.g. on possible shapes of dose-effect relationships, on trends in dose-rate effects, modifications by the presence of oxygen or other chemically active molecules, etc. A few examples of such trend analysis shall now be mentioned.

As mentioned above, the dose-effect curves for somatic late effects depend on a number of variables as e.g. dose, dose rate, time, sex, age, organ, radiation quality, homogeneity of population, etc. Frequently only the dose dependence at a fixed time is considered, and the typical problem of extrapolation to low doses from the dose regions where data are available is indicated in fig. 2. There is some indication that some dose-effect curves (e.g. for leukemia in mice) might follow approximately the relationship

$$y = (a_0 + a_1 D + a_2 D^2) \cdot e^{(-b_1 D - b_2 D^2)}$$

For risk assessment at low doses we are mainly interested
in the magnitude of the respectivelinear terms. If there
are, however, only data available in the region 1 or 2 as
indicated in figure 2 and no further information on the
shape of the curve is known, a linear extrapolation to
zero from the effects at these doses might overestimate
(1) but also possibly underestimate (2) the needed risk
function at low doses. Even more complicated would be the
derivation of an accurate, realistic risk function if the
linear term at low doses is small compared to the quadratic term. This could be the case e..g. for Co-60 irradiations.

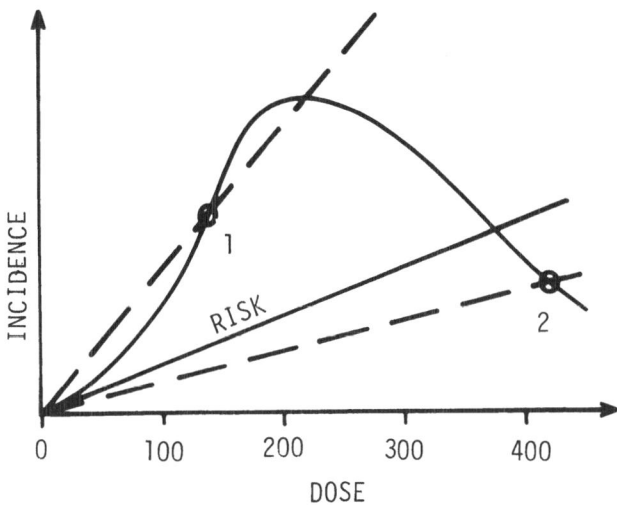

Fig. 2   Possible curve for the incidence of somatic late
          effects as a function of dose. Indicated are also
          possible implications from data in dose regions
          1 or 2 if no further information should be avail-
          able.

For such low LET radiation fields also the normally conservative linear assumption will be applied until further
insights will permit more accurate conclusions.

        The shape of such a dose-effect curve has also to be
expected to be dose-rate dependent. A dependency observed
e.g. for the induction of dicentric chromosome aberrations
and of possibly more general importance especially for low
LET-radiation is indicated in fig. 3. The observations can
be described by a second order polynom whereby only the
quadratic term is dose rate sensitive. The linear, dose
rate independent coefficient would be sufficient to spe-

cify radiation risks of low dose rates. However, frequently
our experience reuslts from single or multiple irradiations
with higher dose rates and thus higher, unknown contributions of the quadratic term.

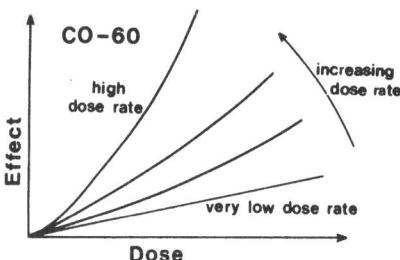

Fig. 3   Observed trend of change in dose-effect curves
          (e.g. for chromosome aberrations) of low LET-
          radiation with dose rate as a parameter.

The shape of relationships for somatic late effects
in addition to dose rate also depend on the time of observation. In some animal experiments, curves for the
number of animals with tumors as a function of time after beginning of chronic exposure with the different dose
rates as a parameter look like those drawn schematically
in fig. 4. These curves often are almost straight lines
when plotted on a logarithmic probability paper as in
this figure. The time τ_{50} after which exactly 50 percent
of the animals have developed at least one cancer can
be taken as one important characteristic of the temporal incidence curves and plotted against the dose rate
(fig. 5). From such a plot, one might find in certain
cases a functional relationship of the form

$$\tau_{50} \sim \dot{D}^{-n}$$

with n roughly of the order of 1/3. The implications of
such a lenghtening of the latency period with decreasing dose rate has been discussed and used in prognostic
calculations e.g. by Mayneord and Clarke (1975) and others.

However, it should be stressed that the human data are too scanty and of too low statistical significancy to permit testing for the general validity of this hypothesis.

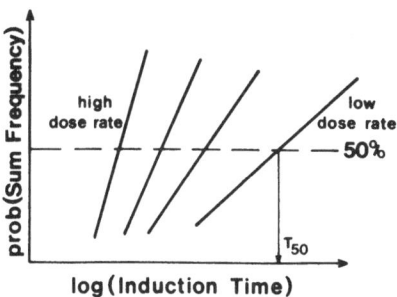

Fig. 4   Observed curves for the sum frequency of animals with tumors as a function of dose and dose rate.

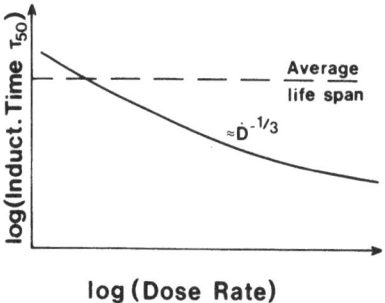

Fig. 5   Possible relationship between mean tumor induction time and dose rate of chronic irradiation.

Some further aspects of the derivation of risk factors from medical data shall at least be mentioned.

The radiation <u>sensitivity</u> of a population is usually
assumed to be approximately uniform. The existence of
considerably higher than average sensitive subgroups
(e.g. suffering from some particular diseases) in certain cases could lead in principle to a steeper than
linear increase of dose effect curves at low doses in a
heterogeneous population. However, the size of such sensitive subgroups must be rather large to find influences
of heterogeneity in a data set that can be resolved with
statistical singificance.

Another question is the determination of the <u>cells
at risk</u>: are all cells of an organ equally likely to
lead to somatic late effects after irradiation or are
there large and location dependent differences in this
respect? It might not be necessary to know the absolute
numbers of cells at risk in an organ if their relative
fraction of all cells is not a function of spatial position in an organ. However, some recent experimental results, (e.g. of Coggle and Peel (1978)), point exactly
into this direction and thus further work is urgently
needed to identify such cells at risk.

Another problem is the <u>organ dependence of the quality factor</u>. From the experience with the atomic bomb
survivors in Hiroshima (considerable neutron component
in the radiation field) and Nagasaki (essentially only
a photon field), some indication arises that the relative efficiency of neutrons for inducing e.g. leukemia
might be considerably higher than for producing cancer
of the breast or lung. However, lacking statistically
better data and sufficient insight into the basic mechanisms, in radiation protection an organ independent quality factor is still being used which depends only on
the organ average LET-spectrum of the radiation field.

In conclusion of this chapter it should be reminded
that - in spite of all these interesting open scientific questions - the radiation risk estimates recommended
by the ICRP for practical application in radiation protection have been derived from human experience taking
such problems into account automatically and that they
are considered as safe estimates in the low dose region.

6 DERIVATION OF RADIATION RISKS FROM EPIDEMIOLOGICAL DATA

The types of problems encounted in analysis of epidemiological data with respect to radiation risk factors of low doses and dose rates have partly been indicated already in the chapter above. However, here we find a number of additional facts which must be realized and taken into account properly when using data from larger populations.

After a careful analysis for possible heterogeneity in radiation sensitivity due to open or hidden diseases, one has to try to match the irradiated group as close as possible with a reference group to check for excess cases e.g. of cancer and leukemia due to radiation and to eliminate possible other causes. The importance of this matching e.g. for the same sex and age distribution can be demonstrated e.g. in the case of breast cancer and thyroid cancer. Women are much more susceptible than men to both types of cancer. In addition, women between, say, 15 and 40 years have a considerably higher breast cancer risk due to radiation as compared to women of other ages.

Much more difficult is matching for the same social status, living habits, occupations, and other parameters which are measured in more qualitative than in quantitative units. One well-known example for the importance of comparing groups having occupational histories as identical as possible is the so-called "healthy worker effect". This effect can clearly be seen e.g. in a study which has been provoking large public attention and many scientific discussions, namely the cancer mortality study of about 25 000 workers at the Hanford plant by Mancuso, Stewart and Kneale (1977). This publication has been causing considerable concern because of the very low dose values (of the order of ten rad) derived from these data, which would double the natural incidence rates of malignant diseases. If the mortality of these workers (about 3600 death certificates were available for analysis) is compared with the age-calendar year-specific mortality rates of white males in the United States, one finds a considerable, general reduction in the mortality of these "healthy" workers (Tab. 2).

However, evidently the cancer death rates are less reduced than the death rates for all causes. This fact can lead to a severe methodological problem wich will be mentioned finally. The standardized mortality ratio method intercompares the death rates of all persons

at risk with that of a comparable, non-irradiated reference group. With this well-established epidemiological
method, according to Marks et al. (1978) no significant
effect of radiation could be detected in the Hanford
mortality data (exept for possibly one or two cases).

Table 2   Standardized Mortality Ratios of the Hanford
           Workers (after Marks, Gilbert, and Breiten-
           stein (1978))

Cause of death	Length of employment at Hanford	
	< 2 Years	≥ 2 Years
All causes	68 %	75 %
All malignant neo-plasms	88 %	85 %
Diseases of the circul. system	87 %	76 %
Accidents, poison-ings, and violence	109 %	75%

On the contrary, the cumulative mean dose method intercompares the mean doses of all those persons who where
exposed and died already. With this method, which is
also frequently used in other contexts and which is
statistically somewhat more sensitive in any respect,
extremely high radiation risk factors have been derived
from the same data set. Besides evident objections
against the meaning of an average dose of a very skew
dose distribution, there are other arguments put forward against the use of the latter concept in this
analysis. One possible artefact of this technique
which is mainly due to the consideration of dead persons only is demonstrated in table 3. Let us assume
that we have two groups of 1000 persons at risk each.
It shall further be asumed that in the reference group
10 percent died from cancers and 40 percent died from
other causes during the period of observation. In the
other group, the healthy worker effect shall reduce
the cancer death rate to 85% of the reference group
(i.e. 8.5 percent) and the death rate from other causes to 75% of the reference group (i.e. 30 percent).

Due to the different degrees of rate reduction, the relative contribution of cancer to the causes of death
naturally will be higher by about 10 percent (22 percent instead of 20 percent) in the latter group independent of the possible existence of additional noxes
in this group.

Table 3   Schematic demonstration of possible results
           from intercomparison of deaths only in the
           presence of a typical healthy worker effect.

	Group 1	Group 2
Cancer Death Rate	100/1000 = 0.1	85/1000 = 0.085
Other Causes Death Rate	400/1000 = 0.40	300/1000 = 0.30
Proportion of Deaths due to Cancer	100 / 500 = 0.20	85 / 385 = 0.22

From the discussion especially of this set of data
it became evident that it is still extremely difficult
to derive dose-effect relationships for somatic late
effects of small doses of ionizing radiation with epidemiological methods even from rather large populations.
Much work remains to be done in collecting data (e.g.
from occupationally exposed persons) as well as in re-
fining the statistical methods of their analysis.

7 CONCEPTS OF MEASUREMENT AND WEIGHTING OF RISKS

Finally a quite different aspect of this topic shall
shortly be mentioned. Above we have only been concerned
with methods and problems of the quantification of the
probabilities for the incidence of or mortality from
somatic late effects of radiation as a function of a
number of parameters. Now the questions remain,
a) with which quantities the individual or collective
detriment due to these health effects shall be measured,
and b) which objective or subjective weight could be
given to a certain amount of detriment e.g. in risk
intercomparison or optimization studies?

At present the frequency of mortality from radiation induced somatic late effects plays a central role
in radiation protection (see e.g. ICRP Report 26).
It is used e.g. in arguments for the recommendations of

numerical values for annual dose-equivalent limits which
should not be exceeded. The Commission, however, is well
aware of certain shortcomings of this concept, which
considers fatalities only, and ideas towards possible
improvements can be found e.g. in ICRP Report 27 (1977).
There are at least two major limitations of this fata-
lity criterion:
- non-fatal diseases are not given weight at all,
- late effects occur long after exposure thus leading
  to a smaller length of residual life span lost as
  compared to not-radiation-related fatal occupational
  accidences.

Especially the latter fact leads to another possibility of the measurement of radiation detriments, namely
the use of the expectation value for the length of
healthy life lost due to a certain exposure. However,
because of severe problems with the derivation of the
necessary functions from the scanty data available,
the agreement on quantitative expressions or even the
implimentation of this concept in radiation protection
cannot be expected for the next future.

The problem of weighting effects of diseases is
closely connected to the second problem, namely of
developing risk preference theories which could possibly be applicable to radiation protection. Although
the term "risk" in many contexts has become an almost
every day word, no generally valid definition of "risk"
exists till now. Nor does any formal theory of risk or
risk acceptance or aversion based on axioms, etc. exist
at all. However, there is a large amount of literature
available on industrial utility theories, sociological
risk preference investigations, etc. which should be
carefully analysed with respect to possible lines along
which the present mortality concept could be improved
which considers only the occurence of one type of adverse events.

One of the problems encountered e.g. in risk intercomparisons, in risk/benefit analyses, in benefit/costs
analyses, etc. is the lacking commensurability of detriments, benefits, etc. which are to be compared.
The unit of detriment can be e.g. "loss of life", "days
in bed due to a temporal sickness", loss of
"life quality", or loss of monetary values. This question is still unsolved, and life or health insurance
related considerations encounter another problem, namely the irrationality of human subjective evaluation of
probabilistic, positive or negative events in ones

life. Especially because of this aspect, optimization
studies in radiation protection, technology assessment
studies, etc. have turned out to be an extremely difficult task.

In conclusion, this lecture could hopefully indicate
that the quantification and evaluation of late effects
of low doses and dose rates, on one hand, is an extremely interesting and important scientific field of basic
and applied character, and on the other hand, that our
present knowledge is almost sufficient to permit an
estimate of stochastic radiation health hazards with an
accuracy acceptable for radiation protection purposes.

REFERENCES

Coggle, J.E., and Peel, D.M., The relative effects of
        uniform and non-uniform external radiation on
        the induction of lung tumours in mice, 1978,
        in: "Proceedings of the Symposium on Late Bio-
        logical Effects of Ionizing Radiation", IAEA/
        STI/ PUB/489, p. 83-94, Vienna
ICRP, Recommendations of the International Commission
        on Radiological Protection, 1977, Report 26,
        Annals of the ICRP, Pergamon Press, Oxford,
        New York, Frankfurt
ICRP, Problems Involved in Developing an Index of Harm,
        1977, Report 27, Annals of the ICRP, Pergamon
        Press, Oxford, New York, Frankfurt
Jacobi, W., and Paretzke, H.G., 1979. Gesundheitsschä-
        den durch Strahlung, in: "Deutsche Risikostudie
        Kernkraftwerke", Bundesministerium für Forschung
        und Technologie, Bonn, Verlag TÜV-Rheinland
Mancuso, T., Stewart, A., and Kneale, G., 1977, Radia-
        tion Exposures of Hanford Workers Dying from
        Cancer and Other Causes, Health Physics, 33, 369
Marks, S., Gilbert, E.S., and Breitenstein, B.D., Can-
        cer Mortality in Hanford Workers, 1978, in:
        "Proceedings of the Symposium on Late Biological
        Effects of Ionizing Radiation", IAEA/STI/PUB/489,
        p. 369-386, Vienna
Mayneord, R.V., and Clarke, R.H., 1975, Carcinogenesis
        and Radiation Risk: A Biomathematical Reconnais-
        sance, Br. J. Radiol. Suppl. 12, Br. Inst. of
        Radiology, London
United Nations Scientific Committee on the Effects of
        Atomic Radiation, 1977, United Nations Publi-
        cations, Sales No. E.77.IV.1, New York

RADIATION RISKS IN MAN AND THE RATIONALE FOR THE ICRP SYSTEM OF DOSE LIMITATION

Giovanni Silini

Division of Radiation Protection
C.S.N. Casaccia
Rome, Italy

INTRODUCTION

The object of the present lecture is to give an outline of the general philosophy on which current evaluation of radiation protection problems is founded. In doing so I shall be following the basic concepts developed by the ICRP in their publication 26 edited in 1977 and in order to make them easier to a non specialized audience, I will also make reference to another publication of the ICRP bearing the number 27 which is an essential tool for interpreting Publication 26. I will also cite some data derived from UNSCEAR Report 1977, where risk estimates in man are very thouroughly analyzed.

The general outline that I will present is the outcome of many decades of study of the theory and practice of radiation protection by a great number of individuals and evaluating bodies who have helped the evolution of the discipline. I would not therefore presume to enter into the details and particularly into the practical aspects of radiation protection. The audience on the other hand, should not presume to have more than a feeling of the subject after this lecture. Reading and meditation of the above publications will be required for a deeper understanding of these concepts.

The present lecture requires some basic knowledge of mammalian radiation biology with special regard to the pathogenesis of the various lesions, to the most important clinical characteristics and their mode of expression in the irradiated individual or within an irradiated population of mammals. Although these subjects have not been considered specifically during the course, many lecturers will make reference to biological effects and I have myself put forward some considerations about the various levels of expression of radiation damage in mammalian systems. It would however be appropriate for me to recall just a few ideas in this respect and to remind you that radiation effects in mammals are usually classified into somatic and genetic, according to whether they are induced on the cells of the somatic or of the germinal line. The somatic damage is expressed within the life of the irradiated individual, the genetic effects are induced on his progeny. Somatic effects are further distinguished into immediate or late if they appear soon or a long time after irradiation.

RADIATION EFFECTS

The immediate somatic effects are expressed in man within a few days or a few weeks from exposure and are caused by damage to one or more of the self-renewing tissues. They may be localized to a given tissue or generalized to the whole body under a specific syndrome. Their clinical severity changes considerably with the dose, the dose-rate, the type and energy of the absorbed radiation, the part of the body irradiated. Immediate effects are also called "functional" because they depend on the inactivation of a great number of functional cells of a given differentiative line. Immediate effects are non-stochastic or deterministic in nature in that they are expected to occur in an individual beyond a given level of absorbed dose. This dose varies from tissue to tissue and is called the threshold dose. It is of the order of a few tens of rads for most functional effects.

The existence of this threshold dose is a most important characteristic of immediate effects from the point of view of radiation protection: it makes it virtually impossible for them to appear after doses that are below the threshold and therefore it makes it possible to avoid them completely by keeping exposure below that dose.

Late somatic effects are those that appear within populations of irradiated individuals and are expressed as leukemias, tumors, and possibly other degenerative

lesions. They are stochastic or statistical in nature,
in so far as it is impossible to identify for them a
causal dose-effect relationship in any single case. The
dose-induction correlation may be shown only on large
populations of irradiated individuals as an extra increase of these diseases over their "natural" background
incidence. Since it is impossible to establish with any
certainty the shape of the dose-induction relationships
at low doses, radiation protection philosophy assumes
that their frequency of occurrence is linear with dose
and without threshold. The clinical severity of these
conditions is variable but radioprotection assumes
conventionally that they are of a uniform and maximum
severity, comparable with the death of an individual.

Genetic effects showing on the progeny may appear
within the first generation from irradiation, in which
case the damage is called dominant, or within subsequent
generations when genes that carry the same mutation in
the male or female genetic complement match into the
genome of the zygote. Clinically, these conditions may
be very trivial (imagine, for example, a condition like
color blindness) or very harmful (mongolism, for example). Radiation protection usually makes reference to
the most severe genetic diseases which are either
incompatible with life or are very disabling for the
individual that carries them and very heavy on his family and the society. The presence of those effects has
never been shown in man although there is no reason to
doubt that at appropriate doses they may in fact be
produced. Radiation protection also assumes that for
this class of effects there may be a linear non-threshold dose-induction relationship.

BASIC ASSUMPTIONS

Any consideration of the specific risks should be
preceded by some discussion of the probable form of
dose-effect relationships. The philosophy adopted by the
ICRP assumes non-threshold linearity for stochastic damage and this is the foundation on which the entire
philosophy depends. It is important to stress the meaning of this assumption. It postulates that there is no
dose, however small, that may be considered safe and
there is no dose increment, however small, which may not
produce a corresponding increase of effect and therefore
of risk. The summation of doses taken as a measure of
total risk and the calculation of collective doses as
expression of the global detriment to the exposed

population could not be justifiable outside the assumption of linearity.

Whatever the assumptions , however, what determines the real added risk due to an increased exposure to radiation is the actual trend of the dose-response relationships and this trend is rarely linear for complex effects in mammals. In recent times the hypothesis is frequently entertained that the form of the dose relationships for significant effects in radiation protection might be of a composite type, for example,

$$E = a D + b D^2 \qquad\qquad\qquad (1)$$

where the effect, E, would be determined by the sum of two component terms, one linear and one quadratic with the dose D, and the constants a and b would be the numerical expression of these components.

In this type of relationships the constant a would naturally prevail at low doses, while b would tend to predominate with increasing dose and dose-rate. To give some value to these terms, one might say that b would become absolutely prevailing above about 100 rad or 100 rad/min. Actually, for higher doses it would be necessary to introduce into the preceding formula another term to account for a decrease of effect due to killing of the potentially transformed cells.

Experimental series showing a trend of the dose effect relationship compatible with the above expression have frequently been described. Recent analyses are showing however that the actual value of a and b would change in absolute terms and relative to one another in different instances. It is also easily appreciated that the lower the ratio a/b and the higher will be the possibility that the interpolation of certain series of data might extimate in excess the value of a. This would naturally be seen in all those cases (which are by far the majority) where the istantaneous dose rate is high, where the exposure is unique and where the total dose is in absolute high.

It must be said that the possibility of overestimating the risk is clearly more acceptable to a radiation protectionist than that of underestimating it However, the consideration should be kept in mind that in the course of a general optimization an overestimate of radiation risks would tend to favour choices that might be in reality worst from the point of view of

public health than those involving a potential
radiological risk.

The above considerations underline the basic importance of the linearity assumption, its relationship with
the practical and experimental situations, and its
limitations. It should be realized that if this principle is adopted, other ideas are also implicitly
accepted. In fact, if there is a linear proportionality
function between dose and induction of stochastic
effects, it should be possible to use the average dose
received by a given organ or tissue as the significant
reference quantity. Thus, it would not be necessary to
consider the variability in the dose distribution
between various parts of the bone marrow or of the liver
if the response of the component cells, taken to be of
uniform sensitivity, is in any case linear with respect
to the dose absorbed. The reactions of the single component cells will sum up to give an overall effect
corresponding to that expected from the mean dose
absorbed by that organ or tissue. This assumption obviously simplifies the solution of many practical cases.

There is a problem with respect to internal
irradiation from point-sources, where the following
considerations apply. Firstly, with regard to stochastic
effects, the cellular inactivation resulting from the
absorption of high doses within microscopic volumes of
tissues should produce less harm than that expected when
the same dose is distributed uniformly within large volumes of the tissue; this is because in the former case
one would expect killing of transformed cells by the
high doses, which would not be possible in the latter.
On the other hand, with regard to non-stochastic
effects, the loss of cells around the zones of highest
dose absorption should not result in significant loss of
tissue or organ function, unless the functional reserve
of the tissue is impaired for other reasons.

One of the most important conceptual developments
that the ICRP have introduced to their new system of dose
limitation is to abandon the old concept of critical
organ or tissue. This was visualized in the past as the
functional unit that was more vulnerable by the lowest
doses under any given exposure condition, because of its
special importance in the general economy of the organism. The concept of critical organ did not allow a
global evaluation of risk taking into account the
effects on the whole body and it did not allow to sum up
all detriments to specific organs and tissues within a

global estimate relative to various types of effect.

The ICRP have chosen instead to give estimates of risk relative to various organs and tissues based on their susceptibility to various types of damage. These risks are weighted relative to each other and the Commission propose that they may be added for each individual or over the whole population, by making use of the assumption of linearity. It is realized of course that the global risk estimates may vary according to the characteristics of the exposed individuals (genetic characteristics, age, sex) or to the structure of the exposed population. But it is also reputed that, at least as a first approximation, an acceptable level of precision might be reached with a single value of risk to be applied to all exposed persons - be they workers or members of the population at large - irrespective of the variability of the single individual or of the various classes.

The criteria for judgement in the evaluation of the risk factors are the probabilities for induction of immediate non-stochastic effects, of late somatic damage and of genetic diseases. Although other forms of damage and of physical and psicological suffering - that we would usually value as less important - could be included in this evaluation, it must be recognized that at the present stage of our knowledge we could not set them into an appropriate perspective with the former ones. The publication 27 of ICRP is specifically concerned with an analysis of this problem and with the development of some objective index of harm. However at present technical as well as ethical difficulties appear unsurmontable for the purpose of providing a more rational and detailed approach to these problems.

There are various points of the ICRP analysis that I would now like to enter into and to discuss in some details because they deserve special attention.

RISK ESTIMATES

The first important point is specifically underlined in the title of the present lecture and concerns the risk estimates. The ICRP state that the assumed risk estimates are derived for the purpose of radiation protection and they explain that the estimates are consistent with those of the UNSCEAR but are adjusted to make them applicable at low doses and low

Table 1. Non-stochastic Effects

Tissue or organ	Type of damage	Estimates from UNSCEAR 1977	Assumptions by the ICRP 26
Testis	Sterility	The 1977 UNSCEAR Report	Temporary sterility at 25 rad, permanent at 250 rad
Ovary	Sterility	has not considered specifically	Variable with age but permanent at 300 rad
Bone marrow	Marrow aplasia	non-stochastic damage. Previous	It is not a limiting effect for doses below 50 rad/year
Lens	Clinical cataract	evaluations are compatible	It is not a limiting effect for doses below 30 rad/year
Skin	Cosmetic lesions	with ICRP assumptions.	These lesions are avoided for doses below 50 rad/year

Table 2. Late Stochastic Effects

Tissue or organ	Type of damage	Estimates from UNSCEAR 1977[*]	Assumptions by ICRP 26[*]
Bone marrow	Induction of leukemia	15-25	about 20
Bone	Tumor induction	2-5	about 5
Lung	Tumor induction	about 50	about 20
Thyroid	Tumor induction	5-15	about 5
Breast	Tumor induction	$10-60/10^6$ females	about $25/10^6$ men and women
Other tissues (stomach, large intestine, salivary glands, liver, other pelvic organs, skin)	Tumor induction	25-50	50 cases as a maximum; no tissues will contribute for more than 1/5 of this value
Embryo and fetus	Induction of leukemia and tumors	200-250 cases/10^6 live-born	The susceptibility to the induction of certain malignancies is higher for irradiation during pre-natal ages

[*] Fatal cases induced per 10^6 persons exposed to 1 rad of low-LET, low-dose-rate irradiation.

Table 3. Genetic Effects

Tissue or organ	Type of damage	Estimates from UNSCEAR 1977	Assumptions by ICRP 26
Male and female gonads	Mutations and chromosomal defects causing clinically important genetic diseases	About 200 important genetic diseases are expected for all future generations per 10^6 live born children for 1 rad of radiation of low LET and low-dose rate received by either parent	About 200 important genetic diseases are expected for all future generations per 10^6 live born children for 1 rad of radiation of low LET and low-dose rate received by either parent. Of these, about 100 cases will appear within the first two generations.

dose-rate. This means that the empirical estimates of risk derived from the observation of special population groups have been somewhat corrected by the Commission to render them applicable to exposure conditions of general significance, such as those at low doses and low-dose-rates. They have also been adjusted to make them applicable to a population of both sexes and all ages. Thus, the resulting ICRP estimates may sometimes appear different from those to be seen in the literature. In all cases they represent average estimates derived only for radiation protection purposes.

I have summarized in Tables 1, 2 and 3 the risk estimates referring to non-stochastic immediate effects, to late stochastic effects and to genetic effects as can be derived from the 1977 Report of the UNSCEAR. Also, for each tissue I have summarized the assumptions of the ICRP which are in a few cases slightly different from the values of UNSCEAR for the reasons that I have alluded to already. There is no need for me to insist here on the major problems referring to the derivation of these estimates, to the difficulties in obtaining them and to the large errors and the consequent limitations by which they are affected. These estimates are essentially to be regarded as indications of the order of magnitude of the risk and there may be doubts on their real value beyond the first significant figure.

There is often controversy over the numbers shown in the Tables and it may be interesting to discuss the rationale for these figures. I will do so by using examples derived from the ICRP publication 27.

One could imagine that the risk of induction of malignancies could be of the order of 100 cases per rad per million of irradiated persons under optimal conditions of expression of damage. Based on experimental evidence we could suppose that about 20 of these cases might be attributed to leukemia and about 80 to other solid tumors. If a population of male workers would be irradiated at a constant rate between 18 and 65 years, some of these potentially inducible tumors would not show as actual causes of death because the older members of the population would die from other causes before the radiation-induced tumors could develop.

Actually, under the reasonable assumptions that latency times for leukemia might be 10-13 years and for solid neoplasms 20-26 years, the curves A and B in Fig.

1 would represent the relative risk of death for leukemia and tumors, respectively in the various age classes. If one multiplies the frequency in each age class (D) for the global risk of leukemia and other tumors at the various ages (C) one obtains the cumulative number of radiation-induced neoplasms for the various age classes (E). It is seen that this number increases with age but never attains the theoretical

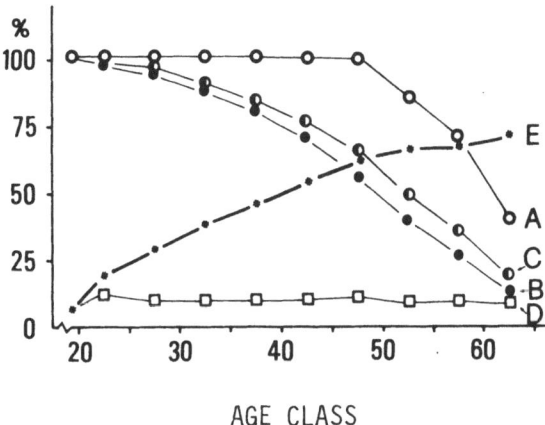

AGE CLASS

Fig. 1.   Effect of the dose distribution in various age classes of an irradiated population on the risk estimates derived for radiation protection purposes (see text) (from ICRP 27).

value of 100 that one would expect for a complete expression of the neoplastic induction. Following the above model, a risk evaluation that would take into account the age distribution of the population within various age classes would only be equivalent to 72% of the postulated maximum risk. The same model could be applied to a female population, in which case the average risk for a population of all ages would be equivalent to 87% of the theoretical risk. The variations between UNSCEAR and ICRP risk estimates are largely explained on this type of argument.

A second point for discussion concerns the evaluation of the genetic risk.   The ICRP start from the
assumption (see Table 3) that a reasonable risk estimate
for genetic diseases within the first two generations
might be of the order of 100 ċases per rem per million
live born children after exposure of either parent
during the fertile age. If this dose is absorbed at a
constant rate over a very long time it is obvious that
only the portion of dose absorbed during the fertile age
might result in possible genetic consequences. In Fig.
2 the curves A (males) and B (females) show the percentage of individuals within various age classes of a working population. The probability of conceiving children
follows   the   curves   C   (males)   and   D   (females)
respectively for each age class . If one multiplies the
percentage of people in each age class and the probability of having children after that age and cumulates the
resulting values one obtains the genetically significant
component   of   dose   for   constant   exposure   rate   as   in
curves E (males) and F (females). Compared to the maximum risk for a complete expression of genetic damage
applying to people of maximum fertility, the actual risk

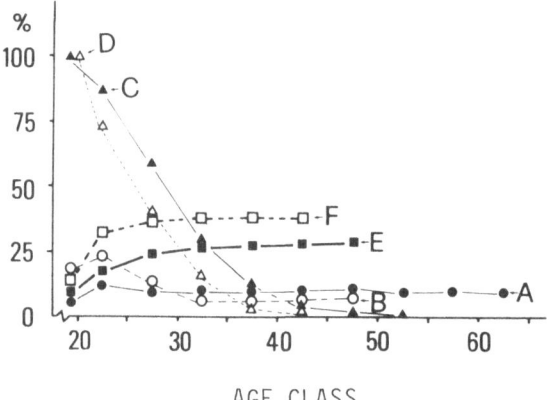

AGE CLASS

Fig. 2.   Estimation of the genetically significant
           component of dose in a population of different
           age distribution (see text) (from ICRP 27).

over a population of all ages would amount to 29% in males and 38% in females.

Similar calculations performed for working populations exposed to radiation under various conditions yielded estimates of the genetically-significant dose component ranging from 10% to 60%. Taking into account the age structure of a population at large the ICRP propose that a ratio of 0.4 between genetically significant dose and dose actually absorbed might be used. This would result in an estimate of the genetic risk over the first two generations of the order of 4×10^{-5} rem^{-1}. The estimate would double (see Table 3) when the risk over the whole population would be involved, in order to include other hereditary damage for all generations after the second. In this case the relevant figure would thus be 8×10^{-5} rem^{-1}.

DERIVATION OF EXPOSURE LIMITS

Going now to the recommendations offered by the Commission, it should preliminarly be stated that their object is two-fold: a) to absolutely avoid non-stochastic effects; and b) to limit the occurrence of stochastic effects to acceptable levels.

With regard to point a) the Commission believe that the object can be met by applying a dose equivalent limit of 50 rem in one year to all tissues except the lens of the eye where a limit of 30 rem is envisaged. These limits are valid both when a given tissue is irradiated alone or in conjunction with other organs and tissues. As I will say later on, they are set in order to further limit exposures that would be compatible with the limits for stochastic effects.

Concerning stochastic damage, the Commission adopt the principle that the risk should be the same when irradiation is delivered uniformly over the whole body or limited to a few tissues. This condition can be met if the annual dose equivalent limit for the whole body ($H_{wb,L}$) is higher or equal to the sum of dose limits applying to each tissue (H_T) multiplied by a weighting factor (w_T) representing the proportion (over the total stochastic risk) of the risk resulting from the irradiation of a given tissue T.

In other words, this condition is met when:

$$\Sigma_T \ w_T \ H_T \ \leq \ H_{wb,L} \qquad\qquad\qquad (2)$$

The Commission give therefore a tabulation of the w_T values, as in Table 4. Before examining these values, let me stress once again their meaning. If one assumes that the global risk of genetic and late somatic damage (this time taken together) resulting from a given dose of radiation uniformly distributed over the whole body is 1, this risk would be subdivided among the various tissues and for the various effects according to the percentages of Table 4.

Consider, for example, the bone marrow. If this tissue were exposed alone to a given dose, the resulting risk would be about 0.12 times that to be expected by exposure of the whole body to the same dose. Therefore the limit of dose equivalent applicable to the whole body must be increased about 8 times for irradiation of the marrow alone, in order to keep the equivalence of risks in the two situations. The values of w_T are at the basis of calculations on which the derived limits for internal exposure are based.

The dose equivalent limits recommended in order to avoid stochastic risk are generally adequate also for the prevention of non-stochastic effects. But for some radionuclides (particularly those that concentrate in tissues having a low susceptibility to cancer induction) the annual dose limiting the stochastic risk could add up to very high figures when integrated over the whole working life and it might therefore produce non stochastic effects. This is, for example, the case of the thyroid. Since its w_T value is only 0.03, in theory a worker could receive over his whole working life:

65-18   x   5 rem/year   x   1/0.03 = 7.833 rem
  years   of   yearly dose equivalent   inverse of w_T
working   life   limit   for workers

This dose is clearly too high and for this reason the limit is set applying to the non-stochastic risk that no worker shall receive more than 50 rem/year at any given tissue or organ.

During the latter part of this lecture I shall try to illustrate the derivation of the exposure limits, according to the ICRP philosophy and I will consider separately the cases of professional exposure and of exposure of the whole population. I should point out that the considerations to follow apply only to the

Table 4. Derivation of the Values of w_T

	Risk estimate (cases x 10^{-5} rem^{-1})	Values of w_T
Gonads (M + F)	4.0	0.25
Breast (M + F)	2.5	0.15
Hemopoietic bone marrow	2.0	0.12
Lung	2.0	0.12
Thyroid	0.5	0.03
Bone	0.5	0.03
Other tissues[*]	5.0	0.30
	16.5	1.00

[*] Five other tissues different from the preceding ones, each with a value of w_T of 0.06

Table 5. Dose-Equivalent Limits

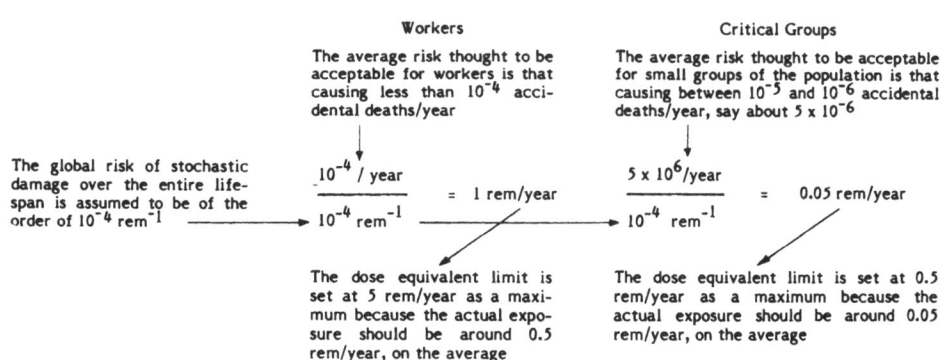

exposure exceeding the natural background and the medical component of the total radiation received.

Concerning exposure of workers, the criterion for
establishing the limit is that of comparing the radiological risk with the risk of other occupations having a
high standard of safety. Through a comparative evaluation of the accidental mortality in the course of
various working activities, the ICRP come to the conclusion that occupations having a yearly accidental death
rate not exceeding 10^{-4} may be considered to be
reasonably safe. In theory all forms of harm, fatal or
not, should be considered for this type of analysis. In
practice however it is observed that radiation does not
produce non-lethal harm within the dose range of concern
for radiation protection. Thus, if a given occupation
involving radiation exposure is judged to be safer than
another occupation without radiation risk only on the
basis of accidental death, it should be even safer with
respect to other non lethal risks.

The second consideration to be born in mind is one
deriving from past experience of radiation protection.
Experience shows that when a dose-limiting system is
applied correctly and this system has an upper exposure
limit at, say, 5 rem/year, the distribution of the doses
received by a large group of exposed persons tends to be
log-normal with an arithmetic mean falling at about 1/10
of the limit, the bulk of the population largely below
this mean and a tail of only a small fraction of the
population at or near the limiting value. On the basis
of the two preceding points the ICRP derive the dose
equivalent limits for workers in Table 5.

Concerning exposure of the population at large
there are two possible cases to be considered. The first
one is exposure of individual members or small groups
frequently referred to as "critical groups" living, for
example, around nuclear plants and for which a protection system must be designed. It is quite clear that
occupational risks should not be taken for a comparison
in this case. Comparisons should more correctly be made
with other forms of risk involving fairly large groups
of people; risks that are only partially dependent on
the acceptance of the exposed persons; risks that persons cannot easily avoid if they decide to live
confortably according to the standards of a modern
industrial society; risks controlled by the
administrative authority that may, directly or indirectly, act upon them to the advantage of the whole civil

community. Typical examples may be the risks from public transport or from the use of electrical home appliances. The ICRP calculate that, with a large range of
variability, these risks entail an incidence of accidental death between 10^{-5} and 10^{-6}/year and that this level
of risk seems to be accepted by single members of the
society. On the above grounds, the ICRP derive the limit
for exposure of single members of the population as in
Table 5.

     Finally one should consider exposure of the popula-
tion at large. In the past the ICRP had suggested (but
never actually explicitly recommended) that a limit of 5
rem in 30 years could be considered in this case. It is
now recognized that this limit might be too high and
perhaps dangerously so, if it would be taken as a
permissible limit. The most recent experience suggests
that the actual value of this limit might be evaluated
in each single case together with other information on
the possible risks (genetic as well as somatic) originating from other causes. In this respect the ICRP recall
that justification and optimization are the principles
to be followed for such evaluation they should be
applied thoroughly before coming to the actual
consideration of limits. There is, as one can easily
appreciate, a great difference between the past and
present position because it is a matter of evaluating
the global risk from all justified exposures from different causes instead of subdividing between the various
causes of damage a limiting value available for appor-
tionment.

     What I pointed out above is, in my view and very
concisely, the essence of the ICRP philosophy of risk.
There would be many other things to be added but they
would require a much longer time to be developed. If
possible, I will point out a few of them in the course of
the discussion.

REFERENCES

International Commission On Radiological Protection,
Recommendations of the ICRP, ICRP Publication 26,
Pergamon Press, Oxford, 1977.

International Commission On Radiological Protection,
Problems involved in developing an index of harm, ICRP
Publication 27,Pergamon Press, Oxford, 1977.

United Nations, Sources and Effects of Ionizing Radia-
tion, UNSCEAR Report to the General Assembly, Official
Records of the General Assembly, Thirty-Second Session,
Supplement No. 40 (A/32/40), New York, 1977.

THE ROLE OF THE THEORY OF DUAL RADIATION ACTION IN RADIATION PROTECTION

Harald H. Rossi

Radiological Research Laboratory, Department of
Radiology, Cancer Center/Institute of Cancer Research
Columbia University College of Physicians and Surgeons
New York, NY 10032

Radiation protection is a very complex subject and the development of an adequate system of radiation protection involves many problems including basic philisophy, concepts, quantities and units, measurement, administrative aspects, and many others. However, the fundamental question that needs to be answered before any of these problems are dealt with concerns the nature and the magnitude of the risk to human populations exposed to ionizing radiation. This is a matter on which there is still a considerable amount of controversy and no doubt this is at least in a large measure due to inadequate information. It is a very difficult problem which almost by definition cannot be solved on the basis of studies of irradiated human populations. There are a variety of such populations which have received a broad range of doses with the evident trend of a decreasing frequency of detrimental effects with decreasing dose. However, it is invariably postulated that the probability of any harmful effects that can be induced by maximum permissible doses must at most be very small. Consequently, doses which cause even moderately discernible effects are likely to be larger than those that are considered acceptable in radiation protection. A further basic difficulty is that all of the harmful effects that can be attributed to radiation also occur in populations that have not been irradiated. Their "natural" incidence varies considerably with age, sex, habitat, socioeconomic status, and a host of other factors. It is very probable that the maximum permissible doses that have been found acceptable in recent years cause incremental frequencies of injury that are even less than these variations.

One aspect of the problem on which there is broad agreement concerns the nature and relative importance of radiation risks.

Although not unequivocally demonstrated in humans, genetic effects
are almost universally believed to be caused by radiation. By
definition, these effects are only observable in offspring of
irradiated individuals and they may linger in populations for many
generations. Pre-natal effects have certainly been observed in
man at least in the form of substantial malformations or mental or
physical retardation. There is also no question as to the existence of somatic effects. These are usually divided into those
for which there is believed to be a threshold (sometimes termed
non-stochastic effects) and those for which the existence of a
threshold is unlikely or uncertain (stochastic effects). Virtually
the only important stochastic effect is the induction of various
forms of cancer.

There has been an increasing consensus that cancer is the
dominant radiation risk because of a belief that maximum permissible doses which pose a generally acceptable risk of cancer constitute risks of other effects that are likely to be at least as
acceptable (ICRP, 1977). It may be well to remember, however,
that this has not always been the general opinion. Some twenty
years ago, genetic risks were considered to be the principal hazard
of ionizing radiation and for reasons which will become apparent,
there is a possibility that this position might be adopted again
in the future.

There is also some uncertainty regarding the magnitude of the
risks attendant to irradiation of the embryo and fetus. It has,
however, been stated that limitation of the dose received during
pregnancy to 5 mGy should result in a very small risk (NCRP, 1977).
Quite possibly there is a threshold for congenital defects and like
non-stochastic somatic effects they will not be considered here.

Because of the paucity of human data, attempts have been made
to utilize animal data as an additional source of information. It
is evident that this must be done in a very critical fashion since
radiation sensitivity does not only vary between animal species,
but frequently also among strains in a given species. A critical
evaluation of animal data requires theoretical considerations of
one kind or another. Theories thus form the third kind of input
to risk estimation and in the following I shall discuss the role
of biophysical theories and specifically that of the theory of
dual radiation action.

As already mentioned, there is very little numerical information relating to human radiation genetics. It has, therefore, been
found indispensable to rely on extrapolations from animal data.
The only two animal species investigated in detail are the fruit
fly and the mouse and there are substantial differences in both
the magnitude and the shape of the dose-effect relation for these
two species. A variety of biological assumptions have to be made

when this information is extended to human populations with consequent substantial uncertainties.

However, while there may be questions as to the magnitude of the risk, theory provides firm information on the shape of the dose-effect curve at low doses. This is so because the cells of the germ plasm may be considered to be <u>autonomous</u>, i.e., the cell response is independent of irradiation of other cells. The response of an autonomous cell must, therefore, depend only on the specific energy received by it. When the dose is decreased to such low levels that cell traversal by more than one particle is improbable, one is dealing with only two kinds of populations: cells that have received no energy at all or those that have been traversed by one particle. The fraction of the latter must be proportional to absorbed dose and consequently any effects must be proportional to absorbed dose. It follows that there can be no threshold for genetic effects and that the incidence of these effects must be proportional to absorbed dose, approximately up to doses of which the mean number of cell traversals is about 1. Such doses are of the order of 1 milligray for low LET radiations if the cell diameter is 10 microns. There is, however, strong indication that the specific energy of importance is not the one in the cell as a whole, but that one should be concerned with subnuclear volumes having diameters of the order of 1 micrometer. This means that linearity for genetic effects should occur for low LET radiations up to doses of the order of 0.1 gray. For high LET radiations this limit can be more than 2 orders of magnitude larger. Since the linear portion of the dose-effect curve is due to single event action, it follows also that there can be no dose-rate effects.

According to the theory of dual radiation action, there is in addition to the linear component a quadratic component in dose which becomes dominant at doses where multiple traversals are more probable. The magnitude of this component should, however, depend on dose-rate and at extremely low dose-rates it should be absent. It would, furthermore, be predicted that at those levels of effect where single event action predominates for high LET radiation and multi-event action predominates for low LET radiation, the RBE of the former relative to the latter should be inversely proportional to the square root of the high LET dose.

Available genetic data for animals are generally in good agreement with these predictions. It should be noted that for certain kinds of genetic damage (e.g., point mutations) dual radiation action may not apply. However, this merely should be reflected by a strengthening by the linear component if overall genetic damage is considered.

The situation is quite different with respect to radiation carcinogenesis. There are a number of studies that have shown a

clear-cut increase of cancer incidence in persons who have received
substantial doses of ionizing radiation. Most of these involve
exposure to low LET radiation and in most instances even the approximate incidence or mortality can be assessed only if the doses
exceed 1 Gy. The studies of Japanese A-bomb survivors furnishes a
very important exception to this. The populations were so large
that excess risks were evident at lower doses and, in addition,
persons exposed at Hiroshima received a substantial neutron dose.

It has been a widely employed practice by epidemiologists to
estimate the low dose risk by a linear interpolation between data
points including the zero dose point which is derived from data on
persons who received essentially no radiation. This practice has
been defended by stating that the available information is not
inconsistent with such a simple model and this is indeed usually
the case. It is, however, also true that it is usually impossible
to distinguish between the goodness of fit of linear relations and
relations that are non-linear.

Studies of radiogenic cancer in animals make it appear unlikely
that any simple function accurately represents the relation between
dose and cancer risk. Figure 1 shows such relations for various
neoplasms induced in mice by low LET radiation (Ullrich, et al.,
1976). This large scale study discloses a variety of shapes of the
dose-effect curves and in 1-b some instances are shown where there
appears to be a reduction of incidence at doses of less than about
1 gray. These observations make it improbable that the adoption of
any reasonably simple dependence of effect on dose can be more than
a rough approximation of the true state of affairs. However, it
should also be noted that none of the curves in Figure 1 are
consistent with the notion of a linear dependence, roughly speaking,
all of them exhibit positive curvature.

The curves in Figure 1 also suggest that radiation carcinogenesis is a very complex process and that it seems hardly realistic to
assume that it is simply due to a transformation of autonomous cells.
It would seem that under these conditions the theory of dual radiation action could not make any useful contributions to the problem
because it certainly has been developed on the basis of comparatively simple considerations of energy deposition in individual cells.
It turns out, nevertheless, that a basic aspect of the theory can
in fact be applied.

This may be illustrated by Figure 2 which shows an incidence
of mammary neoplasms in the Sprague-Dawley rat as a function of
dose of both x-rays and neutrons. In this logarithmic representation and in this relatively restricted range of low levels of effect,
it appears possible to approximate the data adequately by nearly
straight lines. However, the striking finding is that the slope of
the line for neutrons is considerably less than 1 indicating a non-

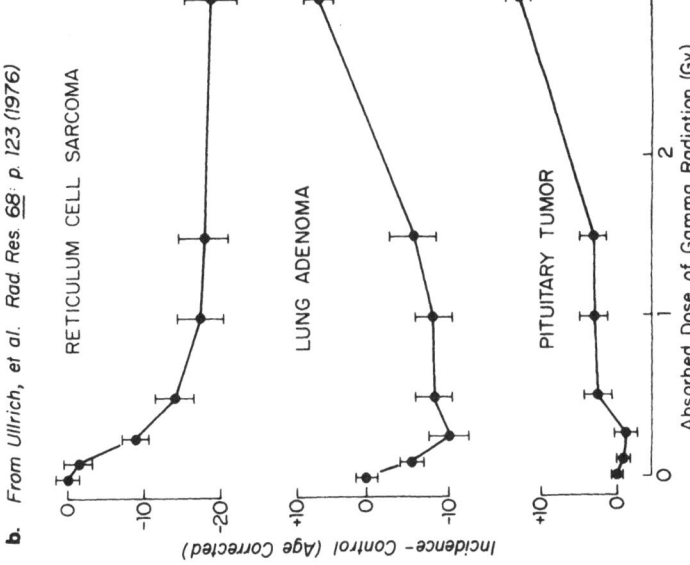

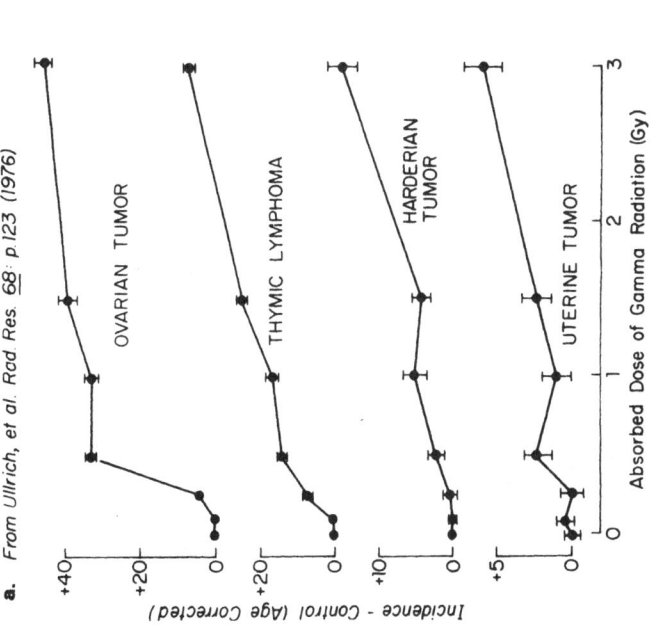

Fig. 1-a. Cancer mortality in mice exposed to gamma radiation.

1-b. shows instances where there appears to be a depression of natural incidence at low doses.

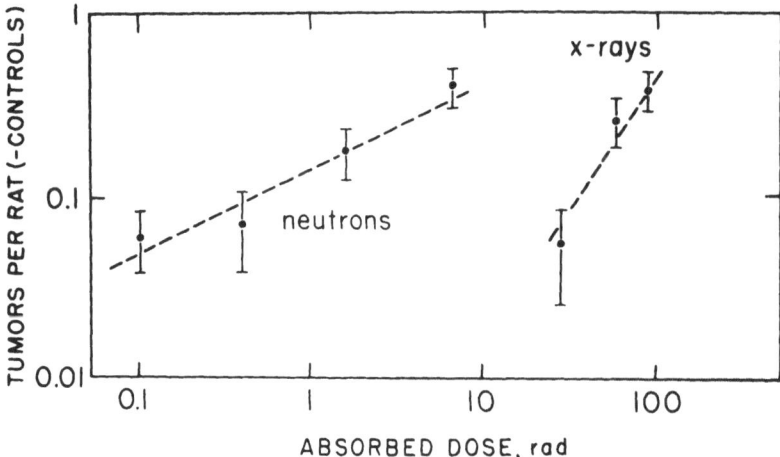

Fig. 2. Logarithm of mammary tumors in Sprague-Dawley rats vs.
logarithm of dose of 430 KeV neutrons and 250 kVp x-rays.

linear dependence of tumor frequency on dose under conditions where
the number of neutron secondaries per cell is considerably less than
1. It has been shown (Rossi and Kellerer, 1972) that this proves
the point which has already been made to the effect that it cannot
be assumed that tumors arise as a result of transformation of
individual non-interacting cells. Nevertheless, as shown in
Figure 3 a logarithmic plot of RBE vs. neutron dose discloses a
relation in which the former is inversely proportional to the square
root of the latter. (This follows from the slope of the line being
equal to $^-0.5$.) This is precisely the relation derived for autonomous cells. A plausible explanation for this agreement is that
cellular interactions which can greatly affect the dose-effect
curve do not affect the dose-RBE curve if they proceed in the same
manner for either type of radiation.

Figure 4 shows a plot of RBE vs. the dose of neutrons having
energies of the order of 1 MeV for a variety of effects where the
same dependence is seen. Some of these effects (such as lenticular
opacifications) are almost certainly quite complex. These findings
indicate that while the shape of the dose-effect curve for radiation carcinogenesis may be complex or unknown they are very
unlikely to have the same shape for neutrons and x-rays and that
one should in fact be able to derive the shape of a dose-effect
curve for either radiation if it is known for the other radiation.

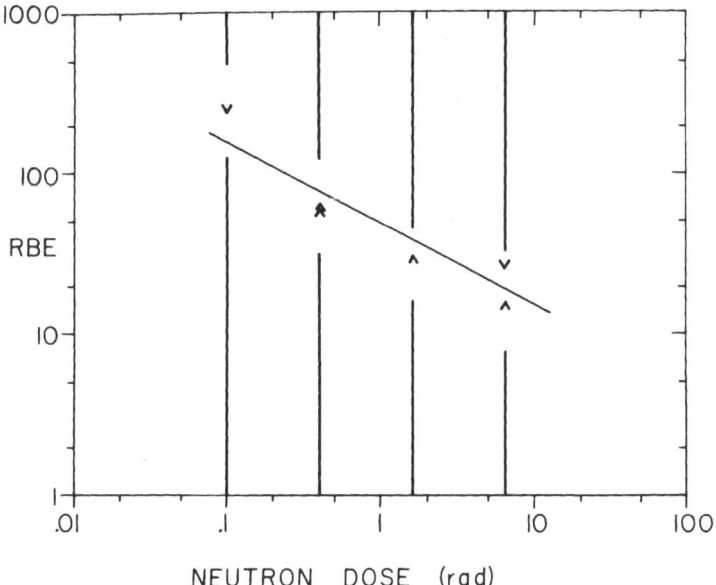

Fig. 3.   Logarithm of RBE vs. logarithm of neutron dose for
          mammary neoplasms for Sprague-Dawley rats. The bars
          delineate values of RBE which are excluded with 95%
          probability. The arrowheads indicate values that are
          excluded with lesser probability.

     There are strong reasons for doubting the validity of the
assertion that there is a linear dose-effect relation for radiation
carcinogenesis by both gamma rays and neutrons. An analysis was
therefore performed which specifically investigated this question
(Rossi and Kellerer, 1974). The malignancy chosen was leukemia
because of its relatively high incidence in irradiated populations
and its relatively low incidence in non-irradiated populations.
The analysis was based on different radiations at Nagasaki and
Hiroshima. In the former city exposures were almost exclusively
to gamma radiation while in Hiroshima a substantial neutron component was also present. It was found that while data from either
Hiroshima or Nagasaki could individually be fitted by linear dose-
effect relations, the possibility that they were both linear could,
in a joint analysis, be rejected with a probability of the order
of 85% because the RBE was found to increase with decreasing dose.
It was also shown that to the extent that the relation between

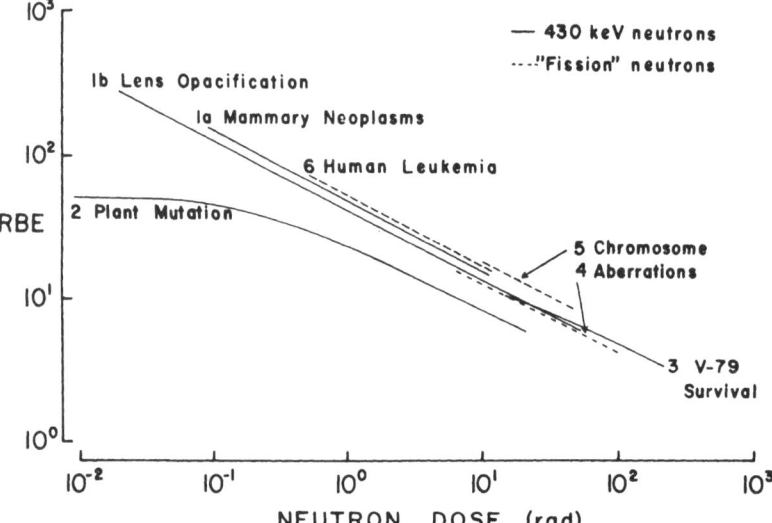

Fig. 4. Logarithm of RBE vs. logarithm of neutron doses for a
         variety of biological effects.

mortality and dose could be approximated by linear and quadratic functions, the former applied to neutrons and the latter to gamma rays. The same conclusion was reached in a somewhat different approach (Rossi and Mays, 1978) which also derived the leukemia risk in terms of the dose to the bone marrow. It was, furthermore, significant that the dependence of RBE on dose was the same as that found for other endpoints and this relation is in fact one of the ones shown in Figure 4. Figure 5 summarizes not only the dose-effect relations which have been derived in these studies, but also indicates that a maximum permissible annual dose of neutrons appears to result in a leukemia risk that is equal to the "natural" annual risk. This implies that a single exposure to the maximum annual

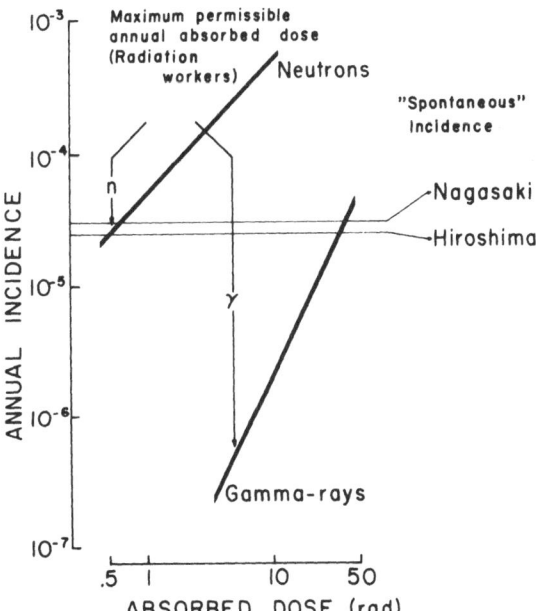

Fig. 5. Annual excess mortality (averaged over a period of 27 years) due to leukemia following various doses of fission neutron and gamma radiation. The maximum permissible annual absorbed doses for radiation workers are also indicated, as well as, "natural" leukemia mortality.

dose of neutrons would double the leukemia risk for a period of the
order of at least 15 years. There is now increasing acceptance of
a curvilinear relation between leukemia risk and dose of low LET
radiation and there is also general agreement as to the magnitude
of the neutron hazard.

The relatively favorable statistics on radiogenic leukemia
might be also expected for the incidence of all forms of cancer.
An analysis has recently been performed (Rossi, 1979) to investigate this subject. Figure 6 shows the annual mortality (1950 -
1974) from all malignant neoplasms at Hiroshima and Nagasaki.
There is an obvious difference between the two cities despite the
fact that in Hiroshima little more than 10% of the total dose to
most organs was due to neutrons. A preliminary analysis indicates
that for the period from 1950-1974 the excess average annual
mortality is 1.7×10^{-8} $(D/rad)^2$ for gamma radiation and 1.3×10^{-4}
(D_n/rad) for neutrons. These are crude numbers that may well be
subject to change, but it seems unlikely that these estimates need
to be modified by a factor of more than 2.

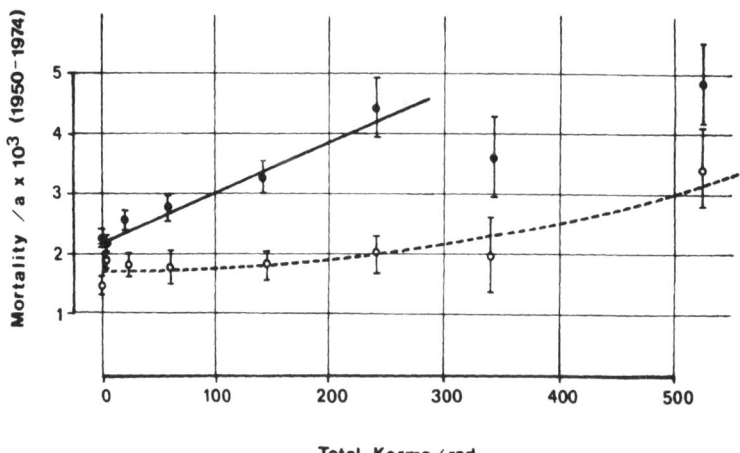

Fig. 6.   Annual mortality from all malignant neoplasms (averaged
          over 24 years) in individuals exposed to various values
          of total kerma at Hiroshima and Nagasaki.

It will be seen that theoretical radiobiology and specifically the theory of dual radiation action can make important contributions to risk estimation. They indicate that as a rule, linear extrapolations from high to low doses are likely to yield sensible estimates of the risks from neutrons, but that this procedure can lead to gross overestimates of the risks from gamma radiation which generally follow quadratic relations. A further and perhaps even more important conclusion is that the hazard from high LET radiations such as fission neutrons is substantially greater than frequently believed because at low doses the RBE can reach values of the order of 100 or more. It, therefore, appears necessary to substantially increase the value of the quality factor Q in order to reduce the maximum permissible dose of neutrons to lower levels (Rossi, 1977).

ACKNOWLEDGEMENTS

This investigation was supported by Contract EP-78-S-02-4733 from the Department of Energy and by Grant Nos. CA 12536, CA 15307 to the Radiological Research Laboratory/Department of Radiology, and by Grant No. 13696 to the Cancer Center/Institute of Cancer Research, awarded by the National Cancer Institute, DHEW.

REFERENCES

Recommendations of the International Commission on Radiological Protection, Adopted January 17, 1977, ICRP Publication 26, Pergamon Press, Oxford (1977).

Review of NCRP Radiation Dose Limit for Embryo and Fetus in Occupationally-Exposed Women, Issued March 1, 1977, NCRP Report No. 53, (National Council on Radiation Protection and Measurements, Washington).

H.H. Rossi, A Proposal for Revision of the Quality Factor, Rad. and Environm. Biophys. 14: 275-283 (1977).

H.H. Rossi (1979) Unpublished.

H.H. Rossi and A.M. Kellerer, Radiation Carcinogenesis at Low Doses, Science 175: 200-202 (1972).

H.H. Rossi and A.M. Kellerer, The Validity of Risk Estimates of Leukemia Incidence based on Japanese Data, Radiat. Res. 58: 131-140 (1974).

H.H. Rossi and C.W. Mays, Leukemia Risk from Neutrons, Health Physics 34(4): 353-360 (1978).

R.L. Ullrich, M.C. Jernigan, G.E. Cosgrove, L.C. Satterfield, N.D. Bowles and J.B. Storer, The Influence of Dose and Dose Rate on the Incidence of Neoplastic Disease in RFM Mice after Neutron Irradiation, Radiat. Res. 68: 115-131 (1976).

DETECTORS FOR RADIATION DOSIMETRY

Victor Perez-Mendez

Lawrence Berkeley Laboratory
Berkeley, CA 94720

INTRODUCTION

In previous years Radiation Dosimetry in Medicine was concerned mainly with measuring doses produced by x-rays, gamma rays and electrons. These radiations have the common feature that their R.B.E. $\approx$ 1, and they can be monitored by the same types of detectors.

More recently, we have had to contend with measuring fluxes and doses produced by the radiations now coming into use in Radiation Therapy. These are heavy and light ions (Argon, protons), negative pions and neutrons.

The heavy ions, light ions and pions have the common feature that they are charged; their R.B.E. is a function of velocity and charge, and they produce some sort of Bragg Peak. The negative pions have the additional complication (and potential) of star formation when captured at low energies or at rest. Neutrons pose different problems since they are uncharged. Their effective R.B.E. depends on the energy, and since it is due to the recoils and secondary interactions of heavy charged particles produced in matter, it depends on the material through which they pass.

Health physicists have had to monitor the doses to personnel working in the vicinity of accelerators and reactors and have developed various detectors for monitoring the fluxes of these particles and measuring their dose to personnel working at the various installations. The material presented here draws heavily on the literature and practice of Health Physics technology. For completeness I will review briefly various detectors used in dosimetry and beam monitoring for x-rays and electrons, since their use also extends, with

143

Table I. Applications of Radiation Detectors

---

1) X-Rays, Gamma Rays, Electron Detectors
   a) Ionization Chambers
   b) Proportional and Geiger Counters
   c) Thermoluminescent Detectors
   d) Radioluminescent Glasses

2) Heavy- and Light-Ion Detectors
   a) Ionization Chambers
   b) Proportional and Geiger Counters
   c) Track Detectors

3) Pion Detectors
   a) Ionization Chambers
   b) Proportional and Geiger Counters

4) Neutron Detectors
   a) BF_3 Counters (Thermal Neutrons), ^3He Counters (Fast Neutrons)
   b) Plastic-Moderated Counters, Bonner Spheres
   c) Fission, Ionization, and Proportional Counters

5) General Detectors
   a) Scintillation Counters
   b) Solid State Detectors

---

slight modifications and different calibrations, to the other forms of radiation.

In Table I I have listed the most widely used detectors for monitoring beam fluxes and their applicability to the various types of radiations.

GAS-FILLED DETECTORS

As noted in Table I, gas-filled detectors, i.e. ionization chambers, proportional counters and Geiger counters, are used in detecting and measuring the dose delivered to various objects by all the forms of radiation discussed here. In various situations the properties of the basic detector, i.e. ionization chamber or proportiona counter, are modified by the gas filling and by the surrounding material to make it suitable for the specific radiation in question.

Ionization Chambers: General Properties

The unit of x-ray intensity, Roentgen, is defined in such terms that it is conveniently measured by an air-filled ionization chamber. One Roentgen = "the quantity of x or gamma radiation such that the associated corpuscular emission per 0.001293 g of air produces in air ions carrying one unit of electrostatic electricity of either sign." (0.00293 g of air = 1 cc at N.T.P.) Note that the definition of Roentgen cannot be used for electrons or other forms of radiation.

Ionization chambers are often used for the following two purposes: 1) to measure and monitor the collimated flux of radiation incident on an object; 2) to sample, with as little perturbation as possible, the energy deposited locally at various points within an object which is being radiated. The electrical principles of operation of ion chambers for these two objectives are the same: the detailed shape, gas filling and materials of construction for these two are quite different.

A typical chamber for beam flux measurement is shown in Figs. 1b and 1c. The chamber is a parallel plate device whose sensitive area is larger than the collimated beam flux cross section. A chamber of this general shape can be used to monitor x-ray, gamma ray, heavy and light ion, pion and neutron beam fluxes. In general, the wall material thickness and gas volume should be thin enough so that the ion chamber, when placed in a beam as shown in Fig. 1a, interacts to a minimal extent with the beam, i.e. does not change its energy distribution or produce appreciable secondary particles.

The electrical characteristics of a chamber used to monitor heavy and light ion beams at LBL are shown in Fig. 1b, which can serve to explain the principles involved, including the use of guard ring electrodes to prevent collection of leakage currents. The thin windows (1) are the enclosure of the gas-filled region which can be filled to a higher pressure than atmospheric, and hence these windows can bulge outwards. The electrical collection part of the chamber consists of the two gold-plated (plated areas 2, 3, 4) plastic foils and the thin metal foil in the center. The high voltage (positive or negative) is connected to the central foil. The sensitive region of the chamber (B) is that between the metallic areas (3) on the plastic foils facing the high voltage foil (see Figs. 1b, 1c). All the metallized areas on the foils are at ground potential; the only ones which are read out are the central areas (3) which define the sensitive volume. The plated areas (2) and (4) at ground potential serve as guard rings to collect various leakage currents. Platings (4) receive the leakage currents across the insulators from the central high-voltage foil to ground. The metallic layers (2) on the plastic foils facing the gas envelope windows collect the ionization current produced in the gas volumes (A).

Fig. 1c shows a further elaboration in the structure of such a chamber: by dividing the sensitive region into a series of

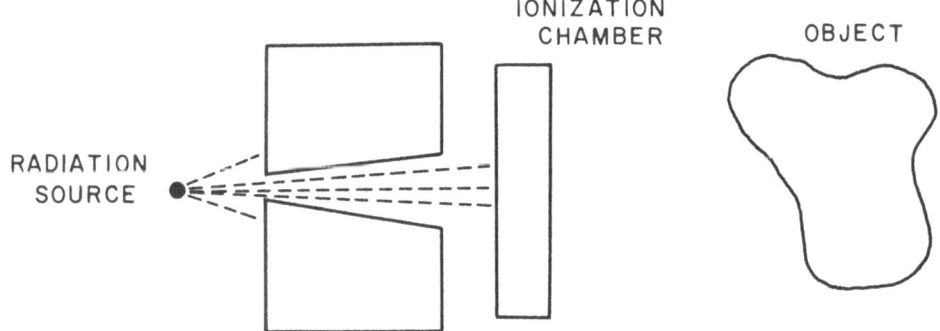

Fig. la.   Beam-monitoring from a collimated source.

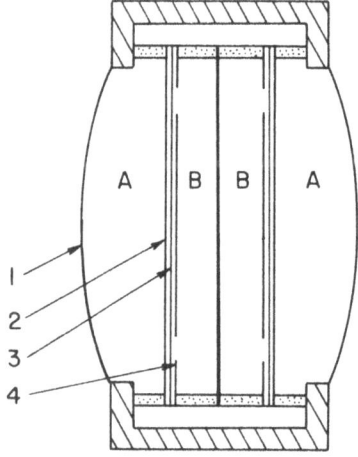

Fig. lb.   Parallel-plate ionization chamber with guard rings.

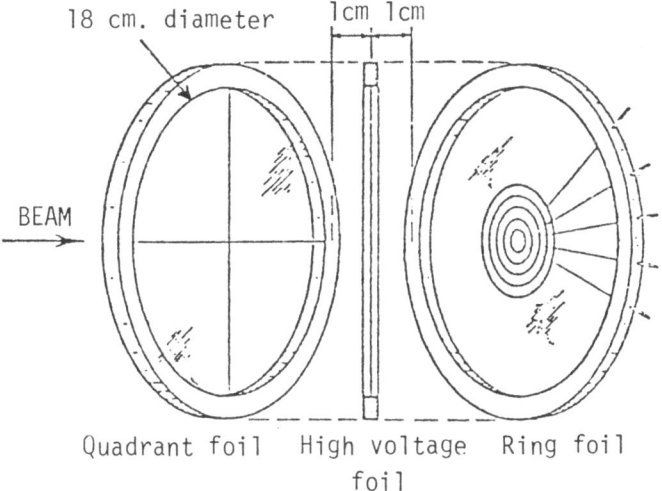

Fig. 1c. Parallel-plate ionization chamber for beam centering with
          quadrant or annular collector foils.

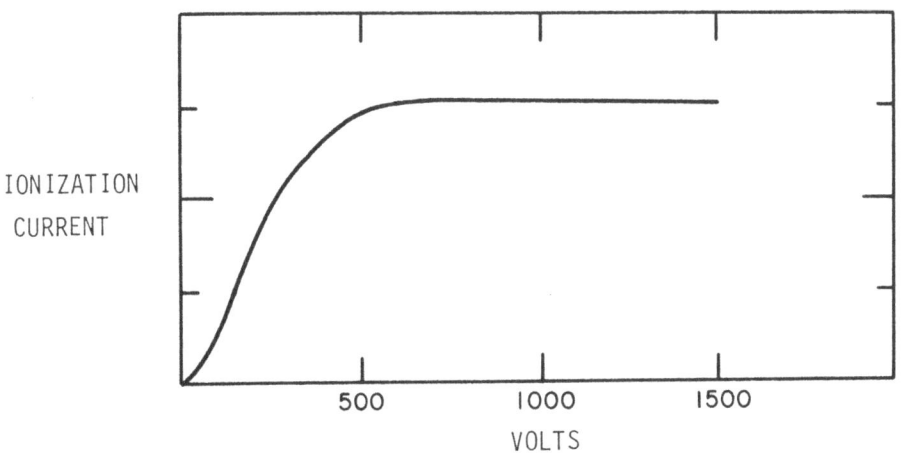

Fig. 1d. Ionization chamber collector voltage plot for chamber,
          showing plateau region.

concentric rings, each connected to a separate electronic charge
integrator, it is possible to monitor continuously the beam distribution profile, assuming it is spherically symmetric. An alternative arrangement, useful for centering the beam, is to divide the
central area into four independent quadrants.

Fig. 1d shows the charge collection voltage plateau typical of
such a chamber. In the plateau region the electric field is high
enough to drift electrons (or negative ions) and positive ions to
their respective collectors with minimal recombination. Since the
drift velocity of electrons is 10-1,000 times faster than the drift
velocity of negative ions, the recombination probability is smaller
in gases which are not electronegative and hence tend to produce
very few negative ions, i.e. noble gases (He, Ne, Ar, Kr, Xe) N_2,
CO_2, CH_4, etc.

At very high beam fluxes, even in these gases, space-charge
effects become sufficiently large, so that electron-ion recombination becomes appreciable and the full charge is not collected. At
these beam levels it then becomes necessary to use other beam monitoring devices, such as the Secondary Electron Monitor (S.E.M.)
described below (Fig. 2). These high fluxes can occur in heavy-ion,
light-ion and pion producing accelerators in which high peak intensity fluxes occur, due to the pulsed characteristics of the machines
when running in low duty cycle conditions. The S.E.M. consists of
a series of metallic electrodes placed in a good vacuum (10^{-6} torr)
container. Charged particles traversing these foils eject secondary
electrons from the foil surfaces. The current produced by these
secondary electrons can be collected and recorded in the same manner
as the ionization chamber currents. For a given beam flux, the
S.E.M. currents are 10-1,000 times smaller than those of a comparable
ionization chamber. In a good vacuum, the only charges present are
electrons, hence problems due to ionic recombinations do not arise
and any non-linearity effects are due to space-charge effects alone.
This problem does not occur in practice, since it is always possible
to raise the collector voltage to a point where the space-charge
non-linearity is reduced to a negligible value for any given peak
beam intensity.

Calibration of Ionization Chambers as Charged Particle Beam Monitors

In principle, a knowledge of the L.E.T. of the incident particles, of W (the energy loss in producing a single-ion pair in the
gas), and of the width of the sensitive region of the chamber is
sufficient to calculate the flux of charged particles traversing
the ion chamber as a function of the collected charge. In practice,
there are a number of additional variables whose specific effect is
difficult to estimate precisely. For instance, the pion beams

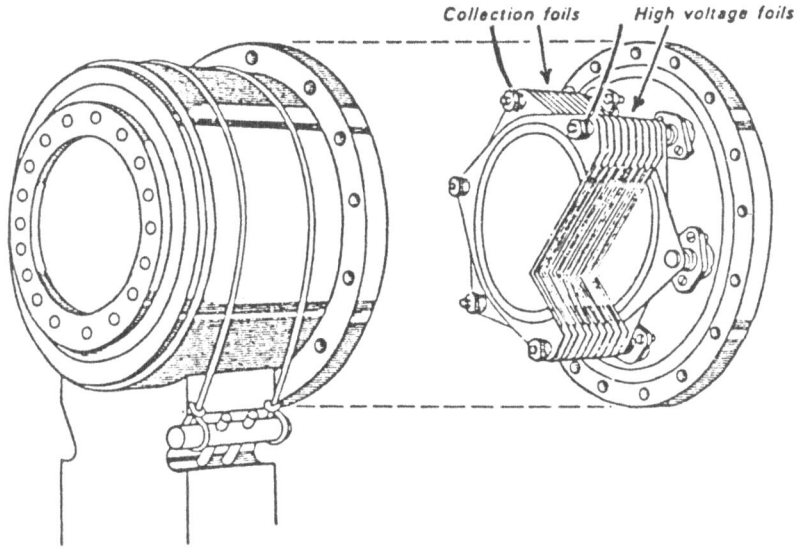

Fig. 2a.   Exploded view of secondary electron beam monitor.

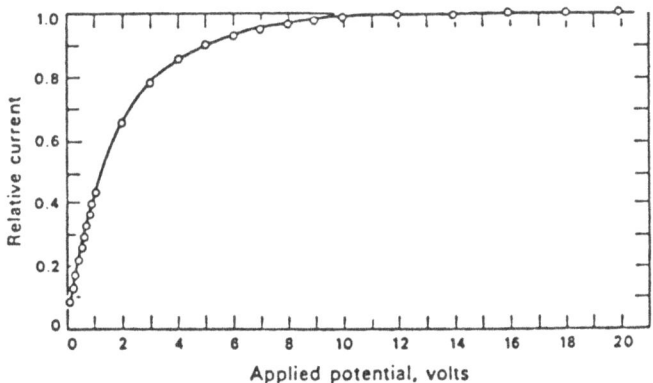

Fig. 2b.   Secondary electron current – collector voltage plot.

Table II. Summary of Measured $^{12}C \rightarrow {}^{11}C$ Cross Section

Particle Energy (MeV/A)	Cross Section (millibarns)	Beam Intensity (ions/pulse)
400	63.5	$2.5 \times 10^4 - 1.2 \times 10^5$
1,050	57.4	$4.0 \times 10^4 - 1.5 \times 10^5$
2,100	60.9	$4.7 \times 10^4 - 1.5 \times 10^5$

always contain variable numbers of electrons and muons which have different L.E.T. than the pions. Again, the heavy ion beams may be accompanied by small numbers of ions in different charge states produced by scattering from the beam slits.

The general procedure has been to calibrate these chambers against some reaction whose cross section is known. Proton beam fluxes can be calibrated by putting a thin foil of CH_2 in the beam and measuring the yield of elastically scattered protons in a coincidence scintillator counter telescope, which detects the incident and recoil protons. Another reaction which has been used to calibrate beam flux detectors is the yield of ^{11}C from ^{12}C targets exposed to various ion beams from protons to Argon ions. The ^{11}C activity is counted with a calibrated NaI crystal. The cross section for Argon ions for this reaction, as measured by the Health Physics Group at the Lawrence Berkeley Laboratory, is shown above in Table II.

Ionization Chambers: X Ray and Gamma Ray Beam Monitors

The beams used in medical practice are those produced by diagnostic x-ray machines (100-250 kVp), ^{60}Co or other radioisotope sources, and Radiotheraphy electron linear accelerators (4-35 MeV).

For x-ray sources in the diagnostic range, it is possible to construct air equivalent ionization chambers which are absolute monitors in the sense that their collection efficiency satisfies the definition of the Roentgen. Two different geometrical approaches are possible, as shown in Fig. 3. In the first case, one has a sensitive volume defined by guard rings with a large column of air in

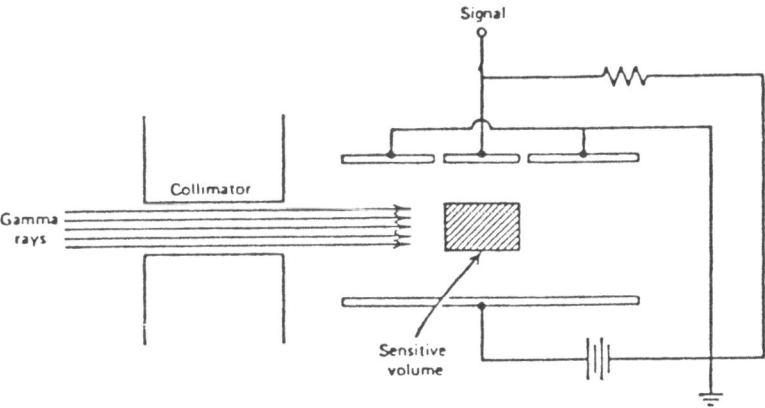

Fig. 3a. Free-air ionization chamber showing guard-ring defined
          sensitive volume.

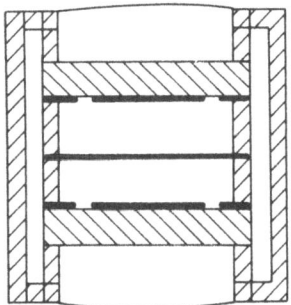

Fig. 3b. Air equivalent ionization chamber with wall thickness
          compensation

Photon Energy (MeV)	Thickness (g cm^{-2})
0.02	0.0008
0.05	0.0042
0.1	0.014
0.2	0.044
0.5	0.17
1	0.43
2	0.96
5	2.5
10	4.9

Fig. 3c. Thickness of wall material (g/cm^2) for establishment of
          electronic equilibrium versus photon energy in MeV.

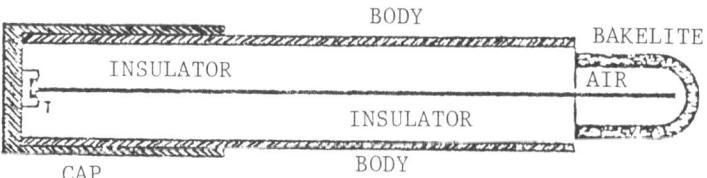

Fig. 3d.   Small-volume (thimble) tissue equivalent ionization
chamber.

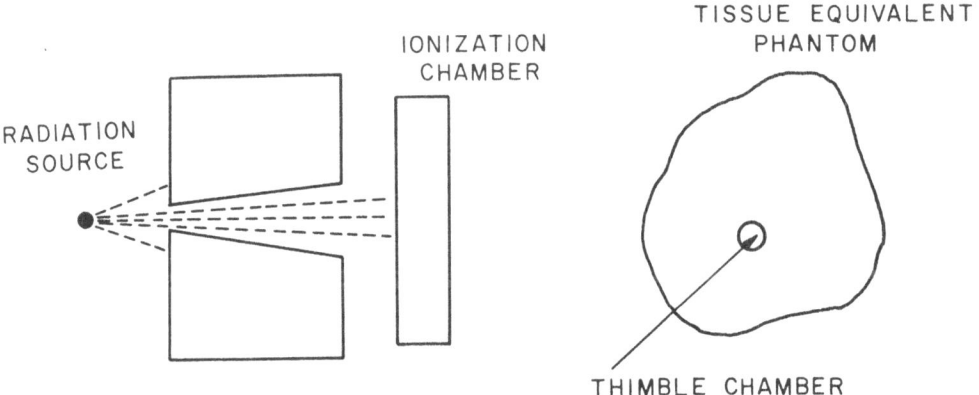

Fig. 4a.   Use of ionization chamber to measure incident radiation
(Roentgens) and tissue equivalent ionization chamber to
measure distribution of absorbed energy in rads.

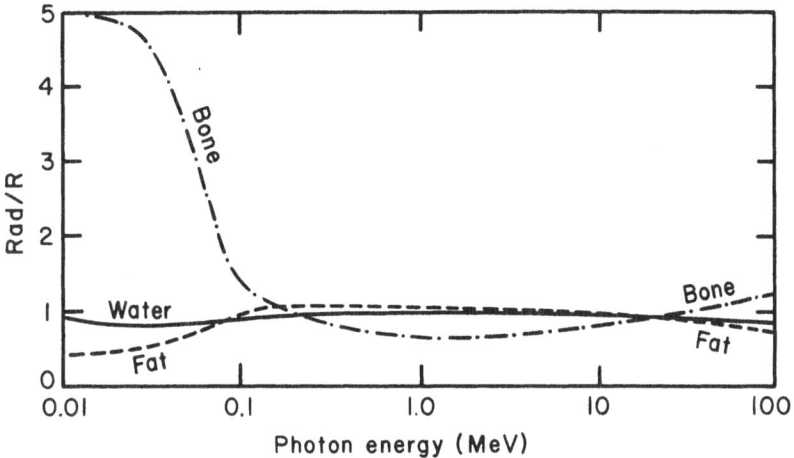

Fig. 4b.   Exposure – dose relations for some materials of biological
           interest.

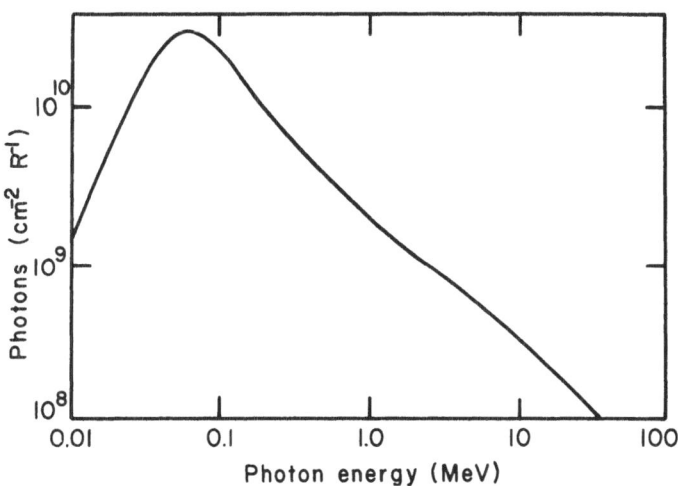

Fig. 4c.   Photon number fluence required to produce one rad as a
           function of photon energy.

front and back so that equilibrium is achieved between the charge produced by x-ray interactions inside and outside the boundaries of the sensitive volume.

An alternative approach to air equivalent chambers is to use the Bragg-Gray prescription, in which a wall material of the appropriate density and appropriate average Z composition is used; this wall is made thick enough so that equilibrium is achieved between the air volume and wall interactions. This is shown in Fig. 3b. The minimum thickness of wall material needed for the equilibrium conditions is given in the table in Fig. 3c. A small-volume ionization chamber used for measuring the dose distribution within a given tissue volume is shown in Fig. 3d. The bakelite wall of the sensitive volume satisfies the Bragg-Gray condition. It is coated with a thin graphite layer on the inside to make it conducting.

Fig. 4 shows the use of a thimble chamber in measuring the dose distribution within a phantom. The total radiation flux incident on the phantom is monitored by a parallel-plate ionization chamber, as discussed previously. Fig. 4b shows the relationship between Roentgens (a unit of fluence) and rads (a unit of absorbed energy) for various materials of biological interest. Fig. 4c shows the number of photons/cm^2 required to produce a fluence of one Roentgen as a function of photon energy.

Ionization Chambers: Beam Monitors for Fast Neutrons

Fast neutrons used in medical radiotherapy treatments have ranged in energy from 5-50 MeV. On the assumption that the incident fast neutron beam is collimated by a suitable entrance channel, it is possible to use appropriate ionization chambers to monitor the neutron flux as shown in Fig. 1. For this purpose, a standard parallel-plate ionization chamber with suitable guard rings can be used when filled with a suitable gas, such as methane at 1-10 atmospheres pressure. The ionization current is produced by the n-p elastic interaction with the hydrogen atoms in the gas. Since the maximum intensity of fast neutron fluxes is approximately 10^{11}-10^{12} neutrons/cm^2/sec, there is little risk of non-linearities due to saturation. In principle, since the elastic n-p cross section is well known at these energies and the recoil spectrum of the protons easily calculable, the energy deposited in the sensitive region of the ion chamber can be determined quite accurately. At the higher neutron energies, the interaction with the walls and the carbon atoms has to be taken into account.

In summary, ionization chambers of the parallel-plate type with suitable precautions to eliminate leakage currents by guard rings and other devices can be used as absolute monitors for collimated beams of the radiations discussed here. In practice, since absolute

devices must be maintained very carefully, including the readout
and charge integrating electronics, it is advisable to maintain some
secondary standards against which their calibration can be checked
periodically.

Gas-filled Detectors: Proportional and Geiger Counters

The ionization chambers described in the previous sections are
suitable for monitoring large fluxes of radiation (more than 10^6
charged particles/sec) by measuring the current produced in the chamber by some suitable electrometer.

The proportional counters and Geiger counters described in this
section are capable of detecting single charged particles and recording the energy deposited/particle in the counter. As beam monitoring
devices, they have to be used with beam fluxes of less than 10^7
charged particles/sec for a proportional counter or less than 10^5/sec
for a Geiger counter.

The ability to detect single particles is due to the internal
gas amplification of the primary charge deposited in the counter
sensitive volume by the passage of a charged particle. Gas amplification occurs in electric field regions which are large enough to
cause primary electrons to ionize successive gas atoms before collection. High electric fields at low voltages are obtained by making
the detectors cylindrical in shape with the anode consisting of a
small diameter wire. Typical dimensions for these counters are:
anode wire 20-50 microns diameter; cathode diameters range from a
few mm to several cm. From electrostatics, the electrical field in
a cylindrical configuration is

$$E(r) = \frac{V}{r \ln b/a} \quad ; \quad \begin{array}{l} V = \text{applied potential,} \\ a = \text{anode wire radius,} \\ b = \text{cathode inner radius.} \end{array}$$

From this formula it can be seen that for applied potentials of
1,000-2,000 volts, electric fields of the order 100-300 KV/cm can
be readily obtained at distances of a few anode wire radii, which
is the region where most of the avalanche gain takes place.

Fig. 5a shows a typical construction of an avalanche gain
detector. The distinction between a proportional and Geiger counter
can be understood from the avalanche gain versus applied voltage
graph of Fig. 6b. At low voltages, the electric field is sufficient
to collect all the primary electrons and ions: this is the ionization regime. At higher voltates there is some avalanche gain which
increases with voltage. At still higher voltages, there are charge
collection saturation effects, and the avalanche gain is no longer

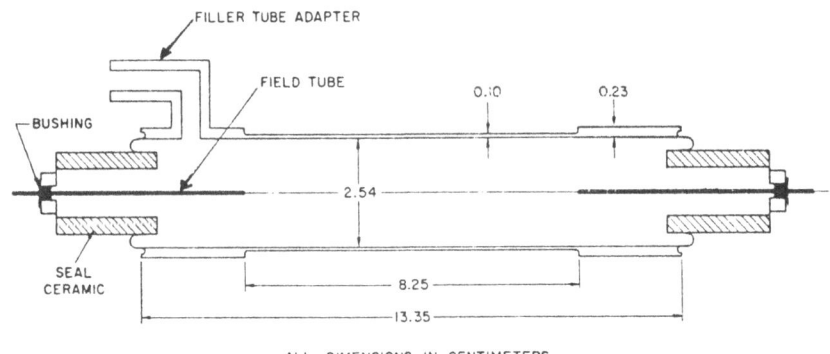

Fig. 5a. Cross-sectional view of proportional chamber showing field
tubes that serve as guard electrodes.

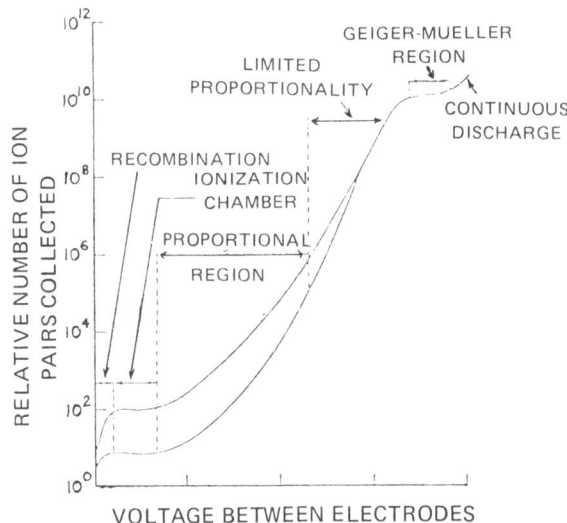

Fig. 5b. Charge multiplication versus voltage for cylindrical
proportional counter showing transition from ionization
chamber to proportional to Geiger to breakdown conditions.

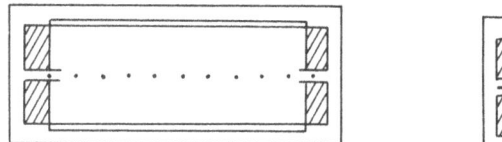

Fig. 5c.   Side view of multiwire proportional chamber.

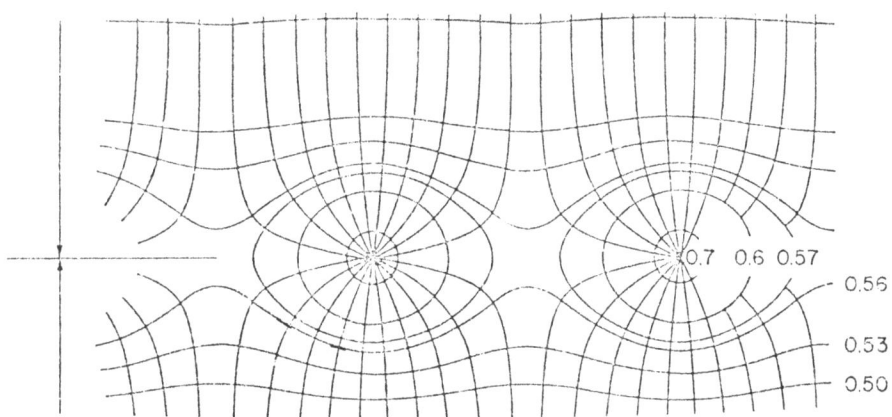

Fig. 5d.   Electric field and equipotential distributions in multi-
           wire proportional chamber.

proportional to the number of primary electrons. A small voltage
region between this point and breakdown is the Geiger region, where
an input pulse of any amplitude produces the same saturated output
pulse of approximately 10^8 electrons.

The characteristics of these counters depend strongly on the
gas filling. Since the avalanche gain is produced by electrons, it
is necessary to use gases that are not electronegative, i.e. do not
capture free electrons to form negative ions. Suitable gases for
this purpose are the noble gases He, Ar, Ne, Kr and Xe. All of
these gases have metastable states which emit photons in the ultraviolet, and therefore the possibility exists that they can start
secondary discharges by ejecting photoelectrons from the cathode
walls. In order to prevent this, it is necessary to add a small
admixture, typically 5-10% of a "quenching gas," which is usually
a non-electronegative polyatomic gas with strong absorption in the
ultraviolet. Typical quenching gases are saturated hydrocarbons,
methane, ethane, isobutane, alcohol and inorganic gases, such as
CO_2.

There is an extensive literature on the properties of proportional and Geiger counters and the detailed mechanisms of the quenching action of the various gas mixtures.

Figs. 5b and 5c show a modern development of the cylindrical
proportional counter, called the MWPC (Multiwire Proportional
Chamber). These devices are used extensively in physics research
when it is necessary to cover large areas with proportional counters
with a minimum amount of absorbing or scattering material. As shown
in Fig. 5d, a set of parallel anode wires produces electric field
equipotentials and gradients in the immediate vicinity of each wire
similar to those in cylindrical proportional counters. Thus, a local
avalanche region is formed at each wire. By good design, it is
possible to ensure uniformity of response over areas of a few square
meters.

For our purposes in this review, we note the following points:

1) For charged particle detection, these counters can be filled
with any noble gas-quenching gas mixture that produces satisfactory
electrical signals.

2) Neutron counters, in which the neutrons are detected by their
interaction with the specific filling of the chamber to yield an
ionizing particle, require special gas mixtures containing ^{3}He or
BF_3. An alternative approach is to coat the inner surface of the
cathode with a boron or lithium compound.

3) Proportional counters are used if there is any need to

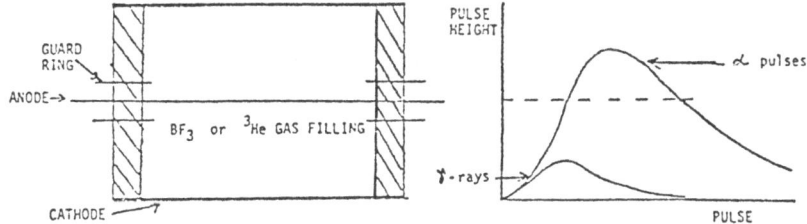

Fig. 6a. Neutron - gamma discrimination by pulse height in BF$_3$ counters.

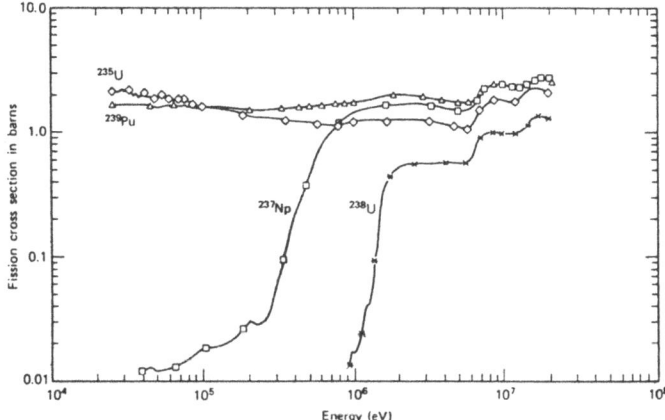

Fig. 6b. Fission cross sections for thermal and fast neutrons for some isotopes used in fission chambers.

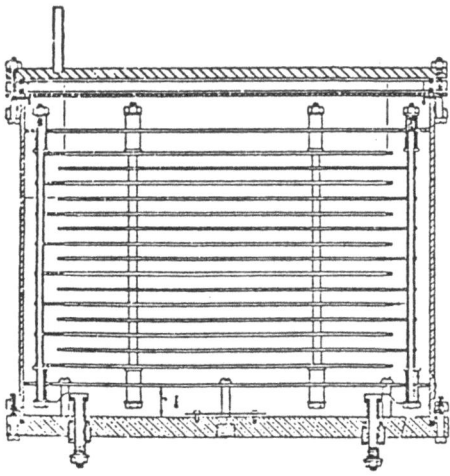

Fig. 6c. Schematic of multiple-plate ionziation fission chamber.

Table III. Properties of Neutrons:
Gas-filled Neutron Detectors

Thermal Neutron: Slow neutron E_n = 0.025 eV
                   Capture in matter → gamma rays

Fast Neutron: Anything above 10 keV
                Radiotherapy range 5-50 MeV
                Capture in matter → charged particles
                                  → gamma rays

Thermal Neutrons: Reactions with high capture probility yielding
                    highly ionizing charged particles: exothermic
                    reactions.

$n + {}^{10}_{5}B \rightarrow {}^{6}_{3}Li + {}^{4}_{2}He$ E_α = 2.79 MeV

$n + {}^{3}_{2}He \rightarrow {}^{3}_{1}He + {}^{1}_{1}H$ E_p = 0.57 MeV

$n + {}^{6}_{3}Le \rightarrow {}^{4}_{2}He + {}^{3}_{1}H$ E_H = 2.73,
                                                              E_α = 2.05 MeV

$n + {}^{235}_{92}U \rightarrow$ Fission   E(fission fragments)
                                             150 MeV

Fast Neutrons: Reactions with interaction probability. Elastic
                 scatter or break-up into charged particles.

$n + {}^{1}_{1}H$ (hydrocarbons) → Hydrogen recoils

$n + {}^{238}_{92}U$ (${}^{232}_{90}Th$) → Fission fragments

---

discriminate between different types of radiation incident on the
chamber by the magnitude of the ionizing energy retained within the
sensitive volume of the counter.

4) Proportional counters can operate at higher speeds than
Geiger counters, typically up to 10^7 cts/sec versus less than 10^5/sec
for the Geiger counters.

5) Geiger counters produce very large uniform pulses which can
be scaled by very simple electronics; hence, they are often used in
survey meters and other portable monitoring instruments.

Gas-filled Detectors: Detection of Thermal and Fast Neutrons

In a previous section we noted that hydrocarbon gas-filled ionization chambers of conventional construction could be used for monitoring large fluxes of fast neutrons. The detection mechanism in that case was the ionization current produced by the proton recoils.

In this section we describe some gas-filled detectors particularly suitable for use as neutron detectors. The type of detector depends on whether it is intended for thermal neutrons (E_n = 0.025 eV) or for fast neutrons. In any application of fast neutrons, such as Neutron Radiotherapy (E_n = 5-50 MeV), it should be remembered that the background radiation flux in any concrete-shielded area is due to the primary fast neutrons and also to the thermalized neutrons from the slowing-down interactions in the walls.

For thermal neutron detection, it is necessary to use a gas filling or wall coating of some element which yields an energetic charged particle with a high interaction probability which then ionizes the counter gas. In Table III we list some thermal neutron capture reactions which are extensively used in various neutron detectors.

Standard thermal neutron proportional counters are made with BF_3 or ^3He gas fillings. The proportionality feature of the counter is used (Fig. 6a) in allowing the scalers, which count the individual pulses, to discriminate between a thermal neutron indirect pulse and a gamma ray pulse from the high flux of gamma rays that is usually present in a thermal neutron ambience.

Another ionizing reaction that is characteristic of neutrons, thermal and fast, is the fission process in heavy elements. Fig. 6b shows the fission cross section for ^{235}U, ^{239}Pu, ^{238}U and ^{237}Np. ^{238}U and other even isotopes of uranium and thorium are fast neutron fission isotopes.

Since the charged fragments from the fission process release an energy of approximately 80 MeV/fission fragment, the ionization pulse is large enough so that avalanche gas gain is not needed and a parallel-plate ionization chamber can be used in a pulse mode. A fission neutron detector is shown in Fig. 6c. It consists of a large number of thin metal plates coated on both sides with a thin deposit of the appropriate uranium or other heavy element isotope. The maximum thickness of the deposit has to be small enough so that the pulse height from a fission fragment which has to traverse the entire thickness of the plating emerges into the gas volume with a sufficient energy to be detected above the amplifier noise and the pulses from alpha particles, which usually occur in the spontaneous

decay of the trans-uranium isotopes. Note that all the high-voltage
plates are mounted on one set of insulating rods and ground plates
on another set: this is a form of guard ring configuration to prevent electrical leakage signals.

In Figs. 7 and 8 we show two configurations of fast neutron
detectors in which the detector element itself is basically a thermal
neutron counter: a BF_3- or ^3He-filled proportional counter. The
fast neutrons are slowed down in a surrounding hydrogenous CH_2
plastic envelope and detected with high efficiency by the central
detector when thermalized.

The counters shown in Fig. 7 are called Bonner spheres: they
are made with various diameter CH_2 spherical blankets. The sensitivity of various Bonner spheres versus neutron energy is shown in
Fig. 7b. These spherical detectors were intended initially for use
in estimating the energy distribution of a flux of fast neutrons by
comparing the counting rates of the various diameter spheres. From
the known sensitivities it is possible to calculate a rough energy
spectrum.

The interest to neutron dosimetry of this type of counter is
illustrated in Fig. 7c, which shows the sensitivity-versus-neutron
energy response of the 10-inch (25.4 cm) sphere and the relative
equivalent dose per neutron; the two curves are sufficiently close
so that dose measurements can be done by use of this particular
diameter Bonner detector.

Another approach to fast neutron detection for an unknown
distribution of neutron energies is to design a hydrocarbon-moderated
detector whose response is the same over a wide range of neutron
energies. Counters of this type, known as long counters, are shown
in Fig. 8. In Figs. 8a and 8b we show an earlier version of a long
counter with a single BF_3 proportional detector at the center. A
flatter response curve is obtained with the more complicated long
counter, shown in Fig. 9a and 9b, using multiple ^3He detectors.

Thermoluminescent (TLD) and Radiophotoluminescent Detectors (RPL)

The interaction of radiation with matter in crystalline and
amorphous forms produces a variety of deformations which can be
used as a measure of the intensity of the incident radiation. Some
interactions with crystalline matter are shown in Table IV. From
this list the thermoluminescent and radiophotoluminescent phenomena
have turned out to be particularly suitable for dose measurements
in the useful medical range, i.e. from a few mrads to thousands of
rads.

Thermoluminescent detectors. The thermoluminescence process

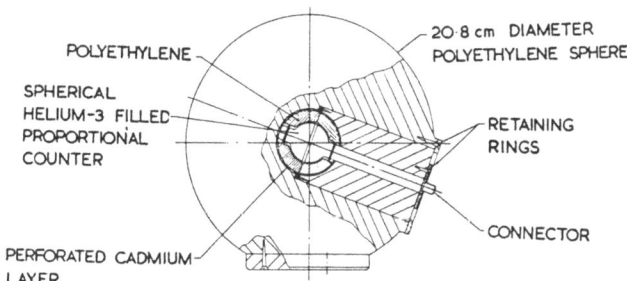

Fig. 7a. Spherical neturon dosimeter based on ^{3}He-filled proportional counter.

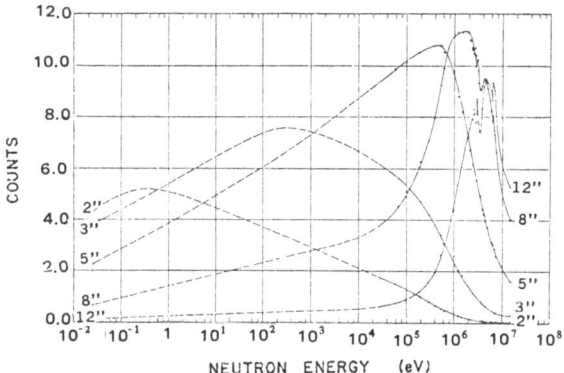

Fig. 7b. Neutron detection efficiency versus neutron energy for
spherical Bonner detectors of various diameters.

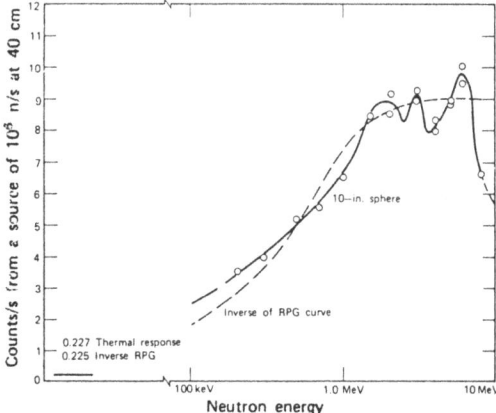

Fig. 7c. Sensitivity of 25.4 cm spherical detector with 4 x 4 mm
central detector versus neutron energy. Also shown is
the relative dose per nuetron labeled as "inverse of RPG
curve."

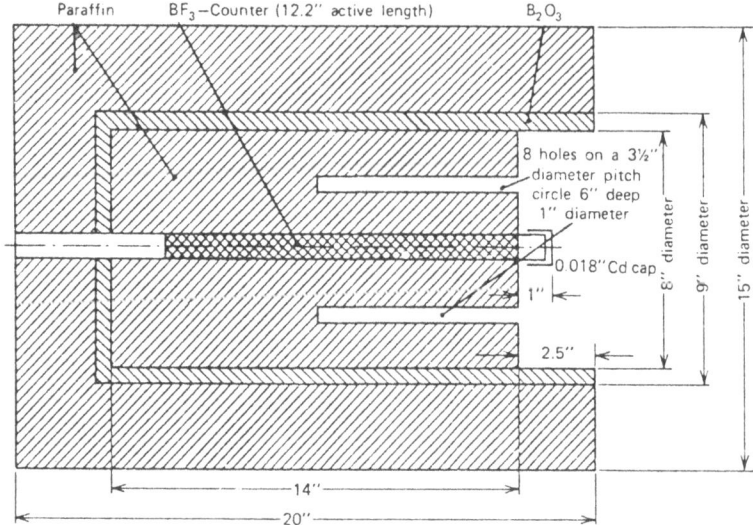

Fig. 8a. Schematic of paraffin-moderated neutron counter with BF_3 proportional chamber for detecting the thermalized neutrons.

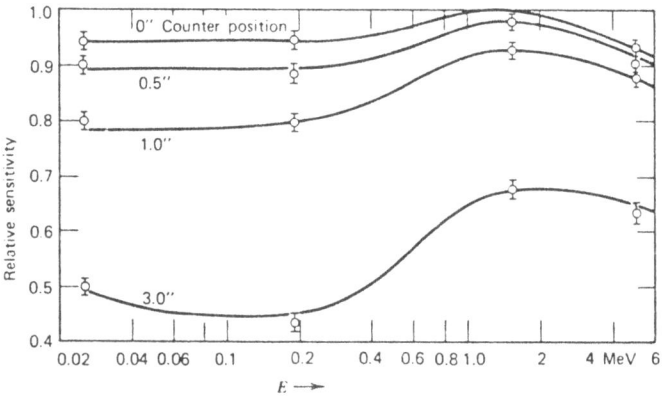

Fig. 8b. Detection efficiency of the counter as a function of neutron energy.

Table IV. Summary of the Types of Solid State
           Systems Used for Dosimetry

Type of centers produced	Measured parameter	Material
Centers stable with respect to readout.	A. Coloration	1. Glasses 2. Plastics 3. Dyes
	B. Radiophotoluminescence	Phophate glass
	C. Degradation of luminescence	Anthracene, etc.
	D. Electron spin resonance	Alanine
Centers destroyed by measurement	A. Thermoluminescence	1. $CaSO_4$:Mn 2. CaF_2:Mn 3. LiF
	B. Infrared-stimulated luminescence	SrS:Eu, Sm

can be understood in terms of Figs. 10a and 10b. Fig. 10a shows
the electronic energy diagram of an ionic crystal. Radiation can
eject an electron from the filled valence band to the conduction
band, resulting in an electron and hole which are both mobile. If
these are collected by external connectors, we have a semiconductor
detector. However, as shown in Fig. 10b, the crystal may have a
small concentration of another element that forms energy levels
which act as trapping centers for both the electrons and holes
released by the radiation. By suitable choice of crystal and impurity element, these trapping centers may, under the action of
heat, cause the electron to be released from its trapped level
and cause recombination with the hole. The energy released is often
in the form of a photon in the visible region.

A variety of inorganic crystals with these suitable properties
exists, as shown in Table V. From this table it can be seen that

all of these crystals can be used for radiation dosimetry in the
medically useful range. Fig. 11a shows the response as a function
of x-ray or gamma ray energy of LiF and CaF_2 doped with Mn. The
flatter response curve of the LiF for incident x-ray energy below
200 KeV is one reason for its wide use as an x-ray dosimeter. The
response curve to other forms of radiation will not necessarily
have the same shape, and hence other TLD crystals may be equally
useful.

LiF can be used as a dosimeter for thermal neutrons by use of
^{6}LiF crystals; the response of the dosimeter is then due to the
ionization produced in the crystal by the exothermic proton from
the neutron capture reaction. The Health Physics group at the
Lawrence Berkeley Laboratory has also used LiF dosimetry in heavy
ion beams and calibrated their response relative to x-rays.

The thermoluminescent material, in the form of a powder, small
rods or powder extruded into teflon plastic discs, is read out by
heating it in an opaque enclosure. A photomultiplier tube built
into the enclosure records the amount of light released, which is
proportional to the integrated flux of radiation to which the crystals have been exposed. Fig. 11b shows a typical light emission
curve (glow curve) of a thermoluminescent crystal (CAF_2:Mn) as a
function of heater temperature. LiF and other crystals have very
similar curves, each peaking at slightly different temperatures.

Radiophotoluminescence (RPL) detectors. Although the RPL
phenomenon has been known since 1912, its theoretical basis is not
completely understood. In silver-activated phosphate glass, ionizing radiation produces two effects: (1) it increases the optical
density over a broad wavelength region in the ultraviolet and the
visible, and (2) it forms stable fluorescing centers that emit orange
light (500-700 nm) under ultraviolet excitation (365 nm). These
fluorescence centers are due primarily to silver deposits in the
form of Ag^{2+} ions formed from the Ag^+ ions of the crystal by loss
of an electron. There are other effects which contribute to the
fluorescence. Fig. 12a shows a schematic diagram of an RPL reader
and Fig. 12b shows typical absorption and luminescence curves of
an RPL glass.

In Table VI we show some properties of commonly used RPL
glasses. Fig. 13a shows, for various RPL glasses, the minimum dose
that can be measured, relative sensitivity, mean atomic number, and
response as a function of gamma ray energy. Fig. 13b shows the
fading effect of the RPL response with time. If we accept a basic
accuracy of $\pm$10%, then all the RPL glasses listed here are satisfactory from a few minutes after exposure to over 100 days.

Both TLD and RPL dosimeters are coming into wide use and dis-
placing film as dosimeters for personnel working in laboratories

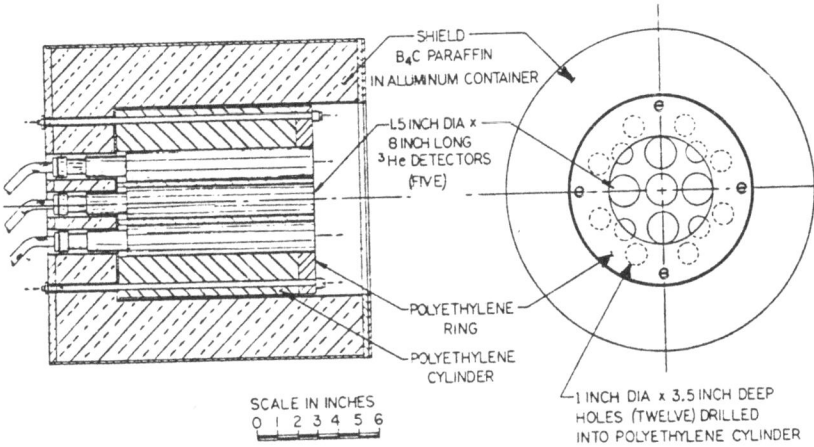

Fig. 9a. Schematic of polyethylene-moderated neutron counter
utilizing multiple ³He-filled proportional counters.

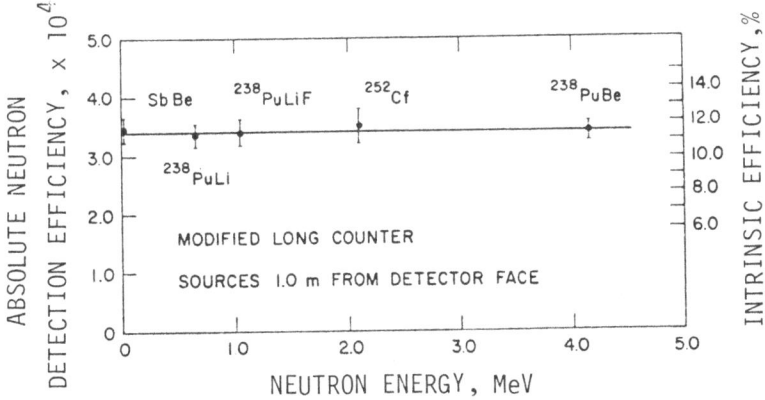

Fig. 9b. Detection efficiency of polyethylene-moderated neutron
counter versus neutron energy.

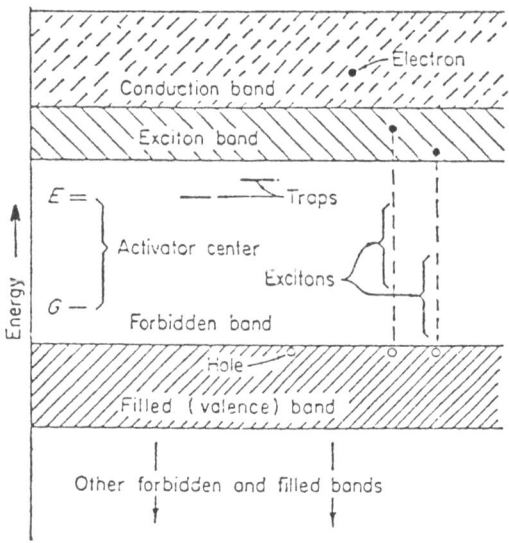

Fig. 10a. Electronic band diagram of an ionic crystal.

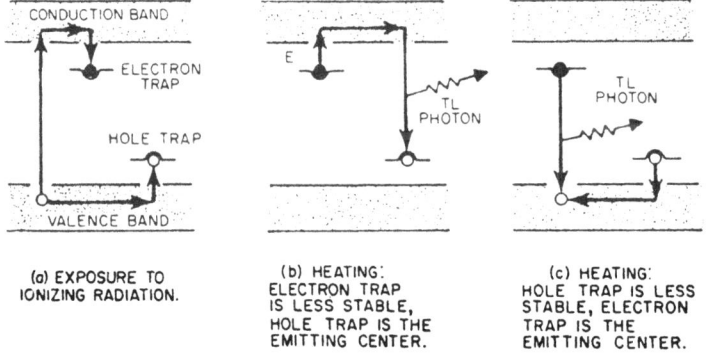

Fig. 10b. Schematic explanation of thermoluminescence.

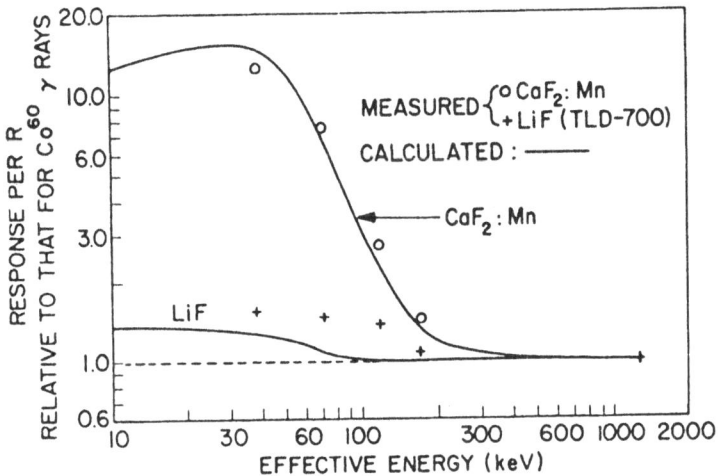

Fig. 11a. Measured and calculated energy dependence of CaF and LiF versus photon energy.

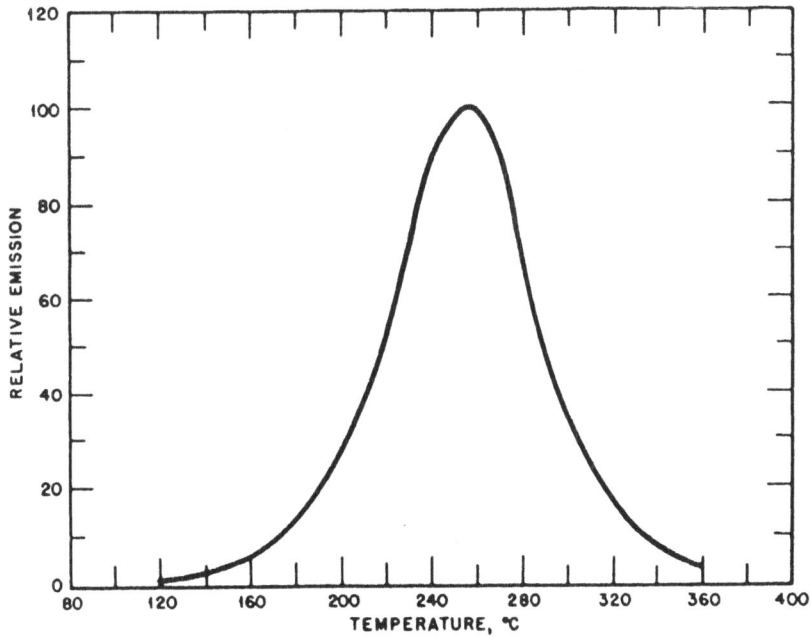

Fig. 11b. Glow curve of CaF dosimeter versus temperature.

Table V. Characteristics of TL Phosphors

Characteristic	LiF	$Li_2B_4O_7$:Mn	CaF_2:Mn	CaF_2:nat	$CaSO_4$:Mn
Density (g/cc)	2.64	2.3	3.18	3.18	2.61
Effective atomic number	8.2	7.4	16.3	16.3	15.3
TL emission spectra (Å) a) range b) maximum	3500-6000 4000	5300-6300 6050	4400-6000 5000	3500-5000 3800	4500-6000 5000
Temperature of main TL glow peak	195°C	200°C	260°C	260°C	110°C
Efficiency at ^{60}Co (relative to LiF)	1.0	0.3	3	~23	~70
Energy response without added filter (30 KeV/^{60}Co)	1.25	0.9	~13	~13	~10
Useful range	mR-10^5 R	mR-10^6 R	mR-3×10^5 R	mR-10^4 R	μR-10^4 R
Fading	Small, <5%/12 wk	~10% in first mo.	~10% in first mo.	No detectable fading	50-60% in first 24 hrs.

Table VI. Composition and Dosimetric Properties of Some Radiophotoluminescent Dosimeter Glasses

Author (Manufacturer)	Composition (% by weight)						Pre-dose in rad γ equivalent (approx.)	Relative γ-sensitivity (approx.)	Effective atomic number	Energy dependence: $\frac{50\ keV}{1\ MeV}$ (calculated)
	Ag	Al	Li	P	O	Others				
Schulman et al. (1951) (Bausch & Lomb High-Z)	4.3	4.7		28.4	44.1	10.8 Ba 7.7 K	10.0	1.0	28.0	32.0
Ginther and Schulman (1960) (Bausch & Lomb Low-Z)	4.3	4.7	1.9	33.7	52.3	3.1 Mg	10.0	1.0	17.6	10.0
Yokota et al. (1961) (Toshiba)	4.2	4.6	3.6	33.3	53.5	0.8 B	0.2	2.2	17.5	10.0
Francois et al. (1965) (C.E.C.)	2.4	3.5	2.5	33.8	52.5	0.5 Be 4.7 Na	0.7	1.9	15.4	7.3

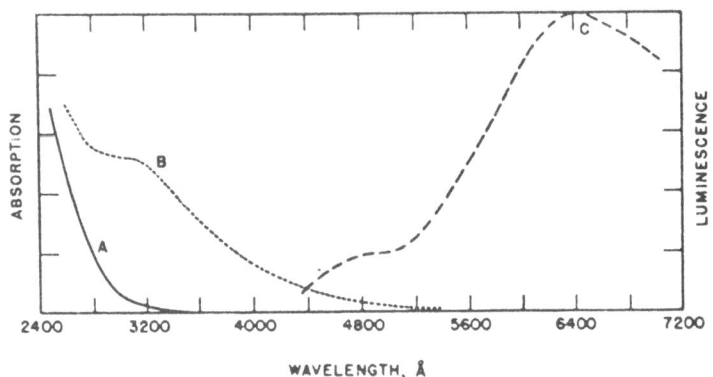

Fig. 12b.   Typical absorption and luminescent curves for RPL glass:
(A) Absorption band: (B) Emission band.

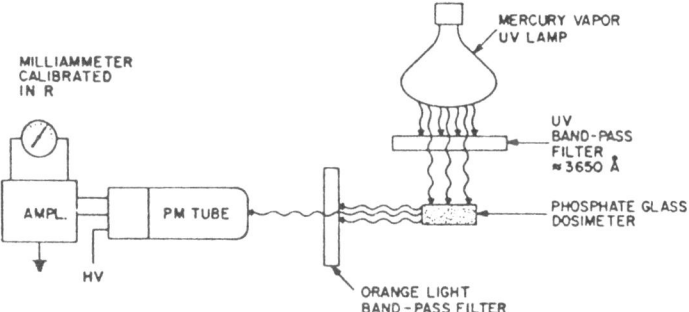

Fig. 12a.   Schematic diagram of basic components in an RPL dosimetry
reader.

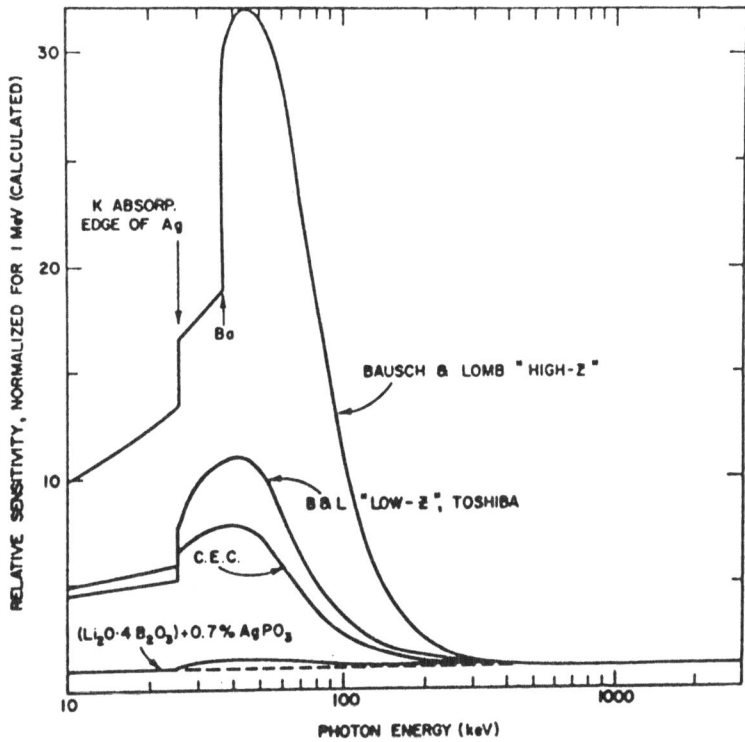

Fig. 13a. Calculated energy dependence of several RPL glasses.

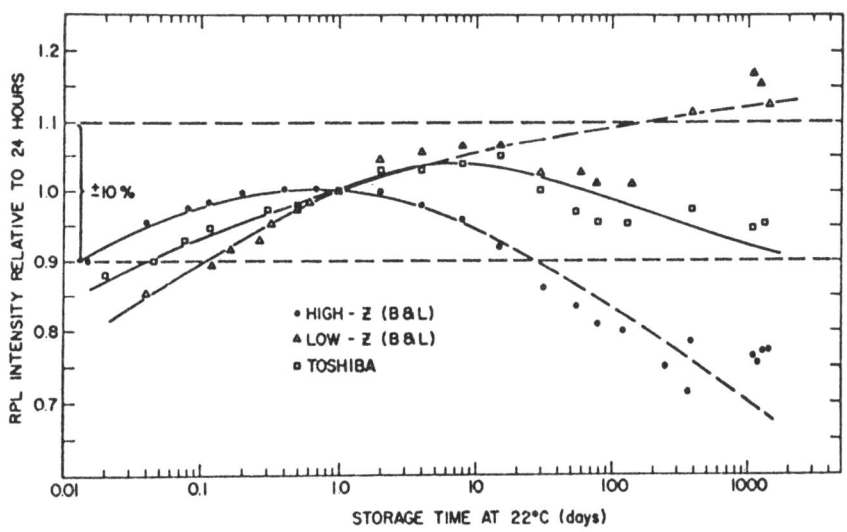

Fig. 13b. Fading characteristics of RPL glasses.

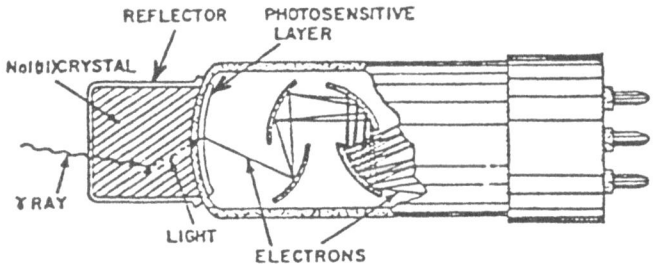

Fig. 14a. Schematic of NaI crystal mounted on photomultiplier.

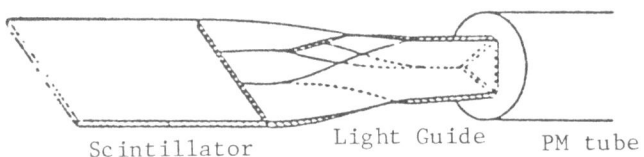

Fig. 14b. Plastic scintillator edge coupled to P-M tube by plastic
           light guides.

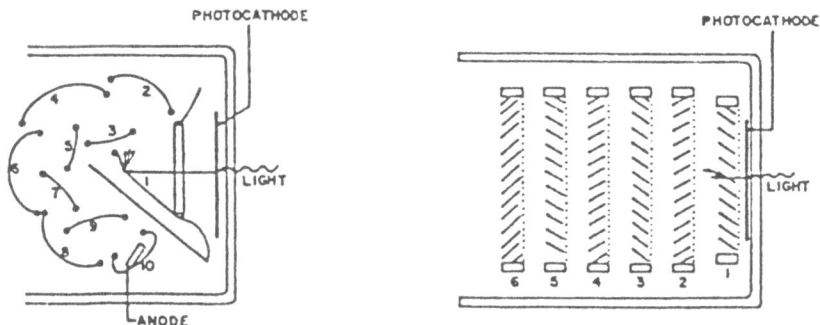

Common photomultiplier tubes. Left, focused
electrode tube, and, right, venetian blind tube.

Fig. 14c. Some types of dynode structure in P-M tubes.

where there is some radiation exposure. They are both used also as patient dosimeters in Radiation Therapy exposures.

SCINTILLATION DETECTORS

The scintillation process was used in the earliest experiments in alpha particle counting by Rutherford and his group in the early 1900's, using ZnS as the scintillator and the human eye as the the detector. After being supplanted in 1919 by the invention of the Geiger counter, it came into use again in particle physics around 1950, following the development of the photomultiplier tube some years earlier, and with the discovery by Kallman that various organic crystals with a cyclic structure that were previously known to fluoresce strongly under ultraviolet radiation were capable of emitting photons in the visible range under the action of ionizing particles. Since then, the scintillation detector has come into wide use in particle physics research, radioactive tracer work, and gamma ray spectroscopy. Scintillation counters are not used very much in radiation dosimetry, primarily because their gain sensitivity is too variable compared to the other detectors discussed here.

In Fig. 14a we show a basic scintillation detector. The scintillator material is mounted directly on the face of the P-M tube. Fig. 14b shows a flat scintillator optically coupled to the P-M tube by polished plastic light pipes in step form. In actual use, the scintillator light pipes and P-M tubes would be enclosed in light-tight wrapping. Fig. 14c shows the dynode structure of other types of P-M tubes than that shown in Fig. 15a. All the P-M tubes have a photocathode whose response is selected to match the light emission of the scintillator material and whose cathode efficiency ranges from 20-30%. The multiplication process takes place in the evacuated glass container, where the dynodes are located. By applying suitable potentials to the dynode, an individual electron from the photocathode is accelerated to the first dynode where it releases a small number (3-5) of secondary electrons. Each of these electrons in turn releases more secondary electrons from the next dynode until the final pulse of electrons is collected in a metallic collection box. Depending on the number of dynodes in a P-M tube and the applied potential across them, gains between 10^5-10^8 can be achieved. The photocathodes are also made with different sensitivity peaking at different wavelengths from 700-300 nm.

Since 1950, many different types of scintillators have been developed for detector applications. The low Z scintillating C_nH_m compounds are generally used in solutions or polymerized into a plastic binder. The liquid scintillators are used extensively for accurate counting of radioactive tracers. The plastic scintillator

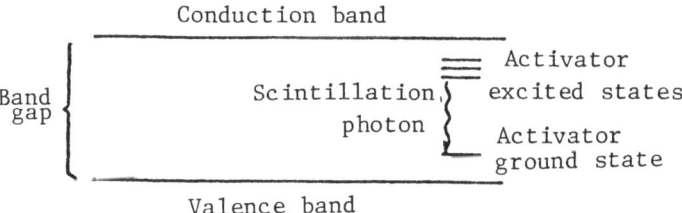

Fig. 15a.   Energy band structure of an activated crystalline
            scintillator.

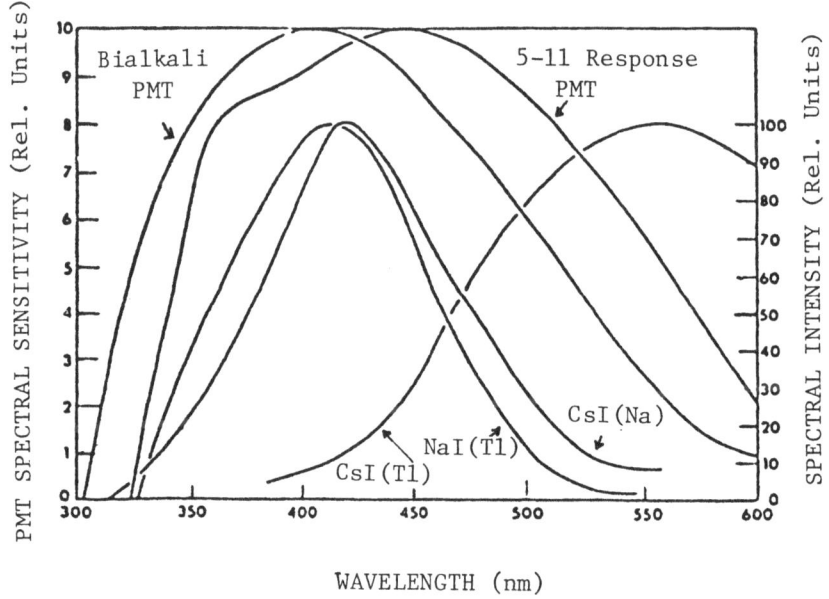

WAVELENGTH (nm)

Fig. 15b.   Emission spectra of several common inorganic scintillators
            compared to spectral response of some P-M tubes.

is the most widely used type for particle counting, for beam monitoring, and for low-energy gamma ray counting. The ability to shape
the plastic to any desired form makes it universally useful. The
general requirement for an organic scintillator, in crystalline,
solvent or plastic polymerized form is that its optical absorption
bands and emission bands should not overlap. Table VII lists various
scintillation materials which can be used as crystals in solutions
or in plastics.

Table VIII summarizes some of the properties of inorganic crystals. The main interest in the inorganic crystals is that they can
contain specific elements of interest for neutron or gamma detection
and spectrometry. Fig. 15a explains schematically the properties
of a scintillating inorganic crystal. Non-overlap between the absorption and emission wavelengths is achieved by adding small amounts
of selected impurities to various transparent cyrstals. These activation compounds create suitable levels in the forbidden gap. The
process is very similar to the thermoluminescence activation. The
main difference is that in these crystals the activator-excited
states release their energy rapidly (10^{-6} sec or faster) in order
to allow the crystals to be used with fast-counting electronics.
These scintillators have to be used in crystalline form; it is not
possible to have the versatility of using them in solution or in
plastics. This is due, of course, to the fact that the crystal
lattice and energy levels are essential for the light emission process. Fig. 15b shows that the light emission bands of the crystals
listed here are a suitable match to the photomultiplier tubes.

For neutron detection, the crystal ^{6}LiI(Eu), with good scintillation properties, is available. ^{6}Li-loaded glasses and ^{10}B-loaded
glasses are available also for neutron detection. The main use of
the inorganic scintillators has been for gamma ray detection and
gamma spectroscopy, where the fact that some of them have high Z
elements means that their interaction probability is many orders of
magnitude higher than for the CH compounds. Until recently, the
almost universally used gamma detector has been the NaI (various
activators) crystal. During the last few years a new crystal,
Bismuth Germanate ($Bi_4Ge_3O_{12}$), has been developed and is coming
into use. The main advantage of this crystal is that it contains
bismuth, whose higher Z, relative to that of the iodine, makes it
more sensitive to gamma ray interactions. It has a defect in that
its light output is approximately 10% that of a comparable NaI
crystal.

In summary, the scintillation counters have been the most
widely used detectors in research in Radiation Physics and applications of radiation to biology and chemistry. Their easily-controlled
gain response, although an asset in a research environment, is probably a drawback for a medical environment where long-term stability
is more important.

Table VII. Organic Scintillators

Scintillator	Density $\rho \left(\frac{gm}{cm^3}\right)$	Effective atomic number Z	Wavelength of max. emission (nm)	Refractive index	Light yield relative to anthracene	Decay time (nsec)
Anthracene	1.25	5.8	445	1.59	1.00	25
Quarter-phenyl	---	5.8	438	---	0.85	8
Stilbene	1.16	5.7	410	1.62	0.73	7
Terphenyl (para)	1.12	5.8	415	---	0.55	12
Diphenyl-acetylene (Tolan)	1.18	5.8	390	---	0.26–0.92	7
Naphthalene	1.15	5.8	345	1.58	0.15	75

Table VIII. Properties of Inorganic Scintillators

Material	Wavelength of Maximum Emission (nm) λ_m	Decay Constant (μs)	Index of Refraction at λ_m	Specific Gravity	γ Scintillation Efficiency Relative to Na(Tl)
NaI(Tl)	410	0.23	1.85	3.67	100%
CsI(Na)	420	0.63	1.84	4.51	85
CsI(Tl)	565	1.0	1.80	4.51	45
^{6}LiI(Eu)	470–485	1.4	1.96	4.08	35
ZnS(Ag)	450	0.20	2.36	4.09	130
CaF (Eu)	435	0.9	1.44	3.19	50
$Bi_4Ge_3O_{12}$	480	0.30	2.15	7.13	8
CsF	390	0.005	1.48	4.11	5
Li glass	395	0.075	1.55	2.5	10

REFERENCES AND ACKNOWLEDGMENTS

     The material for this lecture has been obtained from various
Lawrence Berkeley Laboratory reports written by staff members of
the Health Physics and Biophysics groups. I have also made extensive
use of graphs and information from the following textbooks:

1) F. H. Attix, W. C. Roesch and E. Tochlin, "Radiation Dosimetry"
   (2nd ed., New York: Academic Press, 1966), II and III.
2) W. J. Price, "Nuclear Radiation Detection" (2nd ed., New York:
   McGraw-Hill Book Co., 1964).
3) N. W. Holm and R. J. Berry, "Manual on Radiation Dosimetry" (New
   York: Marcel Dekker, Inc., 1970).
4) G. F. Knoll, "Radiation Detection and Measurement" (New York: John
   Wiley & Sons, 1979).

     The figures listed below are reprinted with permission of John
Wiley & Sons, Inc. from "Radiation Detection and Measurement," G. F.
Knoll, copyright 1979, John Wiley & Sons: Figs. 3a, 5a, 6b, 7a, 7b,
7c, 8a, 8b, 9a and 9b.

     The figures listed below are reprinted with permission of
Academic Press, Inc. and Dr. F. H. Attix from "Radiation Dosimetry"
(2nd ed.), F. H. Attix W. C. Roesch and E. Tochlin, copyright 1966,
Academic Press: Figs. 2a, 2b, 10a, 15a and 15b.

     The figures listed below are reprinted with permission of
Marcel Dekker, Inc. from "Manual on Radiation Dosimetry, N. W. Holm
and R. J. Berry, copyright 1970, Marcel Dekker, Inc: Figs. 10b, 11a,
11b, 12a, 12b, 13a and 13b.

     I would like to take this opportunity to thank for following
members of the Lawrence Berkeley Laboratory staff, whose assistance
is deeply appreciated: R. H. Thomas, J. McCaslin, J. Howard, A.
Smith, P. Wiedenbeck and C. Johnson-Joy.

     This work was supported by the Physics Research Division of
the United States Department of Energy under contract no. W-7405-
ENG-48.

PASSIVE DETECTORS

Ralph H. Thomas

Lawrence Berkeley Laboratory
University of California
Berkeley, California

"They also serve who only stand and wait"

John Milton, 1608–1674
Sonnet XIX "On His Blindness"

INTRODUCTION

In this lecture I shall interpret the term "passive" radiation detector as meaning one that will yield up its information after an irradiation is completed, and often only after some processing of the detector to obtain the data or some considerable data processing. Examples of such detectors would thus be:

(1) Nuclear Emulsion.

(2) Activation Detectors (often imprecisely referred to as threshold detectors).

(3) Integrating Ionization Chambers.

(4) Thermoluminescent Dosimeters.

NUCLEAR EMULSIONS

The use of nuclear emulsion as a detector of ionizing radiations is, of course, as old as the discovery of radioactivity itself. Its use has persisted, particularly because nuclear emulsion provides a graphic presentation of nuclear interactions. It is partly for this reason that nuclear emulsions have played an extremely important role in the discovery of fundamental particles.[1] Whenever a new source of radiation is developed, almost inevitably one of the first exposures made is with some form of nuclear emulsion. As an example, Fig. 1 shows a microphotograph of the tracks of 2.1 GeV/amu nitrogen ions tracks in Ilford G5 emulsion. This emulsion was exposed shortly after heavy ions were accelerated at the Bevatron[2,3] and shows $^{14}N_7$ ion fragmenting into two helium ions and three protons.

Shielding Experiments

Thick nuclear emulsions were used in some of the earlier high-energy shielding experiments.[4-6] Figure 2 shows a photomicrograph of a 10 GeV neutron in 600 μ Ilford G5 emulsion.[7] The use of emulsion facilitated studies of mean energy of the high energy particles as a function of depth in shielding, growth of the neutral component within the shield, angular divergence of

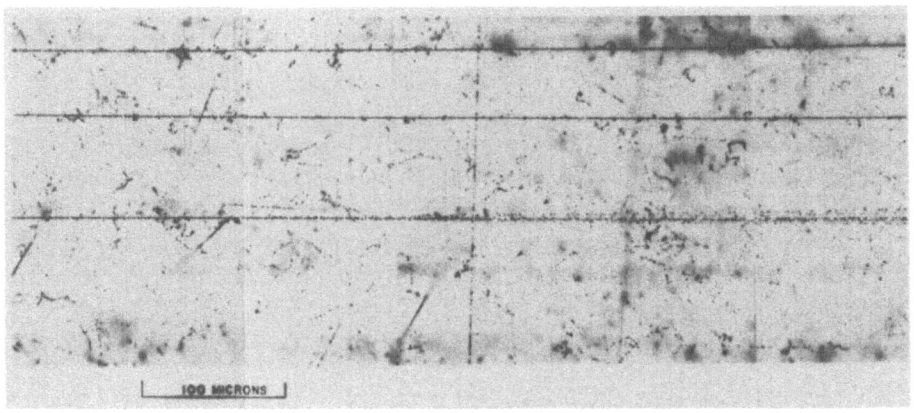

Figure 1. At a point on the left hand side of the figure an incident $^{14}N^{7+}$ ion fragments into two helium ions and three protons. The incident beam energy was 2.1 GeV/amu. and the nuclear emulsion was Ilford G5. (Courtesy of H. Heckmann).

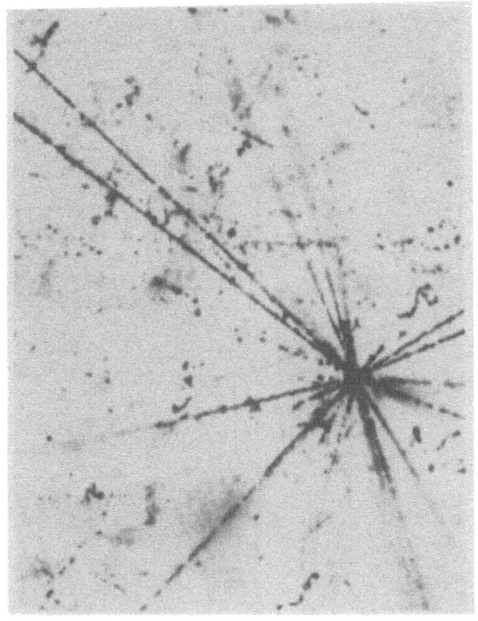

Figure 2. Photomicrograph of 10 GeV/c
neutron interaction in Ilford
G5 emulsion.

the beam, in addition to the overall attenuation of particle flux
with depth. Thus, for example, Fig. 3 shows the fraction of
"neutral" stars as a function of depth in steel irradiated with
an incident 20 GeV/c proton beam.

Absolute Beam Intensity Measurements

Nuclear emulsions may be used for absolute beam intensity
measurements. Smith et al.[8] have used Kodak NTA films to
measure the intensity of carbon, oxygen, and neon ions. The
films were exposed at an angle of 45° to the incident beam and
the large ionization of the heavy ions produced dense tracks
which are very easily identified (Fig. 4). Optimum exposures
produce 40-20 tracks in a 245 μm square field. The upper limit
to track density that could be scanned with an air objective
lens on the microscope (magnification 430x) was about 4×10^5

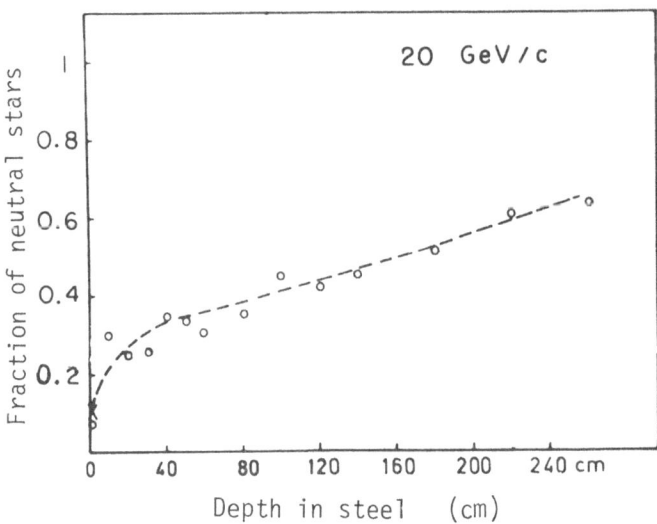

Figure 3. The fraction of "neutral" stars as a
function of depth in still. Incident
proton momentum 20 GeV/c. A "neutral"
star is defined as one with no charged
primary in the backward hemisphere
(after Citron et al.).

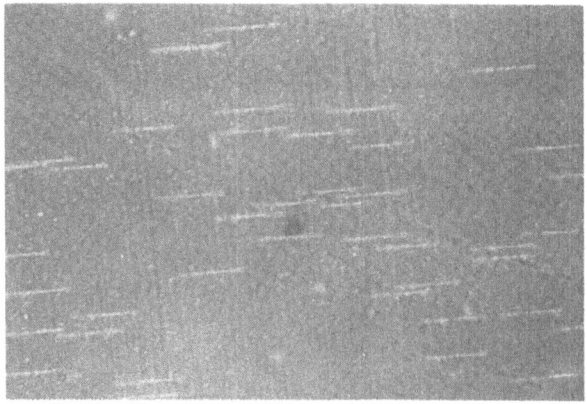

Figure 4. Photomicrograph of carbon ion tracks in Kodak NTA
emulsion (magnification 430x).

tracks cm[-1]. An oil immersion objective would raise this limit
by about a factor of ten. This technique of beam intensity
measurement was applied to the determination of radionuclide
production cross sections,[9] measurement of the efficiency of
thermoluminescent dosimeters,[10-12] and to dosimetry in radiobiological experiments.[13-15]

Personal Dosimetry

   At the present time only three types of personal neutron
dosimeters are commonly used in radiation protection.[16] The
oldest and still most popular personal dosimeter is Kodak NTA
film (or equivalent emulsions produced by other manufacturers),
although in recent years it has become somewhat fashionable to
denigrate the use of this dosimeter.[17] This criticism is based
upon the inherent difficulties of the dosimeter, its instability,
and its energy response. The dosimeter is read by detecting
individual recoil proton tracks in a thin (25 μm) nuclear
emulsion, using an optical microscope. While the technique
seems laborious to those who prefer "gadgetry," it is,
nevertheless, in the hands of a skilled technician extremely
reproducible, quite quick, and inexpensive.

   It has been known for many years that the latent image of the
proton recoil tracks is unstable and fades, particularly at high
temperature and humidities.[18] Much has been made of this
fading[19] so that there has been a decline in the popularity of
NTA film in recent years. This is, in the view of the writer, an
error since it is certainly preferable to provide some personal
dosimetry—rather than none of all! In fact the causes of latent
image fading are well understood and it may be adequately
controlled by careful packaging.[20,21]

   The most serious disadvantage of NTA film is that it is
limited to neutrons in the energy range from 0.5 MeV to about
20 MeV. Figure 5 shows the energy response of NTA film per unit
neutron fluence, compared with the dose equivalent per unit
fluence given by the ICRP. However, despite this limitation NTA
film is of great value of a personal neutron dosimeter around
high energy particle accelerators which the largest fraction of
dose equivalent is due to neutrons from 0.5-20 MeV.[23] Some
years ago the use of thick emulsions (200 μm-600 μm) was proposed
for neutron monitoring at high energies[24,25] but this suggestion
has not been taken up.

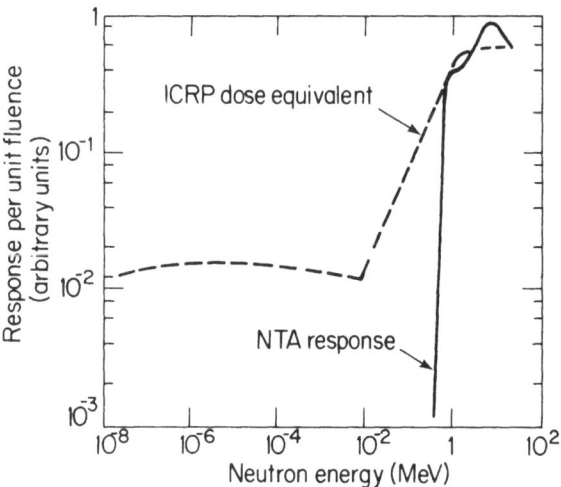

Figure 5. Energy response of NTA film.

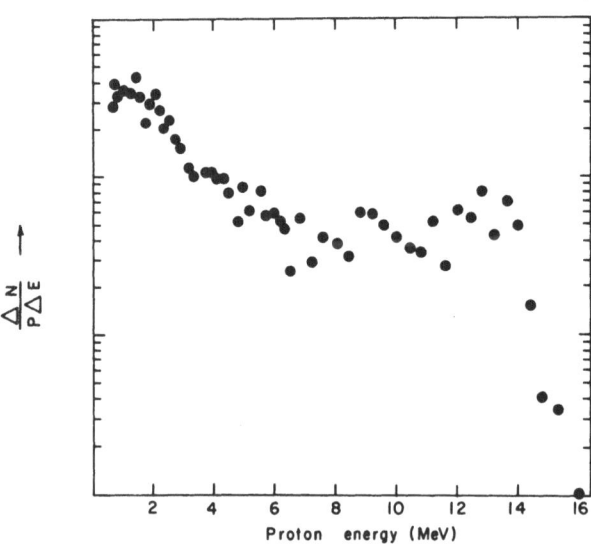

Figure 6a. Proton spectrum measured in an
emulsion exposed to 14 MeV neutrons.

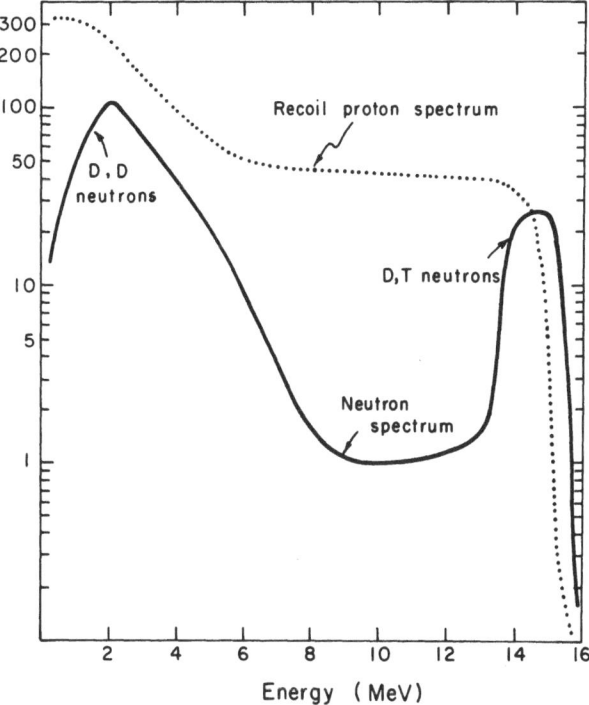

Figure 6b. Generation of a neutron spectrum
from the proton recoil data shown
in Figure 6a. The dotted line
shows the smoothed curve fitted to
the proton recoil scanning data,
while the solid line shows the
derived neutron spectrum.

Neutron Spectrometry

Nuclear emulsions have been used extensively for neutron
spectroscopy and details of the techniques used may be found in
Patterson and Thomas.[26] Two principal methods are used:

(a) Proton recoil spectrum measurement

(b) Star prong counting.

In the first technique the spectrum of recoil protons
produced by neutron interactions in the emulsion is first
measured, and then the neutron spectrum that produced it calculated. Figures 6a and 6b gives an example of the result
obtained.

Figure 7 shows examples of neutron spectra measured by
Lehmann and Fekula around the Bevatron.[27] "The general form of
the neutron spectra (measured between 0.7 and 20 MeV) at eight
locations near the Bevatron is a broad peak in the 0.5–2 MeV
region, followed by a smooth hundred–fold drop in value between
the peak at 12 MeV."[27]

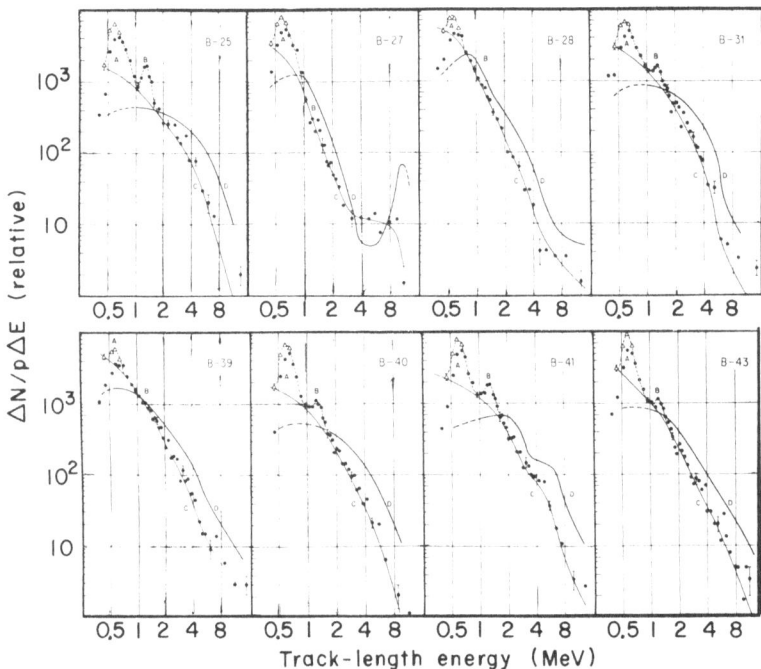

Figure 7. The neutron spectra derived from measurements of
the recoil proton spectra in nuclear emulsions
exposed at several locations around the
Bevatron. In each diagram: A identifies the peak
at 0.6 MeV due to the $^{14}N(n,p)^{14}C$ reaction of
thermal neutrons; B identifies the 1.25–MeV peak
due to α–particles from the decay of the
naturally radioactive constituents of the
emulsion; the curve C shows the smoothed recoil
proton spectrum corrected for background, and the
curve D shows the derived neutron spectrum. The
notations B–25, B–27, etc., identify location of
the emulsion exposure (after Lehman and Fekula,
1964).

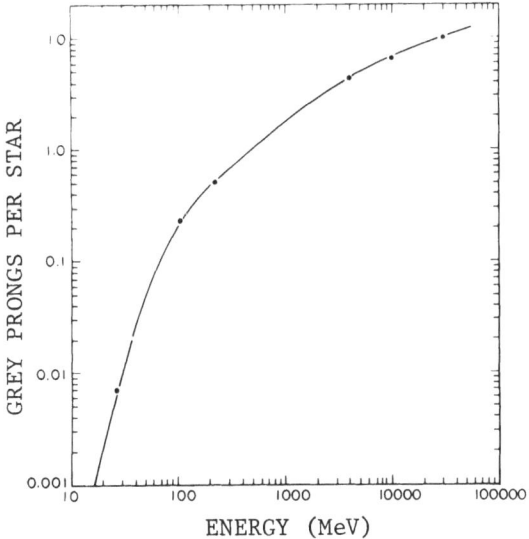

Figure 8. Average number of grey prongs per
star as a function of neutron energy.

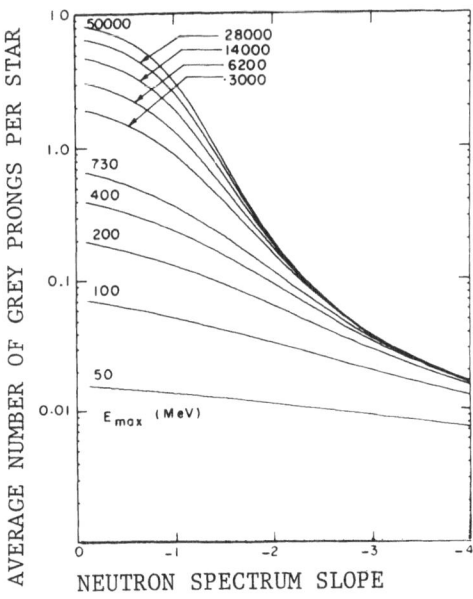

Figure 9. A graph relating the number of
gray prongs per star and different
shapes of neutron spectra characterized by the longarithmic slope, γ,
and the maximum energy of the spectra.

Table 1. Spectral Indices Obtained from Measured Values of the
Average Number of Gray Prongs per Star.

Location	E_{max} (MeV)	n_g	Spectral index, γ
184-inch cyclotron between Bays 10 and 11	730	0.44	0.75
Bevatron west tangent tank shielding wall	6,200	0.55	1.50
Bevatron Col. 7, main floor	6,200	0.32	1.68
Bevatron mezzanine	6,200	0.27	1.78
CERN PS	14,000	0.29	1.80
CERN PS	14,000	0.21	1.95
CERN PS	28,000	0.45	1.68
White Mountain 12,000 ft. altitude	(50,000)	1.07	1.32
White Mountain 14,000 ft. altitude	(500,000)	1.04	1.35

The second application of emulsions to neutron spectroscopy
is to determine the average number of gray prongs, n_g, produced
in the nuclear interactions. For monoenergetic neutrons there is
a strong relationship between the average number of gray prongs
per star and the incident neutron energy. This relationship has
been determined by Remy[28] and by Patterson et al.[29] and is
shown in Fig. 8. Patterson et al.[29] showed how to use this
relationship to determine the characteristics of hard neutron
spectra found around, for example, high energy accelerators.
Under the assumptions that these spectra are smooth and may be
characterized by an index,* γ, that the cross section for star
production is independent of neutron energy and that a maximum
neutron energy may be defined, Patterson et al. calculated the
relationship between γ and $\bar{n}_g$ shown in Fig. 9, where several
curves are shown for differing maximum energies. Thus, under the
assumptions of the method, if $\bar{n}_g$ is determined, E_{max} known
(e.g., as for example at an accelerator) then γ may be determined. Experience has shown that around accelerators γ has
values usually in the range 1.3-20 (see Table 1).

Where independent checks have been possible--for example when
spectra have been determined by alternative techniques, e.g.,
activation detectors--quite good agreement was obtained.[30]
Patterson et al. concluded that "The study of stars in nuclear
emulsion extends the usefulness of emulsion methods in nuclear
spectroscopy to energies much higher than the upper limits of
recoil proton techniques."[29]

---

*The neutron differential energy spectrum $d\phi/dE$ is written:
$d\phi/dE = E^{-\gamma}$.

Table 2. Properties of Some Commonly Used Threshold Detectors

Detector	Reaction	Half-Life	Energy range	Typical minimum flux density measurable[a] $n\,cm^{-2}s^{-1}$	Remarks
BF$_3$ proportional counter	$^{10}B(n,\alpha)^7Li$	–	Thermal		
Gold foil	$^{197}Au(n,\gamma)^{198}Au$	2.7 days	Thermal	10^2	
Indium foil	$^{115}In(n,\gamma)^{116m}In$	54 min	Thermal	1	
Moderated BF$_3$ counter	$^{10}B(n,\alpha)^7Li$	–	Thermal–15 MeV	10^{-2}	Energy range and sensitivity depends upon moderator size ~15 cm dia. values quoted.
Moderated gold foil	$^{197}Au(n,\alpha)^{198}Au$	2.7 days	Thermal–15 MeV	10^2	
Moderated indium foil	$^{115}In(n,\alpha)^{116m}In$	54 min	Thermal–15 MeV	1	
Thorium fission counter	Th(n,fiss.) fission products	–	>2 MeV	1	
Sulphur	$^{32}S(n,p)^{32}P$	14.3 days	>2.5 MeV	10^4	
Aluminum	$^{27}Al(n,\alpha)^{24}Ha$	15 h	>6 MeV	1	
Aluminum	$^{27}Al(n,spall.)^7Be$	53.4 days	>25 MeV	10^4	
Polystyrene: plastic scintillator	$^{12}C(n,2n)^{11}C$	20.4 min	>20 MeV	1	
	$^{12}C(n,spall.)^7Be$	53.4 days	>30 MeV	10^4	
Bismuth fission chamber	Bi(n,f) fission products	–	>50 MeV	1	
Mercury	Hg(n,spall.)^{149}Tb	4.1 h	>600 MeV	10	

[a]Based on 1 h measurement.

ACTIVATION DETECTORS

The use of threshold detectors in neutron dosimetry is a well understood and universally accepted technique in radiation physics. Their use has found widespread application at most high-energy particle accelerators and has been described in several articles, reviews and text books.[23,26,30-34]

Threshold detectors may be both active[35] and passive and both types of detector are frequently used in tandem in radiation surveys or other measurements.[36] We are particularly concerned here with threshold detectors that utilize their own induced radioactivity to monitor radiation.

Table 2 summarizes some of the threshold reactions commonly used as accelerator laboratories. Column 5 indicates the typical sensitivity which may be readily achieved for these detectors. Sensitivity is, however, clearly a function of detector size and the precise experimental techniques employed, and the values indicated are intended only as a general guideline. They indicate the order of magnitude of minimum flux density that may be detected after a measurement lasting one hour. For precise details, the reader is referred to the original sources. Furthermore, Table 2 is not intended to be comprehensive but to indicate the reactions in common use. Particular laboratories may have their own preferred specialties that they have perfected.

It may be seen from Table 2 that threshold detectors are of high sensitivity available over the entire energy range normally of interest at accelerators (0.1-100 MeV). No details of the shape of the neutron spectrum below about 1 MeV will be obtained using only one size of moderator with a thermal neutron detector. Fortunately such detail is not often required for two reasons: (1) because the dose-equivalent contribution is not large, and (2) because, below 10 keV, the dose-equivalent per unit fluence is independent of neutron energy. In principle, should more detailed information of the spectrum be required in the energy region from $\sim 10^{-8}$ to 1 MeV, several moderators of different size could be used.[37]

At high radiation intensities ($\geqslant$10 rem/h) several less sensitive reactions provide additional information. Figures 10a and 10b show the variation of sensitivity with energy for the reactions listed in Table 2.[23]

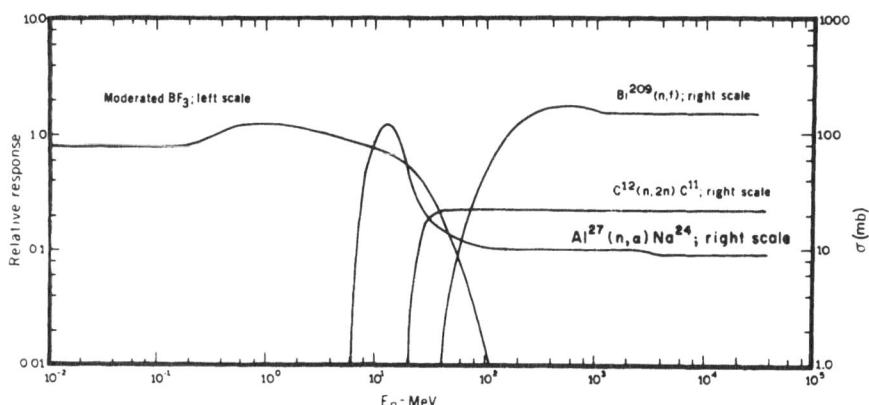

Figure 10a. Response functions of four threshold detectors used to
            determine neutron spectra in well-shielded locations
            at high-energy accelerators. (After Thomas Ref. 23).

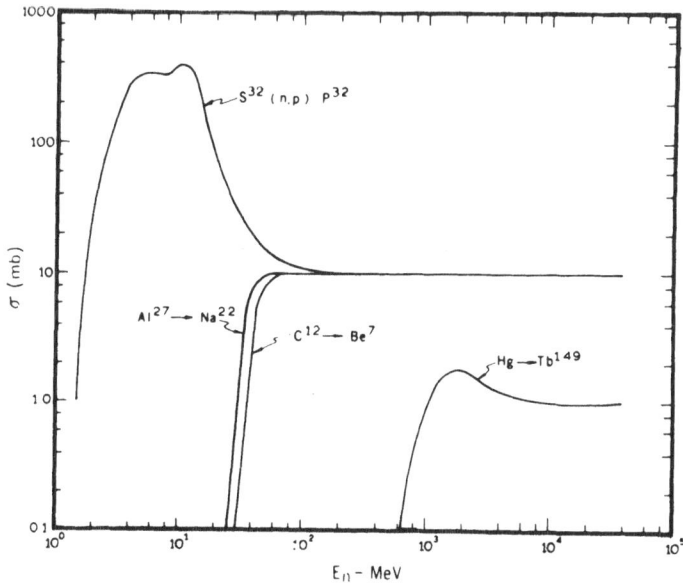

Figure 10b. Response functions for additional detectors used to
            determine neutron spectra in high radiation levels at
            high-energy accelerators. (After Thomas Ref. 23).

Activation detectors are widely used radiation measurements at accelerators. Typical uses include:

(a) Shielding Experiments

(b) Neutron Spectrum Measurements

(c) Radiation Surveys

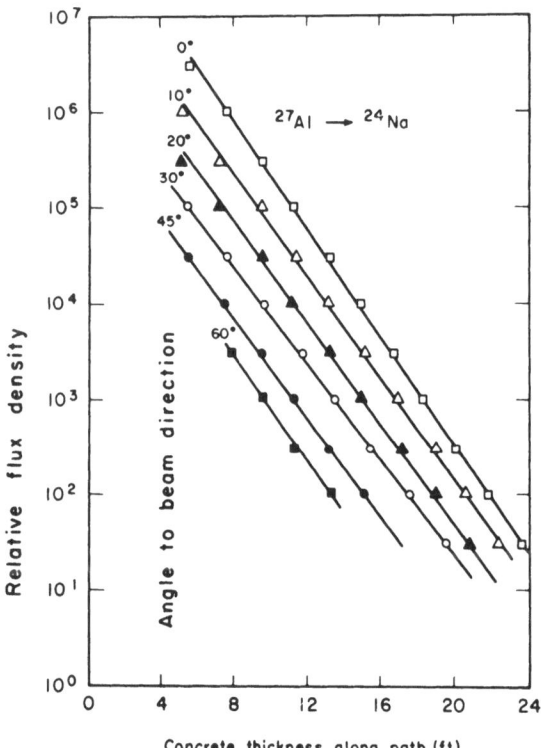

Figure 11. Relative flux density distribution measurements along paths drawn at several angles to the line of incidence of the proton beam, from the point of entry on a concrete shield. Measurements made with the $^{27}Al \rightarrow {}^{24}Na$ reaction. Incident proton energy 6 GeV. (After Smith et al. Ref. 34.)

Shielding Studies

A great many shielding studies at high energy accelerators have been made using activation detectors. Many of these measurements have been reviewed by Patterson and Thomas.[38] Figures 11 and 12 show typical examples of the data that may be obtained. Figure 11 shows the relative flux density distributions measured in a concrete shield bombarded by 6 GeV protons. The reaction used was the production of ^{24}Na from ^{27}Al (6 MeV threshold).

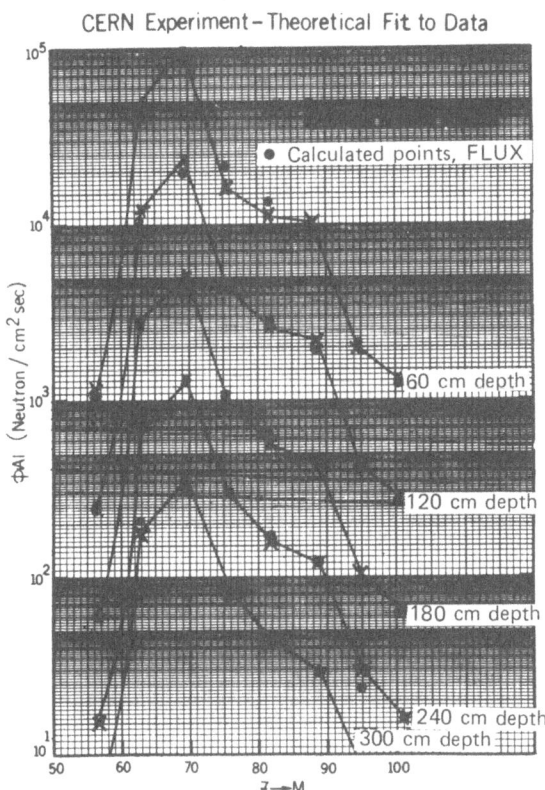

Figure 12. Measurements of neutron flux density in the earth shield of the CERN 28 GeV proton synchrotron. The crosses show experimental data points measured with the ^{27}Al$\rightarrow^{24}$Na reaction while the circles show calculated values. (Gilbert et al. Ref. 30).

Figure 12 shows experimental data using the same reaction obtained in the earth shield of the 28 GeV CERN Proton Synchrotron. The neutron flux density measured is plotted as a function of distance from the proton beam (depth) and along the beam direction (z). Measured values are in excellent agreement with those calculated using a computer program FLUXFT.[30]

To determine the spectrum a solution for the neutron spectrum $\phi(E)$ is sought from a set of activation equations of the form

$$A_j = C_j \int_{E_{min}}^{E_{max}} \sigma_j(E) \, \phi(E) \, dE \quad \text{for} \quad j = 1, 2, \ldots m \qquad (1)$$

where

A_j is the saturation activity of the jth detector,

$\sigma_j(E)$ is the cross-section for the appropriate reaction at energy E,

C_i is a normalizing constant between activity and flux density, and

E_{min}, E_{max} are the minimum and maximum neutron energies in the spectrum.[30]

Gilbert et al.[30] have described the use of an interactive technique that employs on-line facilities of a CDC-6600 computer for the determination of neutron spectra from a few threshold detectors: "TELLY." The operator indicates to the computer his best estimate of the neutron spectrum which will match his experimental data. This is done by drawing the spectrum with a light pen on the screen of a CRT display. The computer then calculates the detector responses and presents them for comparison with the experimental data. The operator then systematically modifies his suggested spectra to the computer until, after a few iterations, the detector responses are matched with an accuracy reflecting the experimental errors. TELLY was found to work well, avoiding many of the pitfalls of more "sophisticated" methods of spectrum analysis. Its only drawback is that it is somewhat difficult to use in a systematic manner when many detectors with overlapping regions of sensitivity are used.

Equation (1) is a degenerate case of a Fredholm integral of the first kind. Formal methods of solution are not applicable when, as is the case with activation detectors, the A_j's are known only as a set of discrete points.[39]

Routti[39] has critically reviewed the numerical techniques commonly used for solution of such first-order Fredholm equations, and the interested reader is referred to his paper for a detailed account.

Early attempts to obtain neutron spectra from activation detector data were frustrated by difficulties such as non-uniqueness or an oscillatory (and even negative) character to the solutions to the Fredholm equations. Some of these problems arise from the mathematical characteristics of the equations to be solved, while others are related to the specific method of solution adopted.

Routti suggests that a suitable method of solution must be able to combine the information contained in the measured data with any already existing information of the neutron spectrum. Such prior information is almost always available on physical grounds. Thus, for example, the solution must be non-negative and zero beyond a given maximum energy. In addition, the spectrum of radiation penetrating thick shields constructed of a complex material such as concrete may be assumed to be smooth. Some information on intensity or shape may be available from previous measurements. It is important that all this prior information be properly taken into account in the solution technique selected. However, care must be taken to ensure that the consequent additional constraints imposed on the spectrum do not prevent it from matching the measured response or from assuming any physically acceptable shape.

Any appropriate solution must fulfill two basic requirements

(a) The neutron spectrum which is found to be a solution to the activation equations must accurately match the detector responses; and

(b) if many solutions are found that fulfill condition (a), there should be a flexible way to apply physical prior information on the solution so that the most appropriate solution may be selected.

It is important that any solution method be tested to ensure that it meets all these requirements. This is most conveniently done by computing the response of the system to test spectra. The resolutions of the system and the influence of experimental errors or uncertainties in the detector response functions may then be systematically studied.

Routti has applied a generalized least-squares method to solve the activation equations. In his technique, the solution is forced to be non-negative and prior information on the

spectrum can be incorporated in a very flexible way. The
technique and the computer program LOUHI, written to perform the
analysis, have been subjected to the tests described in the
previous paragraph. These tests show that the method meets the
two basic requirements for an appropriate solution.

Considerable experience has now been obtained with LOUHI
which has been found extremely reliable and capable of calculating neutron spectra with adequate accuracy for radiation
protection purposes. A desirable feature of LOUHI is that, in
addition to providing activation detector data, it may be used to
determine neutron spectra from Bonner sphere or nuclear emulsion
data.

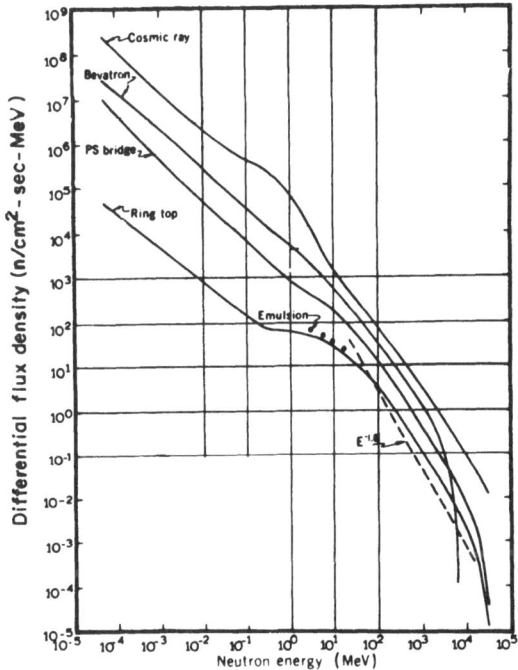

Figure 13. Typical high energy accelerator neutron spectra.
The spectra are arbitrarily displaced on the
vertical scale to show the differences between
the spectra. The measurements shown here were
made at the Bevatron and the CERN 28-GeV Proton
Synchrotron — above concrete ("Bridge") and
earth ("Ring Top"). In the case of the ring,
top measurements were made with activation
detectors (solid line), Ilford L4 emulsion (dots),
and by star prong counting (dashed lines). The
Hess cosmic ray spectrum is shown for comparison.

It is of interest to compare neutron spectra obtained using threshold detectors with those obtained from nuclear emulsions. This was possible in an experiment carried out at CERN where neutron spectra outside the shielding of the 28 GeV CERN proton synchrotron were determined by several methods.[30]   Figure 13 shows several neutron differential energy spectra measured outside accelerator shields.   In the case of one (labelled "Ring Top"), which was measured above the earth shield of the CPS neutron spectrum measurements, were made with activation detectors, proton recoil spectrum and gray prong number.   The three spectra obtained are shown in the figure and are in reasonable agreement.

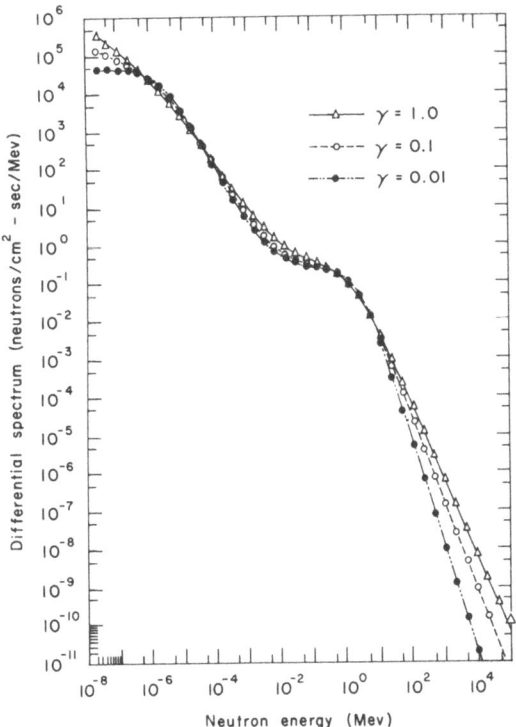

Fig. 14.   Neutron differential energy
           spectra unfolded from data
           obtained in an aircraft flying
           at 41,000 ft, geomagnetic
           latitude 20°N, for various
           values of the smoothing para-
           meter γ. (Stephens et al.
           Ref. 40).

Stephens et al.[40] have investigated the uncertainties in
spectra unfolded by the LOUHI routine for the particular case of
Bonner Spectrometer data.   The LOUHI unfolding routine permits
varying constraints to be placed on the rapidity and magnitude of
fluctuations in the solutions it generates.   This is achieved by
the use of a smoothing parameter, γ, which normally is given
values between 0 the 1.0.   The higher value of γ, the more con-
strained are fluctuations in the solution generated and the lower
the ability to resolve sharp structure in the spectrum.
Figure 14 shows solutions for values of γ = 0.1 (moderately
damped) and γ = 0.01 (slightly damped).   Significant differences

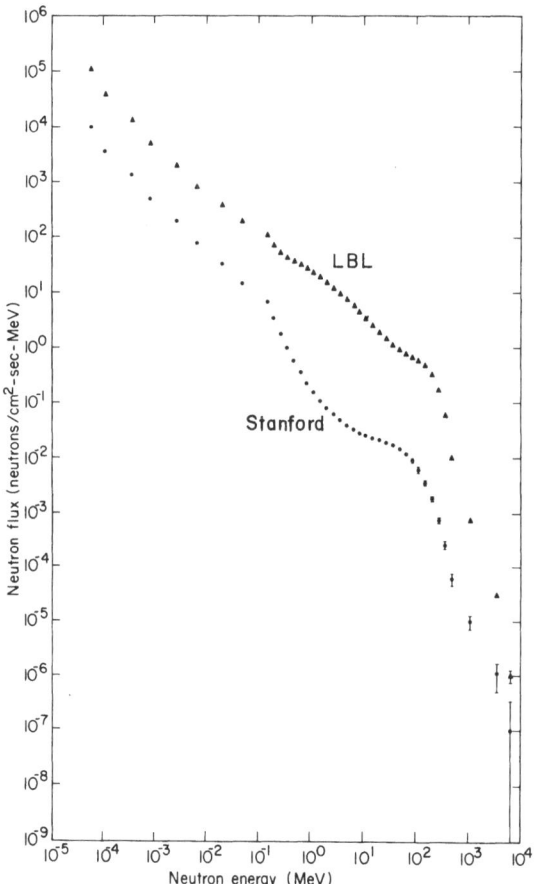

Fig. 15.   Differential Neutron Energy
           Spectra measured at the
           Bevatron (LBL) and the Stanford
           20 GeV electron linear acceler-
           ator (McCaslin et al. Ref. 41).

in the calculated spectra are evident, but it is not entirely
clear whether this inconsistency is due to fluctuations of
statistical origin or systematic errors introduced by the use of
two detectors of different size and a mixed set of moderators.
However, these authors also showed that the addition of data
taken with aluminum activation detectors significantly limit the
uncertainty in derived neutron spectra.

For some years there had been some difference in opinion
between accelerator health physicists as to whether there was any
significant dose equivalent contribution from neutrons with
energy greater than 20 MeV. Activation detectors were applied to
this problem.[41] Figure 15 shows the neutron differential
energy spectra measured outside shielding at the 6 GeV proton
sychrotron at Berkeley (the Bevatron) and the Stanford 20 GeV
electron linear accelerator. From these spectra, the integral
space dose equivalent curves may be calculated (Figs. 16 and
17). The integral dose equivalent spectra are quite similar and
clearly show the presence of a significant high-energy neutron
fluence at the Stanford Linear Accelerator, confirming theoretual
predictions by de Staebler.[42]

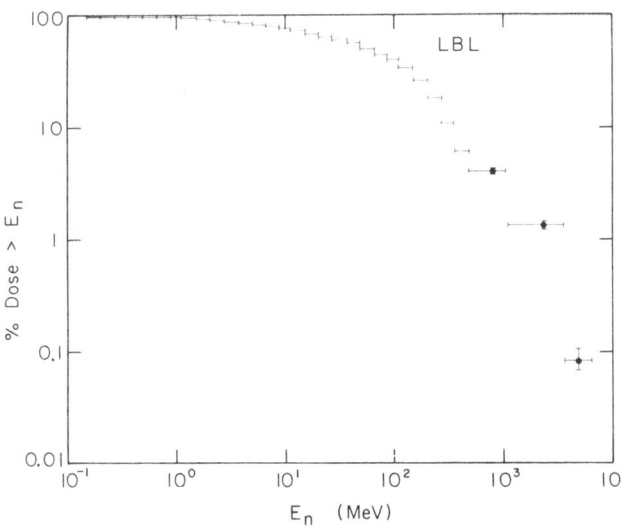

Figure 16. Integral dose equivalent spectrum (LBL).

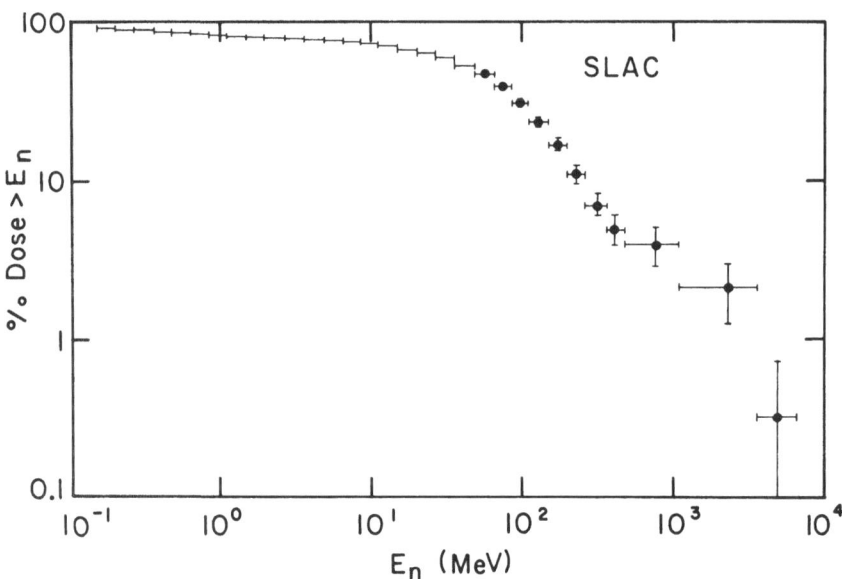

Figure 17. Integral dose equivalent spectrum (SLAC).

Despite the difficulties in measuring neutron spectra with threshold detectors the method may be used with some confidence as evidenced by the reasonable agreement between the experimental measurements of the cosmic-ray neutron spectrum by Hess et al.[43] made in 1958 and the relatively recent calculated values of Armstrong et al.[44] These workers used a Monte Carlo code to comput the production of protons, charged pions, and neutrons by the incident galactic protons, and the subsequent transport of these particles down to energies of 12 MeV. The calculated production of neutrons of energy ⩽12 MeV calculated by the Monte Carlo code was used as input to a discrete-ordinates code to obtain the low-energy neutron spectrum. Figure 18 shows the results of these calculations and an absolute comparison with the experimental data of Hess et al.[43] at atmospheric depths of 200 and 1,033 g/cm^2. The calculated and measured spectra differ somewhat at lower energies, but are in very good agreement at high energies.

The Use of Activation Detectors to Measure High-LET Radiation

Activation dosimeters have been used with great success in the absolute determination of particle fluences at proton

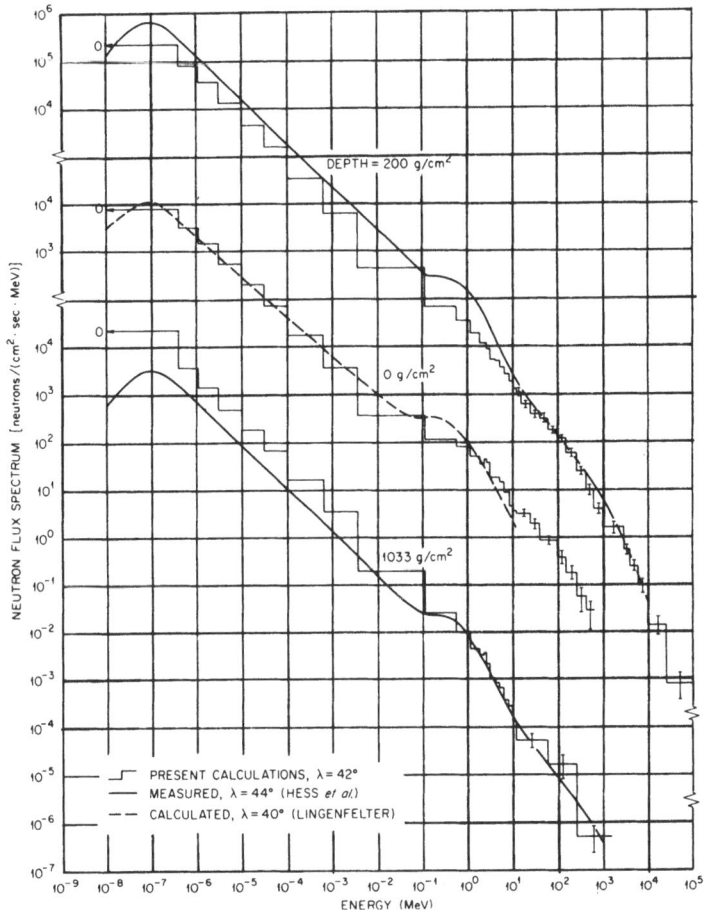

Figure 18. Neutron flux spectra at various depths
in the earth's atmosphere produced by
galactic protons near solar minimum.
These calculations are compared with
calculations of Lingenfelter and the
measurements of Hess et al. (λ =
geomagnetic latitude) [Armstrong et al.
Ref. 44].

accelerators and there is no reason to doubt that they might be used to monitor high–energy heavy–ion beam intensities.

The specific advantages of activation dosimeters are:

(1) Suitable reactions give high activity enabling relative data of high statistical precision to be obtained. Such reactions may therefore be used as a standard with which the reproducibility and linearity of all other dosimetric systems may be compared.

(2) The use of several activation detectors can give information on incident beam purity.

Smith[45] has reported initial tests of aluminum activation detectors irradiated by 375 MeV/u Ne^{10+} ions. The induced ^{24}Na activity was measured using a NaI(Tl) γ-spectrometer situated in the Health Physics Department Low Background Facility.[46]

Sufficient activity is induced in 40 g of aluminum to easily obtain a statistical precision of ±1 percent for exposures corresponding to 30 rad (the efficiency of the spectrometer used was ~35 percent for ^{24}Na). Competing radionuclides produced were ^{11}C(20.3 min) and ^{18}F(110 min). The production of ^{18}F in aluminum is capable of comparable statistical precision to that obtained counting ^{24}Na, but correction for the production of ^{11}C (which also emits positrons) is an added complication.

Table 3 summarizes some possible activation reactions. As a general rule, those reactions that produce readionuclides with half–lives comparable to the length of irradiation will prove to be most convenient.

In radiobiological experiments it will often be convenient to use activation targets of composition similar to that in tissue, such as polyethylene, lucite, or polystyrene, in which the pro-duction of ^{11}C from ^{12}C is a convenient reaction. For absolute dosimetry, the reaction cross section must be known. Measurements for the production cross section of ^{11}C by 375 MeV/u Ne^{10+} ions in carbon give a value 75 ± 7 mb.[9] A reasonable compromise between energy loss in the target and sensitivity puts the lower limit of sensitivity for this reaction at an absorbed dose of ~10 rad in tissue.

Activation detectors would be more acceptable in some experiments if they could be placed behind the irradiated bio-logical sample to determine the particle fluence leaving the experiment, but it is then necessary to study the production of radionuclide to be measured as a function of depth in an

Table 3. Some Useful Target Activation Reactions that Produce
Radionuclides with Halflives Longer than about 10 Min.

Target material	Reactions useful for dosimetry				Competing and Reactions
1. ^{27}Al in aluminum	$^{27}Al \rightarrow {}^{24}Na$	15.0 hr	β−	1368,2754 keV	
	$^{27}Al \rightarrow {}^{22}Na$	2.60 yr	e-capt, β+	511, 1275 keV	
	$^{27}Al \rightarrow {}^{18}F$	109.7 min	β+	511 keV	
	$^{27}Al \rightarrow {}^{13}N$	9.93 min	β+	511 keV	
	$^{27}Al \rightarrow {}^{11}C$	20.34 min	β+	511 keV	
	$^{27}Al \rightarrow {}^{7}Be$	53.6 day	e-capt	478 keV	
2. ^{9}Be in beryllium	$^{9}Be \rightarrow {}^{7}Be$	53.6 day	half-life, e-capt	478 keV	
3. ^{12}C in graphite, polystyrene, or polyethylene	$^{12}C \rightarrow {}^{11}C$	20.34 min	β+	511 keV	
4. ^{19}F in Teflon	$^{19}F \rightarrow {}^{18}F$	109.7 min	β+	511 keV	$^{12}C \rightarrow {}^{11}C$
	$^{19}F \rightarrow {}^{13}N$	9.93 min	β+	511 keV	$^{12}C \rightarrow {}^{7}Be$
	$^{19}F \rightarrow {}^{11}C$	20.34 min	β+	511 keV	
	$^{19}F \rightarrow {}^{7}Be$	53.6 day	e-capt	478 keV	
5. ^{14}N in boron ^{7}Be nitride	$^{14}N \rightarrow {}^{13}N$	9.93 min	β+	511 keV	$^{11}B, B^{10} \rightarrow$
	$^{14}N \rightarrow {}^{11}C$	20.34 min	β+	511 keV	
	$^{14}N \rightarrow {}^{7}Be$	53.6 day	e-capt	478 keV	
6. ^{16}O in water	$^{16}O \rightarrow {}^{13}N$	9.93 min	β+	511 keV	
	$^{16}O \rightarrow {}^{11}C$	20.34 min	β+	511 keV	
	$^{16}O \rightarrow {}^{7}Be$	53.6 day	e-capt	478 keV	
7. ^{16}O in beryllium oxide	$^{16}O \rightarrow {}^{13}N$	9.93 min	β+	511 keV	
	$^{16}O \rightarrow {}^{11}C$	20.34 min	β+	511 keV	
	$^{16}O \rightarrow {}^{7}Be$	53.6 day	e-capt	478 keV	$^{9}Be \rightarrow {}^{7}Be$
8. $^{28}Si, {}^{29}Si, {}^{30}Si$ in fused quartz					

irradiated sample. Smith[47] has reported some preliminary
studies of the distribution of ^{18}F and ^{24}Na in aluminum
irradiated by 375 MeV/u Ne[10+] ions. Figure 19 shows the
distribution of ^{18}F in aluminum. The initial portion of the
^{18}F curve is caused by the production of ^{18}F fragments from
the original neon ion beam.[48] Such fragments have a range of
~17 g cm^{-2} in aluminum and are thus seen superposed on the
continuous distribution of ^{18}F due to target activation.
Beyond the range of the primary ions (~16–17 g cm^{-2}), ^{18}F is
produced by the reaction of lighter fragments with the target
aluminum (residual activation). The total ^{18}F due to stopped
fragments is approximately equal to that due to target activation. Residual activation is not insignificant being nearly of
the same magnitude as that due to target activation at the
entrance of the stack. The total quantity of residual ^{18}F
activity is, however, only a few percent of that due to target
activation because it rapidly decreases with increasing depth,
beyond the primary ion range.

The distribution of ^{24}Na is shown in Fig. 20 and exhibits
the typical buildup of activity as a function of depth observed
in high-energy irradiations. Beyond the range of the primary
ions there is a sharp reduction in ^{24}Na production. Residual
activation is seen to be higher beyond the neon ion range than
the target activation at the target entrace. Since ^{24}Na is

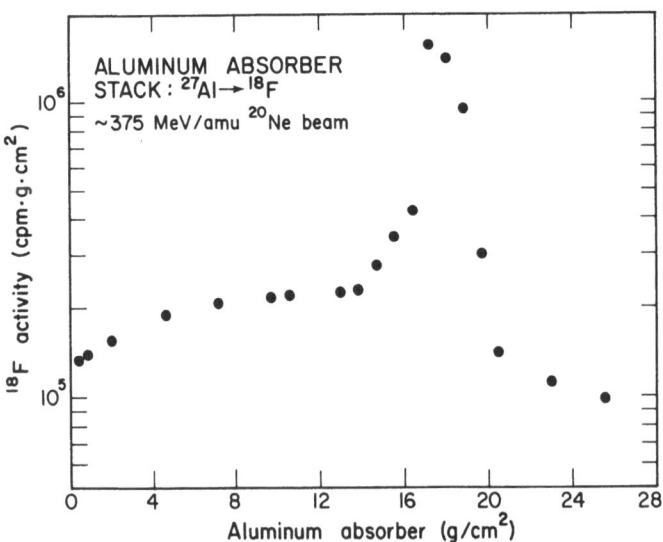

Figure 19. The distribution of ^{18}F along an aluminum
target bombarded by 375 MeV/amu Ne[10+] ions.

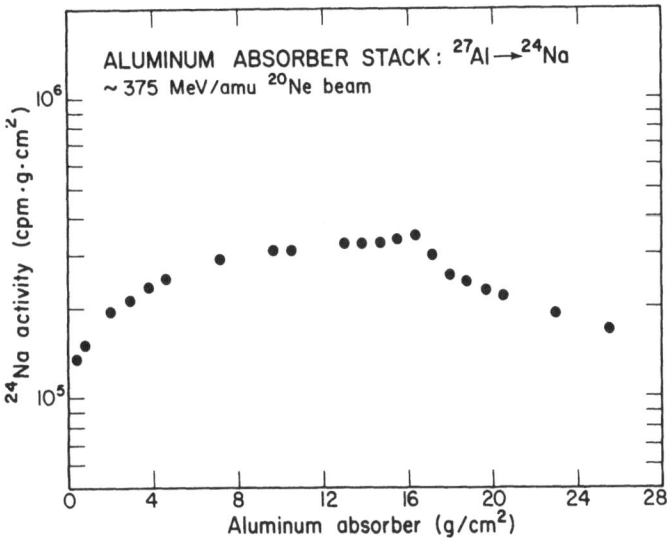

Figure 20. The distribution of ^{24}Na activity
along an aluminum target bombarded
by 375 MeV/amu Ne^{10+} ions.

heavier than the primary ions (^{20}Ne), autoactivation is not
feasible.

These preliminary studies indicate the need to study the
three mechanisms of activation in some detail if activation
techniques are to be used for absolute dosimetry behind considerable thickness of material. Smith[49] has reported approximate
cross sections for the production of ^{18}F and ^{24}Na in several
materials irradiated by 375 MeV/u Ne^{10+} ions. Table 4
summarizes these values.

Recently, work at Los Alamos has used the production of
^{24}Na in ^{27}Al for pion in-vivo dosimetry. They report that
the ^{24}Na activity is primarily produced by stopping π^-
mesons, but that 15-25 percent of the activity results from
neutrons. Comparison of the induced activity distributions with
the high-LET absorbed dose measured by a silicon detector and a
Rossi chamber show that the ^{24}Na activity may be used as a good
measure of the high-LET dose.[50]

IONIZATION CHAMBERS

The charge Q, collected under conditions of electronic
equilibrium as a result of the passage of a number N of particles

Table 4. ^{11}C Production Sections—375 MeV/u Ne^{+10} Ions.

Target	Reaction	Production cross-section (m barn)
Al	^{24}Na	63
Si	^{24}Na	15
S	^{24}Na	11
Al	^{18}F	40
Si	^{18}F	36
S	^{8}F	16

across the plates of a parallel-plate ionization chamber placed normally to a uniform, parallel charged particle beam, is related to the average energy required to create an ion pair, w, by the equation:

$$Q = \rho e \frac{N}{\bar{w}} \Delta E \qquad (2)$$

where w is measured in eV and

ρ is the density of nitrogen in the ionization chamber, $g\ cm^{-3}$;

e is the electronic charge, coulomb;

ΔE is the energy absorbed in the gas, eV.

If the energy of the incident particles is high, the energy loss in passing through the gas is small and equation (2) may be written:

$$Q = 10^6\ se\left(\frac{dE}{dx}\right)_{N_2} \cdot \frac{N}{\bar{w}} \qquad (2a)$$

where s is the separation between the collection plates, cm;

$\left(\frac{dE}{dx}\right)_{N_2}$ is the mass stopping power of the particles in the nitrogen within the chamber, $MeV\ g^{-1}\ cm^2$.

Equation (2a) may be related to the absorbed dose in tissue, D, by the equation:

$$D = 10^5 \ Q \ \frac{\bar{w}}{m} \ S'$$

(3)

where m is the mass of irradiated gas in the chamber (in grams)

S' is the ratio of stopping powers of tissue to gas for the incident particles.

The Biology and Medicine Division of the Lawrence Berkeley Laboratory has designed large parallel-plate, nitrogen-filled ionization chambers for dosimetry in radiobiological experiments. These chambers are constructed so as to present a minimum of absorbing material ($\sim$0.05 g cm^{-2}) in the heavy ion beam

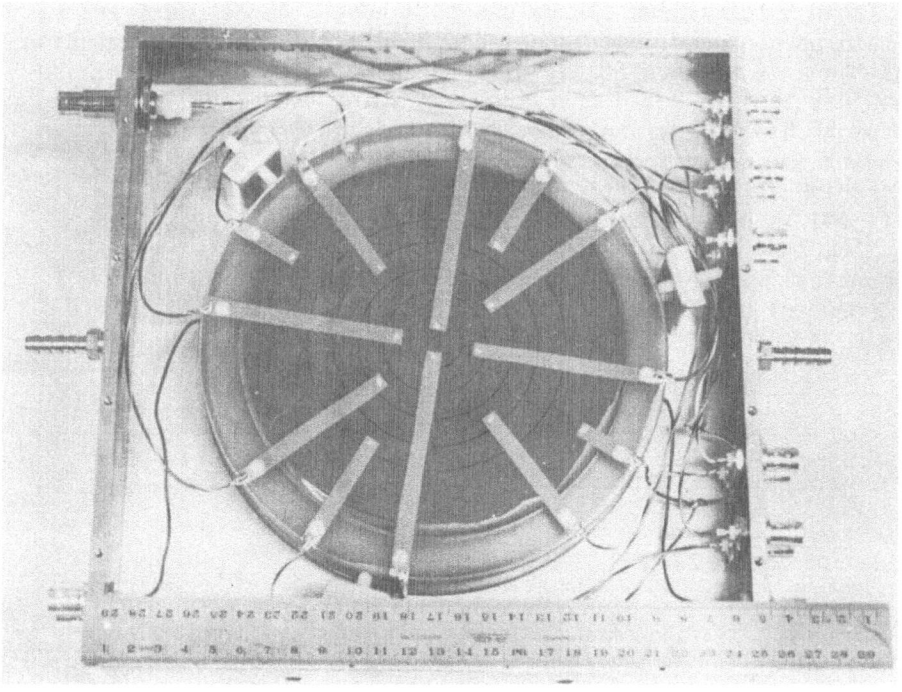

Figure 21. Photograph of a prototype annular ionization chamber used for Biomedical dosimetry.

path. The electronic equilibrium established in the air path through which the beam passes before entering the chamber is essentially maintained as the beam passes through the chamber. The collecting electrodes of the chambers are circular in cross section, and are spaced 1 cm from the high voltage electrode and placed at right angles to the incident beam direction.

Many radiobiological experiments at the Bevalac utilize rather large irradiation fields (typical beam dimensions might in some cases be a fullwidth, half maximum of 10-12 cm). Chambers have been constructed with collecting electrodes up to 18 cm in diameter to make measurements in such radiation fields. Each collecting electrode is divided into several regions which make it possible to use the chambers to explore the uniformity of the radiation fields used in the experiments (Fig. 21).

For accurate dosimetry, values of $\bar{w}$ must therefore be known. There were until recently no values of $\bar{w}$ published in the literature for high energy ions (a few hundred MeV/amu). Thomas and his colleagues have reviewed the literature and described measurements of $\bar{w}$ for 250 MeV/amu $^{12}C^{6+}$ ions, 375 meV/amu $^{20}Ne^{10+}$ ions and 479 MeV/amu $^{40}Ar^{18+}$ ions.[51,52]

Table 5 summarizes the values obtained. The absolute errors are $\partial\pm5$ percent and just large enough to allow the interpretation that there is no significant difference between one another, and even that they do not differ from the value used for electrons $\bar{w}_\beta$, of 34.6 ± 0.3 eV/ion pair.[53] If this interpretation of the data is chosen then the suggestion by Bakker and Segre that $\bar{w}$ for high velocity particles would tend to the value $\bar{w}_\beta$ (Ref. 54) is confirmed. This suggestion was, however, based upon the empirical evidence available at that time (1961). Theoretically, there are grounds for expecting the value of $\bar{w}$ to be dependent upon LET. As the LET of the ion increases, there will be a correspondingly greater energy density around the

Table 5. Values of w for Heavy Ions.

Ion Species	Energy (MeV/amu)	w (eV)
$^{12}C^{6+}$	250	36.3 ± 1.9
$^{20}Ne^{10+}$	375	35.3 ± 1.5
$^{40}Ar^{18+}$	479	34.6 ± 0.9

particle trajectory, which in turn increases the probability of ionic recombination, leading to higher values of $\bar{w}$. However, comparison of the published values of $\bar{w}$ shows this variation is probably not larger than about 10-15 percent.[55,56]

The data of Thomas et al.[52] do indeed show a trend of decreasing $\bar{w}$ with increasing charge state of the ion and increasing energy. Their accuracy, however, does not permit definitive conclusions and additional and more accurate measurements are needed.

The values of $\bar{w}$ obtained thus far give estimates of absorbed dose in tissue which are in good agreement with values obtained by other experimental techniques. Thus, Patrick et al.[13] have compared measurements of the entrance absorbed dose in soft tissue irradiated by 250 MeV/amu C^{6+} ions using a nitrogen filled ionization chamber and ^{7}LiF thermoluminescent dosimeters. Table 6 summarizes the results. The error for the ionization chamber measurements were ±5 percent. Only statistical errors are included for the TLD measurements in the table. The absolute uncertainty was about ±3 percent. The two sets of absorbed dose estimates are seen to be in good agreement.

Table 6.   Comparison of Entrance Absorbed Dose in Issue Irradiated by 250 MeV/u Ions

Irradiation No.	Group No.	Entrance Absorbed Dose in Tissue (red.) Ionization chamber[a]	(LiF) Thermoluminescent dosimeter[b]	Ratio ionization chamber to TLD
1	a	64.0 ± 3.2	61.2 ± 0.5	1.045
1	b	96.6 ± 4.8	93.6 ± 0.9	1.032
2	a	128 ± 6.4	124 ± 2	1.032
2	b	193 ± 10	186 ± 4	1.038

[a]Using a value W = 36.6 eV/ion pair. Errors ±5 percent.

[b]Using a value e = 0.89. Statistical errors only.

Table 7. Characteristics of TL Phosphors.

Characteristic	LiF	$Li_2B_4O_7$:Mn	CaF_2:Mn	CaF_2:nat	$CaSO_4$:Mn
Density (gm/cc)	2.64	2.3	3.18	3.18	2.61
Effective atomic no.	8.2	7.4	16.3	16.3	18.3
TL emission spectra(Å) range maximum	3500-6000 4000	5300-6300 6050	4400-6000 5000	3500-5000 3800	4500-6000 5000
Temperature of main TL glow peak	195°C	200°C	260°C	260°C	110°C
Efficiency at Cobalt-60 (relative to LiF)	1.0	0.3	3	23	70
Energy response without added filter (30 keV/Cobalt-60)	~1.25	0.9	13	13	10
Useful range	mR-10^5 R	mR-10^5 R	mR-3 x 10^5R	mR-10^4 R	R-10^4 R
Fading	small, 5 /12 wk	10 in first month	10 in first month	no detectable fading	50 -60 in the first 24 hrs
Light sensitivity	essentially none	essentially non	essentially none	yes	yes
Physical form	powder, extruded, Teflon-embedded, silicon-embedded, glass capillaries	powder, Teflon-embedded	powder, Teflon-embedded, hot-pressed chips, glass capillaries	special dosimeters	powder, Teflon-embedded

THERMOLUMINESCENT DOSIMETERS

Thermoluminescent dosimeters are now widely used as personal dosimeters,[57] for measurements of environmental radiation[58] and in radiological physics.[59]

The basic technique has been known for more than three hundred years and was first used to measure ionizing radiation by Weidmenn and Schmidt in 1895.[60]

In the past twenty years several phosphors have been developed but easily the most widely used is Lithium fluoride. Table 7 shows the characteristics of several thermoluminescent phosphors and Fig. 22 shows that relative response of several phosphors as a function of photon energy.

Lithium Fluoride Phosphors

It is the rather flat energy response of LiF that has made it so popular for photon dosimetry (the difference in response between 40 keV and 1 MeV is only 30 percent). However, LiF phosphors present

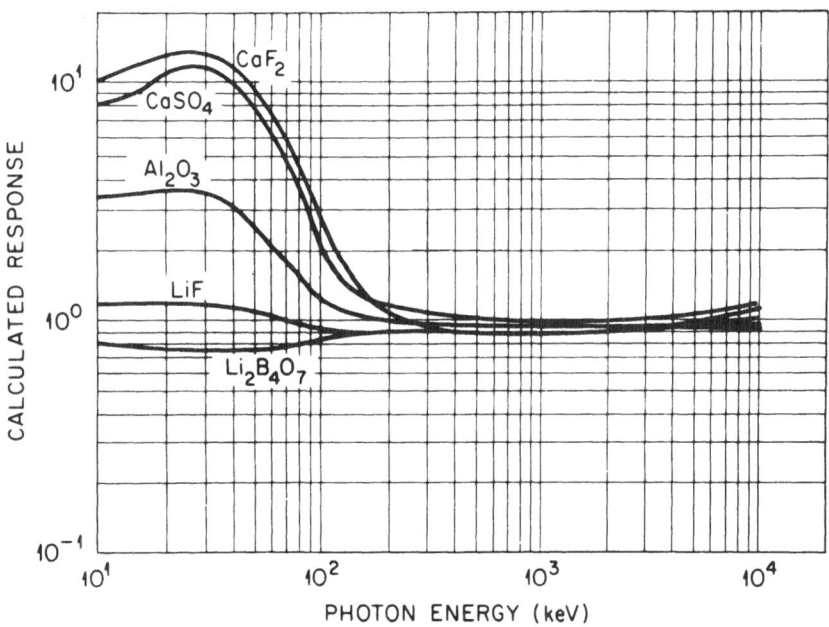

Figure 22. Theoretical sensitivity of several thermoluminescent phosphors.

some difficulties, demanding very careful technique if they are
to be used for accurate dosimetry.

The glow curve is complex and far from ideal, as may be seen
from Fig. 23 which shows a typical glow curve after the phosphor
has been annealed at 1 hr at 400°C, irradiated to 100 R, and
then immediately read.

The glow curve is complex showing 5 peaks. These peaks decay
at room temperature with approximate half-lives of 5 minutes,
10 hours, 0.5 years, 7 years, and 80 years. Peaks 4 and 5 are
the most suitable for dosimetry. Peaks 1 and 2 may be removed by
various combinations of pre- and post-irradiation annealing.
Post-irradiation annealing for 10 minutes at 100° is usual.

Annealing procedures can have a significant influence on the
sensitivity of LiF phosphors and carefully controlled technique
is required for consistency.[59] The light output from LiF
phosphors is linear up to an exposure of about 1000 R as may be
seen in Fig. 24 and the output is photon dose-rate independent up
to dose rates of at least 2×10^{11} rads s^{-1}. LiF in its pure
form exhibits little thermoluminescence, the phenomenon depending

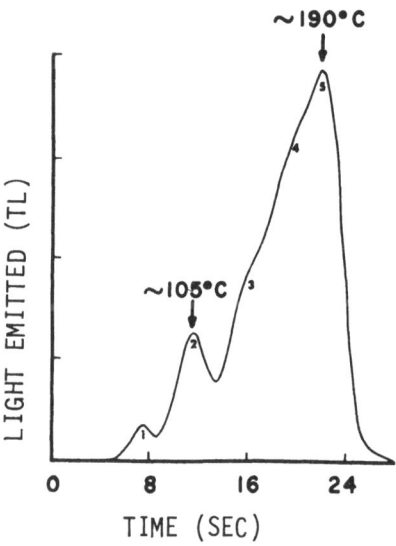

Figure 23. Typical glow curve of LiF
(TLD-100) after phosphor
phosphor has been annealed
1 hour at 400°C read soon
after irradiation to 100 R.

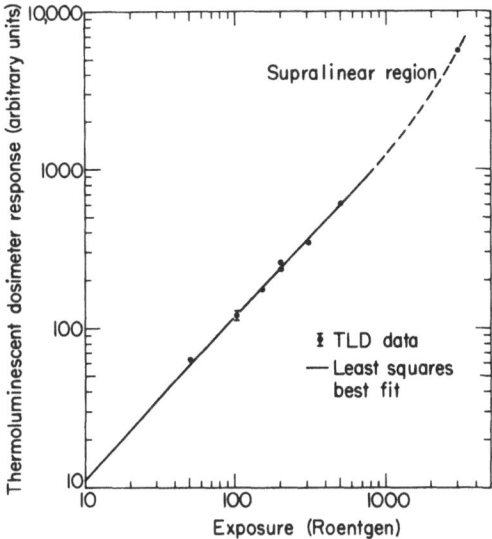

Figure 24. ^{7}LiF thermoluminescent
dosimeter response to ^{60}Co
γ-rays. One representative
error bar is indicated.

upon the correct mix of impurity ingredients. There is no
general agreement as to what these are but Mg, Eu, Y, Zr, and Ce
have been reorted in various descriptions of techniques of
production. (Harshaw manufactures LiF: Mg, Ti under the
trade-names TLD-100, 600 and 700, and the composition of these
phosphors is shown in the Table 8.)

The main dosimetry peak is around 200° which is favorable
for rapid heating. The emission spectrum from LiF lies in the

Table 8.   Isotopic Composition of Some
LiF Phosphorus.

Isotope	TLD-100	TLD-600	TLD-700
^{6}Li	7.5	95.6	0.01
^{7}Li	92.5	4.4	99.99

blue region enabling infra-red emission from the heater to be
eliminated by the use of a red/infra-red filter.

Neutron Detection

The high thermal neutron capture cross section of Li^6 has
led to the use of ^6LiF phosphors for neutron detection both in
personal dosimetry and to measure fast neutrons by surrounding
them with moderator—as for example, Bonner spheres.[62] Since
TLD 600 contains 95.6 percent ^6LiF, TLD 700 99.99 percent
^7LiF, a pair of dosimeters can be used to measure thermal
neutron exposures by difference.

Cameron et al.[60] found that TLD-100 gives the following
light output:

$$10^9 \text{ neutrons cm}^{-2}(\sim 1 \text{ rem}) \cong 37R \ ^{137}Cs \ \gamma\text{-rays}$$

On the basis of energy absorbed thermal neutrons produce ~1/7 the
amount of light produced by γ-rays.

Because of the high thermal neutron capture cross section of
^6Li some care must be taken in measurements using LiF phosphors
when neutrons are present. It should be noted that:

(a)   Traces of ^6Li in TLD-700 may give errors.

(b)   TLD-100 and TDL-600 have large capture cross sections
      for thermal neutrons. Samples must therefore be thin:

      (~0.1 mm of TLD-100 absorbs 5-10 percent incident
      thermal neutrons. ~0.1 mm of TLD-600 absorbs 50 percent
      of incident thermal neutrons.)

(c)   Fast neutrons may be moderated by the human body to
      produce thermal neutrons.

(d)   The glow curves for photons and thermal neutrons are
      slightly different.

(e)   Supralinearity is less for thermal neutrons than for
      photons.

Response of LiF Phosphors to Charged Particles

LiF phosphors are sensitive to electrons, protons alpha
particles and heavy ions. In general the light output per unit

of energy absorbed decreases with the LET of the incident radiation.

Jähnert[60] has reported measurement of the efficiency of the light output of TLD-700 (^{7}LiF) using protons from a few to 20 MeV and alpha-particles emitted by radioactive materials. He found decreasing efficiency with increasing LET.

Jähnert has proposed two alternative theoretical models (which differ in the number of electron traps assumed), both of which fit his experimental data quite well in the range of LET from 0.02 to 300 keV/ m, when the accuracy of the experimental data is considered. The one-trap theory seems to fit the data better and may be used to interpolate values of ε. Jahnert's work has been continued using energetic heavy ions by Thomas and his colleagues.[10-12]

The Bevalac is capable of producing heavy ions as heavy as Ar^{+18}, up to energies as high as 2.1 GeV/amu.[61] ^{7}LiF dosimeters were exposed to carbon, oxygen, neon and argon ions of energies shown in Table 9. The experimental technique used was to expose the dosimeter simultaneously with nuclear emulsions (Kodak type NTA), AgCl crystal detectors[62] or activation detectors to determine the particle fluence, ϕ. The absorbed dose in the irradiated dosimeters may then be calculated from their known stopping power, $(dE/dx)_{LiF}$, for the incident ions. Finally, the dosimeter response to heavy ions, L, is compared with the dosimeter response to ^{60}Co photons for an exposure of 1 R, τ.

Table 9. Measurements of ε.

Ion Speciex	Energy (MeV/amu)	(dE/dx) in ^{7}LiF (MeV g^{-1}cm^2)	
H^{+1}	798	1.89	1.08 ± 0.08
C^{+6}	252	116	0.89 ± 0.02
O^{+8}	300	112	0.90 ± 0.05
O^{+8}	1050	186	0.82 ± 0.05
Ne^{+10}	372	259	0.73 ± 0.03
A^{+18}	447	770	0.52 ± 0.02

It may then be readily shown that

$$\epsilon = \frac{5.05 \times 10^7}{\left(\dfrac{dE}{dx}\right)_{LiF}} \cdot \frac{L}{\phi} \qquad\qquad (4)$$

when L and τ are light outputs in arbitrary units, and

$$\left(\frac{dE}{dx}\right)_{LiF} \text{ is measured in MeV g}^{-1}\text{ cm}^{-2},$$

$$\phi \text{ is measured in particles cm}^{-2}.$$

Figure 25 summarizes the data of Jähnert[63] and Thomas et al.,[10-12] together with theoretical predictions.

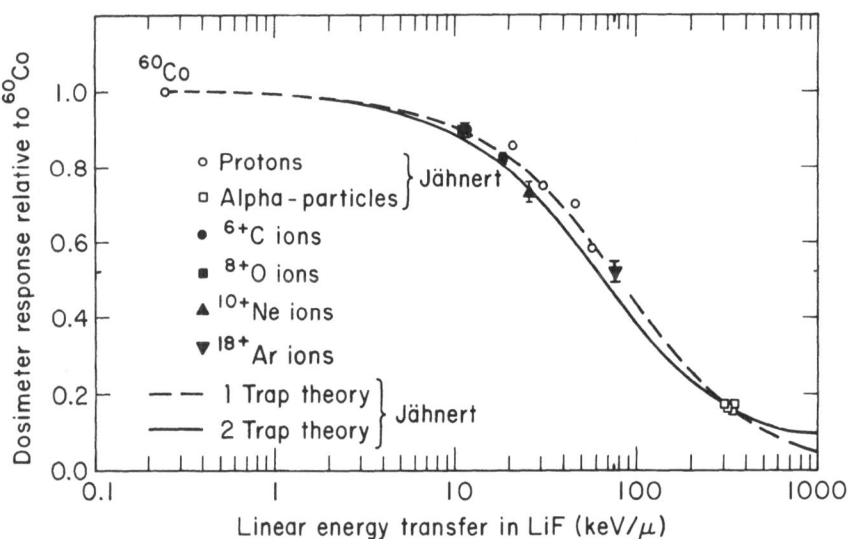

Figure 25. Measured values of ^{7}LiF TLD efficiency relative to ^{60}Co γ-rays and comparision with theoretical models.

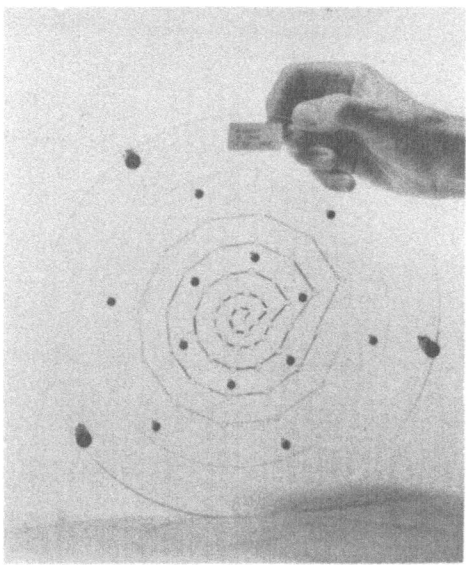

Figure 26. Lucite assembly used to determine spatial distribution
of absorbed dose in heavy ion irradiation fields.

These studies have enabled ^{7}LiF phosphors to be used for
accurate heavy ion dosimetry (few percent) but more fundamental work remains to be done. The light output as a
function of absorbed dose due to heavy ions can be inferred to
be linear from a few rads to about 500 rads by the work of Thomas
et al.,[10-12] but controlled experiments are needed. Dose rate
phenomena need to be investigated and details of the glow curves
as a function of ion species, absorbed dose and dose, rate must
be studied. Nevertheless LiF phosphors are extremely valuable
tools having a wide range of application in particle dosimetry.

Measurement of the Spatial Variations of Radiation Fields

^{7}LiF dosimeters are useful in field intensity mapping.
Smith et al.[8] have reported measurements using 0.125 in. x
0.125 in. x 0.035 in. mass ~25 mg phosphors, manufactured by the
Harshaw Chemical Company. Figure 26 shows a dosimeter assembly
that may be placed in the radiation field, while Fig. 27 shows
the spatial distribution of absorbed dose across a beam of 380
MeV/amu[10+] Ne ions.

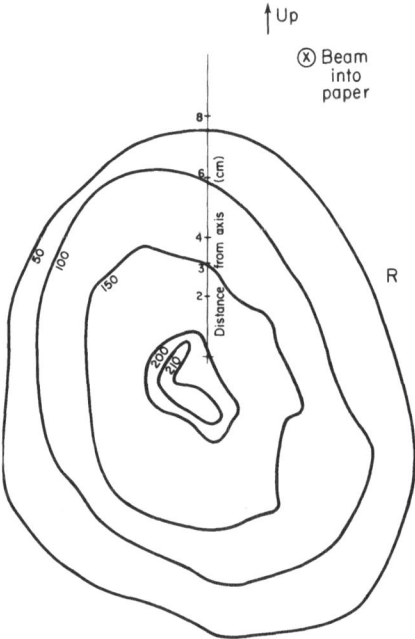

Figure 27. Spatial distribution of absorbed dose around a $^{10+}$Ne
         beam used to irradiate mice. Determined using ^{7}LiF
         thermoluminescent dosimeters.

Patrick et al.[13] report reasonable agreement between beam
profile measurements made using ^{7}LiF dosimeters and Du Pont
NDT-45 X-ray film, as may be seen by inspection of Fig. 28.

Absorbed Dose Distribution Studies

Thermoluminescent dosimeters are also extremely useful for
absorbed dose distribution studies. Patrick et al.[13] have
reported measurements of the spatial distribution of absorbed
dose in a lucite cylinder (used to simulate irradiated mice)
irradiated by a wide beam of 251 MeV/amu $^{6+}$C ions. Figure 29
shows both the dosimeter response and the calculated absorbed
dose distribution using energy-loss data of Steward et al.[63]

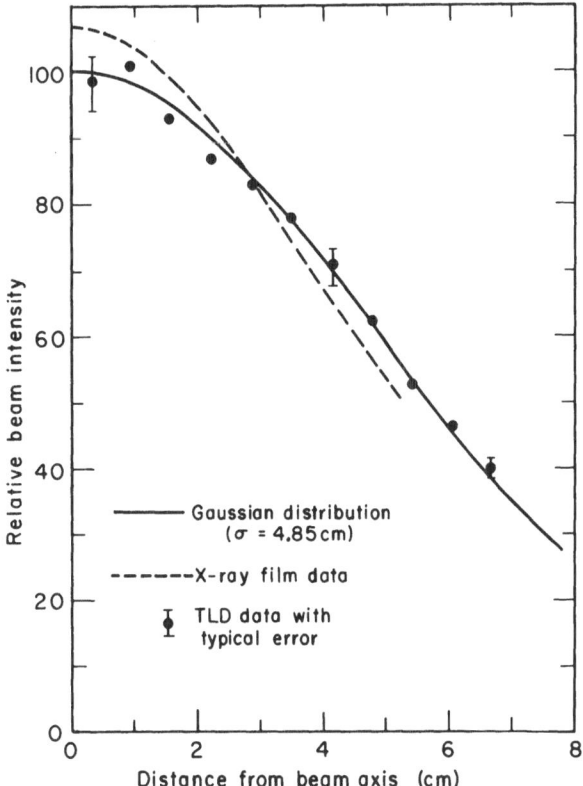

Figure 28.   Comparison of a carbon ion beam profile measured with
X-ray film and TLD.

Secondary particles resulting from primary particle interactions
are not taken into acount in the calculated curve. Comparison
between the calculated curve and dosimeter readings is not precisely possible because the dosimeter efficiency is a function of
LET and changes along the length of the phantom and these effects
must be taken into account for precision. However, the general
agreement between the two sets of data is gratifying.

ACKNOWLEDGEMENT

    This work was supported by the U. S. Department of Energy
under contract No. W07405-ENG-48.

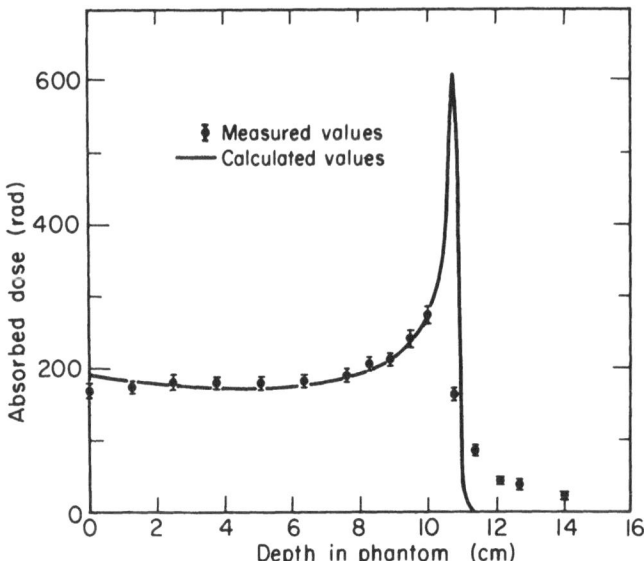

Figure 29. ^{7}LiF Thermoluminescent Dosimeter responsence as a
function of depth in a lucite phantom irradiated by a
broad beam of 251 MeV/amu $^{6+}$C ions. The measurements are compared with calculated values of the energy
deposition from ionization energy losses.

REFERENCES

1. C. F. Powell, 1959, "The Study of Elementary Particles by the
   Photographic Method," Pergamon Press, Oxford.
2. R. H. Thomas, 1972, High Energy Heavy Ions, Phys Bull. 23:
   143.
3. H. A. Grunder et al., 1971, Acceleration of Heavy Ions at the
   Bevatron Science, 174: 1128.
4. A. Citron, L. Hoffmann and C. Passow, 1961, Nucl. and Meth.
   14: 97.
5. R. H. Thomas (ed.), 1963, Report of the Shielding Conference
   held at the Rutherford Laboratory, Sept. 26-27th, 1962,
   Rutherford Laboratory Report NIRL/R/40.
6. H. Bindewald et al., 1965, Shielding Studies in Steel with 10
   and 20 GeV/e Protrons, Parts 1-V, Nucl. Inst. and Methods 32:
   45.
7. R. L. Childers, C. D. Zerby, C. M. Fisher and R. H. Thomas,
   1965, Measurement of the Nuclear Cascade at 10 GeV/c with
   Emulsions, Part III op. cit Ref. 6.

8. A. R. Smith, L. D. Stephens and R. H. Thomas, 1961, Dosimetry for Radiobiological Experiments, Health Physics, 32: 343 (1977).

9. A. R. Smith and R. H. Thomas, The Production of ^{11}C by the Interaction of 375 MeV/amu Ne^{10+} ions with carbon, Nucl. Inst and Methods. 137: 459.

10. J. W. Patrick, L. D. Stephens, R. H. Thomas and L. S. Kelly, 1975, The Efficiency of ^7LiF Thermoluminescent Dosimeters for Measuring 250 MeV/amu ^{6+}C ions, Health Physics 28: 614.

11. J. W. Patrick, L. D. Stephens, R. H. Thomas and L. S. Kelly, 1976, The Efficiency of ^7LiF Thermoluminescent Dosimeters to High-LET particles, relative to ^{60}Co γ-rays, Health Physics 30: 295.

12. A. M. Henson and R. H. Thomas, 1978, Measurement of the Efficiency of ^7LiF Thermoluminescent Dosimeters to Heavy Ions, Health Physics 34: 389 (1978).

13. J. W. Patrick, L. D. Stephens, R. H. Thomas and L. S. Kelly, 1975, The Design of an Experiment to Study Leukemogenisis in Mice Irradiated by Energetic Heavy Ions, Rad. Res. 64: 492.

14. L. S. Kelly, S. J. Daniels, and R. H. Thomas, 1977, Life Shortening and Leukemogenic Effects of a High-Energy Carbon-Ion Beam I: Irradiation of Young RF Mice, Lawrence Berkeley Laboratory internal report, HPN 92, September 1977 (unpublished).

15. L. S. Kelly and R. H. Thomas, 1977, Life Shortening and Leukemogenic Effects of a High-Energy Carbon-Ion Beam II: Irradiation of Middle-Aged RF Mice, Lawrence Berkeley Laboratory internal report, HPN 93, September 1977 (unpublished).

16. R. V. Griffith, D. E. Hankins, R. B. Gammage, L. Tommasino and R. V. Wheeler, 1979, Recent Developments in Personnel Neutron Dosimeters, A Review, Health Physics 36: 235.

17. K. Becker, 1973, "Solid State Dosimetry," CRC Press, Cleveland, OH.

18. J. Jasiak and T. Musialowicz, 1973, "Personnel Monitoring in Poland," in Proc. I.AE.A. Symp. on Neutron Monitoring for Radiation Protection Purposes, Vol. 1: 191, I.AE.A., Vienna.

19. K. Becker, 1973, Long Term Stability of Film, TLD and other Intergrating Dosimeters in Warm and Humid Climates, Oak Ridge National Laboratory Report ORNL-TM-4297.

20. P. N. Krishnamoorthy, G. Venkataramon, D. Singh and Dayashankar, 1973, "Comparison of Various Neutron Personnel Dosimeters" in Proc. Symp. Neutron Monitoring for Radiation Protection Purposes, Vol. II., p. 343, IAEA, Vienna.

21. A. Knight, 1974, Fading of Proton Recoil Tracks in Nuclear Emulsions, Health Physics, 27: 606.

22. A. M. Sayed and E. Piesch, 1974, Study of the Latent Fading of the NTA film and Track Etching Detectors at Various Temperatures and Humidities, Kernforschungszentrum, Karlsruhe, Internal Report KFK-2032.

23. R. H. Thomas, 1973, "Neutron Dosimetry at High-Energy
    Particle Accelerators in Neutron Monitoring for Radiation
    Protection Purposes," Vol. I, p. 327, IAEA, Vienna.
24. H. W. Patterson et al., 1972, An Evaluation of the Potential
    of Thick Nuclear Emulsions for Use as a High Energy Neutron
    Personal Dosimeter, Health Physics Group, Lawrence Berkeley
    Laboratory Internal Report No. 160.
25. R. H. Thomas, 1973, Personnel Neutron Dosimetry Studies at
    the Lawrence Berkeley Laboratory, Lawrence Berkeley
    Laboratory Report LBL-2057.
26. H. W. Patterson and R. H. Thomas, 1973, Radiation
    Measurements, in "Accelerator Health Physics," Academic
    Press, New York.
27. R. L. Lehmann and O. M. Fekula, 1964, Energy Spectra of Stray
    Neutrons from the Bevatron, Nucleonics 22 (No. 11): 35.
28. R. Remy, 1965, Neutron Spectroscopy by the Use of Nuclear
    Stars from 20-300 MeV (M. S. Thesis) Lawrence Berkeley
    Laboratory Report UCRL-16325.
29. H. W. Patterson, H. Heckmann and J. T. Routti, 1969, "New
    Measurements of Star Production in Nuclear Emulsions and
    Applications to High-Energy Neutron Spectroscopy," in Proc.
    Sec. Conference Acc. Dos. Stanford, California. Nov. 1969.
    USAEC Report - CONF-691101 p. 750.
30. W. S. Gilbert, et al., 1968, 1966 CERN-LRL-RHEL Shielding
    Experiment. Lawrence Berkeley Laboratory Report UCRL-17941.
31. A. H. Sullivan, 1969, "The Present Status of Instrumentation
    for Accelerator Health Physics," in Proc. 2nd Int. Conf. on
    Accelerator Dosimetry and Experience, Stanford, California,
    Nov. 1969, CONF-691101.
32. D. R. Perry, 1967, Neutron Dosimetry Methods and Experience
    on the 7 GeV Proton Synchrotron, Nimrod, in "Neutron Moni-
    toring, Neutron Monitoring," Proc. Symp. Vienna, 1966,
    p. 355, IAEA, Vienna.
33. A. R. Smith, 1965, Threshold Detector Applications to Neutron
    Spectroscopy at the Berkeley Accelerators, in "Proc. First
    Symp. Acc. Rad. Dos." Brookhaven National Laboratory USAEC
    Report, CONF-651109, p 224.
34. A. R. Smith, 1965, Some Experimental Shielding Studies at the
    Bevatron, op. cit. ref. 33, p 365.
35. V. Perez-Mendez, 1979, Instrumentation--Active Detectors
    Lecture No. 9 in "Advances in Radiation, Protection and
    Dosimetry in Medicine," Course Proceedings, International
    School of Radiation Damage and Protection, Ettore Majorana
    Centre for Scientific Culture, Erice, Italy.
36. R. H. Thomas, 1979, Radiation Monitoring, Lecture No. 26, in
    Radiation Protection and Dosimetry in Medicine," Course
    Proceedings, International School of Radiation Damage and
    Protection, Ettore Majorana Centre for Scientific Culture,
    Erice, Italy.

37. R. L. Bramblett, R. I. Ewing, T. W. Bonner, 1960, A New Type of Spectrometer, Nucl. Instrum. Meth. 9 (1960); 1.

38. H. W. Patterson and R. H. Thomas, 1971, Experimental Shielding Studies--A Review, Particle Accelerators 2: 77.

39. J. T. Routti, 1969, "Mathematical Considerations of Determining Neutron Spectra from Activation Measurements," in Proc. Second Int. Conf. on Acc. Dos. Stanford, USAEC Report CONF 691101, p 494.

40. L. D. Stephens, J. B. McCaslin, A. R. Smith, R. H. Thomas, J. E. Hewitt, L. Hughes, 1978, Ames Collaborative Study of Cosmic Ray Neutrons, II Low and Mid-latitude Flights, Lawrence Berkeley Laboratory Report LBL-6738.

41. J. B. McCaslin, A. R. Smith, L. D. Stephens, R. H. Thomas, T. M. Jenkins, G. T. Warren and J. W. Baum, 1977, An Inter-comparison of Dosimetry Techniques in Radiation Fields at High-Energy Accelerators, Health Physics 33: 611.

42. H. de Staebler, 1979, Similarity of Shielding Problems at Electron and Proton Accelerators. op cit Ref. 33, p 429.

43. W. N. Hess, H. W. Patterson, R. W. Wallace, E. L. Chupp, 1959, The Cosmic Ray Neutron Energy Spectrum, Phys. Rev. 116: 445.

44. T. W. Armstrong, K. C. Chandler and J. Barish, 1972, Calculation of the Neutron Spectrum in the Earth's Atmosphere Produced by Galactic Cosmic Rays, Neutron Physics Division Annual Rep., ORNL-4800, p. 63.

45. A. R. Smith, 1974, Activation Element Monitoring for Mouse Irradiations (Lola Kelly Bevalac Run, 11/28/74), Lawrence Berkeley Laboratory, Health Physics Department Internal Note, HPN 21.

46. A. R. Smith and H. Wollenberg, 1966, A Low Background Counting Enclosure, Health Physics 12: 53.

47. A. R. Smith, 1974, Distribution of Fluorine-18 and Na-24 in Thick Aluminum Stack Irradiated in 380 MeV/N Neon Beam, Lawrence Berkeley Laboratory, Health Physics Department Internal Note.

48. C. A. Tobias, A. Chatterjee and A. R. Smith, 1971, Radioactive Fragmentation of N^{7+} Ion Beam Observed in a Beryllium Target, Physics Letters, 37A, α2.

49. A. R. Smith, 1975, Lawrence Berkeley Laboratory, Private communication.

50. Private Communication.

51. L. D. Stephens, R. H. Thomas and L. S. Kelly, 1976, A Measurement of the Average Energy Required to Create An Ion Pair in Nitrogen by 250 MeV/amu C^{6+} Ions, Phip. Med. Biol. 21: 570.

52. R. H. Thomas, J. T. Lyman, and T. M. de Castro, 1979, A Measurement of the Average Energy Required to Create an Ion Pair in Nitrogen by High-Energy Ions, Lawrence Berkeley Laboratory Report, LBL-6710, Rev. 2.

53. G. N. Whyte, 1963, Energy per Ion Pair for Charge Particle in Gases, Radiat. Res. 18: 265-271.

54. C. J. Bakker and E. Segre, 1961, Stopping Power and Energy Loss for Ion Pair Production for 300 MeV Protons, Phys Rev. 81: 489-492.

55. M. N. Varma, J. W. Baum and A. V. Kuhner, 1977, Radial dose, LET and w for ^{16}O Ions in N_2 and Tissue Equivalent Gases. Radiat. Res. 70: 511-518.

56. M. N. Varma, J. W. Baum and A. V. Kuhner, 1975, Experimental Determination of w. for Oxygen Ions in Nitrogen, Phys. Med. Biol. 20: 955.

57. K. Becker, 1973, "Solid State Dosimetry," CRC Press, Cleveland.

58. G. de Planque and T. F. Gesell, 1979, Second International Intercomparison of Environmental Dosimeters. Health Physics, 36: 221.

59. J. R. Cameron, N. Suntharlingham and G. N. Kenney, 1968, "Thermoluminescent Dosimetry," University of Wisconsin Press, Madison.

60. B. Jähnert, 1972, The Response of TLD-700 Thermoluminescent Dosimetry to Protons and Alpha Particles, Health Physics, 23: 112-115.

61. H. Grunder, ed., 1973, Heavy Ion Facilities and Heavy Ion Research at Lawrence Berkeley Laboratory, Lawrence Berkeley Laboratory Report LBL-2090.

62. E. Schopper, et al., 1972, Review an AgCl Crystals as Visual Detectors of Nuclear Particles, in Proc. 8th Int. Conf. on Nuclear Photgraphy and Visual Detectors, Bucharest.

63. P. G. Stewart, 1968, Stopping Power and Range for Any Nucleus in the Specific Energy Interval 0.01 to 500 MeV/amu in Any Non-Gaseous Material, Lawrence Berkeley Laboratory Report, UCRL-18127.

NEW SOURCES OF RADIATION

Walter Schimmerling

Biology and Medicine Division, Lawrence Berkeley
Laboratory, Berkeley, CA 94720

INTRODUCTION

Man, in all cultures, has worshipped radiation. The most stirring words ever written are the words used in Genesis to describe the Creator focusing His Will that there be light. Now, approximately twenty billion years later, that initial flash of light has cooled to approximately $4^{\circ}K$, but still retains all of its mystery.

In the last one hundred years or so, our knowledge of radiation has increased considerably beyond the visible part of the electromagnetic spectrum discussed in Genesis. Means of producing new and different kinds of radiation have sprung forth from the ingenuity of scientists and engineers, and have been applied in elegant ways to the study of nature and to some of the most pressing problems of society. This process has been so intense and productive that very little remains that is "new" in the sense that it has not been proposed or studied—if not embodied already in an operating device. To that extent, therefore, there are very few "new" sources of radiation; as is well known, there is little, if anything, new under the sun. If this seems regrettable, you should take heart from another human endeavour, the institution of marriage, whose enduring charm it is, precisely, to visit endless renewal upon the known.

In that spirit, we shall consider as new not only the novelty of a device per se, but the novelty of its interaction with the world at large. "New sources," then, is to be understood as new sources brought to bear on old problems as well as old sources brought to bear on new problems.

It is impossible to cover the enormous range of endeavours that deal in the production and use of radiation within a single lecture, or even a single course. Indeed, to encompass radiation and its entwinement in the fabric of our civilization would require several careers!

Most of us here will be interested in one particular aspect of radiation: its interaction with matter. It is that knowledge which we wish to expand in order to heal, to diagnose, and to predict, prevent, and assess damage. It should be remembered, however, that the importance of any given source for any given application cannot be foretold. Until lasers became almost household appliances, visible light was not a major hazard; until recently, radiologists operating an x-ray machine in their office did not have to worry about modulation transfer functions for CT scanners, and most physicists innocently thought that pi mesons were nothing but the carriers of nuclear force.

The present discussion is an attempt to select examples of radiation sources whose application may make new or unconventional demands on radiation protection and dosimetry. A substantial body of knowledge about high energy facilities exists and, partly for this reason, the great high energy accelerators will be mentioned only briefly. The textbook by Patterson and Thomas (1973) is recommended for those interested in further details. In addition, many excellent and complete descriptions of the new high energy physics facilities have been published and are easily available to the interested student (Cole and Donaldson, 1977; Hendrickson, 1979).

SOURCES AND SOURCE CHARACTERISTICS

General Features

The sources of radiation to which we shall refer are mostly based on accelerators. The radiation produced still consists, to a large extent, of the familiar charged particles: electrons and protons, as well as neutrons. However, there are several noteworthy developments that will become apparent in the course of this discussion, and attention is called to them here:

1. Accelerators have changed, and can no longer be conceived as a single machine speeding protons or electrons from a simple ion source to a hole in the wall, with perhaps a couple of quadrupoles and a bending magnet thrown in for good measure. New particle accelerators are complex <u>systems</u> of accelerators, beam transport and beam storage elements, each of which performs a specialized function in a carefully optimized region of phase

space. The name of the facility often is only a reflection of its most important application. Thus, a "storage ring" or a "synchrotron radiation facility" may both be based on the same accelerator system, even though emphasis is given to the main purpose of application.

The engineering insouciance associated with this drastic change in the scale of operations is not restricted to the more exotic high energy physics machines. For example, the University of Western Ontario, in Canada, has built a variable energy racetrack microtron for therapy that includes a full-fledged electron linac instead of an accelerating cavity. In a later design, called a "shuttle microtron," electrons are accelerated back and forth along a linac, and steered back into the accelerator by an appropriately shaped magnetic field (Froelich et al., 1973; 1977). (The design of these machines was motivated by the search for small, inexpensive, variable energy machines for radiation therapy in the 30-MeV region.) Similarly, new designs for megavolt electron microscopes have discarded the old-fashioned electron gun, and use symmetrical cascade generators, i.e., an electron accelerator (Reinhold and Gleyvod, 1973).

2. The above developments have come about, to a large extent, as a consequence of the advances in the theoretical understanding of the physics of particle beams. This has made possible acceleration cycles where antiprotons, made on a tungsten target with 80 GeV/c protons at Fermilab, will be collected for injection into the booster synchrotron and decelerated to 200 MeV, for transfer into a storage ring and electron cooling (i.e., reduction in the spread of transverse velocities) before being accelerated to 400 GeV. At the high beam currents necessary for storage rings, when the particles can no longer be treated as approximately independent, the theory of these machines overlaps considerably with the physics of plasmas. Even at lower intensities, one could reasonably ask whether, for example, the shuttle microtron is not really a magnetic mirror machine. Indeed, a recent textbook on charged particle beams provides such a unified treatment of ion sources, accelerator beams, and plasmas (Lawson, 1977).

3. A further development that has played an important role in the design, operation, and use of the new sources of radiation has been the availability of high-speed computers. These have contributed to advances in the theoretical understanding of particle beams and plasmas by making sophisticated calculations possible. Computers provide fast and extremely complex control functions and data acquisition and analysis are unthinkable without them. The same is becoming true of radiation therapy

and diagnosis. The availability of microprocessors is expected
to have a similarly revolutionary effect on the field.

4. Superconductivity is quickly becoming an established
technology. Projects under way to achieve the highest beam
energies, the Energy Doubler at Fermilab and the ISABELLE
colliding beam facility at Brookhaven, are based on superconducting magnets. The quantities involved (e.g., 516 dipoles
and 372 quadrupoles in the case of ISABELLE) are already on an
industrial scale. At the lowest beam energies, a superconducting storage ring has been built for very cold (10^{-6} eV)
neutrons, using the "ultracold" neutron beam of the High Flux
Reactor at the Institut Laue-Langevin in Grenoble (Kugler
et al., 1979). This machine achieves beam bending by coupling
the 3.5 T guiding magnetic field to the magnetic moment rather
than the (inexistent) charge of the neutron. For this reason,
one order higher multipole magnetic fields are required than for
electrically charged particles, and quadrupole magnets must be
used for bending while sextupole magnets are used for focusing.
Thus, super-conductivity may be expected to become as much a
part of new radiation sources and their applications as, e.g.,
RF engineering.

5. Secondary radiations (e.g., neutrons, synchrotron
radiation) were often considered a nuisance in the past because
they interfered with experiments and required extensive shielding. They have now become the source of some of the most interesting applications. It is perhaps a reflection of the
Zeitgeist that the recycling of "waste" radiation has become one
of our more productive efforts. Intense pulsed neutron sources
are at various stages of planning or operation in Canada, Great
Britain, Japan, the U.S., and the U.S.S.R., mainly based on
accelerators. Use of these sources has become one of the most
general experimental methods in condensed matter research,
yielding information that, in many cases, is inaccessible by any
other technique. Applications spanning biology, chemistry,
physics, and materials research are constantly increasing. The
use of neutrons from a high energy proton linear accelerator
incident on a molten-lead target to produce fissile fuel from a
surrounding blanket of U-238 or Th-232 has also been proposed in
the accelerator breeder concept (Steinberg et al., 1977). This
idea has the notable advantage that, if depleted fuel elements
are irradiated, reprocessing steps (and the concurrent risks of
diversion) are minimized. The most spectacular use of "waste"
radiation is, of course, that of synchrotron radiation, to be
discussed in somewhat more extent below. Here, it should be
pointed out that the design of the latest storage rings, such as
PEP, actually requires synchrotron radiation as a mechanism for
damping undesirable oscillations.

6. In the case of charged particles, the <u>charge state</u> has become an increasingly important parameter that requires attention but is also a means for great design flexibility. Negative ion sources, especially H⁻, are more and more common at high energy accelerators and provide the energy resolution necessary to study nuclear energy levels. These sources have long been an intrinsic part of tandem accelerators, which now play such a prominent role in the new generation of heavy ion machines. An understanding of charge exchange is also vital for the improvement of neutral beam injectors in magnetic fusion devices. These are, in a sense, <u>neutral beam</u> accelerators. If this seems odd, the neutron storage ring discussed above is a similar example, pointing out that current advances require great care in applying conventional thinking to new sources of radiation.

Source Parameters

A new source of radiation, to be new, must have a quality of excess to meet the name; it must do something, at least, better than any other device. Whether this requirement, that a machine give evidence of miracles before it is technologically canonized, is a psychological quirk or not, it is based on the reasonable need for certain desirable design characteristics.

These design parameters arise because there are time and space scales associated with the systems with which the radiation interacts. In addition, there is also a "truth scale," which determines the significance of the interaction, and is usually referred to as "statistics." More appropriately, the information content to be derived from the interaction is also called the "signal-to-noise ratio."

The spatial extent of the interacting system determines the necessary energy of the radiation. At the quantum-mechanical level, the wavelength of the radiation must be comparable to the dimensions of the structure being studied, whether a quark or a crystal, and this specifies the energy or momentum. Macroscopically, the range of heavy charged particles in matter is determined by their energy.

Fluorescence decay or charge collection times in detectors influence the desired time scale of the beam, as does the doserate dependence of biological systems and the immobilization time of a patient.

Most of the effects due to radiation have a small probability of occurrence. In order to measure the effect reliably, it is necessary to have a large flux of radiation or a large number of detectors, or both. The large signal-to-noise ratio of life is sustained by nature, using solar energy and a great number of detectors—also known as "plants." Clinical trials are a similar means of achieving high information content.

Radiation generally comes in beams that are not parallel. The brightness is a measure of the source flux density per unit solid angle. It is inversely proportional to the square of the emittance (Lawson, 1977), an all-important quantity describing the extent ("beam spot") and divergence of a beam, as well as its momentum spread and relative timing. The emittance contains all the information about the beam and, accordingly, the beam entropy can be defined as the logarithm of the emittance in units of the area of a phase-space cell. In the case of colliding beams, the intensity-related quantity is called luminosity, and is proportional to the product of particle densities in each beam and the interaction volume. One of the most important consequences of the advances in accelerator theory has been the design of accelerator optics capable of focusing a maximum intensity of particles into the volumes compatible with required source dimensions.

APPLICATIONS AND EXAMPLES

A limited sampling of applications of both old and new radiation sources is given in Table 1, under headings that may seem exaggerated only upon a first examination. In the remainder of this lecture, we shall comment briefly upon a few selected examples.

In the life-and-death category, the greatest impact on medicine can be expected from meson factories and heavy-ion accelerators, which may well end up being complementary rather than competing modalities. The possible therapeutic advantage of these two types of radiation derives from their dose distribution in matter and the high rate of energy deposition (LET) in a selectable depth of material.

It should be emphasized that the direct benefits from the application of these types of radiation to therapy are not the only medical application and may not even be the most important one. Technology does not progress in isolation, and the development of meson factories has already resulted in the incorporation of the side-coupled electron linac into most clinical units used in the United States (Rosen, 1971). Radioactive secondaries from nuclear interactions of heavy ions have been refocused into radioactive beams at the BEVALAC (Alonso et al., 1979), and implanted noninvasively in test animals. The usefulness of beams of radioactive iodine, gallium or technetium,

Table 1. Some Applications of Old and New Radiation Sources

---

1. Life and Death

> Radiation therapy
> Radiation biology and biochemistry
> Radiology
> Radiopharmaceuticals
> Radioisotope implantation
> X-ray diffraction (synchrotron radiation)
> Neutron diffraction
> Electron microscopy

2. War and Peace
> Weapons neutron research

3. Energy
> Inertial fusion
> Neutral beam injection
> Magnetic confinement fusion
> Well-logging
> Ion implantation (solar cells)
> Spallation breeder

4. History and Origin of the Universe
> Simulation of big bang with heavy ions
> Radioisotope dating
> Cosmic rays
> Nucleosynthesis

5. The Fundamental Laws of Nature
> Nuclear physics
> High energy physics
> Radiation chemistry
> Nuclear chemistry

6. Technology and Civilization
> Ion implantation
> Paint curing
> Microlithography
> Analysis of materials
>> Neutron activation
>> Induced x-ray emission
>> Backscattering
> Wear and corrosion studies
> Criminology analysis
> Crystallography

---

that can be made to stop inside any desired organ without the need to inject voluminous pharmaceuticals and circumvent the blood-brain barrier, can be easily visualized. Finally, the sophisticated diagnostic and therapy techniques required to take full advantage of these facilities are likely to have a revolutionary impact on medicine as a whole.

Some of the salient features of the meson factories are summarized in Table 2. These machines have been built primarily for physics research, and hence the emphasis on duty factor, variable energy and energy resolution, which will allow very precise studies of effects associated with nuclear energy levels.

The use of H^- beams is the reason for the recently achieved energy resolution of TRIUMF, as well as for some of the problems that this facility had to solve. The binding energy of the electron in H^- is only 0.75 eV, so that any collision, even the slightest, will remove this electron and the residual hydrogen atom will be lost from the beam during acceleration. In the rest frame of the H^-, however, the magnetic guiding field B appears as an electric field of strength 0.3 $\beta\gamma B$. Therefore, the maximum field that allows for an H^- lifetime comparable to the acceleration cycle is ~5 kG, leading to a much larger machine.

The advantage of this sensitivity of H^- to collisions is that, in a knife-edge, the neutral H atom traversing a very small thickness of material will emerge, while at greater thicknesses it is stripped to H^+ and bent away. Thus, beams with very small radial emittance (correspondingly well-defined in energy with respect to the acceleration cycle) can be produced. Such microbeams may also be of great interest for possible applications to biology and materials science.

Figure 1 is a schematic of the Los Alamos beam areas. Note the large fraction of beams that are dedicated to applications. It is interesting to note that the Weapons Neutron Research Area is not intended to serve aggressive purposes. In fact, the director of the facility has argued very eloquently that the availability of such facilities to the major powers is an important factor in achieving a comprehensive test ban treaty. Figure 2 is a picture of the Swiss Institute of Nuclear Research machine. The ring cyclotron is another instance of the imaginativeness of modern accelerator designers, where the conventional distinction between a cyclotron and a synchrotron has become slightly blurred. Figure 3 is a schematic diagram of the TRIUMF beam lines. Here, the thermal neutron facility, which provides thermal neutron fluxes of 10^{-12} $cm^{-2}s^{-1}$ is not only competitive with a nuclear reactor, but actually compensates for the lack of one in Western Canada.

Table 2. Meson Factories

Laboratory	Accelerator Type	Maximum Proton Energy (MeV)	Design Current (μA)	Maximum Achieved Current (μA)	Duty Factor	Comments
Clinton P. Anderson Meson Physics Facility, Los Alamos USA	Proton linac (800 m long)	800	1000	500	6 %	Full energy in 1972. Can accelerate H$^+$ and H$^-$ simult. Cockroft-Walton injectors, drift-tube linac to 100 MeV, side coupled linac to 800 MeV. $\Delta p/p \simeq 0.25\%$.
Swiss Institute for Nuclear Research, SIN, Villigen, Switzerland	Ring cyclotron	590	100	112	100 %	Full energy, Jan. 1974. 72 MeV sector focused cyclotron injector; separated 8-sector cyclotron with 4 RF cavities to 590 MeV. $\Delta E/E \simeq 0.07\%$.
TRIUMF, Vancouver, Canada	Sector-focused H$^-$ cyclotron	500	100 (500 MeV) 300 (450 MeV)	100 (at 1 % duty factor)	100 %	Full energy Dec. 1974. Large radius (310 m), 4000-ton magnet to keep H$^-$ together. $\Delta E/E \simeq 0.01\%$ (173±12) keV at 200 MeV. Variable energy 180-520 MeV.
Institute for Nuclear Research Moscow, USSR	Proton linac	600	500	—	1 %	Under construction. Low duty factor for high instantaneous intensity (e.g., neutrino experiments)

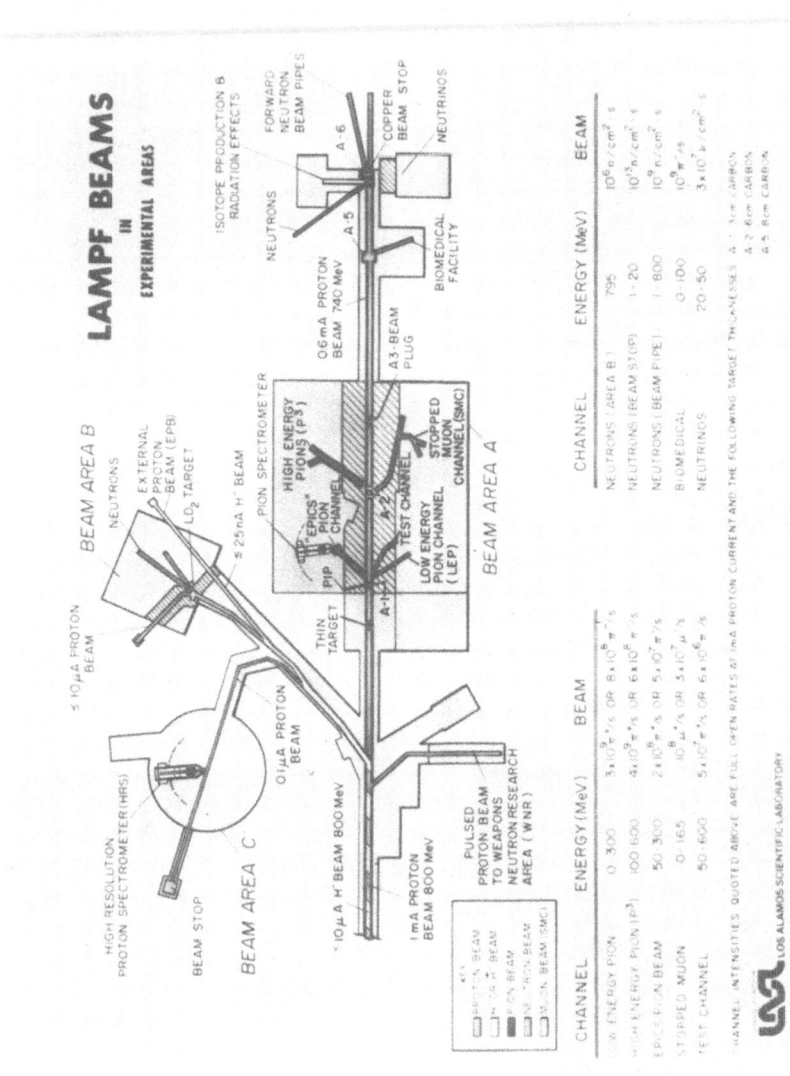

Fig. 1. Schematic diagram of beam areas at the Los Alamos Scientific Laboratory Clinton P. Anderson Meson Physics Facility. Courtesy of J. DiCello Los Alamos Scientific Laboratory.

Fig. 2. A view of the ring cyclotron at the swiss Institute of
Nuclear Research. Courtesy of Dr. Hans Blattman, SIN.

Heavy ions, i.e., beams of atomic nuclei from helium to
uranium, are potentially the most versatile new sources of radiation. They are currently undergoing clinical trials at the
BEVALAC in Berkeley (U.S.) together with an extensive program in
radiation biology and chemistry. The use of radioactive beams
has already been mentioned, and radiography with heavy ions is
also being actively studied. In the energy category (Table 1),
heavy-ion beams are a promising contender for inertial fusion,
and various studies are also being pursued in that direction.
Ion implanation by low energy beams of boron is now a known
technology, and holds some promise in developing solar cells with
efficiencies that may make direct solar energy conversion econom-
ically competitive. Figure 4 is a view of the Mark I device
developed by the Western Electric Company in the U.S. (Rodde
et al., 1974). This 300 kV ion implantation device produced maximum currents of 60 μA $^{11}B^+$, 90 μA $^{31}p^+$, and 110 μA of
N_2 , and has since been replaced by more advanced models giving
throughputs of 200 two-inch wafers/hour for doses up to
$2 \times 10^{14}/cm^2$.

Deductions about the origin and the confinement time of
cosmic rays in the galaxy depend upon a knowledge of heavy ion

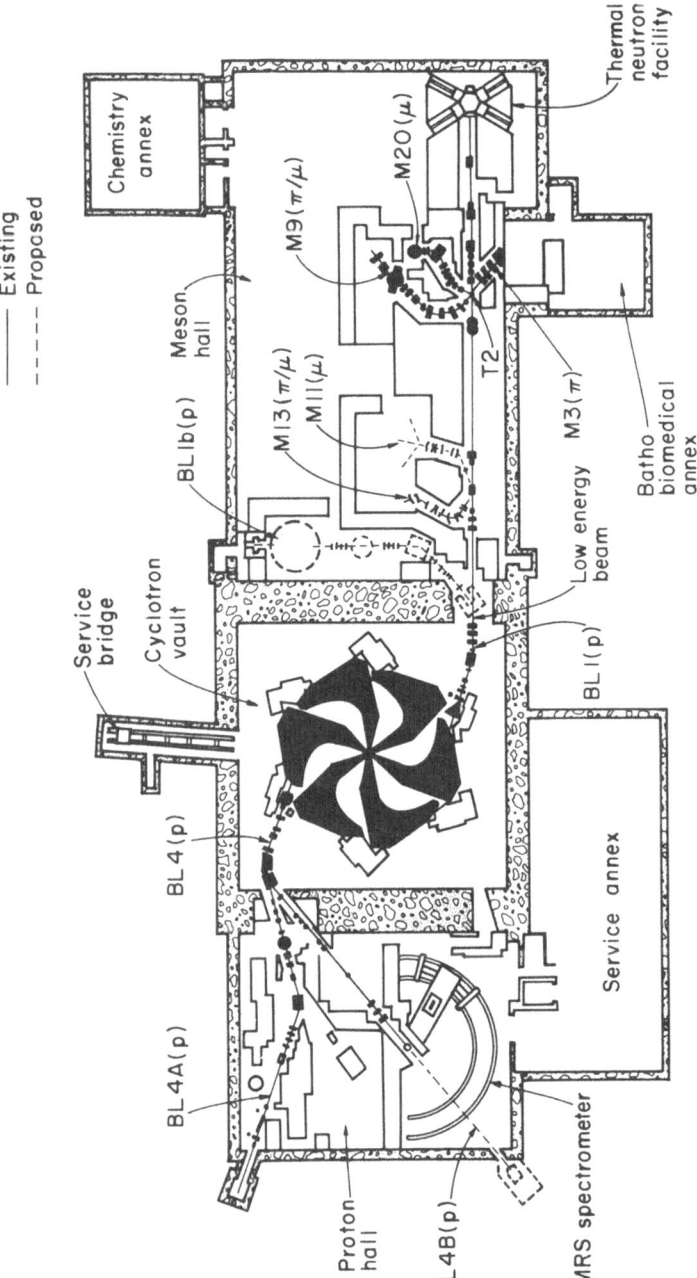

Fig. 3. Schematic diagram of beam areas at the TRIUMF facility. Courtesy of TRIUMF Laboratories.

Fig. 4. Mark I ion implantation device developed by the
        Western Electric Company. Courtesy of Western
        Electric.

cross sections and their role in depleting the observed cosmic
ray fluxes in the interstellar material. On a shorter time
scale, the use of heavy-ion accelerators, such as the 88-inch
Cyclotron at Berkeley, as a mass spectrometer for radioisotope
dating has opened up an entirely new field of research (Muller,
1977). Finally, heavy-ion beams are being used in studies of
nuclear matter, where entirely new phenomena, such as pion
condensation and the formation of quark matter, have been
predicted for velocities of the incident heavy nucleus sufficiently high to compress the target nucleus to several times its
normal density.

As a consequence of this, there are more than sixty proposed
and existing heavy-ion facilities in the world at present. Most
of these projects are for heavy-ion machines with energies below
approximately 100 MeV/A (Ball, 1977), and an excellent recent
review of the field may be consulted for further details
(Grunder and Selph, 1977). The energy per nucleon to be
achieved at the planned facilities is plotted as a function of
atomic mass in Fig. 5 for the low-energy facilities. Of

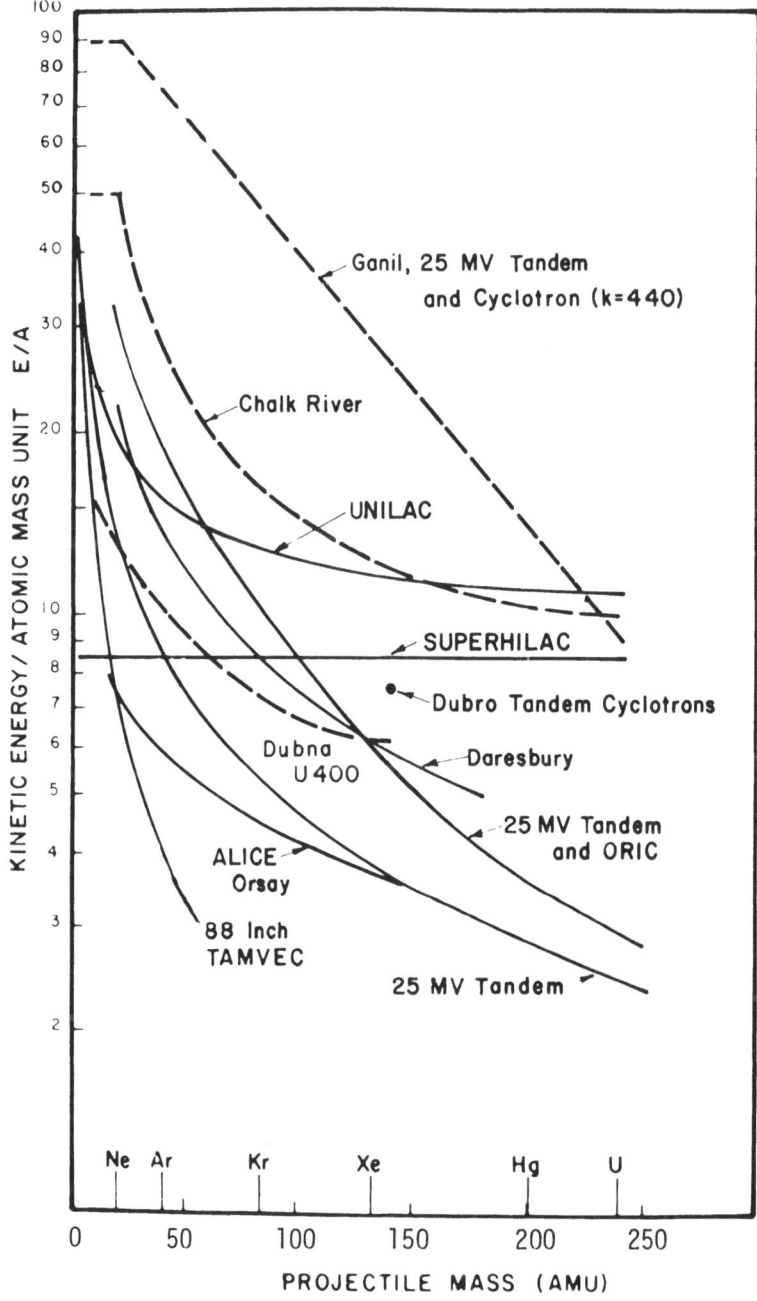

Fig. 5. Energy per atomic mass unit as a function of projectile mass for low-energy heavy ion facilities. Courtesy of J. Ball, Oak Ridge National Laboratory.

these, the recently completed Phase I at Oak Ridge is a good typical example. It consists of a "super-tandem," a 25 MV Pelletron (Herb, 1971) constructed by the National Electrostatics Corporation, and the Oak Ridge Isochronous Cyclotron (ORIC), a combination known as the Holifield Heavy Ion Research Facility (HHIRF). This machine uses a charging chain of metal cylinders, rather than a belt, as shown in Fig. 6. A schematic of the HHIRF machine is shown in Fig. 7. It constitutes a major departure from the traditional tandem configuration in that the

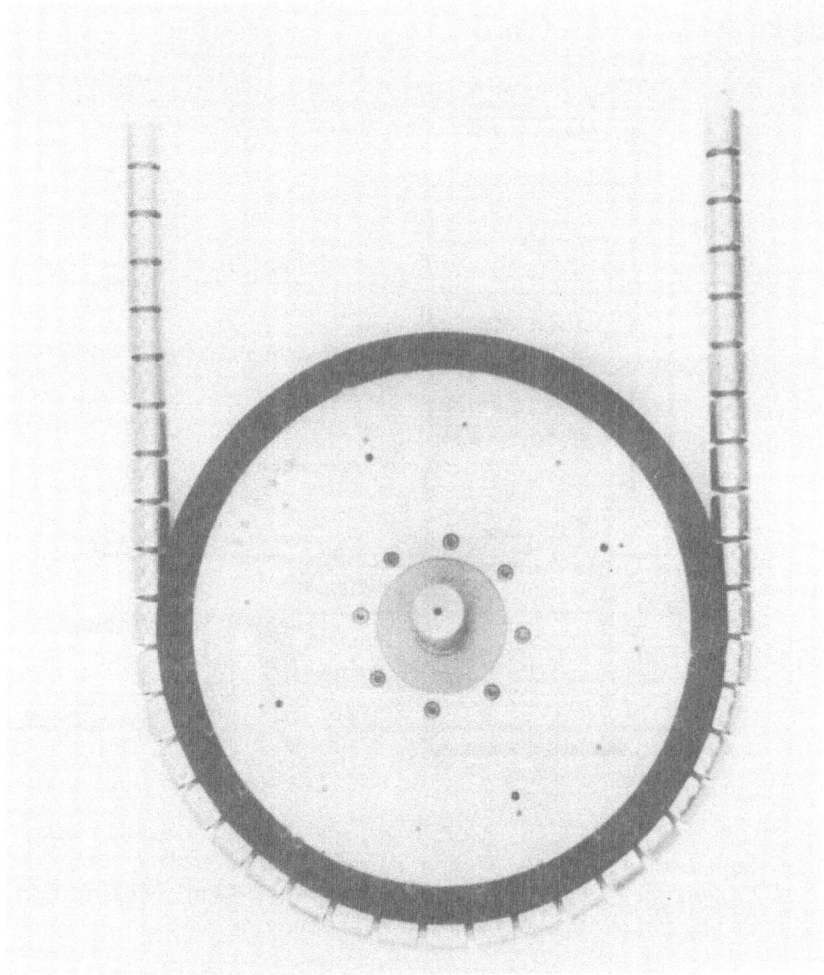

Fig. 6. Charging chain of metal cylinders used
in the Pelletron electrostatic accelerator. Courtesy of G. Norton, NEC.

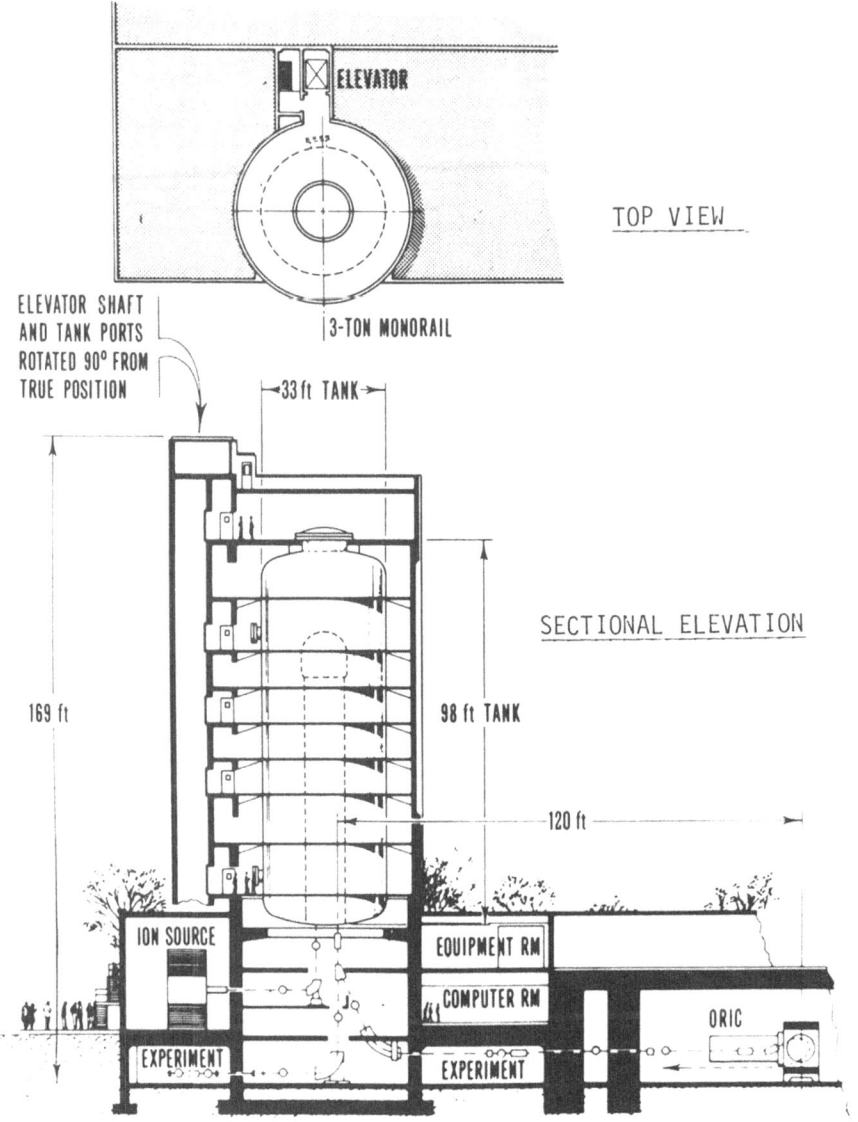

Fig. 7. Schematic diagram of the Holifield Heavy Ion
         Research Facility in Oak Ridge, TN. Courtesy of J.
         Ball, Oak Ridge National Laboratory.

accelerator has a "folded" structure, with both low and high energy acceleration tubes in the same column. Negative ions are accelerated upward into the HV terminal, stripped to positive ions, and accelerated downward to the final potential. The pressure vessel is 30 m high and 10 m diameter, making such an arrangement possible. At 7 atm, it requires 100,000 kg of SF6. Beam energies will be 25 MeV/A for light ions and 6 MeV/A for A up to 160. A new booster cyclotron to raise this energy to 500 q^2A^2 is planned in Phase II of the design.

The use of heavy-ion accelerators in the region just around 100 MeV/A is being seriously considered for driving inertial confinement fusion (Godlove, 1979). This approach to fusion aims at compressing a deuterium–tritium pellet by a factor of 10^4 by implosion with a short (~10 nsec) pulse of radiation, in order to obtain the high plasma densities required for ignition. The use of lasers and light charged particles, also pursued vigorously at this time, may result in preheating (and consequent expansion) of the pellet due to Bremsstrahlung. Heavy ions are attractive because they do not present this problem, and also because the high energy of the particles and their large stopping power reduce the peak current requirements from megamperes to kiloamperes. Table 3 shows some of the characteristics that such a driver might have. Current thinking envisions a three-stage program, consisting of an Accelerator Demonstration Facility (ADF) to perform the necessary research and development, followed by a Heavy Ion Demonstration Experiment (HIDE) and a final stage for initial studies of reactor design, an Engineering Test Facility (ETF).

The prospect of heavy-ion fusion, as well as many other applications depending on intense, high energy pulsed beams, are closely related to progress in pulsed power technology. One of the more significant concepts in this regard is that of the linear induction accelerator (Faltens et al., 1977; Leiss, 1979). A possible configuration is shown schematically in Fig. 8. In this configuration, an electromagnetic pulse produced by a switched high-voltage generator is used to accelerate the beam through the cavity. In other configurations, known as "core-type," a rapidly changing magnetic flux is used to accelerate the charged beam. When a large number of such independently phased modules are threaded by a charged particle beam, they can be thought of as a linear betatron. Such linacs have been built and operated (at lower power levels than required for fusion) for many years, with great reliability. The modular construction makes the induction linac attractive because it places relatively modest demands on each module, which results in greater reliability and lower cost. Approximately 10^4 such modules are envisioned in a 5-km long accelerator for a power plant igniter system.

Table 3. Heavy-Ion Driver Characteristics

("Uranium" Ions, Charge = +1 to +4)

Beam Energy	1 MJ
Beam Power	100 TW
Kinetic Energy	5 to 25 GeV
Range	0.1 to 1 gm cm^{-2}
Specific Energy	20 to 100 MJ/gm
Target Radius	1 to 5 mm
No. of Beams or Clusters	2 to 4
Beams/Cluster	1 to 5
Current/Beam	2 to 7 kA

Physics

- Energy deposition profile understood (classical)
- Beam propagation focussing tractable

Technology
- Mature
- Techniques for high current exist but need demonstration

All fusion reactors, of whatever type, should produce significant numbers of neutrons. What may not be immediately apparent is that most of the magnetic confinement experiments will use high power neutral beam injectors to heat the plasma (Kunkel, 1979), and that these "neutral" beam injectors are themselves sources of substantial fluxes of neutrons (Berkner et al., 1979).

A schematic of a typical neutral beam injection system is shown in Fig. 9. The most critical item in these systems is the ion source, which has to supply tens of amperes of ions more or less continuously, so that well-collimated beams can be formed in simple electrostatic accelerating structures. These are a

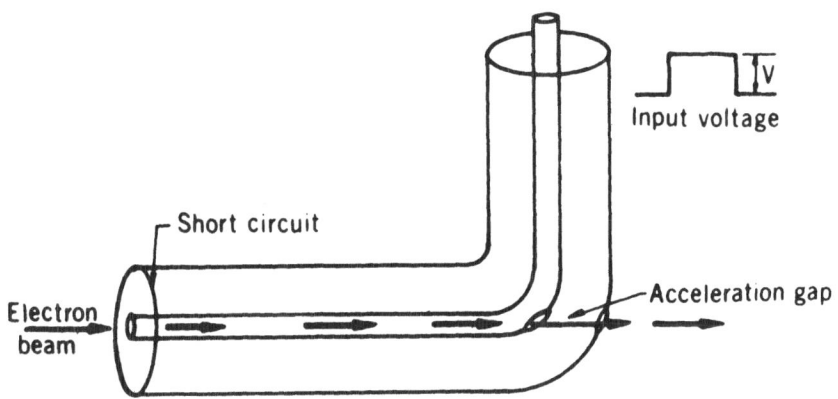

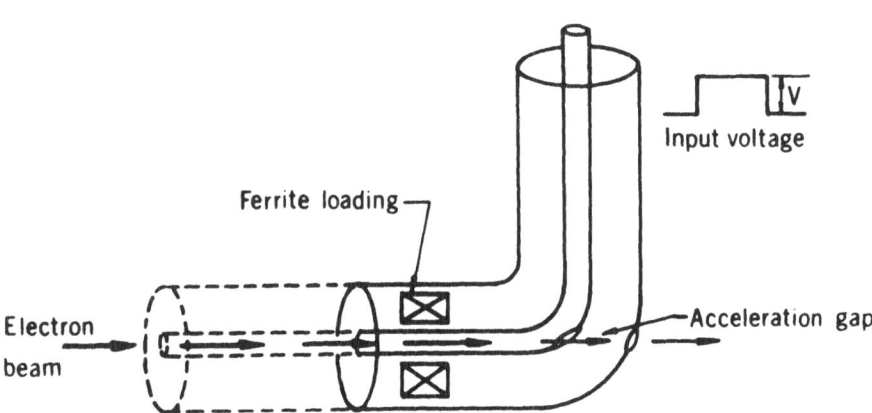

Fig. 8.   Evolution of induction accelerating cavity.   Courtesy
          of D. Keefe, LBL.

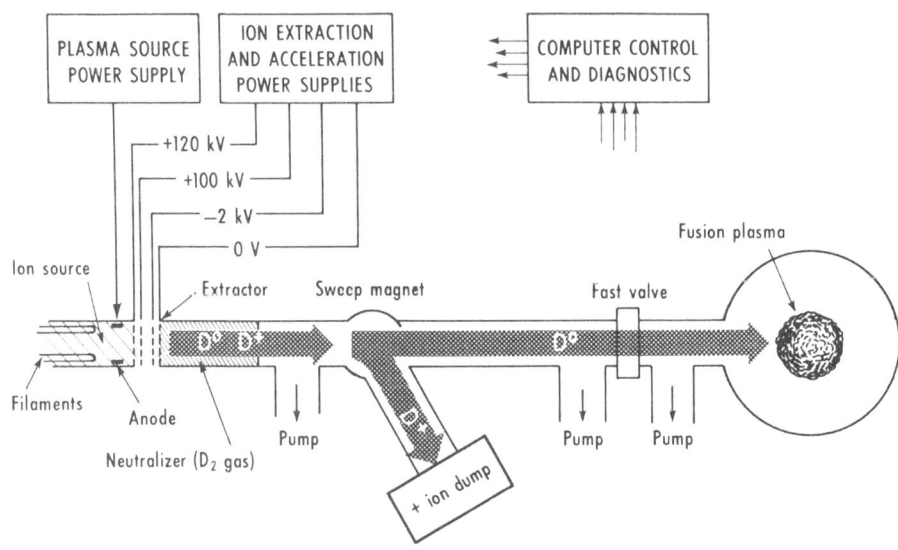

Fig. 9. Schematic diagram of a typical neutral beam injection
         system for plasma heating. Courtesy of R. Pyle, LBL.

set of grids with aligned apertures in the case of the Berkeley
systems. Positive ions are currently used and neutralized in a
gas. The efficiency for producing neutral ions is rather small,
and efforts to produce negative ion sources with the required
intensities are under way.

The large amounts of neutralizer gas must be pumped out to
avoid reionizing the neutral beam emerging from the sweep
magnet. Accordingly, the Berkeley facility, shown in the photograph of Fig. 10, has a large (170,000ℓ) vacuum system. The
spherical chamber seen in the photograph is part of this system,
serving to expand and lower the pressure of residual gas. This
facility has produced 1 MW of power at 120 keV energy (Berkner
et al., 1977). Four such beam lines are envisioned for the
Tokamak Fusion Test Reactor (TFTR) currently under construction
at Princeton.

Most of these injectors will operate with deuterium, and
thus will generate neutrons from the d-d interaction between the
beam and the neutralizer. Deuterons in the beams (both charged
and neutralized) will become imbedded in materials that they
strike, and will thus become high-density targets for following
beam particles, resulting in more neutrons. A measurement of
the absolute yield of neutrons at various shaping currents and a

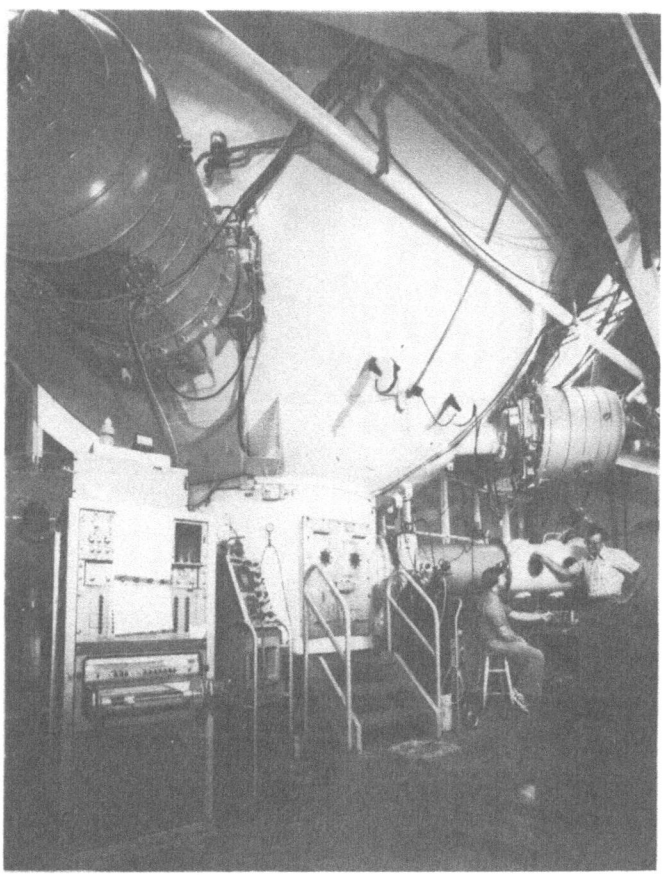

Fig. 10. Neutral beam injection test stand at
         the Lawrence Berkeley Laboratory.
         Courtesy of R. Pyle, LBL.

comparison with calculations are shown in Fig. 11, taken from
Berkner et al. (1979).

The only facility currently producing relativistic heavy
ions for research and biomedical applications is the Berkeley
Bevalac, which can accelerate heavy-ion beams with charge-tomass
ratios of 0.5 up to 2.6 GeV/u. A schematic view of the facility
is shown in Fig. 12. Its injector system consists of two
Cockroft-Walton accelerators, one air-insulated at 750 kV and
the other pressurized at 2.5 MV, either of which can inject into
the SuperHILAC, an Alvarez-type linac of 8.5 MeV/u. The beams
from this machine are then transported via a 250-m long transfer

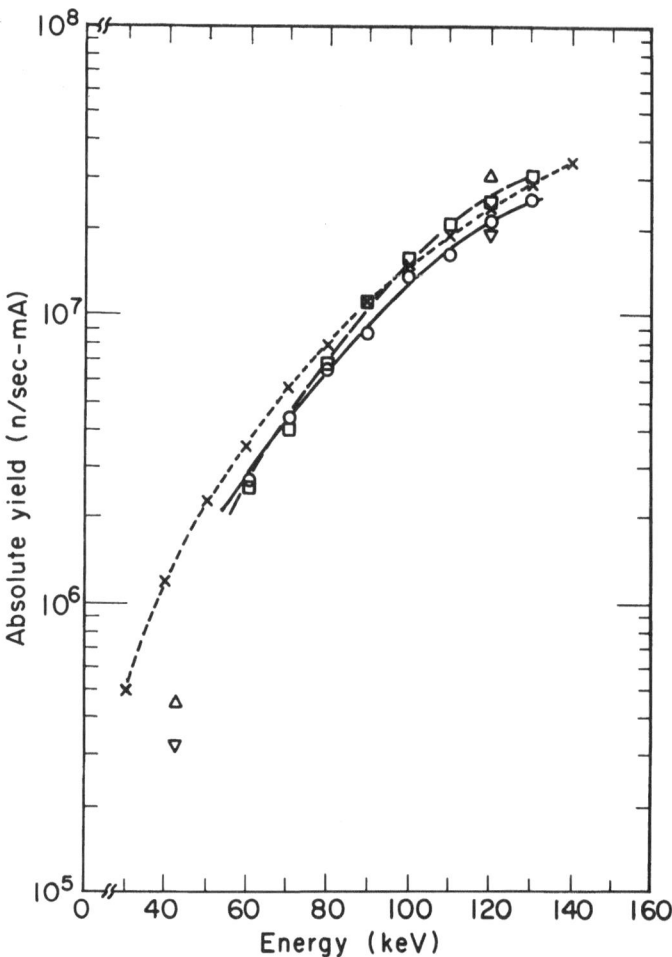

Fig. 11.   Neutron yield as a function of
deuteron energy in a neutral beam
injection line. Courtesy of J.
McCaslin, LBL.

line to the Bevatron, a weak focusing synchrotron. The Bevatron
vacuum of 2 x 10^{-7} Torr allows only acceleration of fully
stripped beams. An improvement program, involving the installation of a high-vacuum liner shown schematically in Fig. 13, is
planned, and will allow acceleration of uranium and partially
stripped ions.

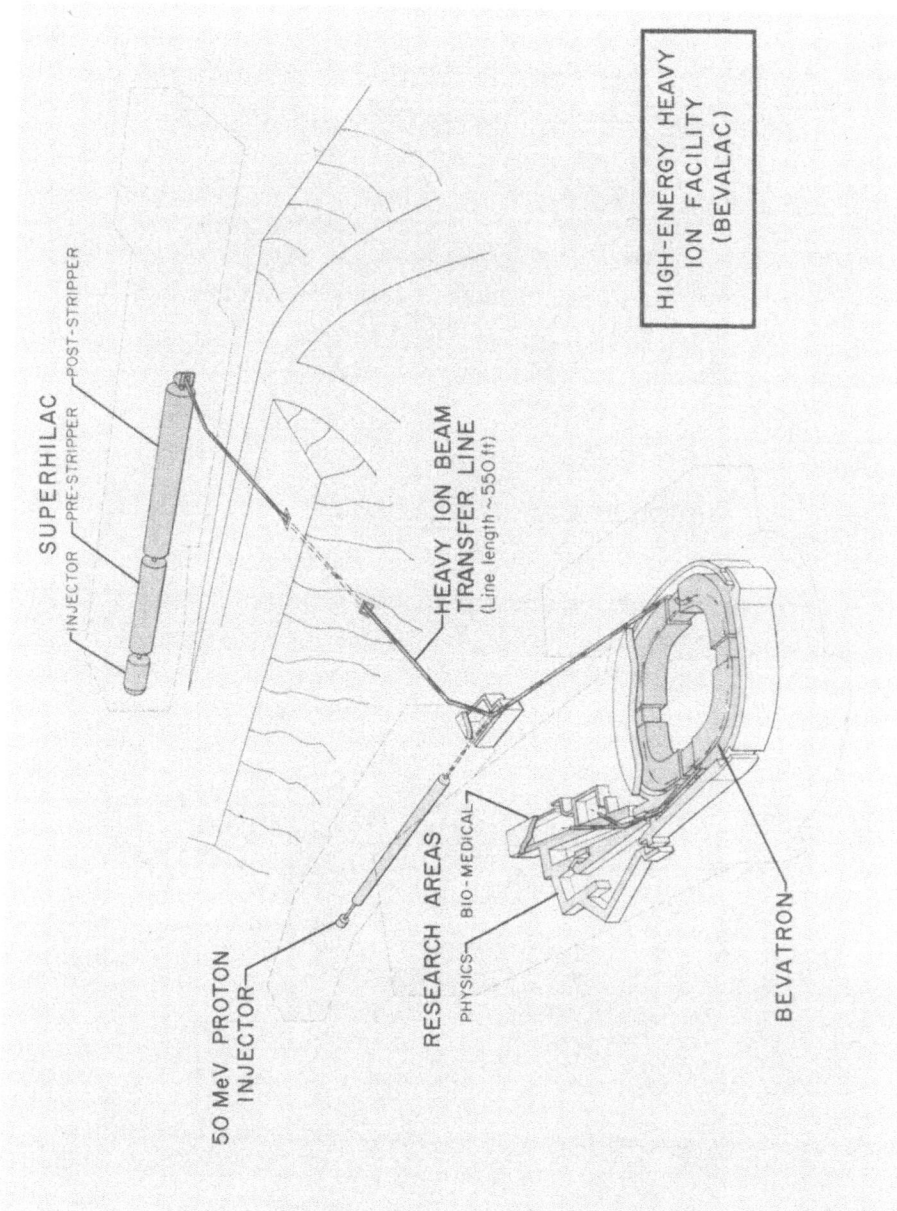

Fig. 12. The BEVALAC: A high-energy heavy-ion facility at LBL.

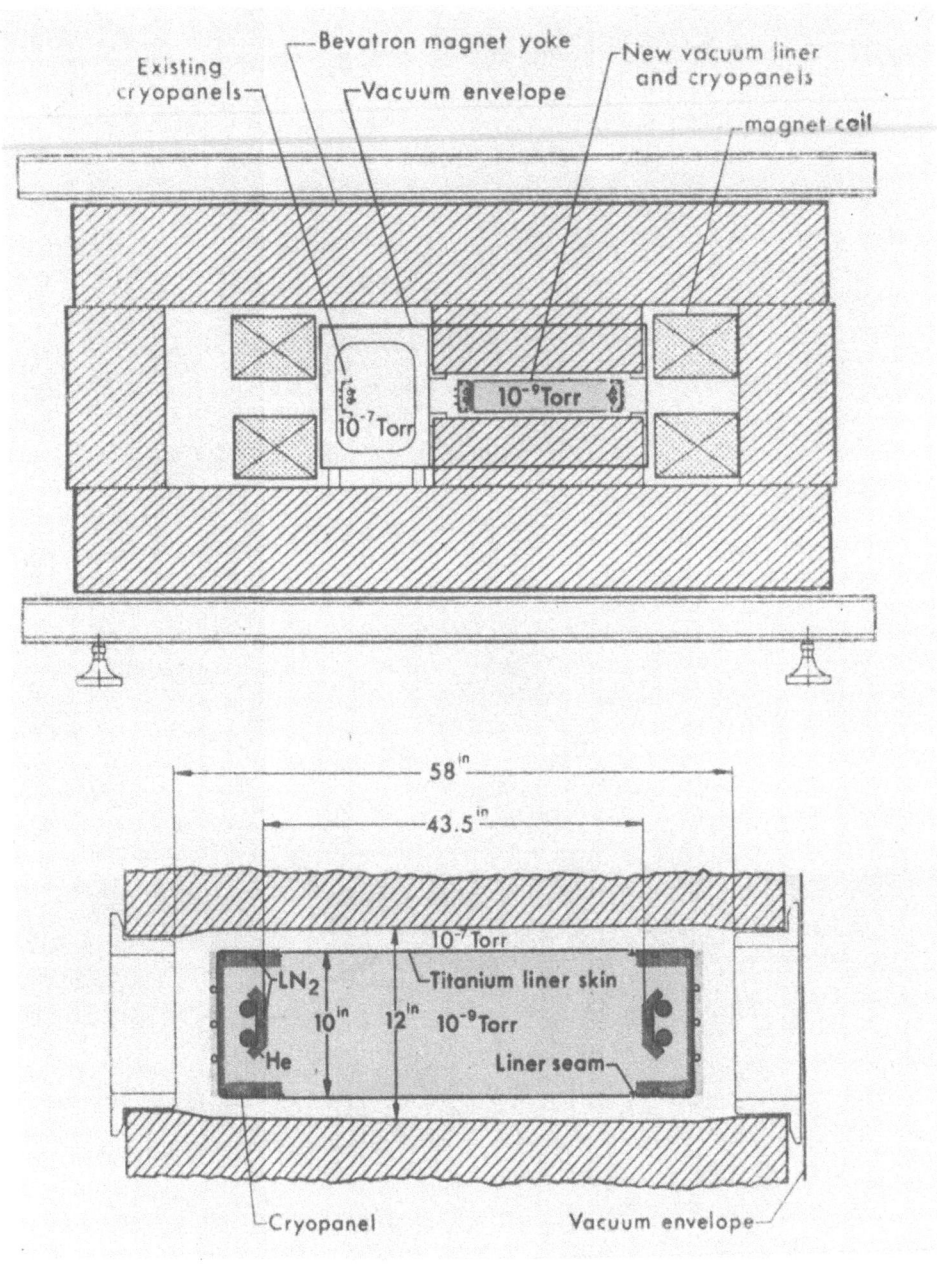

Fig. 13. Proposed vacuum chamber liner for uranium acceleration at the LBL BEVALAC. Courtesy of H. Grunder, LBL.

Several projects to build facilities for the production of multi-GeV heavy ion beams are currently in the proposal and design study stage. At the Gesellschaft fuer Schwerionen-forschung (GSI) in Darmstadt (Federal Republic of Germany), the project to accelerate uranium to energies up to 10 GeV/u, is envisioned as a three-part program summarized in Table 4. In the first stage, their present accelerator, the UNILAC (Fig. 14), will be upgraded by inserting new acceleration cavities in the Alvarez linac section.

The second stage involves the construction of a strong-focusing, separated function synchrotron, the SIS-100, shown schematically in Fig. 15. It will accelerate ions with q/A between 0.042 (corresponding to U^{10+}) and 0.5. The final energy will be variable between 20 MeV/u and 14 GeV/u. The machine will have a mean radius of 125 m and a circumference of 785 m. Use of superconducting magnets is not planned in order to ramp the magnetic fields at rates up to 2 T/s. There will be 12 RF accelerating cavities operating on a frequency range of 0.83 to 7.6 MHz with acceleration at harmonic numbers of 20 or 40. The vacuum in the machine will be approximately 10^{-11} Torr.

The final stage of this proposal would involve design and construction of a high intensity preinjector for the UNILAC to take advantage of the high currents produced by sources for single and double-charged ions.

Table 4. SIS Project

Energy	Intensity (sec^{-1})		Injection energy (MeV/u)	Target Date
	Neon	Uranium		
2 to 20 MeV/u	10^{13}	2×10^{11}	-	1981-1982
20 to 140 MeV/u	10^{11}	10^{10}	1.4	
0.1 to 14.1 0.1 to 7.3 GeV/u	5×10^{9}		5.9	1984
0.5 to 14.1 0.5 to 8.8 GeV/u	3×10^{10}	5×10^{8}	20	1986

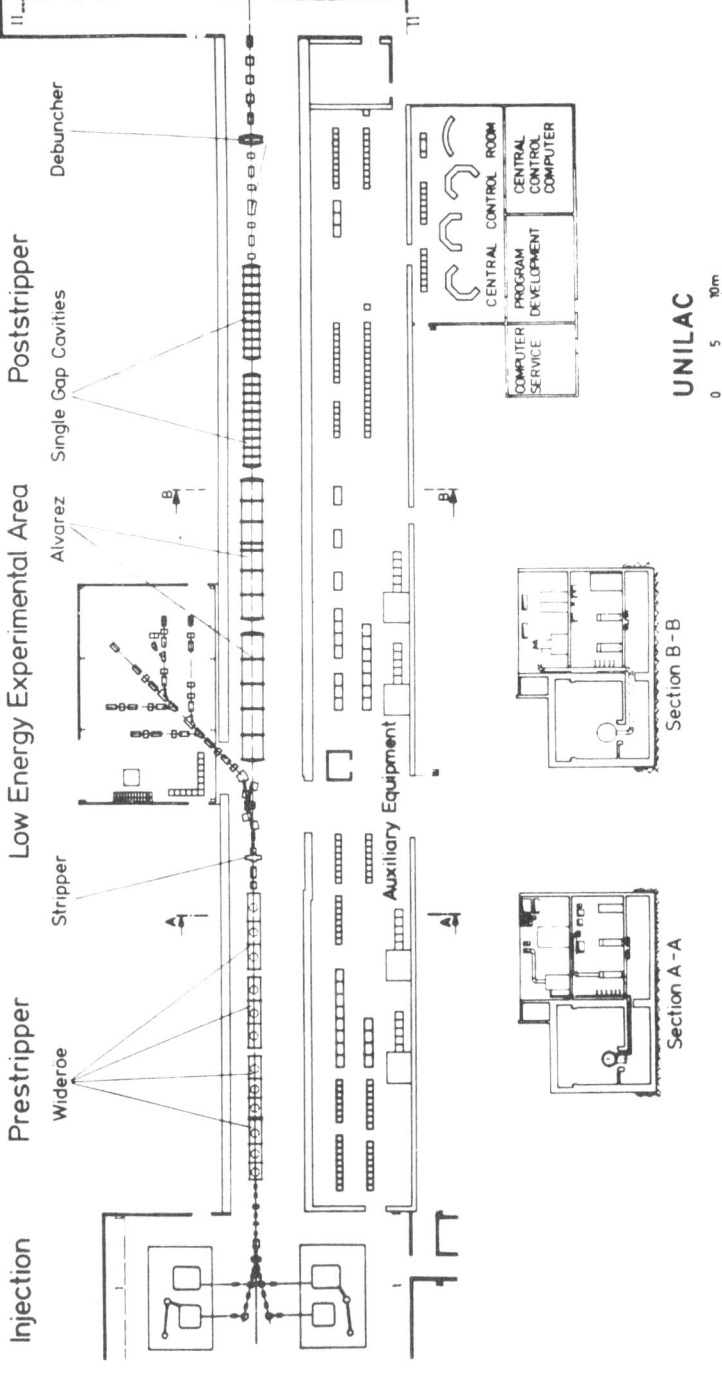

Fig. 14. Schematic diagram of UNILAC experimental beam areas. Courtesy of UNILAC.

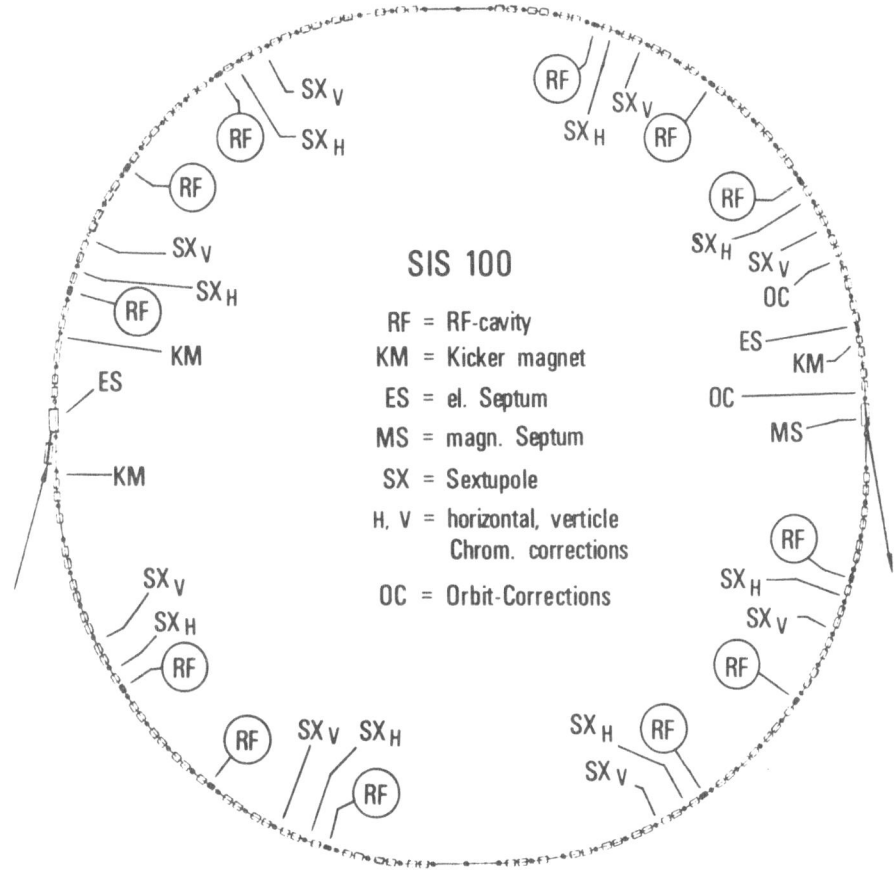

Fig. 15. Schematic diagram of SIS–100 synchrotron planned at
          Gesellschaft fuer Schwerionenforschung in Darmstadt,
          West Germany. Courtesy of that laboratory.

Operation of the SIS project is planned for 1986, at an
estimated cost of 190 million German marks.

The VENUS (Variable Energy Nuclear Synchrotron) project
presently under study at the Lawrence Berkeley Laboratory in the
United States, is intended to satisfy the requirements outlined
in Table 5. An approach that is expected to fulfill these is to
combine an accelerator and storage ring in a single facility
without sacrificing the performance of either component. The
scheme proposed is shown in Fig. 16, and consists of two
identical superconducting rings, located inside a single tunnel,

Table 5. VENUS Capabilities and Design Criteria

---

1. Intense ion beams of all masses, from protons through uranium.
2. Low energies to 40 MeV/A, overlapping the range of the 88-inch Cyclotron.
3. Intermediate energies at intensities significantly greater than those at the BEVALAC.
4. Highest beam energies well above the BEVALAC range, i.e., to 20 GeV/A for the heaviest ions.
5. Center of mass energies E_{CM}/A = 40 GeV/2A or more with colliding beams of equal mass nuclei (i.e., the total CM energy available is 20 GeV times the number of nucleons).
6. Construction and operation should be as economical as possible, with minimum power consumption and staffing requirements.
7. Flexible for adaptation to future research and operating criteria.

---

with the SuperHILAC as injector. One ring would serve as SuperHILAC booster to accelerate the 8.5 MeV/A injected ions to about 1 GeV/A. At this energy they can be stripped without significant losses and transferred to the second ring for acceleration to a maximum of 20 GeV/A for the heaviest ions, up to 25 GeV/A for light ions, and 50 GeV for protons.

The S-shaped reinjection line is used for storage ring-colliding beam operation: half the particles at the desired energy would be split from one ring and reinjected in the opposite direction (reversing the magnetic field) into the other ring. Two different heavy ion beams can also be stacked in one ring and separated subsequently for colliding beam experiments. Both rings are to consist of the same configuration of superconducting magnets, and are referred to as Ring 1 and Ring 2.

In the colliding beam mode, approximately 100 pulses would be accumulated. Three interaction regions are presently planned. In these, as in all storage rings, it will be necessary to have small transverse beam dimensions to maximize the luminosity. For the heaviest beams, 200 particle-milliamperes seem to be a reasonable expectation for the attainable currents. The luminosity for the heaviest ions at 10 GeV/u has been estimated at 10^{29} cm^{-2} s^{-1}. This is somewhat less than that of high-energy physics storage rings, but the cross sections for heavy ion reactions are expected to be higher, so that comparable event rates will be obtained. At 10^{-11} Torr

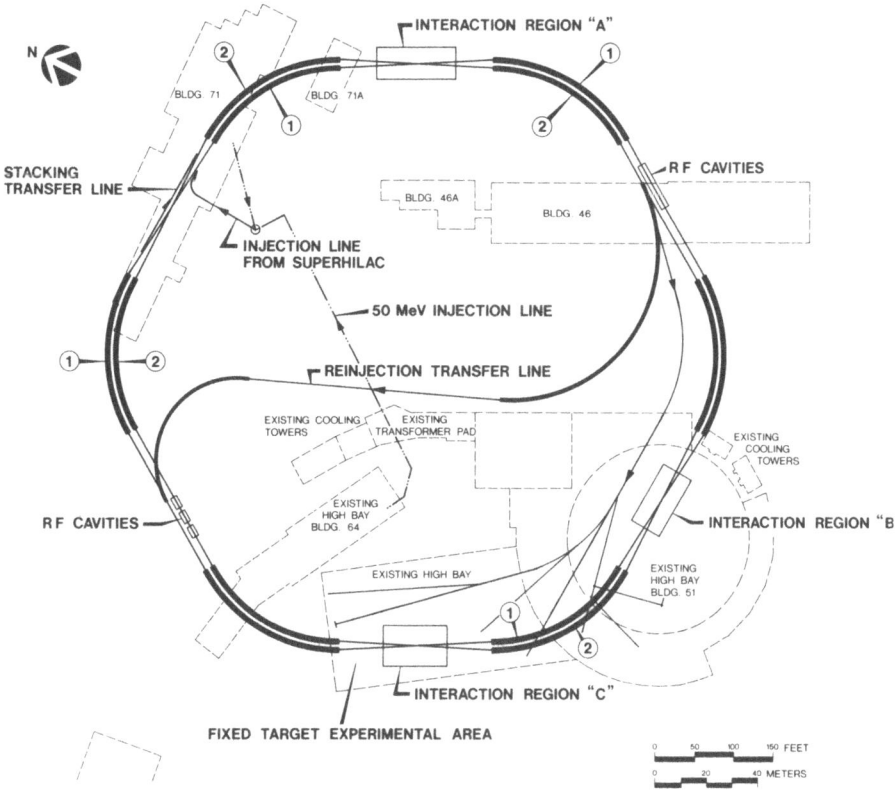

Fig. 16. VENUS: A relativistic ion synchrotron and storage
          ring, LBL.

vacuum, good beam quality and intensity should be maintained for
a few hours.

The ring tunnel elevation will be below the present Bevatron
and its experimental hall. For fixed target operation, the beam
would be extracted from Ring 2 and transported 10 m up to the
level of the present Bevatron experimental halls. A vertical
section through the projected ring is shown in Fig. 17. Beams
would be injected vertically down from the SuperHILAC, which
will be 57 m above the VENUS rings.

An aerial view of the Lawrence Berkeley Laboratory, with the
proposed VENUS layout superimposed, is shown in Fig. 18, showing

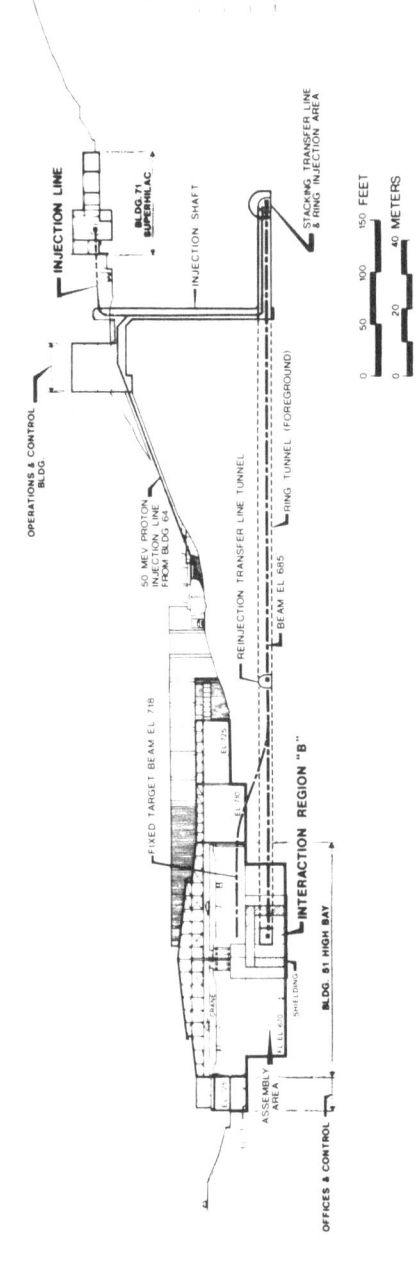

Fig. 17. VENUS: A relativistic ion synchrotron and storage ring, LBL. Section through the ring at the injection line and interaction region B.

Fig. 18. Photograph of LBL with proposed VENUS layout
          superimposed.

the desired maximum utilization of existing facilities. Half
the tunnel is estimated to be cut-and-fill, and the other half
will be bored tunnel, similar to construction of the PEP
tunnel. There are no known earthquake faults going through the
LBL site, but the design is planned for the maximum earthquake
stresses that can be expected.

The facility will require four separate RF systems to:
(1) accelerate to 1 GeV in Ring 1; (2) bunch beams after stack-
ing in Ring 2 (one RF system will be needed per ion species
simultaneously accelerated); (3) capture the beam after reinjection in Ring 1 (for colliding beam operation); and
(4) acceleration to final energies in Ring 1 and 2. For this
purpose, two types will be designed: a 70-kV system with a
large, 1:7.5 frequency swing, and a high voltage, 250-kV system
requiring only a 10% frequency swing.

Superconducting magnets will be of the type tested at
Berkeley in an experimental superconducting accelerator section,
with a maximum field of 5 T and a rate of rise of 1 T/s. Superconducting quadrupoles will also be used, since their smaller
fields place fewer demands on the technology.

The project contemplates authorization for construction in
1983 and completion in 1987, at a cost of 100 to 150 million
dollars (1979).

In France the recently completed Saturne II accelerator at
Saclay is beginning a program to produce proton and eventually
heavy-ion beams. Proton intensities will be 2.5×10^{12} per
pulse. The heaviest nuclei that can be accelerated with the
existing vacuum will be neon ions at intensities estimated to be
10^8 per pulse. The injector system is at present a 750 kV
pressurized Cockroft-Walton accelerator, followed by a 20 MeV
linac. A Cryogenic Electron Beam Ion Source (CRYEBIS) has been
built at Orsay and will soon be installed. It is designed to
provide 200 keV/u beams of fully stripped heavy ions. Saturne II
itself is a 3 GeV, strong-focusing synchrotron. A view of the
ring tunnel is shown in Fig. 19.

Several other high energy heavy-ion projects are under way
in Japan (NUMATRON) (Hirao, 1979), and the U.S.S.R. (the
recently operational U-400 Cyclotron, with energy of
725 q^2/A^2 in Dubna (CERN, 1979b), and the planned adaptation
of the Dubna synchrophasotron to produce beams up to uranium
with energies of 3.4 GeV/A (Baldin et al., 1979).

The high energy physics facilities now in operation,
construction or planning stages are shown in Table 6 for
completeness (Richter, 1979). A summary of their sophisticated
design features and the fundamental insights expected from their
use, that would do them justice, cannot be given in the space
available here.

Synchrotron radiation is possibly the fastest-growing new
field centered around electron accelerators and electron-
positron colliding beam facilities. This radiation arises from

Fig. 19. View of SATURNE II ring tunnel in Saclay, France.
Courtesy of the Institut Gustave-Roussy.

particle, this distribution is folded into a small forward cone
by the transformation to the laboratory frame of reference, as
shown in the lower part of Fig. 20. In principle, all charged
particles emit synchrotron radiation in magnetic fields. The
radiated power varies approximately with the inverse fourth
power of the mass, and, until recently, the only synchrotron
radiation seen came from electron beams. However, even though
the proton mass and energy are such that synchrotron radiation
from protons would not be expected in observable amounts, even
at SPS and Fermilab machines, the magnetic field discontinuities
inevitably present near magnet edges have resulted in observable
proton synchrotron radiation at CERN, at energies above 350 GeV
and for beam intensities as low as 10^{11} per pulse (CERN,
1979a).

     The power emitted by synchrotron radiation is substantial,
on the order of 6 MW in PEP, and must be made up by continuous
acceleration of the electron and positron beams stored inside
the radial acceleration imparted to a charged particle by the
magnetic field. In the rest frame of the circulating particle,
this radiation has the well-known dipole radiation distribution

Table 6. High Energy Facilities[*]

Type	Name (Laboratory)	Energy (GeV)	Status
e^+e^- colliding beams	PETRA (DESY) PEP (SLAC/LBL)	18x18	Operational 1979
e^- beam	SLAC	35	Operational
p beam	FNAL SPS (CERN)	450	Operational Operational
pp colliding beams	ISR (CERN)	31x31	Operational
p beams	DOUBLER (FNAL)	1000	1982
p colliding beams	AA (CERN)	300x300	1982
pp colliding beams	ISABELLE (BNL)	350x350	1986
e^+e^- colliding beams	LEP (Europe)	80x80	After 1988
p beam	UNK (USSR)	3000	After 1988
pp colliding beams	TEVATRON (FNAL)	1000x1000	1985

[*]Adapted from Richter, 1979.

shown in the upper part of Fig. 20. For a relativistic the rings. This power is radiated in a continuous spectrum characterized by a critical energy $\epsilon_C = 2.2\ E^3/R$, where E is the total energy and R is the radius of curvature (Winick, 1975). Figure 21 shows a typical spectrum obtained using the SPEAR storage ring at Stanford. Specific wavelengths from this continuum are selected using precision tunable monochromators. This is the only known means of obtaining intense sources of electromagnetic radiation over the entire spectrum ranging from 0.1Å to the visible.

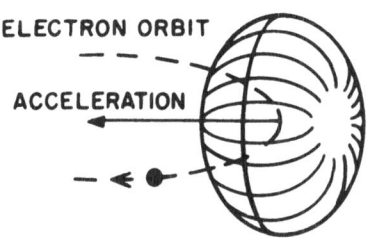

ELECTRON ORBIT

ACCELERATION

CASE I : $\frac{v}{c} \ll 1$

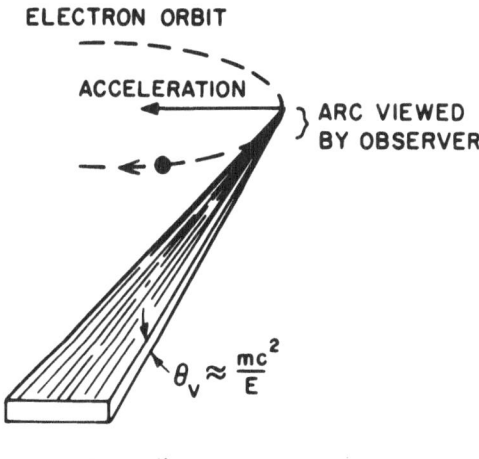

ELECTRON ORBIT

ACCELERATION

} ARC VIEWED
BY OBSERVER

$\theta_v \approx \frac{mc^2}{E}$

CASE II : $\frac{v}{c} \approx 1$

Fig. 20. Radiation emission pattern by
electrons in circular motion.
Courtesy of H. Winick, Stanford
Synchrotron Radiation Laboratory.

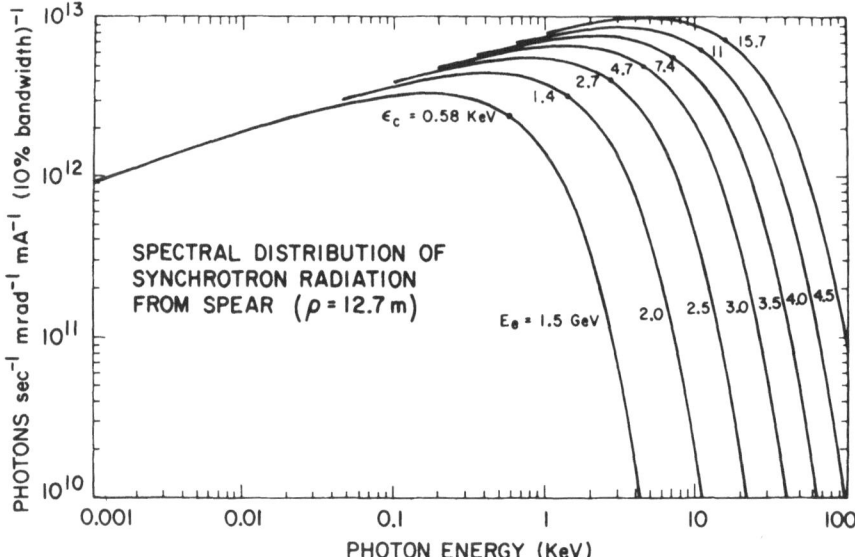

Fig. 21. Typical spectrum obtained using the SPEAR storage
          ring at Stanford. Courtesy of H. Winick, Stanford
          Synchrotron Radiation Laboratory.

As seen in Fig. 20, the entire radiation pattern of the beam
is collimated into an angle ~$1/\gamma$ perpendicular to the plane of
the beam orbit. The photon source size, given by the beam cross
section, is typically less than 1 mm^2. The brightness of
synchrotron radiation sources is thus expected to be between
10^4 and 10^7 times greater than any conventional x-ray or
line-discharge source. In addition, synchrotron radiation is
pulsed, with a time structure dependent on the electron bunch
length, typically nsec, and polarized. These characteristics
have resulted in an explosion of research in physics, chemistry,
and biology, of which the study of muscle cells in vivo and
spectroscopy of proteins may be most interesting in the context
of this course. A recent review of synchrotron radiation
applications has been given by Bienenstock (1979).

Synchrotron radiation sources in operation and under
construction are listed in Table 7 (van Steenbergen, 1979). All
new dedicated sources are built as electron storage rings to
take advantage of the greater source stability and 100 duty
factor. An artist's view of the Stanford Synchrotron Radiation
beam line arrangement is shown in Fig. 22. It shows how five

Table 7. Synchotron Radiation Facilities[+]

IN OPERATION

		GeV	λ_c
USSR:	VEPP4 (NOVOSIBIRSK)	7.0	$0.27 (1.1)_w$
	VEPP3 (NOVOSIBIRSK)	2.0	$2.9 (1.0)_w$
	VEPP2M (NOVOSIBIRSK)	0.7	23
	*ARUS (EREVAN)	4.5	1.5
	*SIRIUS (TOMSK)	1.4	9.4
	*PAKHRA (MOSCOW)	1.3	10
	*FIAN, C60 (MOSCOW)	0.7	28
	N-100 (KARKHOV)	0.1	3100
GERMANY:	PETRA (HAMBURG)	18.0	0.18
	DESY (HAMBURG)	7.5	0.4
	DORIS (HAMBURG)	5.0	0.5
	*BONN I (BONN)	2.5	2.7
	*BONN II (BONN)	0.5	77
USA:	SPEAR (STANFORD)	4.0	$1.1 (1.6)_w$
	**SURF II (WASHINGTON, D.C.)	0.25	344
	**TANTALUS I (WISCONSIN)	0.24	258
FRANCE:	DCI (ORSAY)	1.8	3.4
	**ACO (ORSAY)	0.54	39
JAPAN:	*INS. ES (TOKYO)	1.3	10.1
	***SOR (TOKYO)	0.4	95
ITALY:	ADONE (FRASCATI)	1.5	$8.3 (4.6)_w$
SWEDEN:	*LUSY (LUND)	1.2	11.8

IN CONSTRUCTION

	GeV	λ_c
***JAPAN (TSUKUBA), PH. FACT.	2.5	$3.0 (0.6)_w$
***UK (DARESBURY), SRS	2.0	$3.9 (0.9)_w$
***GERMANY (BERLIN) BESSY	0.8	20
PEP (STANFORD) 1979	18.0	0.16
CESR (CORNELL) 1979	8.0	0.35
***ALADDIN (WISCONSIN) (1980)	1.0	11.6
***NSLS (BROOKHAVEN) (1981)	2.5	$3.0 (0.6)_w$
***NSLS (NAT'L LAB) (1981)	0.7	31

+   Adapted from van Steenbergen, 1979.
*   Synchrotron
** Dedicated to synchrotron radiation research
*** Designed for synchrotron radiation research
w   Wavelength shifter

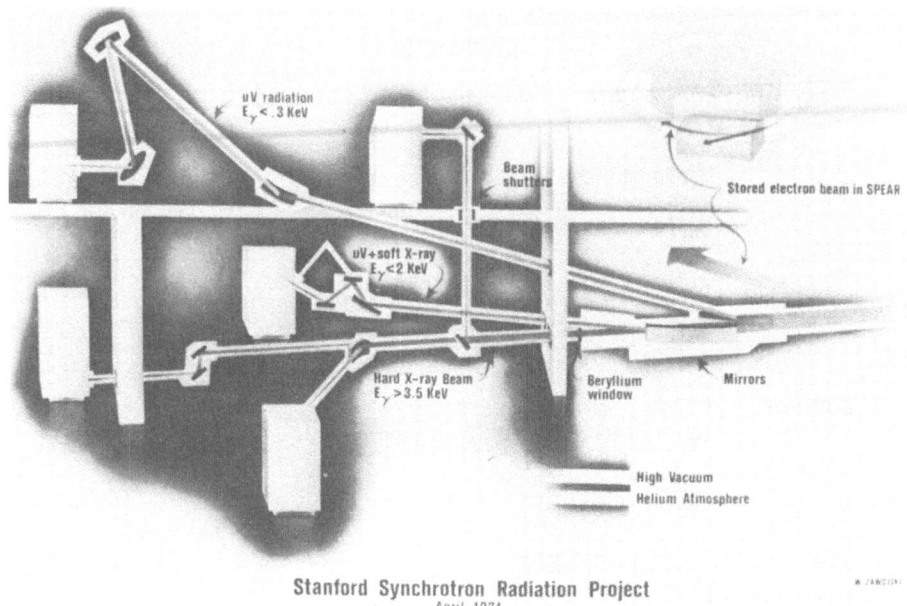

Fig. 22.   Artist's view of the Stanford Synchrotron Radiation
            beam line arrangement.   Courtesy of H. Winick.

simultaneous users can share a single beam line. The UV and
soft x-ray beams are split off by grazing incidence reflection
on polished metal surfaces and continue to grating monochromators. The high vacuum environment of the source is of great
importance to users of soft x-rays and the region below 500Å
(vacuum ultraviolet).

The critical energy of synchrotron radiation, as well as the
power radiated, are inversely proportional to the radius of
curvature of the beam. This feature is used to produce higher
energy synchrotron radiation by means of "wiggler" magnets,
which consist of several short sections of magnetic fields of
alternating polarity, the integrated effect of which does not
result in a net orbit deflection. Recently, a wiggler was
operated for the first time in the SPEAR storage ring. It
consists of a seven-pole device, 1.25 m long, which can be

powered up to fields of 1.8 T, and resulted in an increase of a factor of 6 in radiation intensity. Sequential arrays of wigglers, known as "undulators," have also been proposed, with some expected advantages due to excitation of coherent oscillations.

In the rest frame of the circulating electron, the periodic magnetic field of a wiggler or similar structure appears as a plane electromagnetic wave travelling toward the electron. Compton scattering between the virtual photons of this electromagnetic wave and the electrons in the beam bunch results in real backscattered photons going forward in the laboratory frame. From another point of view, the equivalent electron energy has been transferred to the energy contained in the wiggler. Mirrors that do not interfere with the electron beam can be added at each end of the wiggler to create a resonant cavity. When sufficient photons are produced in phase, their amplitudes add and the intensity increases above the laser threshold. Such a laser is called a "free electron laser," since it is due to stimulated radiation between an upper level consisting of a free electron and a virtual photon and a lower level consisting of a scattered electron of less energy and a scattered photon. Theoretical treatments have been given by Madey (1971) and Pellegrini (1979). Laser action has been observed at Stanford and the U.S. Naval Research Laboratory. These lasers can be tuned by changing the electron beam energy. The possibility of obtaining a continuously tunable, high power and high efficiency laser, unrestricted by properties of a material medium has stimulated active development at many United States laboratories, as well as at the University of Trento and Frascati in Italy (Lubkin, 1979).

We thus come full circle to where the accelerators acquired in the course of studying the nucleus are used to generate visible light. Imitation has been called the sincerest form of flattery. If so, the unconscious reenactment of creation to which we seem bound may be well received. It is only to be hoped that our endeavors will also merit the verdict accorded to the original creation: "And He saw that it was good."

ACKNOWLEDGEMENTS

The author greatly appreciates the graphical material generously provided by J. Ball, J. DiCello, H. Grunder, H. Gutbrod, D. Keefe, J. McCaslin, G. Norton, R. Pyle, W. Samaroo, R. Stock, and C. Weber. The competent secretarial help of Grace Walpole is also gratefully acknowledged, as well as the assistance of R. Stevens and the Lawrence Berkeley Laboratory Photography Department. Some of this work was supported by the Office of Health and Environmental Research of the U.S. Department of Energy under Contract No. W-7405-ENG-48.

REFERENCES

Alonso, J., Chatterjee, A., and Tobias, C. A., 1979, IEEE Trans. Nucl. Sci., NS-26: 3003.

Baldwin, A. M. et al., 1979, IEEE Trans. Nucl. Sci., NS-26: 4294.

Ball, J. B., 1977, IEEE Trans. Nucl. Sci., NS-24: 969.

Berkner, K. H., Cooper, W. S., Ehlers, K. W., and Pyle, R. V., 1977, Lawrence Berkeley Laboratory Report LBL-6384. Berkner, K. H. Cooper, W. S. McCaslin, J. B., and Pyle, R. V., 1979, Lawrence Berkeley Laboratory Report LBL-6366 (unpublished).

Bienenstock, A., 1979, IEEE Trans. Nucl. Sci., NS-26: 3780.

CERN Courier, May 1979a, p. 111.

CERN Courier, June 1979b, p. 158.

Cole, F. T., and Donaldson, R., eds., 1977, Proceedings, 1977 Particle Accelerator Conference, Chicago, Illinois, IEEE Trans. Nucl. Sci., NS-26.

Faltens, A., Judd, D. L., and Keefe, D., 1977, Lawrence Berkeley Laboratory Report LBL-6768.

Froelich,H. R., 1977, IEEE Trans. Nucl. Sci., NS-24: 1022.

Froelich, H. R., Thompson, A. S., Edmonds, D. S., and Manca, J. J., 1973, IEEE Trans. Nucl. Sci., NS-20: 260.

Godlove, T. F., 1979, IEEE Trans. Nucl. Sci., NS-26: 2997.

Grunder, H., and Selph, F. B., 1977, Ann. Rev. Nucl. Sci., 27:353.

Hendrickson, R., ed., 1979, Proceedings, 1979 Particle Accelerator Conference, San Francisco, CA, IEEE Trans. Nucl. Sci., NS-26.

Herb, R. G., 1971, IEEE Trans. Nucl. Sci., NS-18:71.

Hirao, Y., 1979, IEEE Trans. Nucl. Sci., NS-26: 3736.

Kugler, K. J., Paul, W., and Trinks, U., 1979, IEEE Trans. Nucl. Sci., NS-26: 3152.

Kunkel, W. B., 1979, IEEE Trans. Nucl. Sci., NS-26: 4166.

Lawson, J, D., 1977, "The Physics of Charged-Particle Beams," Clarendon Press, Oxford, England.

Leiss, J. E., 1979, IEEE Trans. Nucl. Sci., NS-26: 3870.

Lubkin, G. B., ed., July 1979, Search and Discovery, Phys. Today, p. 170.

Madey, J. M. J., 1971. J. Appl. Physics, 42: 1906.

Muller, R. A., 1977. Science, 196: 489.

Patterson, H. W., and Thomas, R. H., 1973, "Accelerator Health Physics," Academic Press, New York.

Pellegrini, C., 1979, IEEE Trans. Nucl. Sci., NS-26: 3791.

Reinhold, G., and Greyvod, R., 1973, IEEE Trans. Nucl. Sci., NS-20: 378.

Richter, B., 1979, IEEE Trans. Nucl. Sci., NS-26: 4261.

Rodde, A. F., Jackson, J. H., McCullum, J. G., Mostek, P. J., Robertson, G. I., Weissman, B., and Williams, N., 1974, Extended Abstracts of the Electrochemical Society Spring Meeting, San Francisco, CA.

Rosen, L., 1971, Science, 173: 490.

Steinberg, M., Takahashi, H., Powell, J. R., and Kouts, H. J. C. 1977, Brookhaven National Laboratory Report BNL-50731. (unpublished).

van Steenbergen, A., 1979, IEEE Trans. Nucl. Sci., NS-26: 3787.

Winick, H., 1975, Stanford Synchrotron Radiation Project Report No. 75/07.

RADIOISOTOPES PRODUCTION BY ACCELERATORS

CLAUDIO BIRATTARI

Istituto di Fisica – Università degli Studi
Via Celoria, 16 – Milano – Italy

HISTORY

It may be said that one of the most significant advances in
nuclear science was made in 1930, when E.O. Lawrence and his associates built the first operational cyclotron at the University of
California, Berkeley. With his cyclotron Lawrence was able to demonstrate nuclear transformation, a discovery that was to result in
the large-scale use of radioisotopes in medicine, industry, agriculture and science. By 1939, cyclotron were being used in many
parts of the world for the production of a wide range of radioisotopes. One of the main limitations of the cyclotron, however, is
its inability to produce large quantities of radioactive material.
Only one or two radioisotopes are made at a time, and this results
in high production costs. Thus, the development of the nuclear reactor in 1942 by Fermi and his colleagues at the University of Chicago
was seen by many to be the answer to the problem of limited radioisotopes production. Using a reactor, it is possible to make large
quantities of several different radioisotopes simultaneously, and
at a relatively low cost compared with those made on a cyclotron.
As with the cyclotron, so too with nuclear reactor, which has been
developed so that now are used extensively for radioisotope production.

At about the same time as the invention of the cyclotron, Van
de Graaff accelerator was developed. Although not as powerful as
the cyclotron, it is nevertheless capable of producing small quantities of radioisotopes.

The cyclotron, the nuclear reactor, and to a lesser extent
the Van de Graaff accelerator, are used for the production of

269

radioactive isotopes.

INTRODUCTION

Until a few years ago, nearly all radioisotopes for current use were produced by reactors, both as primary production (reactions induced by neutrons in the reactor pipes), and as secondary products in re-processing operations (fission products). However, in recent years, and especially within the framework of medical and pharmacological research, increasing interest has been shown in radioisotopes that could be produced with cyclotrons.

This is easy to understand, bearing in mind the fact that, on the one hand, the cyclotron's function is complementary to that of reactors, since it produces mainly radionuclides in the "neutron-deficient" region, scarcely accessible to reactors, while, on the other, it makes possible a wider choice of isotopes of a particular element, and therefore selection, both from the point of view of half-life and of the types of radiation emitted.

An immediate advantage of production by means of the cyclotron is that the radioisotope thus formed can be separated chemically from the matrix. Thus, the use of reactions of the (p, xn), (d, xn), or (α , xn) type leads to the formation of radio elements different from the element bombarded. Carrier-free products can therefore easily be obtained.

A second advantage consists in the fact that it is very often possible to choose a particular reaction and a particular energy of the bombarding particle in such a way as to maximize production of the desired isotope as compared with others that interfere, and therefore also to reduce the presence of contamination by other radioisotopes.

From the foregoing it is apparent that here we may speak not just of the ability to compete with reactors, but rather of genuinely qualitative differences, and therefore of production of isotopes with cyclotrons on a very different level and with very different aims. From the quantitative angle, the reactor has so far remained the main producer, among other things because, as is known, the production of radioisotopes can be carried on hand in hand with other uses.

REQUIREMENTS

With the employment of radio nuclides in nuclear medicine for diagnostic purposes, there is a tendency, on the one hand, to reduce the dose given to the patient and, on the other, to increase the degree of spatial resolution of the scintigraphic images. In addition, in functional studies, it is necessary not to disturb biolo-

gical systems by introducing significant quantities of non-physio
logical elements.

Reduction of the dose given to the patient in diagnostics with
radioisotopes is a problem that can be approached both by using isotopes with a half-life shorter than that at present in use, and by
using isotopes with simple decay schemes, or isotopes that have gamma emissions with energy more suited to detector systems and that
do not involve emissions of charged particles, such as, for example,
decay with electron capture.

The second problem may be approached by improving computation
statistics, by improving collimation systems, or by using special
detection techniques such as computerized axial tomography, applied
to detection of the coincident gamma rays, such as, for example,
those emitted following annihilation of the positrons.

Improving the computation statistics means increasing the quantity of isotopes administered, and this is possible, without running
counter to the first problem, by using isotopes with a short half-
life.

Collimation may also be improved by using isotopes with simple
decay schemes, with a single gamma emission of suitable energy.

Annihilation gamma rays are emitted only by neutron-deficient
isotopes that decay with emission of β^+

Biological systems are not disturbed provided use is made of
non-physiological radioisotopes whenever the latter are carrier-free.
The quantities of isotopes such introduced into the organism do not
produce imbalance, and do not therefore alter the metabolic functions under study.

Short-lived isotopes, simple decay schemes, decay through electron capture or β^+ emission, and carrier-free forms are therefore
what medicine currently requires to isotope producers.

Production with cyclotrons will make it possible to meet some
of these requirements, while, in other cases, it will favour such
implementation.

Nor must it be forgotten that a cyclotron can always produce
any nuclide obtainable by a reactor with neutronic capture reactions,
since a cyclotrons is also an intense neutron source.

For all these reasons, a symbiosis of a cyclotron and a nuclear
medicine centre would undoubtedly be an advantage, although it would
call for grater effort on the part of users -- an effort that, however, might well be made worthwhile by the increasing amount and

quality of information obtainable through the use of radionuclides
produced in this way.

PROPERTIES OF USEFUL MEDICAL ISOTOPES

From a general point of view, the properties required in a radioisotope to be employed for clinical applications or functional
studies may be summarized as follows:
1) – Gamma-ray energy:
    a) – Adequate penetration in tissues with minimum scattering
    b) – Maximum sensitivity for available detecting systems
    c) – Maximum directionality for best spatial resolution
    d) – Minimum penetration in the collimating system for best
       umbra
2) – Physical half-life: minimum value, compatible with the physio-
    logical half-life of the phaenomena under study
3) – Minimum beta or conversion electron emission
4) – Maximum "merit ratio"

Obviously, a radionuclide cannot be used for studying a physiological process that is much longer than the half-life of the nuclide
itself.

From the point of view of keeping to a minimum the dose of
radiation, the optimal half-life required for a radionuclide in studying a physiological phenomenon in a living organism should be
0.693 times the time interval between administration of the substance and measurement of the radiation emitted.

Since the physiological processes under study are mostly rapid,
there is a strong interest in considering radioisotopes with short
half-lives, less than 12 hours. These nuclides also offer the advantage to allow measurements to be repeated in the same biological
system without the presence of residue activity.

Only X-rays, gamma rays, and radiation from the annihilation
of positrons can be used for detection outside a living organism.

On the other hand, beta particles, conversion electrons, and
low-energy photons are absorbed in the tissue without supplying
information, on the contrary, contributing to the dose given the
patient.

Detection systems are more sensitive for a few hundreds of
KeV gamma ray energy. With this amount of energy, collimation is
easier, and the picture better defined, especially if there are no
other higher energy emissions.

The merit ratio, defined as the ratio between the energy utilized in detection and the total energy, however it may be emitted in

the decay processes, is thus an index of the quality of the decay scheme of an isotope from the point of view of its use for diagnostic purposes.

PRODUCTION STUDIES

Having therefore, on the basis of the parameters explained above, identified an isotope that may be used for clinical applications, the main task is to deal with the problem of determining the nuclear reaction to be employed, of the best bombardment energy, and of the thickness of the target - the object of all this being to maximize production and minimize the presence of radiactive contaminants.

The problem is solved by studying the excitation functions and determination of the yield, or else of the parameter that quantitatively characterizes the production of an isotope.

This parameter, which is analogous to the cross-section, is expressed in μCi/ μAh MeV, and represents the activity obtainable at the end of a bombardment of 1 hour with a current of 1μA upon a target in which the incident particles lose 1 MeV of energy. For this it is not possible to use the cross section, since several reactions may combine simultaneously in the production of an isotope.

To determine yield, use may be made both of the technique of irradiating several thin targets in sandwich mode, and by irradiating individual targets. The letter technique, although more tedious, is nevertheless more accurate, and in certain cases the only one, when, for example, it is a question of determining the yields of elements in gazeous form.

Determination of yield for targets of natural chemical composition makes it possible to establish the energy interval that is most suitable for production, the percentage of contaminants present, and the need, if any, to resort to enriched targets and the degree of enrichment. Fig. 1, 2, 3 and 4.

The subsequent phase of the study consists in the separation of the radio isotope produced by the matrix with chemico-physical techniques.

These separations generally have different characteristics from the conventional techniques that can be applied to stable isotopes; in fact, the quantities, in weight, for radioisotopes to be separated from the matrix are less than one millionth of the weight of the matrix itself.

These separation techniques have to have high yields and to be extremely selective vis-à-vis the matrix element, in order, on the

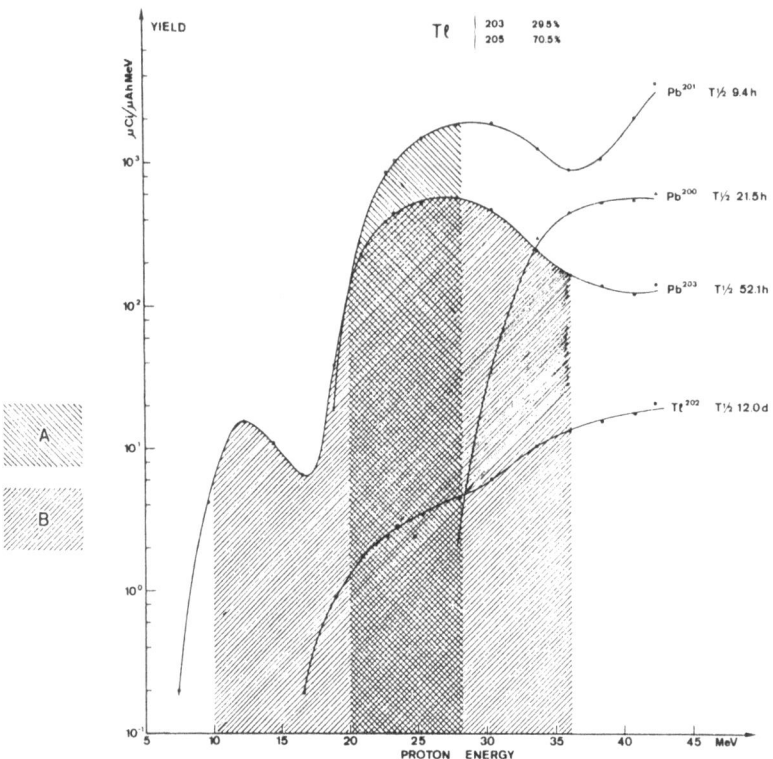

Fig. 1 Excitation function for the production of lead radioisotopes
in a natural thallium target for 7 to 42 MeV protons.
A = proton energy range for maximize Tl^{201} (via Pb^{201} decay)
production and minimize contamination by Pb^{200} Tl^{200}.
(first chemical separation 3 h after end of bombardment,
second chemical separation 32 h after first chemical separation)
B = proton energy range for maximize Pb^{203} production.
(chemical separation 24 h after end of bombardment)

one hand, to ensure high specific activity and, on the other, absence of any elements that might be toxic or in some way interfere
with the biological processes to be studied with the radioisotope
produced.

The production of carrier-free high-specific-activity radio-

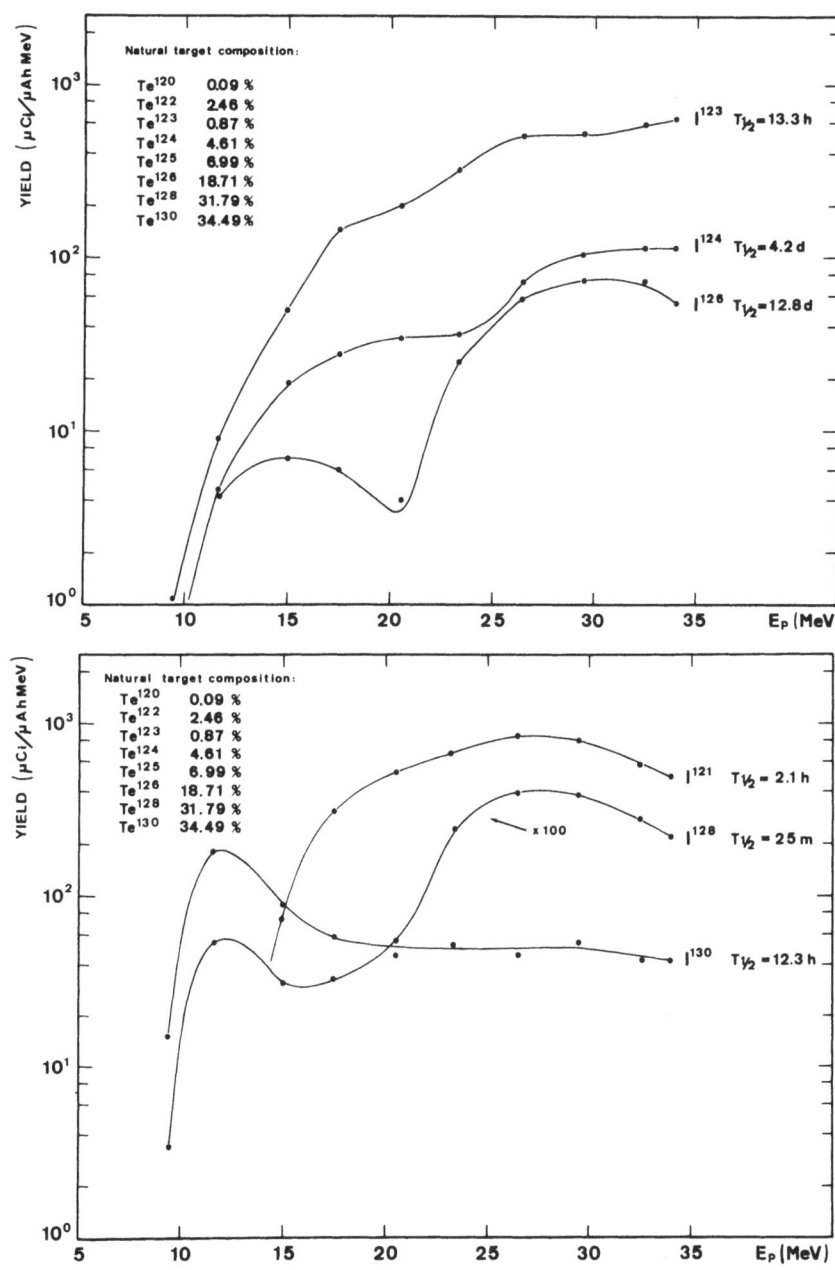

Fig. 2-3 Excitation function for the production of iodine radioisotopes (123, 124, 126, 121, 128,130) in a natural tellurium
target for 10 to 35 MeV protons.

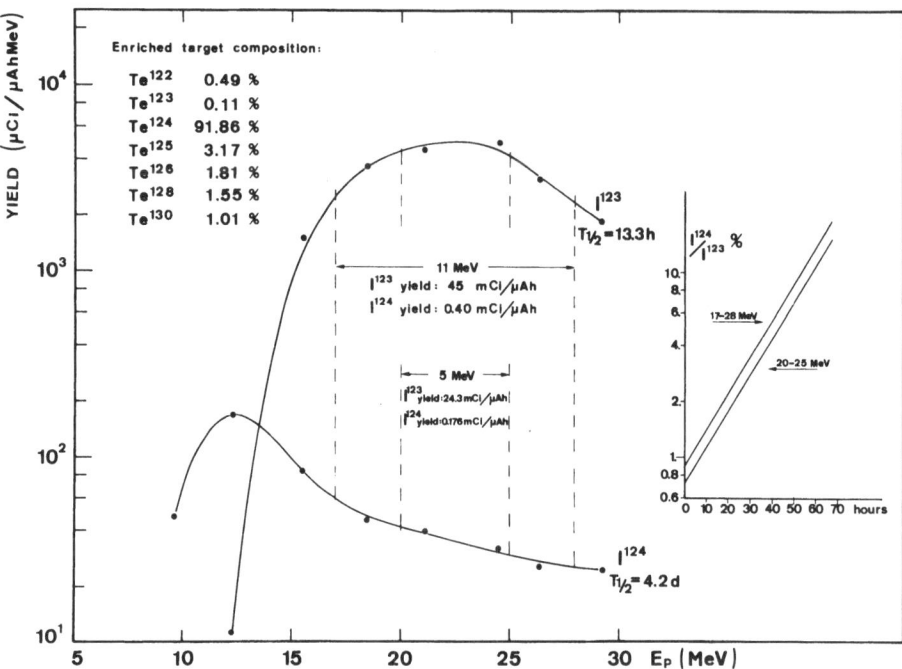

Fig. 4 Excitation function for the production of I^{123}, I^{124} in an enriched tellurium target for 10 to 30 MeV protons. Percentage ratio of I^{124} to I^{123} as a function of the time elapsed after irradiation for two different target thicknesses of 11 and 5 MeV rispectively.

pharmaceuticals generally exploits knowledge already acquired; however, in this case too, a study is also called for, to sift the results and make techniques more suitable for the specific cases under examination.

TARGETS

The setting-up of production techniques requires the study of particular irradiation methods, or the study of devices to contain the targets, of cooling systems for the latter during irradiation, and of recovery, under safe conditions, given the high levels of exposure that occur at the end of bombardment.

The devices may be subdivided into two categories: for internal bombardment on the cyclotron, or for external bombardment, on beam pipes. The former can be used with metal targets with a high melting point and good heat conductivity.

The advantage of internal bombardment lies in the higher

ROTATING PROBE

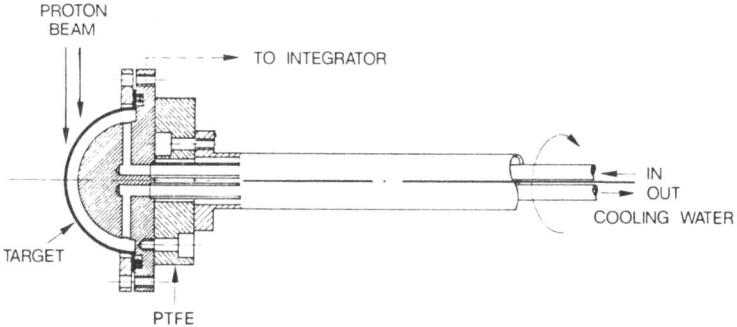

STATIC ANGULATED PROBE

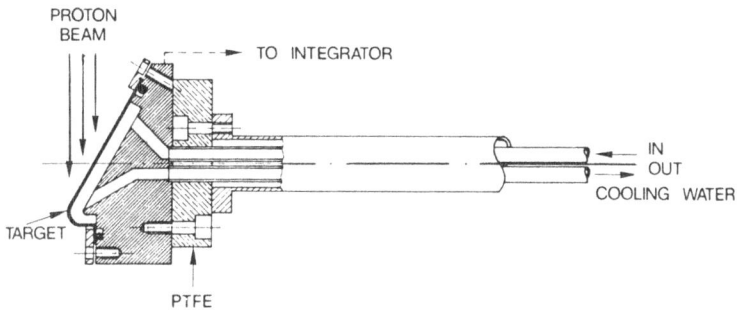

Fig. 5 Internal targets imployed to increase heat dissipation.
   1) - rotating probe
   2) - angulated probe

utilizable beam intensity, the higher beam density, and, lastly, in the better target utilization.

External bombardment, on the other hand, has to be considered in the event of low heat conductivity, when it is necessary to distribute the beam over larger surfaces, or in the case of bombardment of targets in gaseous form.

It is practically impossible to indicate a general method

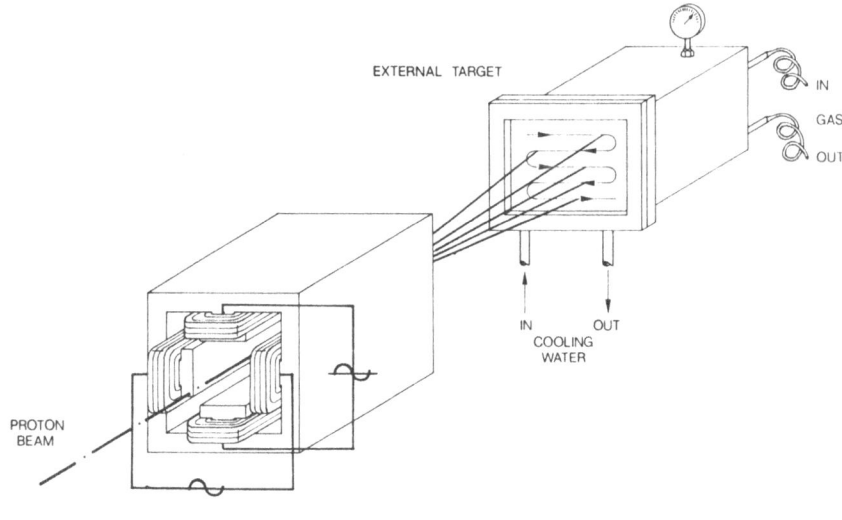

Fig. 6 External targets, a sweeping magnet may be employed to increse proton heated surface.

for designing a target for massive bombardment, given the fact that each case has to be treated individually. This is due to various factors, which have to be taken into consideration, such as:
1) - The nature of the material of which the target is made.
2) - The need for adequate cooling, to avoid damage or melting.
3) - The means used to recover the bombarded material, depending on the half-life of the isotope produced.
4) - The thickness of the target, which is based on maximalization of the isotope to be produced and minimization of the interfering isotopes.
5) - The physical state of the material to be bombarded.

Generally speaking, whenever possible, there is a tendency to increase the impact surface of the beam, mainly by the following three methods:
a) - Oblique incidence
b) - Rotating targets   Fig. 5, 6
c) - Sweeping magnets
the object of this being to reduce power dissipation per unit surface.

CYCLOTRON ISOTOPES IN CURRENT MEDICAL USE

Although it may be said that any radioisotope is potentially usable in clinical applications, only a few are currently in use.

TABLE I — Cyclotron produced Radionuclides in current medical use

Radio nuclide	Half-Life	Principal Radiation	Employed reactions
C^{11}	20.4 m	β^+, γ 511 KeV	$B^{10}(d,n)$; $B^{11}(d,2n)$; $C^{12}(p,pn)$
N^{13}	10.0 m	β^+, γ 511 KeV	$C^{12}(d,n)$
O^{15}	2.1 m	β^+, γ 511 KeV	$N^{14}(d,n)$
F^{18}	110 m	β^+, γ 511 KeV	$O^{16}(\alpha,pn)$; $Ne^{20}(d,\alpha)$
K^{43}	22.4 h	β^+, γ 373,619 KeV	$Ar^{40}(\alpha,p)$
Cr^{51}	27.8 d	E.C. γ 320 KeV	$V^{51}(p,n)$
Fe^{52}	8.2 h	β^+, γ 165 KeV	$Cr^{50}(\alpha,2n)$; $Mn^{55}(p,4n)$
Ga^{67}	77.9 h	E.C. γ 184, 296 KeV	$Zn(p,xn)$
Ga^{68}	68 m	β^+, γ 511 KeV	$Zn(p,xn)$
Br^{77}	57 h	E.C. γ 240, 520 KeV	$As^{75}(\alpha,2n)$
Kr^{81}	13 m	I.T. γ 190 KeV	Rb^{81} decay
Rb^{81}	4.7 h	β^+, γ 190, 450 KeV	$Br^{79}(\alpha,2n)$; $Kr(p,xn)$
Rb^{83}	83 d	E.C. γ 530 KeV	$Kr(p,xn)$
In^{111}	2.8 d	E.C. γ 170, 250 KeV	$Cd(p,xn)$; $Ag^{109}(\alpha,2n)$
In^{113m}	100 m	I.T. γ 393 KeV	$Cd(p,xn)$
I^{123}	13.3 h	E.C. γ 159 KeV	$Te^{124}(p,2n)$; Xe^{123} decay
Cs^{129}	32.4 h	E.C. γ 385 KeV	$I^{127}(\alpha,2n)$
Tl^{201}	73 h	E.C. γ 167 KeV	Pb^{203} decay; $Hg(p,xn)$

The chief isotopes, produced with cyclotrons, at present employed for diagnosis are listed in Table I, with their half-lives,
the principal emissions, and the nuclear reactions normally used
for their production.

The purpose of this table, which is deliberately concise, is just to list those isotopes that can be used today in normal routine at a centre equipped with a cyclotron.

Since it was not desired to go into too many details on the clinical applications of these radioisotopes, it seemed appropriate to classify their uses into four groups, presenting for each group some of the isotopes utilizable for specific clinical applications and also indicating the activities administered, the critical organs, and the dose absorbed by it.

This information has been summarized in the four attached tables. Tables 2, 3, 4 and 5

PHYSIOLOGICAL ISOTOPES

As will be seen, an important part in all these studies is played by the isotopes that I would define as physiological par excellence: C^{11}, O^{15} and N^{13}.

These isotopes are some of those that have the shortest half life, and they are produced by methods that make possible their continuous recovery, as radioactive gases, on line. In addition - and this is an aspect we shall be dealing on afterwards - they can be produced following nuclear reactions induced by low-energy particles.

The half life of these isotopes, which is very short (20.4 min, 122 sec, and 10 min), while on the one hand it represents a definite advantage, on the other hand makes it necessary to use very special techniques to achieve the best utilization.

TABLE II- Measurement of Blood flow

Radio nuclide	Amount administ.	Critical organ	Radiation dose to critical organ	Application
$^{81}Kr^{m}$	20 mCi/l for 5 min	Lungs	95 mrads	Regional pulmunary and cerebral perfusion
^{81}Rb	50 μCi	Whole body	35 mrads/μCi	Myocardial blood flow
^{15}O	1 mCi	Brain	50 mrads/mCi	Cerebral blood flow and metabolism
^{201}Tl	1 mCi	Whole body	70 mrads/μCi	Myocardial blood flow

TABLE III - Study of organ function

Radio nuclide	Amount administ.	Critical organ	Radiation dose to critical organ	Application
$^{81}Kr^m$	20 mCi	Lungs	0.6 mrad/μCi	Regional lung ventilation
^{81}Rb	200 μCi	Spleen	16.5 rads/mCi	Estimation of rate of destruction of red cells by the spleen
^{123}I	500 μCi	Bladder	1 mrad/μCi	Regional renography
^{123}I	20 μCi	Thyroid	42 mrads/μCi	Tryroid function in children
^{123}I	70 μCi	Lungs		Cardiopulmonary function
^{15}O	5 mCi	Lungs Trachea	14 mrads/mCi	Lung function studies. Organ blood flow and oxygen consumption
^{11}C	5 mCi	Lungs	44 mrads/mCi	Assessment of cardiac shunts and lung function studies in children
^{13}N	5 mCi	Lungs	11 mrads/mCi	Lung function studies

TABLE IV - Localisation studies

Radio nuclide	Amount administ.	Critical organ	Radiation dose to critical organ	Application
^{11}C	500 μCi	Foetus	12 mrads/mCi	Diagnosis of placenta praevia
^{18}F	1.5 mCi	Bone	0.4 rad/mCi	Detection of breast metastases in bone
^{81}Rb	200 μCi	Spleen	16.5 rads/mCi	Localisation and assessment of degree of enlargement
^{123}I	40 μCi	Thyroid	42 mrads/μCi	Detection of thyroid nodules in children
^{67}Ga	2.5 mCi	Whole body	340 mrads/mCi	Soft tissue tumours
^{52}Fe	100 μCi	Spleen	30 mrads/μCi	Investigation of bone marrow

TABLE V - Estimation of body spaces

Radio nuclide	Amount administ.	Critical organ	Radiation dose to critical organ	Applications
^{11}C	30 μCi	Whole body	0.033 mrads/μCi	Estimates of red cell volume
^{43}K	50 μCi	Whole body	0.6 mrads/μCi	Exchangeable potassium in multiple electrolyte studies
^{77}Br	30 μCi	Whole body	0.31 mrads/μCi	Exchangeable bromide in multiple electrolyte studies

Carbon 11, used mainly in the forms C^{11} and $C^{11}O_2$, whether directly or in processes of synthesis of other lebelled compounds, is produced by bombarding B^{10}, B^{11}, C^{12}, and N^{14} with deutrons, protons, and He^3.

The oxygen 15, produced by bombarding N^{14} with deutrons, is used as molecular oxygen, or to label CO_2 and water. Its use is particularly interesting due to the combination of two factors: the first stems from the fact that oxydation is the basic phenomenon in the life of superior organisms, and that the metabolism of a tissue depends on the speed of oxydation; the second factor is that metabolic oxydation is a process that, in most of its phases, has a duration comparable to the half-life of oxygen 15. For this reason, oxygen 15 is employed mainly in studies on metabolic phases.

Nitrogen 13, produced following the reaction C^{12} (d,n) is used as molecular nitrogen and to label ammonia radicals (N-H_3).

TOMOGRAPH

These isotopes, whose use is particularly interesting and cover a wide range of clinical applications, decay mainly by positron emissions, and therefore present the phenomenon of the emission of two annihilation gamma rays in coincidence.

Today this phenomenon is widely exploited for more accurate localization of the point of emission with tomographs.

It should be remembered that the emission tomograph, which utilizes the detection of the annihilation gamma rays, is a device designed as a logical consequence of the use of isotopes produced by a cyclotron, which, being neutron-deficient, may decay either through electron capture or by positron emission.

The device, which employs techniques for reconstructing images, can supply information on the density of accumulation of the isotopes administered in various anatomic sections, variously oriented in space.

The localization, which is established by utilizing the emission coincidence of two gamma rays, at $180°$, makes it possible to improve the spatial resolution of the scintigraphic images, or to obtain more anatomic details.

This technique can obviously be used for all positron-emitter isotopes, and among these we may recall Fluorine 18, Iron 52, Zinc 62, and Cesium 128.

GENERATORS

Another group of isotopes that can be used, even long away from a production centre, consists of the generators, or a couple of isotopes, one of which, even if its half life is short, can be eluted at intervals defined by the half-lives of the two constituants, parent and daughter.

This category, which is well known today for the couple Mo^{99}-Tc^{99m}, presents a very interesting couple in the field of cyclotron isotopes imployed for various clinical purposes.

This couple consists of the generator $Rb^{81} - Kr^{81m}$. (fig. 7)

The Kr^{81m}, which is a gaseous element, has a half life of 13 sec and a single emission at 190 KeV. These characteristics make it particularly usable for studies on pulmonary efficiency, in which it supplies information that could not otherwise be obtained on the ventilated areas of the lungs.

However, its use is also being extended to studies on pulmonary and cerebral perfusion; these studies are of particular interest, because they make possible the external analysis of any cerebral occlusions or by pass.

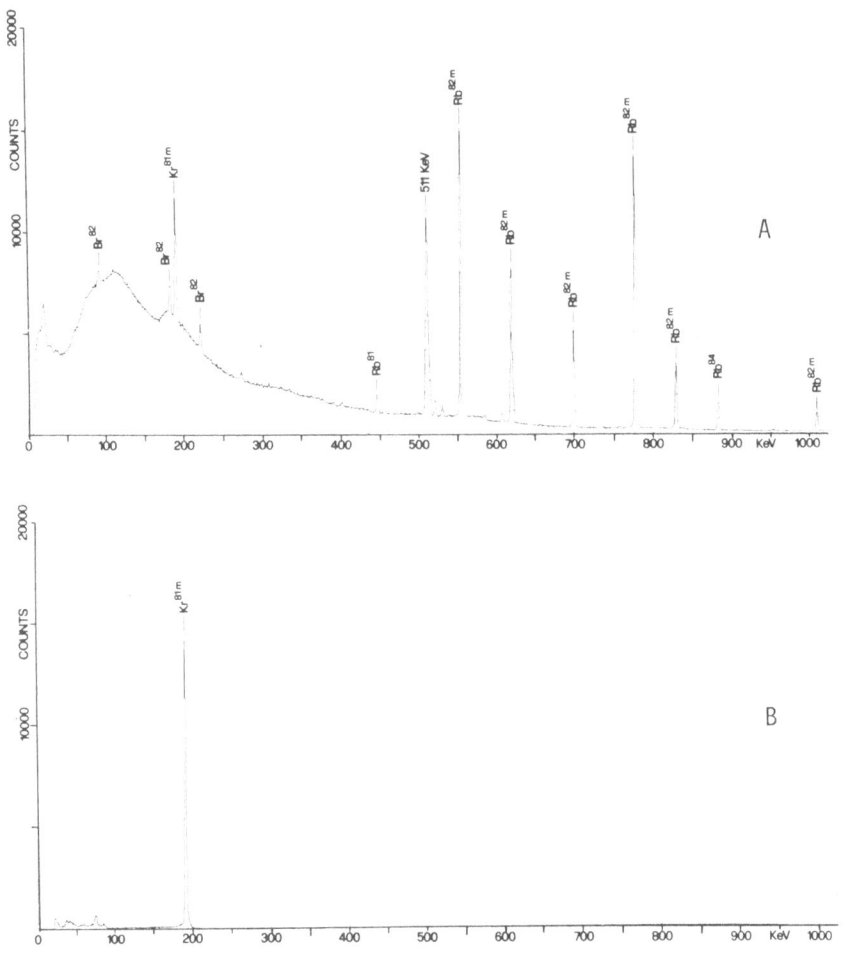

Fig. 7   A - Ge(Li) gamma rays spectra of a Rb^{81} - Kr^{81m} generator
        B - Ge(Li) gamma rays spectra of Kr^{81m} eluted in gas form
        from a generator Rb^{81} - Kr^{81m}.

CONSIDERATIONS ON MEDICAL CYCLOTRONS

There is no point in listing the characteristics of further
isotopes, because, among other things, as has already been said,
any element is potentially usable in medicine.

The difficulty lies in producing the isotope, whose characteristics are of particular interest from the point of view of its application, and producing it with a high degree of radio-chemical purity and in the most suitable form.

Obviously, from the point of view of production, it would be ideal to have a cyclotron with the maximum possible energy, capable of accelerating all the particles with very high intensity.

However, this ideal situation is clearly unrealistic, for reasons of cost, size, and reliability.

A medical cyclotron must be compact, inexpensive, and reliable.

In producing isotopes for medical purposes, it needs, above all, to be clearly established whether the accelerator is to serve only the centre in which it is installed or whether it is also to serve other centres, and therefore to evaluate their distance from the point of production. This fact especially affects the intensity characteristics of the beam, e.i. the amount of produced radioactivity.

However, it is advisable to have available accelerated particles such as proton, alpha, deutons and He^3 , in order to be able to "reach" isotopes of interest through different types of reaction, to minimizing radioactive contaminants.

The maximum energy that may realistically be required should not exceed 30-40 MeV.

In recent years, many cyclotrons have been installed in medical or industrial centres to produce isotopes for medical purposes, and the logic presented in the foregoing has in most cases been followed.

It is universally recognized that credit for this approach goes to the cyclotron and to the staff of Hammersmith Hospital in London.

This accelerator, which was installed almost 25 years ago, was the first medical cyclotron, and was always a point of reference for installations of this type.

However, it is clear that centres with these characteristics, although they can make available to outlying centres isotopes of relatively long life, or generators, will never be able to supply those isotopes, such as C^{11}, O^{15}, N^{13} which, as we have seen, are of particular interest, expecially in functional studies.

In this connection, it is worth reiterating that the use of

these isotopes must be immediate, on line.

MINI CYCLOTRONS

It is interesting to observe that reactions that are utilizable for the production of these isotopes, whose life is extremely short, have very low thresholds and reasonably good yields, even with low energy of the incidente particles.

For this reason, a very favourable view should be taken of the present tendency, on the part of cyclotron manufacturers, to market so-called mini-cyclotrons, or cyclotrons that accelerate several particles (p, α , d, He^3) with considerable intensities (50, 100 μA), but whose maximum energies are limited: 16 MeV for protons and for alpha, 8 MeV for deutrons.

These machines, which are self-shielding and can be installed in limited space, are particularly economical, both to buy and to rum.

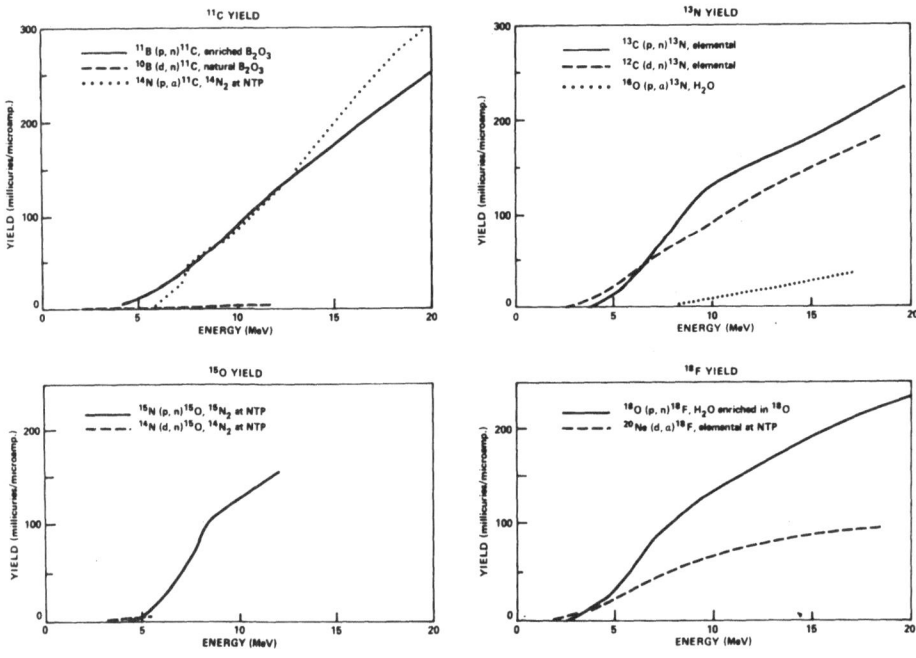

Fig. 8 Yield in millicuries/microampere at saturation for C^{11}, O^{15}, N^{13}, F^{18} production.
The yield at saturation is obtained with a irradiation of about 6 time the half life of isotope produced.

Production in terms of C^{11}, O^{15}, N^{13}, and F^{18} is of the order of 50-100 mCi/μA at saturation, more than sufficient for the purposes for which they are intended. (Fig. 8)

CONCLUSIONS

It is believed that the wider use of this type of accelerator, if installed in accordance with careful planning, at a few specialized centres, and coupled with adequate instrumentation, such as emission tomographs, may in future provide valid support for the development of nuclear medicine.

REFERENCES

J.C. Clark et al.
"Using Cyclotron-produced isotopes at Hammersmith Hospital"
Nucleonics 25, 54 (1967)

A.S. Gerlbard
"Recent aspects of Cyclotron production of medicallu useful radionuclides"
Proceedings of IAEA/WHO – Symposium on new developments in radiopharmaceuticals and labelled compounds
(Copenhagen, March 73) Paper SM-171/93

H.I. Glass et al.
"Cyclotrons in nuclear medicine"
Br. J. Radiol. 43, 589 (1970)

S. Silver
"Uses of radioisotopes in Medicine"
Nucleonics 23, 106, (1965)

D.J. Silvester
"Accelerator production of medically useful radionuclides"
Proceedings of IAEA/WHO – Symposium on new developments in radiopharmaceuticals and labelled compounds
(Copenhagen, March 73) Paper SM-171/6

D.J. Silvester
"Biomedical Cyclotron in a Medical Centre: Capabilities and Problems"
Int. Symposium on Radiopharmaceuticals Atalanta – Georgia – USA February 12-15 – 1974

M.M. Ter-Pogossion et al.
"A new look at the Cyclotron for making short lived isotopes"
Nucleonics 24, 50 (1966)

R.S. Tilbury et al.
"Cyclotron Production of Radioactive Isotopes for Medical Use"
Seminars in Nuclear Medicine <u>4</u> 245 (1974)

CHARGED PARTICLES IN MEDICAL DIAGNOSIS

George T. Y. Chen and William R. Holley

Biology and Medicine Division
Lawrence Berkeley Laboratory
Berkeley, CA 94720

INTRODUCTION

Conventional x-ray radiography is based on the detection of normal and diseased states by recording the attenuation of photons. The attenuation of photons is dependent upon both the electron density and atomic number integrated through the x-ray path. Although the use of contrast agents is of considerable help, one of the important limitations of x-ray radiography has been to detect the size, shape, and position of soft tissue tumors accurately. This is due in part to the similar atomic composition of soft tissue tumors and normal surrounding tissue. Prior to CT x-ray imaging, the presence of large abdominal masses was frequently determined by the extent of distortion and displacement of adjacent anatomical structures filled with contrast media rather than by direct imaging of the mass.

In such cases, a much more sensitive detection of this slight difference in specimen composition may be accomplished by using the variation of the stopping power of the specimen to accelerated charged particles rather than through the variation in x-ray attenuation coefficient. This increased resolution stems from the fact that a monoenergetic heavy charged particle beam has a well-defined range, and is sensitive to the presence of tissues of varying density. Furthermore, nearly all of the particles will traverse the radiographed object intact, given sufficient beam energy. On the other hand, conventional diagnostic x-rays are exponentially attenuated and must be able to penetrate the entire specimen and still retain adequate intensity to sufficiently expose the film without giving an unduly high radiation dose to the entrance side of the tissue. In this case the distance for a large variation in the x-ray radiation intensity must be the order of the thickness of the specimen. Consequently, heavy particle radiography affords a sensitivity increase of the order of 10- to 100-fold in cases where the percentage change in the particle stopping distribution is the same as the percentage change in the transmitted flux.

In recent years, various centers have explored the use of charged particles in diagnosis as a complement to conventional radiography. Although the types of particles used and methods of detection may differ from center to center, the principle advantages of charged particles over x-rays in medical diagnosis are the same: (1) energetic charged particles produce radiographs with high contrast and high depth resolution, and (2) the dose required to produce these radiographs is less than that required by conventional photon radiography. Proposed medical applications for charged particles in diagnosis include the detection of soft tissue tumors, skeletal radiography, mammography, tissue densitometry for charged particle treatment planning in radiotherapy, and charged particle computerized tomography.

PHYSICAL PRINCIPLES

The basic mechanism involved in charged particle radiography is the variation in the energy loss rate due to structures in the radiographed object. These changes may be due to variations in the electron density or the average Z (logarithmic dependence) along the particle path. As described by the Bethe-Bloch equation, the principle contribution to the relative stopping power in biological specimens is through the electron density term.

The transmission characteristics of charged particles compared with photons are shown in Fig. 1. There is very little attentuation of the protons, which have a well-defined range; the amount of range straggling of protons is equal to about 1 percent of the total range. Because of the steep flux gradient at the end of range, a detector near the beam stopping point will detect large differences in signal due to a small absorber thickness change. In contrast, x-rays are attentuated exponentially, and the small change in additional absorber will not produce as large a difference in the detected photon signal. This effect is due to the more gradual slope of the photon attentuation curve. These differences result in the increased sensitivity to small density differences in the radiographed object.

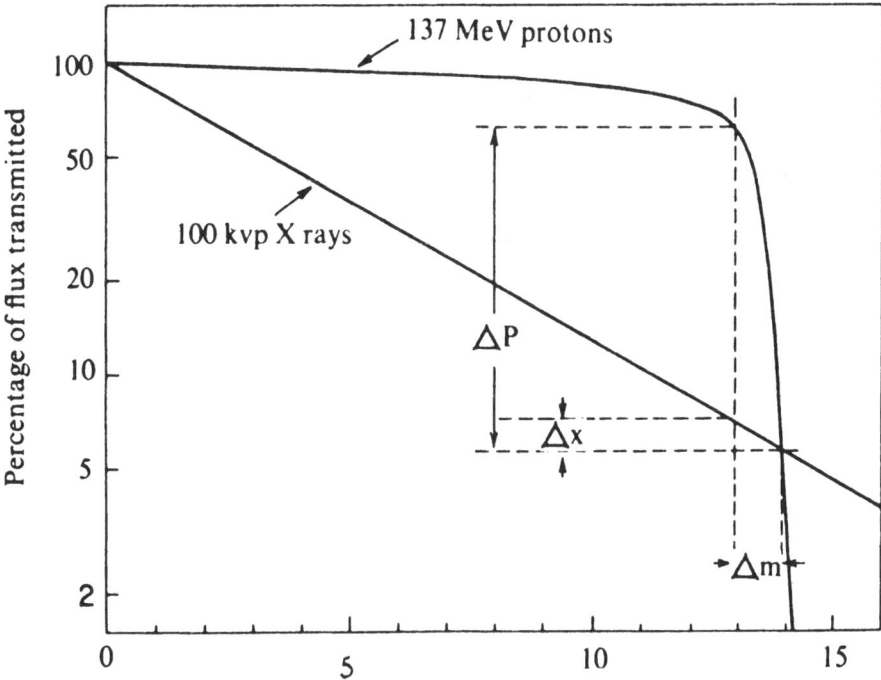

Fig. 1. Transmission characteristics of charged particles versus photons. Reprinted, by permission, from Steward et al. (1975b).

The precision in determining the variation in the radiographed object is governed primarily by the ability to establish the stopping point of the particles. Since the energy loss process is a stochastic one, in which the energy loss of the charged particle beam is governed by statistics, the stopping point distribution of the beam has a finite width. The variation in stopping point is known as range straggling, and may be described as:

$$\sigma_r = 0.012 \ R^{0.95} \ A^{-0.5} \ ,$$

where A is the atomic mass number of the incident beam and R is the range in cm of water. The range straggling for various charged particles as a function of maximum range is shown in Fig. 2a. For a typical range of 30 cm, the range straggling of protons would be about 0.3 g/cm^2, while for neon ions the straggling would be less than 0.07 g/cm^2. The improved depth resolution for heavier particles, however, is somewhat compromised by an increased dose per particle for the heavier ions. For a given dose, fewer particles must be used per subject area with heavier ions, resulting in poorer statistics and increased graininess. This suggests the need for a compromise between maximum depth resolution and dose.

Particles traversing the object are subject to multiple Coulomb scattering, which limits the spatial resolution capability of the end of range method. For a beam of parallel particles entering a water bath with a residual range of R cm and then striking a feature which lies a distance S from the stopping point of the particles, the standard deviation of the Gaussian point spread function as calculated by Henke (private communication) is approximately given by:

$$\sigma_x = 0.091 \ Z^{-0.21} A^{-0.4} [(1.0/S^{0.21}) - (0.9/R^{0.21})]^{0.5} \ S$$

Figure 2b is a graph of the beam deflection due to multiple small-angle scattering for protons and heavier ions as a function of range penetration in water. Here, the object of interest is on the upstream side of the water bath. For a typical range of 30 cm, the mean beam deflection for protons is about a factor of five worse than for neon ions.

Heavy ions (i.e., carbon and neon) are also subject to beam fragmentation, which becomes increasingly important as Z increases. Secondary nuclear fragments produced by heavy ions contribute to the absorbed does without a corresponding contribution to the useful signal. The mean free path for neon-20 ions is about 22 cm as compared with a mean free path of 129 cm for protons.

EXPERIMENTAL METHODS

A variety of experimental techniques (Fig. 3) has been developed for charged particle radiography. The first method utilized (Fig. 3b) was developed by Koehler (1968), who placed a single sheet of photographic film at the distal gradient of a proton beam and adjusted the range to obtain maximum contrast. The first proton radiograph of a test object is shown in Fig. 4. The image shows a pennant shaped aluminum foil with a thickness of 0.035 g/cm^2 inserted in an 18 g/cm^2 thick aluminum absorber. Although the edges of the pennant are blurred by scattering, the presence of this additional material, which represents only 0.2 percent of the total proton range, is clearly visible.

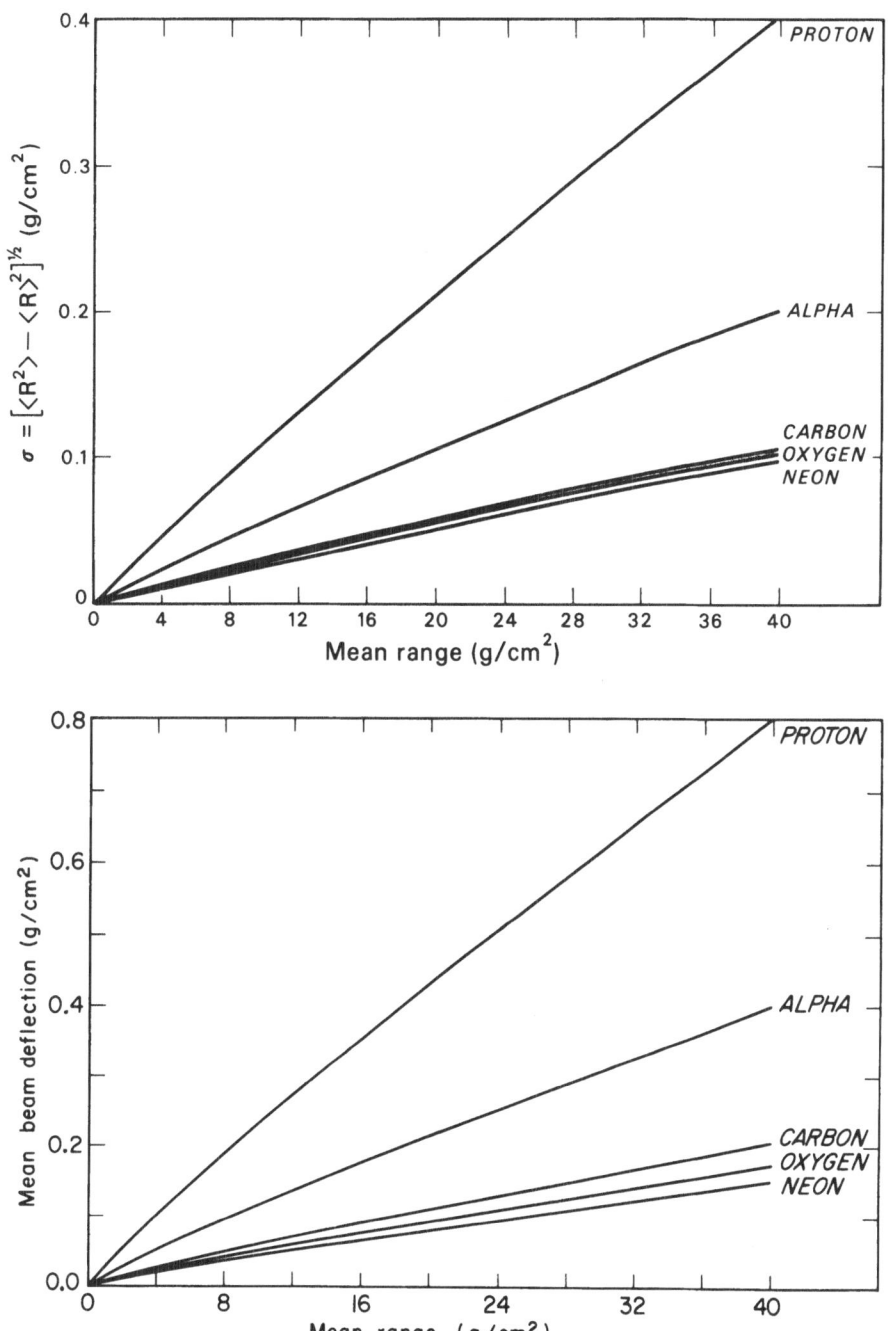

Fig. 2. (a) Range straggling for various ions as a function of range. (b) Beam deflection for various ions as a function of range. Reprinted by permission from C. A. Tobias (1978).

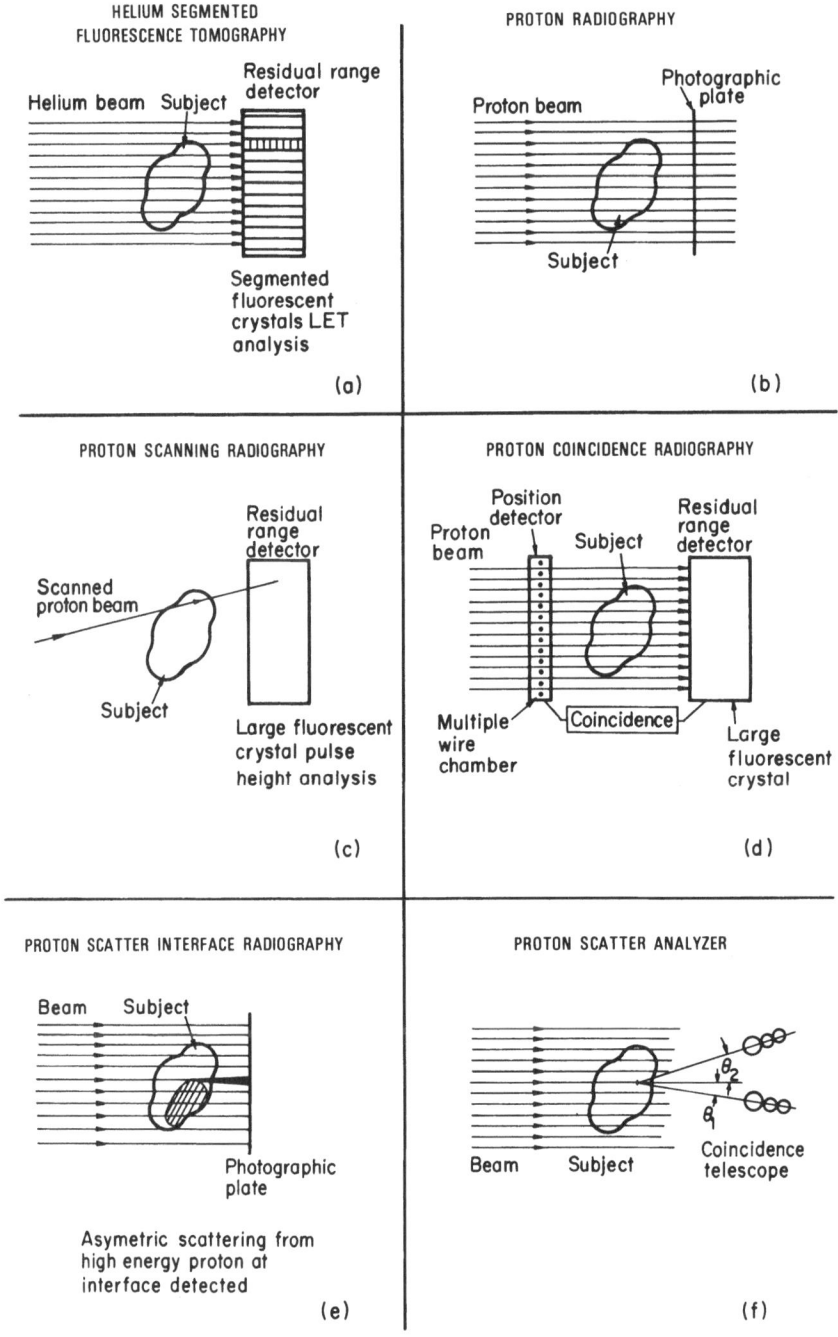

Fig. 3. Various techniques of charged particle radiography.
Reprinted, by permission, from Tobias et al. (1978).

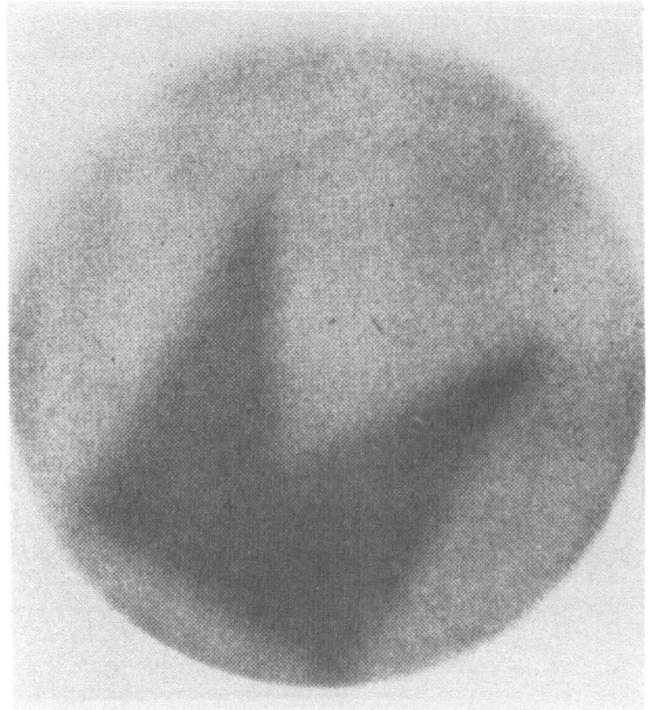

Fig. 4. First proton radiograph of a test object.
The pennant shaped wedge represents only
a 0.2 percent increment in range.
Reprinted, by permission, from Koehler
(1968).

Subsequently, techniques were developed to measure the
residual range of the particle beam after traversing the
sample. Tobias and Benton (Benton et al., 1973, 1975; Tobias
et al., 1977, 1978) developed heavy-ion radiography at LBL with
the use of plastic nuclear detectors. The method is schematically illustrated in Fig. 5a. A nearly parallel beam of heavy
particles crosses the object to be studied and stops in the
stack of plastic foils. These foils are nuclear track detectors and were originally developed to study heavy primary
cosmic rays. The foils are insensitive to the passage of low-
LET particles; high-LET particles cause damage in the plastic
which can be developed with warm sodium hydroxide, resulting in
tiny conical holes. After development (Fig. 5b), each sheet is
illuminated obliquely (which preferentially brightens the
etched areas), and digitized with a video camera and analog-to-
digital converter. A cube of image data is thus generated from

a)

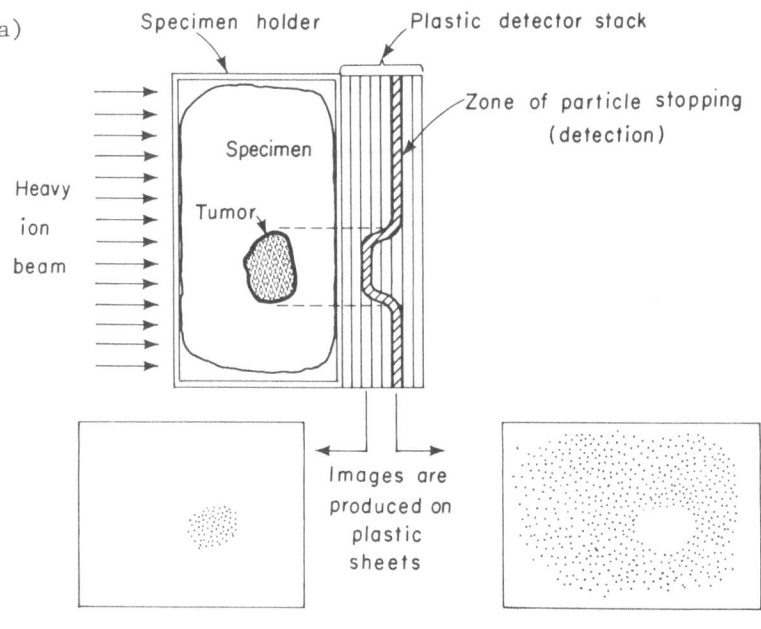

b)

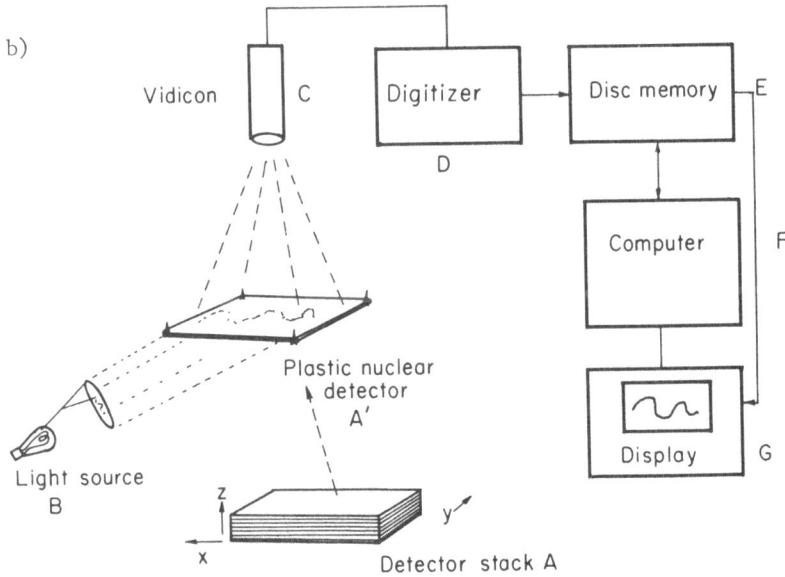

Fig. 5. (a) Experimental method used at LBL for heavy-
ion radiography. (b) Schematic of data
processing at LBL.

sequential sheets. A digital image may be synthesized from
this cube, where each pixel of the image is assigned a value
proportional to the stopping point through that pixel. The
most common representation is as a grey level image, as seen in
Fig. 6. Figure 6a is an x-ray of the foot, while Fig. 6b is a
heavy-ion radiograph of the specimen, and shows in addition to
bone, detail of soft tissue components, including tendons and
muscles. Another representation of the image is seen in
Fig. 7, which shows a surface view in perspective. An isopene-
tration contour plot may also be made from the digital data, as
shown in Fig. 8.

Kramer et al. (1979) at Argonne National Laboratory have
used a scanned proton beam and a large fluorescent crystal with
pulse height analysis for residual range determination. This
all electronic on-line technique has the potential for rapid
image generation, although at the present time is hardware
limited. Currently, rather than using magnetic deflection to
scan the proton beam across the object, the Argonne group has
performed pilot studies which translate the object across the
pencil beam. The total amount of time required to scan a
25 cm x 25 cm field is 10 to 20 minutes, and is limited by the
pulse rate of the accelerator. In the future, with a longer
pulse accelerator and a magnetically scanned beam in both
directions, a scan could be completed in 0.1 seconds.
Figure 3a and 3d show other techniques to simultaneously
measure particle position and residual range electronically.

The methods described thus far depend upon the integrated
density along the particle beam path. Saudinos et al. (1975)
and Charpak et al. (1976) have a different approach to charged
particle radiography. Their method allows, in principle, for
the three-dimensional reconstruction of an object by irradiation from a single direction. Nuclear scattering radiography
(NSR) exploits the wide angle scattering of an incident proton
beam by the nuclei of the target material. Each scattered
particle is detected by multiwire proportional chambers
surrounding the subject. The entry and exit angles are used to
determine the position of the target nucleus. By amassing
sufficient data from these scattering events, the nuclear
density of the target is obtained. NSR has the ability to
distinguish between different types of nuclei by detecting
subtle differences in scatter behavior. NSR measures a
quantity proportional to the density divided by the atomic
number for each volume element. Since this radiography technique detects scattered particles, the dose required may be
rather high. For 1 percent precision in the measured nuclear
density, the dose needed for a 1-mm^3 cubic volume is 50 rad,

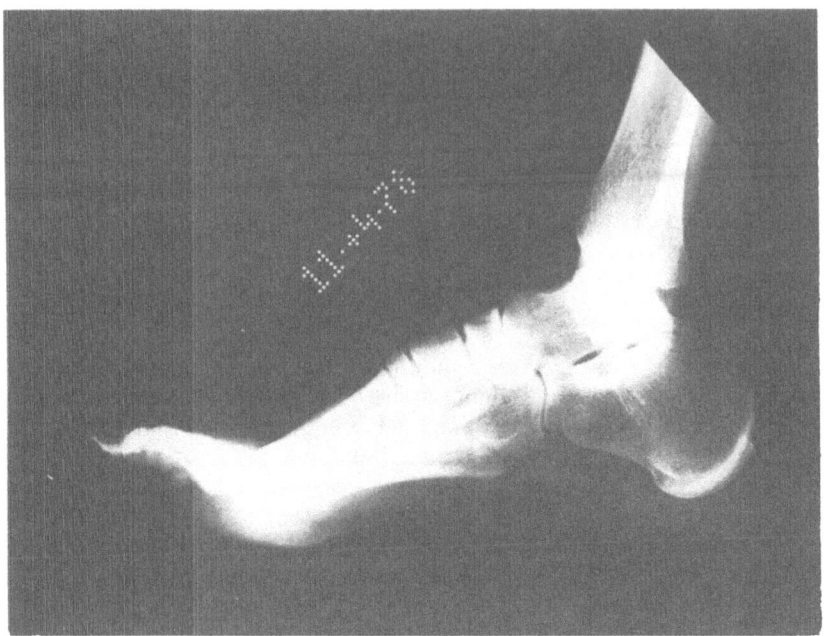

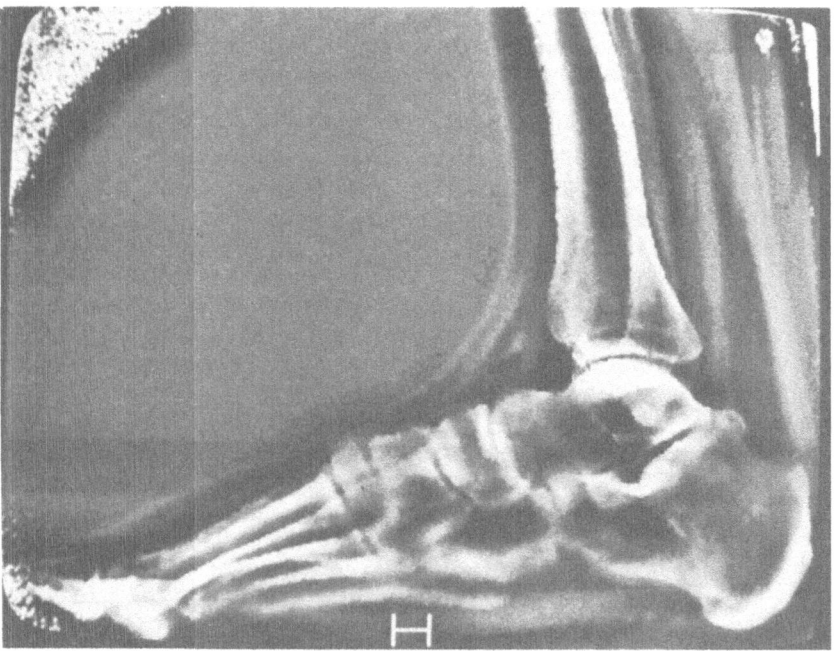

Fig. 6. (a) X-ray of a human foot. (b) Heavy-ion radiograph
(grey level display) of a human foot. Courtesy of C. A. Tobias.

Fig. 7.   Perspective graphical display of Fig. 6.

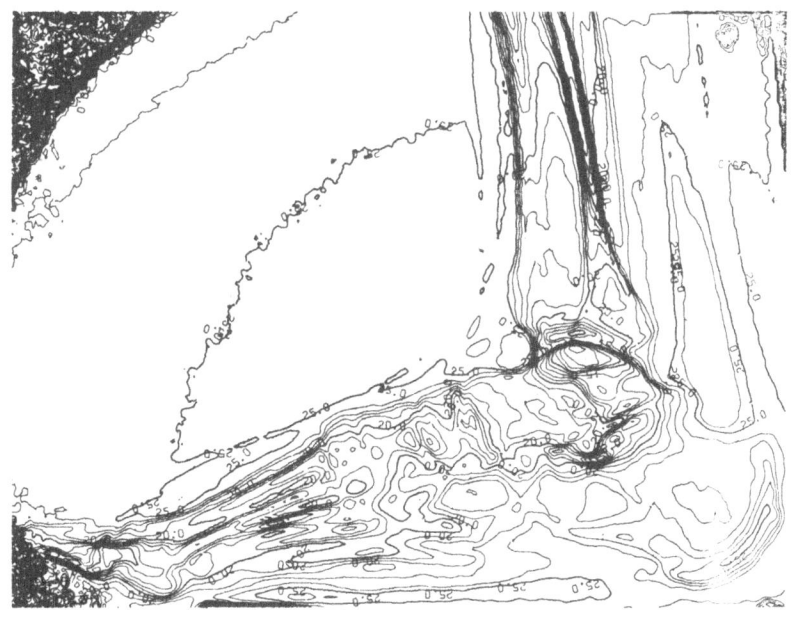

Fig. 8.   Isodensity contour map of Fig. 6.

while for 10-mm^3 volume elements, the dose needed is 5 rad.

Williams and Leith (1978) at SLAC have obtained radiographic images using high-energy pion beams, a magnetic spectrometer, and spark chambers in a similar technique to NSR. Outgoing particles from inelastic reactions such as: $\pi^-A \rightarrow \pi^-\pi^+A'$ or $\pi^-A \rightarrow \pi^-\pi^-\pi^+A$ (where A represents an atomic nucleus in the target) are used to determine the interaction vertices. Radiation dose to the target is relatively high (comparable to the NSR exposures of Saudinos and Charpak), but good spatial resolution should be obtainable because of the 5 GeV/c incident π^- beam momentum.

STUDIES WITH PHANTOMS

The potential advantage of heavy particle radiography over x-rays lies in the superior density resolution attainable at a given dose. This should permit better visualization of tumors differing only by a small amount in density from surrounding tissues. To test the relative abilities of heavy particle radiography and x-rays, the LBL group performed experiments with a phantom representing the body, in which a number of objects simulating tumors with different densities were suspended (Tobias, private communication; Sommer et al., 1978). The phantom is shown in Fig. 9a, and consisted of a rectangular lucite box 7.5-cm thick filled with water. Five 2.5-cm diameter latex rubber spheres containing sucrose solutions of varying specific gravities, from 1.005 (upper left) to 1.030 (lower middle) were suspended in the water bath. The sixth sphere at the bottom right contained monobasic potassium phosphate, a substance of mean Z close to bone. The phantom was radiographed with 250 MeV/amu carbon ions, with the dose to the center of the phantom equal to 19 mrad. Figure 9b is the image from one sheet of the detector, showing that all six spheres are visible. In contrast, the x-ray of the phantom taken with 155 mrad using a screen film combination of 60 kVp and 16 mAS is shown in Fig. 9c. Here, only the balloon filled with the potassium phosphate is unambiguously visible. For further comparison, an x-ray CT scan was taken on an EMI 5000 whole body scanner. The sphere with the 1.010 g/cm^3 is rather ambiguous and the lowest density sphere is not visible. The dose in this scan was about 1 rad. This experiment established that there is a very significant dose advantage in the use of heavy-ion radiography over x-rays in the resolution of tumor-like objects differing only slightly from surrounding tissues. The carbon radiograph taken with 19 mrad mean phantom dose was superior to the CT scan delivering 1 rad to the scan section and much superior to the plain radiograph using 155 mrad mean phantom dose.

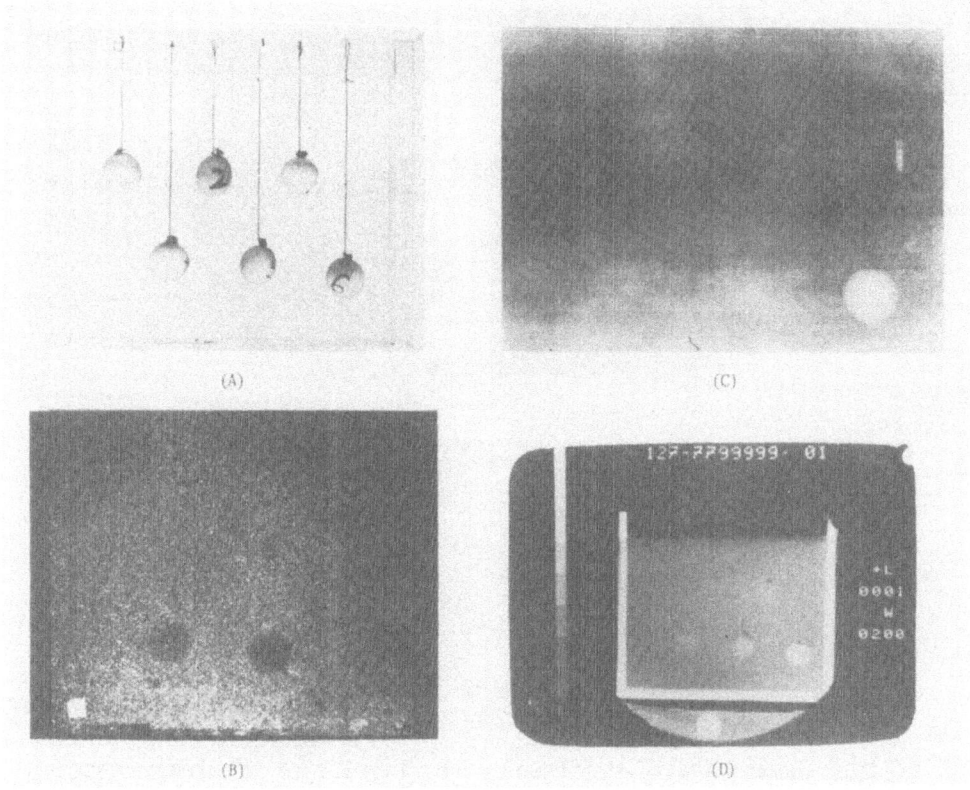

Fig. 9. (a) Photograph of phantom with latex balloons filled
with solutions; (b) heavy-ion radiograph of phantom;
(c) x-ray of phantom; (d) CT scan of phantom. Reprinted,
by permission, from Tobias (1978).

TISSUE SPECIMEN RADIOGRAPHY

Steward and Koehler (1975b) demonstrated the possibility of
visualizing cerebral vascular lesions (strokes) with charged
particle radiography. A blood clot in the brain is 2 to
3 percent denser than healthy brain tissue while a brain
infarct (death of brain tissue due to obstruction of a blood
vessel) is about 5 percent less dense than normal brain
tissue. The proper diagnosis between these two different
conditions would lead to different treatments. Figure 10 shows
an image of a brain with an intracerebral hemorrhage, with the
dark region corresponding to a denser region as compared with
the normal brain. Figure 11 illustrates a cerebral infarction,
due to a recent thrombotic occlusion of the left middle cerebral artery. The lighter regions in the left hemisphere
correspond to the region of the infarct.

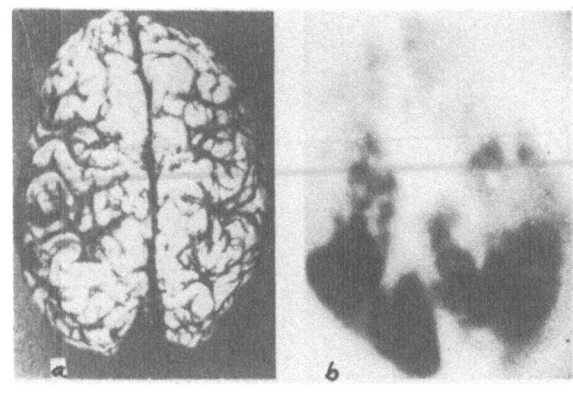

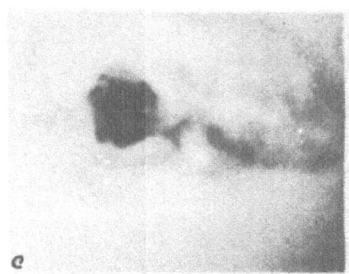
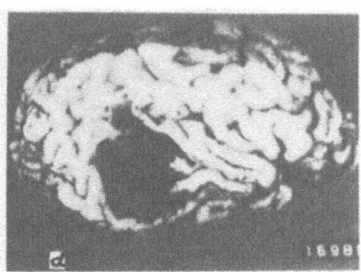

Fig. 10. Human brain specimen with intracerebral hemorrhage.
(a) photograph, superior view; (b) proton radiograph,
superior view; (c) proton radiograph, lateral view;
(d) photograph, lateral view. Reprinted, by
permission, from Steward and Koehler (1975).

     In 1972, Steward and Koehler (1973a) also showed that
proton contact radiography could be used to detect density
differences in tumors in the brain. The specimen consisted of
a human brain with a metastasis from a pancreatic carcinoma.
Figure 12a is a medial view of a formalin-fixed right hemisphere, which contained the metastasis. Figure 12b shows the
proton radiograph of the hemisphere with the lesion visible
below the midpoint of the superior border. Figure 12c is an
x-ray radiograph of the hemisphere in the water box, with the
brain and metastasis not distinguishable. Figure 12d is an
x-ray radiograph of the brain in air, and Fig. 12e a photograph
of a slice taken through the tumor.

     Figure 13 is a photograph and a carbon radiograph of a
fresh liver specimen containing numerous deposits of metastatic
oat cell carcinoma (Sommer, et al., 1978). The light region
surrounding the area representing the full thickness of the

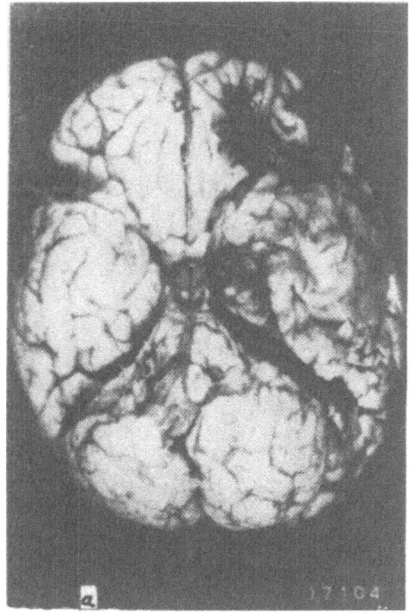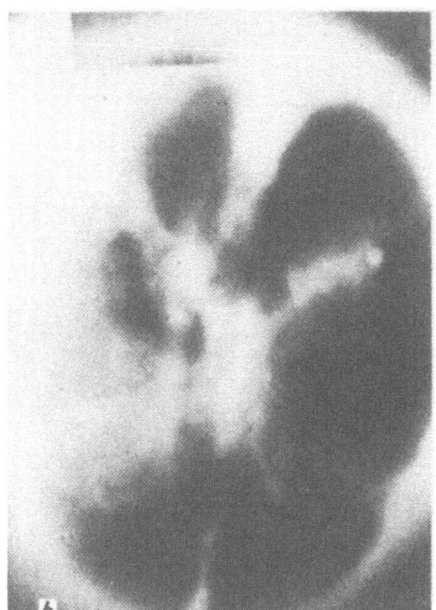

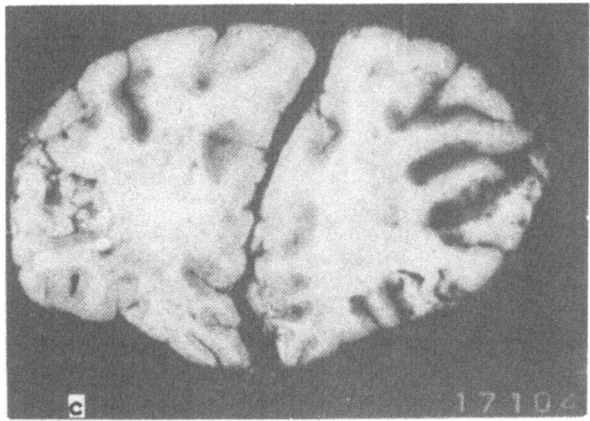

Fig. 11. Human brain specimen with cerebral infarction.
(a) Photograph, inferior view; (b) proton radiograph;
(c) photograph, coronal section. Reprinted, by
permission, from Steward and Koehler (1975b).

tissue specimen is due to the wedge shape of the specimen. The
small light areas within the specimen are due to trapped air in
the vascular and ductal structures. Several metastatic deposits within the liver are easily identified. The corresponding
x-ray radiograph failed to demonstrate the location of the
tumor.

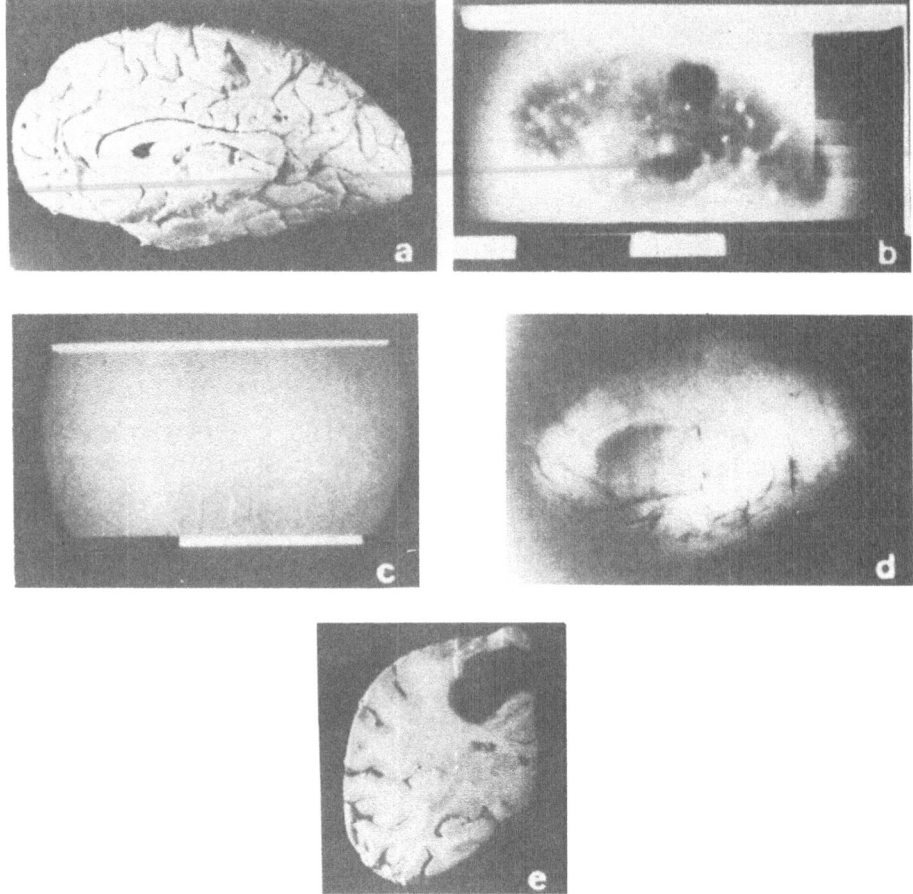

Fig. 12. Metastatic tumor in the brain. (a) Medial view
          photograph; (b) proton radiograph; (c) x-ray in water
          bath; (d) x-ray in air; (e) photograph of a slice of
          the brain taken through the tumor. Reprinted, by
          permission, from Steward and Kohler (1973a).

     The proton radiograph of a freshly excised human heart
(Kramer et al. (1979) by the beam scanning technique is shown
in Fig. 14. The image is approximately in the anatomically
correct position for an anterior view of the chest. This
radiograph shows clear sepa- ration of the right and denser
left ventricles, the atrioven- tricular groove, the auricle of
the right atrium, and the pulmonary trunk and aorta. The less
dense epicardial layer of the heart is also seen.

     These radiographs of tissue and organs demonstrated that
organs with structures of slightly different densities (tumors

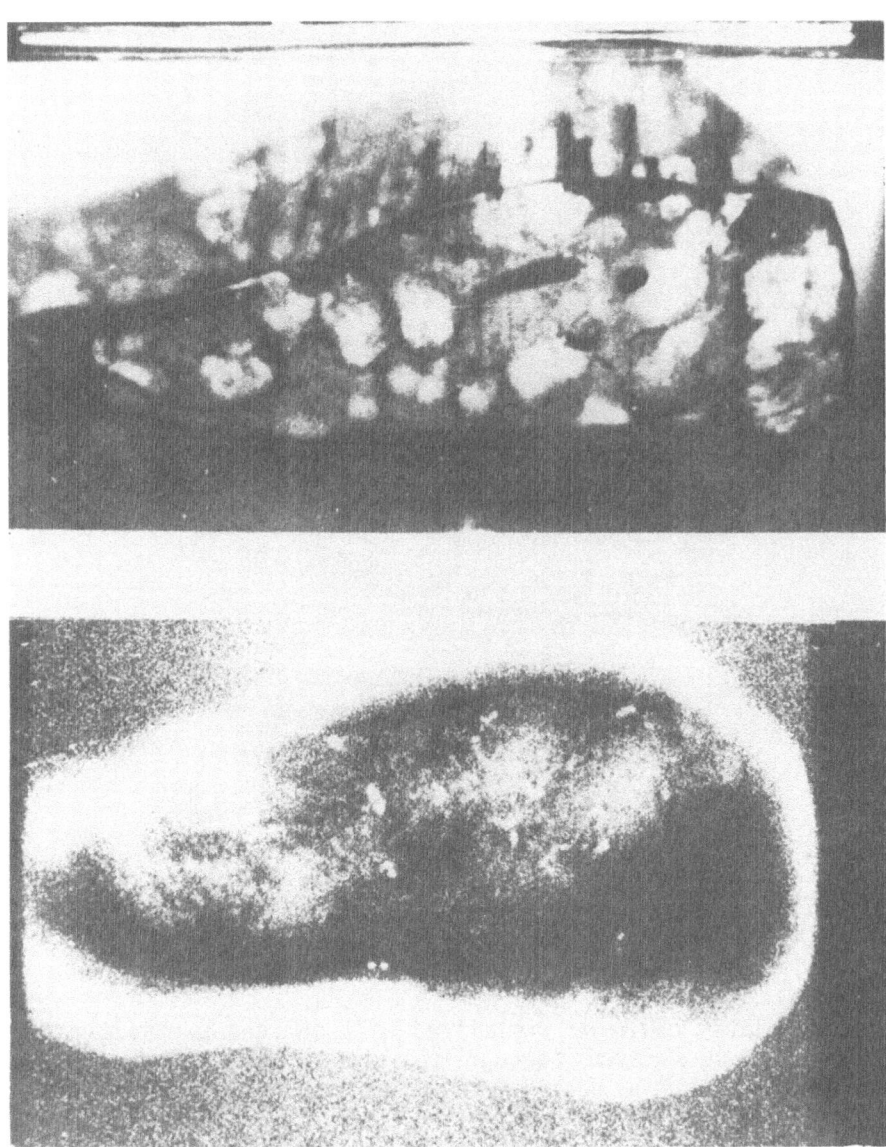

Fig. 13. (a) Photograph of liver specimen containing oat cell
carcinoma; (b) carbon ion radiograph of same
specimen. Reprinted, by permission, from Tobias
et al. (1978).

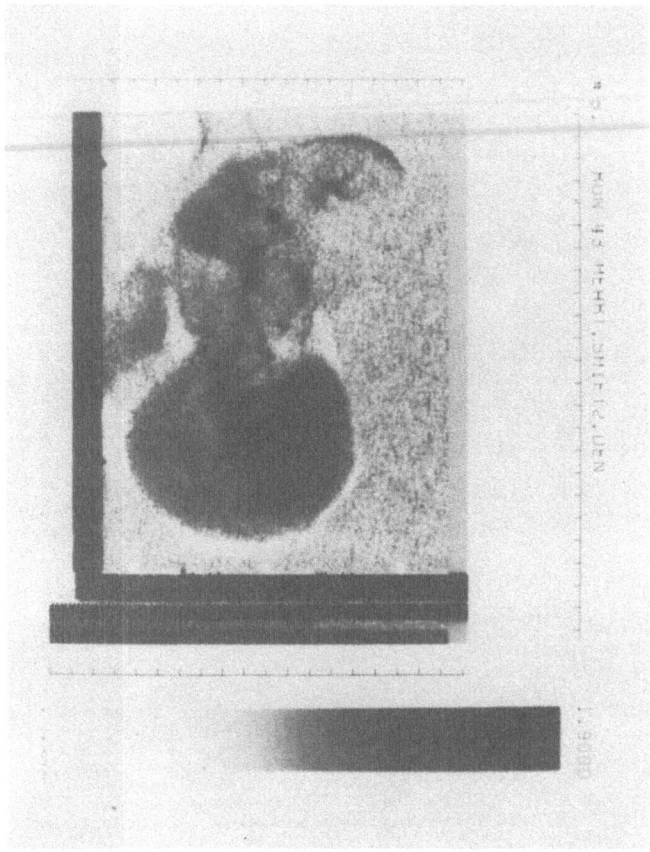

Fig. 14. Proton radiograph of human heart. Reprinted, by
          permission, from Kramer.

or blood clots) could be visualized with charged particle radiography. When x-rayed through the water box containing the
specimen, very little detail was detectable.

HUMAN STUDIES

Most of the studies involving patients both at Harvard
University and LBL (C. A. Tobias, private communication;
Steward and Koehler, 1974) have centered around the use of
charged particle radiography in detection of breast cancer.
The LBL work in this area is now described. Heavy-ion radiography of the breast may prove to be a unique advancement in
quantitative noninvasive diagnostic imagining (C. A. Tobias,

private communication). Its potential lies in augmenting and improving the cancer detection rate of very small lesions with very low radiation dose to the breast.

Studies at LBL on the stopping power values of normal and neoplastic tissues derived from heavy ion radiography (Sommer et al., 1978) of specimens show the promise of more accurate diagnosis of benign and malignant tumors of the breast by diagnostic imaging and accurate quantitative densitometry of the stopping power distribution in tissue. From mastectomy specimens, some stopping power values relative to water have been measured: normal breast tissue, 0.957 ± 0.004, mammary dysplasia, 1.016 ± 0.011, and infiltrating ductal carcinoma, 1.064 ± 0.003. The large difference in stopping power between mammary dysplasia and carcinoma may lead to a quantitative method for differentiating between benign disease and breast cancern using heavy-ion radiography.

Advances in breast imaging with ionizing radiations have depended on dose reduction factors which can be achieved without impairing image quality. New x-ray film screen combinations have succeeded in reducing the dose to the breast by a factor of four. However, the combination of heavy-ion and plastic detector stacks has resulted in diagnostic mammographic images with a dose reduction factor of 25 to 250. Conventional x-ray mammography with soft x-rays often results in a dose per film between 0.5 and 3 rad. A three-view mammographic study could result in a dose to each breast as high as 1.5 to 10 rad. With the new film screen combinations, a complete examination would half this dose to 0.75 to 5 rad per breast. Xeroradiography can now be done at a dose of about 0.5 to 2 rad per film. Heavy-ion radiography of the breast provides a diagnostic procedure to investigate early detection of breast cancer not only because of its ability to detect and quantitate very small differences in tissue density but because it can be done with doses as low as 20 mrad per projection.

The actual radiography is carried out in the following manner. A patient couch is positioned over the beam line. A water bath with parallel walls filled with warm water and a wetting agent contains the breast as the patient lies on the couch. A single beam pulse (less than 1 sec) is used in the irradiation. About 25 sheets of plastic detector are placed downstream of the specimen box. Two views are taken by rotating the couch relative to the beam line. A total of 34 patients have received carbon mammography examination. In all patients, whenever x-ray mammography detected abnormal densities within breast tissue, the carbon ion images demonstrated the same abnormal densities. In addition, carbon ion radiographs (Fig. 15) demonstrated the presence of abnormal densities

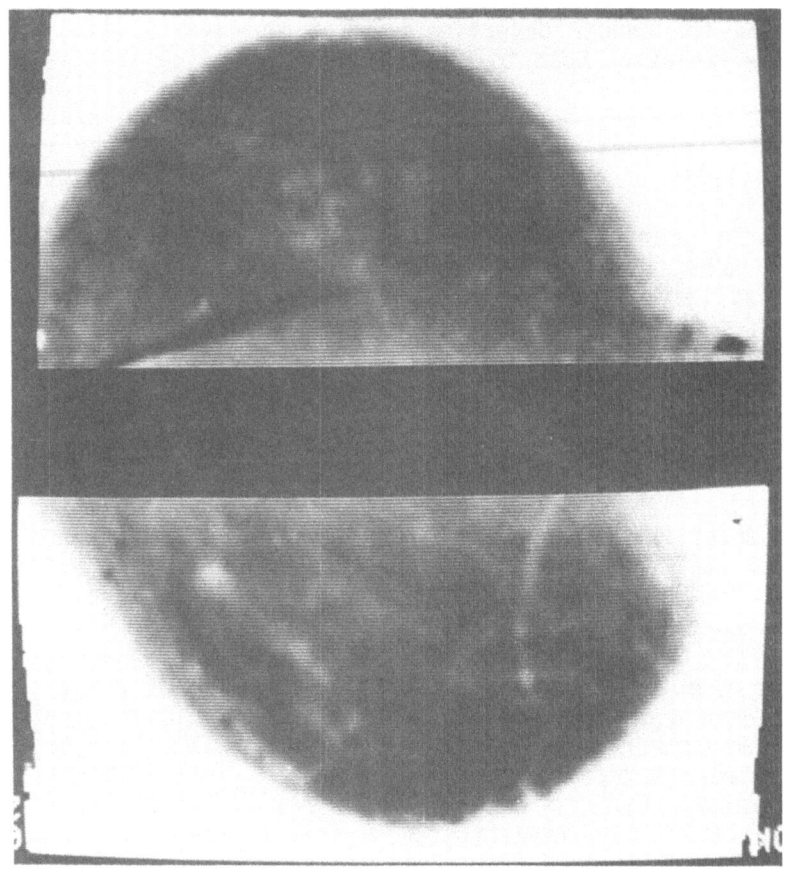

Fig. 15. Carbon-ion mammography of a patient. Upper, normal
          breast; lower, breast with a nonpalpable tumor.

in the breasts of two patients not detected by mammography.
These were nonpalpable masses and measured less than 1 cm on
the heavy ion images. A second phase clinical trial at LBL
will be started to continue this study.

CHARGED PARTICLE COMPUTERIZED TOMOGRAPHY

    The use of charged particles in diagnostic radiology may
also be extended to include axial tomography. In 1975, Crowe
et al. (1975) produced axial scans of the head with 900 MeV
helium ions. Their experimental setup is shown in Fig. 16a. A

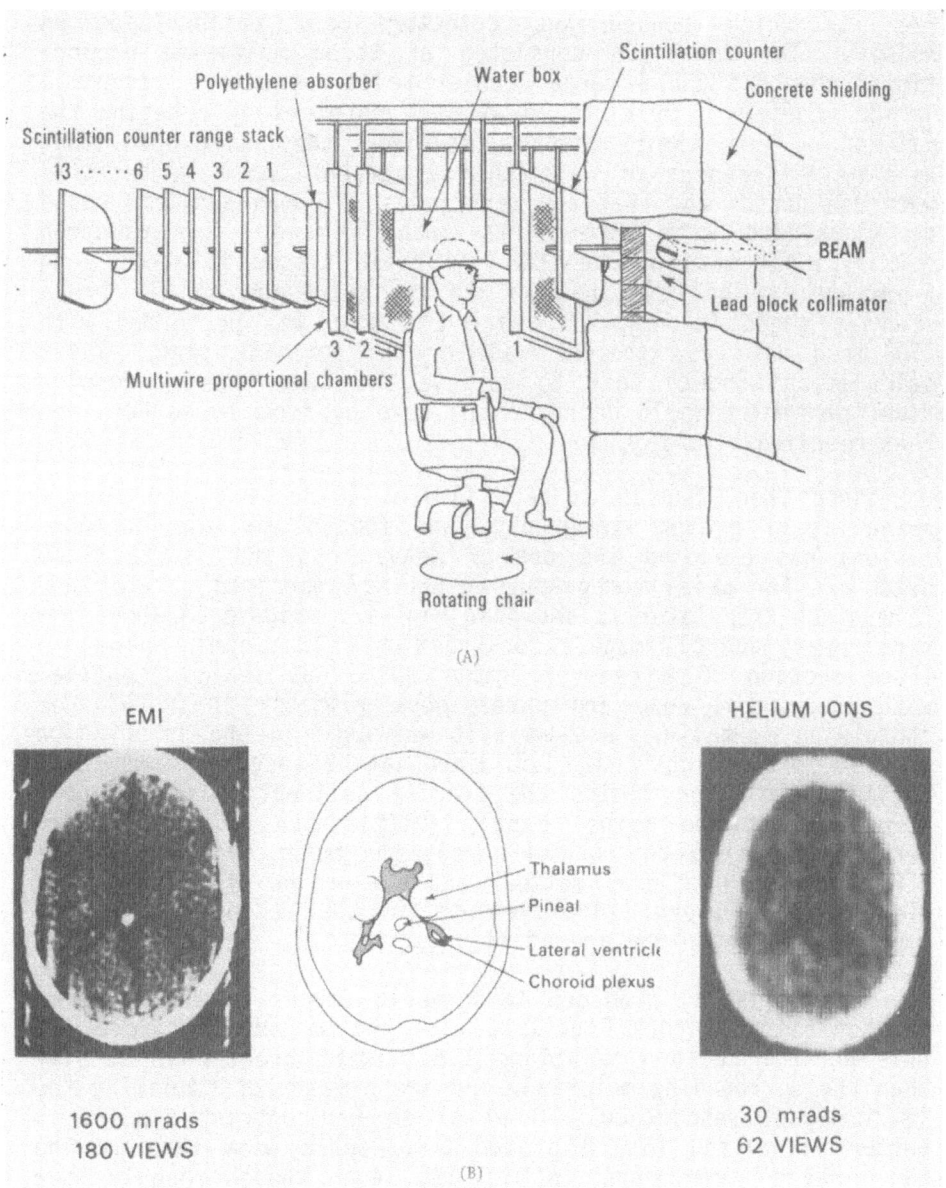

Fig. 16. Helium—ion axial tomography. (a) Experimental setup,
(b) EMI and particle axial tomograms. Reprinted, by
permission, from Crowe et al. (1975).

beam of helium ions was directed down a beam channel to a
facility which housed the detectors and patient/specimen
holder. The detectors consisted of three multiwire proportional chambers and a range counter telescope with 13 scintillation counters. Multiple views were obtained by rotating the
patient. From their phantom studies, they concluded that
density differences of less than 2 percent could be detected
with low doses, and that the ventricles in human subjects could
be visualized with doses less than 50 mrad. The electron
density difference between CSF and brain tissue is about 3 to
4 percent. A helium ion scan and the corresponding EMI brain
scan is shown in Fig. 16b. The EMI scan was performed with
1600 mrad dose vs. the 30 mrad for the particle scan. Their
calculations showed that to achieve the same density resolution, particles could decrease the dose by from 10 to 50 times
that required by x-ray CT.

Since then, Hansen at LASL (1978a, 1978b) has studied the
potential of proton tomography, and Tobias (private communication) has explored the use of heavy ions and pastic stack
detectors for axial tomography at LBL. A specimen rotator and
film stack translator is shown in Fig. 17. Figure 18 shows the
first heavy ion CT image reconstruction of the human spine. A
15-cm section of a cadaver lumbar spine was sealed (together
with the spinal cord and paraspinous muscles) in a cylinder
containing formalin. A 2-mm slit was used in the irradiation
and 90 projections were obtained at 2 degree intervals.
Spatial resolution limits the precise delineation of internal
structures of the spinal canal. Differences in density are
detectable within the spinal canal, the outer cylinder of the
CSF, the region of gray matter, and the region of white matter
of the spinal cord. Improvements in the reconstruction and
experimental technique are still under way.

Hansen (1978B) produced tomographic images of a resolution
phantom using protons Fig. 19a. The bottom set of circles in
this phantom contain solutions 1.8 percent greater in density
than the surrounding material, and are a means of comparing the
low contrast resolution. This set of low contrast circles is
better visualized in the proton tomographic scan than in the
EMI x-ray CT scan shown in Fig. 19b, even though roughly four
times the dose was required for the EMI scan. Another feature
of the proton scan is that there is no variation of the CT
number across the image as found in the x-ray image. The x-ray
image is subject to a cupping effect, where the CT numbers in
the center of the circular phantom are lower than at the edges
(a result of beam hardening of a polychromatic x-ray beam).

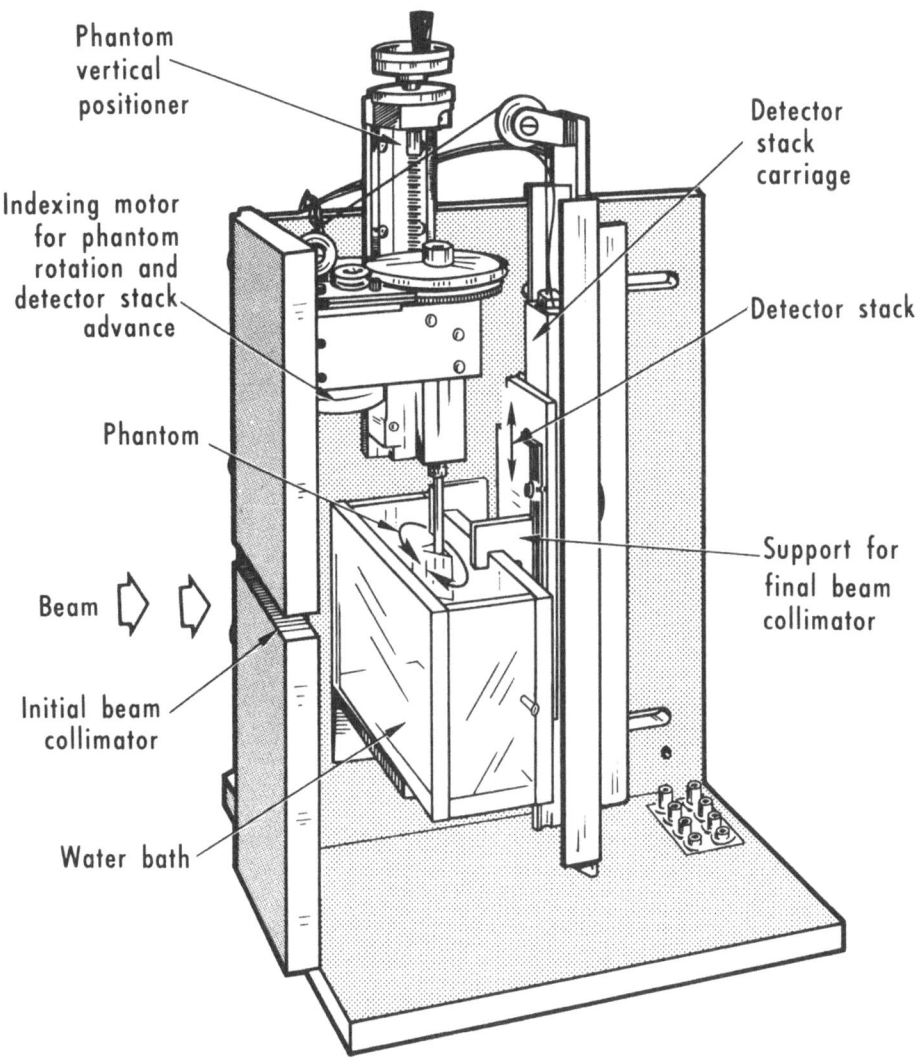

Fig. 17. Device for heavy-ion CT scans of specimens using plastic detectors (courtesy of C. A. Tobias).

SUMMARY

The use of charged particles in diagnostic radiological medicine is being investigated at several centers through the world. These studies are directed toward the development of a technique with improved density resolution and reduced dose as compared with conventional radiography. While physical

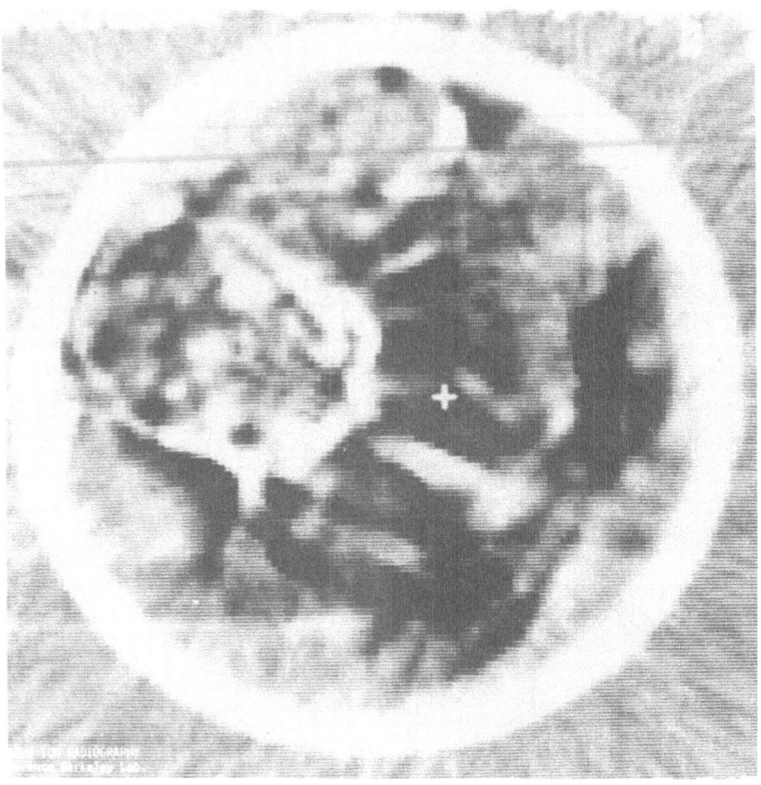

Fig. 18. Particle reconstruction of a human spine specimen using carbon ions. Reprinted, by permission, from Holley (1979).

tomography indicate a substantial improvement in certain areas over conventional methods, further investigations and clinical trials will be needed to establish their clinical efficacy and practicality.

ACKNOWLEDGEMENTS

The authors wish to acknowledge the generous use of figures and photographs from Drs. C. A. Tobias and J. I. Fabrikant of LBL, Drs. E. V. Benton and R. Henke of the University of San Francisco, Dr. V. Steward of the University of Chicago, Dr. A. Koehler of the Harvard Cyclotron Laboratory, Dr. S. Kramer of Argonne National Laboratory, and Dr. K. Crowe of LBL. Some of these studies were supported by the National Cancer Institute (YO1–40302 and CA 19138), and the U. S. Department of Energy under Contract NO. W–7405–ENG–48.

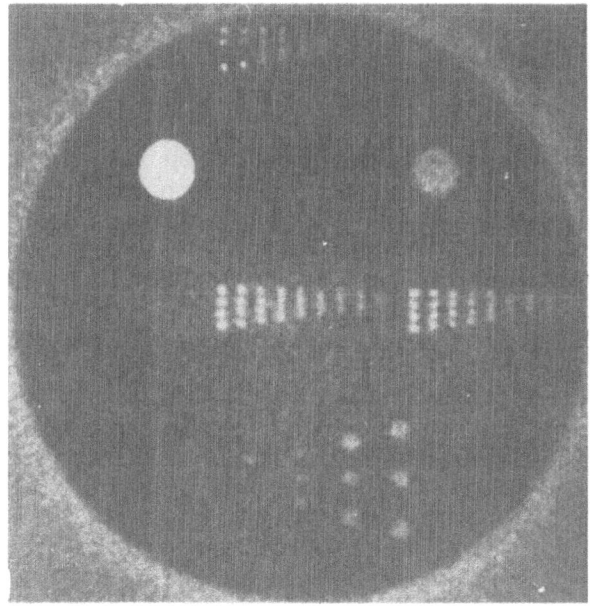

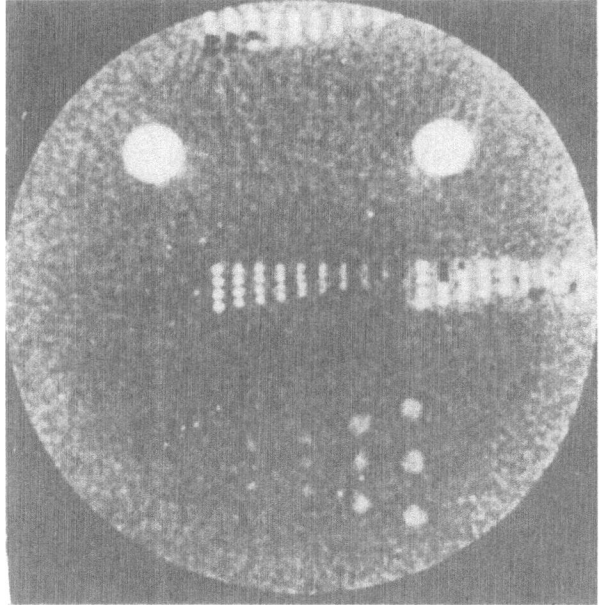

Fig. 19. (a) proton CT image of phanton;
          (b) x-ray scan of phantom.

REFERENCES

Benton, E. V., Henke, R. P. and Tobias, C. A., 1973, Heavy particle radiography, Science, 182, 474.

Benton, E. V. Henke, R. P., Tobias, C. A., and Cruty, C. R., 1975, Radiography with heavy particles at Donner Laboratory. Lawrence Berkeley Laboratory Report LBL-2887.

Charpak, G., Majewski, S., Perrin, Y., Saudinos, J., Sauli, F., Townsend, D., and Vinciarelli, J., 1976, Further results in nuclear scattering radiography, Phys. Med. Biol., 21, 941.

Crowe, K. M., Budinger, T. F., Cahoon, J. L., Elischer, V. P., Huesman, R. H., and Kanstein, L. L., 1975, Axial scanning with 900 MeV alpha particles, IEEE Trans. Nucl. Sci., NS-22, 1752.

Hanson, K. M., 1978a, Proton computed tomography. Los Alamos Scientific Laboratory Report LA-UR 78-1827.

Hanson, K. M., 1978b, Development of a proton radiographic system for diagnosis and localization of soft tissue abnormalities. Los Alamos Scientific Laboratory Report LA-7107-MS.

Holley, W. R., Henke, R. P., Gauger, G. E., Jones, B., Benton, E. V., Fabrikant, J. I., and Tobias, C. A., 1979, Heavy particle computed tomography, Proceedings, Sixth Computer Radiology.

Koehler, A. M., 1968, Proton radiography, Science, 160, 303.

Kramer, S. L., Moffett, D., Martin, R. L., Colton, E. P., and Steward, V. W., 1979, Proton radiography application to medical imaging, submitted to Radiology.

Saudinos, J., Charpak, G., Sauli, F., Townsend, D., and Vinciarelli, J., 1975, Nuclear scattering applied to radiography in medical diagnosis, Phys. Med. Biol., 20, 890.

Sommer, F. G., Capp, M. P., Tobias, C. A., Benton, E. V., Woodruff, K. H., Hinke, R. P., Holley, W., and Genant, H. K., 1978, Heavy-ion radiography: density resolution and specimen radiography, Invest. Radiol., 13, 163.

Steward, V. W., 1976, Proton (and other heavy charged particle) radiography in medical diagnosis, IEEE Trans. Nucl. Sci., NS-23, 5770.

Steward, V. W., and Koehler, A. M., 1973a, Proton beam radiography in tumor detection, Science, 179, 913.

Steward, V. W., and Koehler, A. M., 1975b, Proton radiographic detection of stroke, Nature (London), 245, 38.

Steward, V. W., and Koehler, A. M., 1974, Proton radiography in the diagnosis of breast cancinoma, Radiology, 110, 217.

Tobias, C. A. Benton, E. V., Capp, M. P., Chatterjee, A., Cruty, M. R., and Henke, R. P., 1977, Particle radiography and autoactivation, Int. J. Radiat. Oncol. Biol. Phys., 3, 34.

Tobias, C. A., Benton, E. V., and Capp, M. P., 1978, Heavy ion radiography, in "Recent Advances in Nuclear Medicine," vol. 5, J. H. Lawrence and T. F. Budinger, eds., Grune and Stratton, New York.

Williams, S. H., and Leith, D. W. G. S., 1979, Use of elementary particle interactions for radiological imaging, SLAC-PUB-2417.

ADVANCES IN DIAGNOSTIC INSTRUMENTATION

Victor Perez-Mendez

Lawrence Berkeley Laboratory
Berkeley, CA 94720

University of California, San Francisco
San Francisco, CA 94143

INTRODUCTION

The art of medical diagnosis is a continuously evolving process which often owes its progress to technical developments in other scientific fields.

The major advance in Radiology during the last ten years has been the development of Computerized Tomography. Although many of the principles of this technology have been known for considerably longer periods of time, the actual implementation became feasible only with the development of versatile high speed small computers with large memory storage capacities at modest prices. Other technological developments which made C. T. a reality were (a) advances in Applied Computational Mathematics such as the Fast Fourier Transform algorithms, as well as graphic display programs, and (b) developments in new photon detectors.

These advances in diagnostic x-ray imaging have had their influence in other branches of Radiology, such as Nuclear Medicine and Ultrasonography.

Nuclear medicine provides complementary information to x-ray diagnosis in that it provides information on the vascular and physiological processes that distribute a given radioisotope-labelled compound within the desired organ. The possibility of obtaining tomographical information on the concentration distribution of the radioisotope distribution was recognized even earlier than in the x-ray case. Its implementation is now taking place, although at a slower rate than with the Transmission Computerized Tomography.

Ultrasonography is an imaging technique which is also undergoing rapid new developments. A major impetus to its expanded use
is the fact that it does not involve ionizing radiation, and at the
intensity levels which are under use, is not known to create any cell
damage. The present techniques, which are almost duplicates of older
Radar imaging concepts, provide a Tomographic image in the well known
B-scan approach. However, Radar imaging with the aid of more elaborate electronics to provide faster and more accurate pattern recog-
nition is now providing the technological basis for advances in
Ultrasound Imaging.

Lastly, the well known x-ray tube itself has recently under-
gone improvements in design that permit higher quality transmission
x-ray pictures. In order to detect smaller structures or more
detail in a transmission image, it is necessary to magnify the image
either in the x-ray projection or by enlarging the resultant film.
Principles of electron beam focussing taken from electron accelerator technology have played a role in this development.

Table I lists the topics covered in this paper. Other diagnostic imaging concepts involving Nuclear Magnetic Resonance (NMR),
Thermography (infrared and microwave), as well as Electrical Imped-
ance C. T., are areas in which there is a lot of research activity
at present.

Table I. Developments in Medical Imaging

1) Fine Focus X-ray Tubes
   a) Structure of Electron Gun
   b) Resolution Patterns

2) Computerized Tomography: X-ray Transmission
   a) Configurations of C. T. Machines: 2π Geometry
   b) Fast C. T, Machines: Limited Fan Angle Geometry
   c) Principles of Limited Angle Reconstructions

3) Emission Tomography: Radioisotope Imaging
   a) Single Photon Emission Tomography
   b) Positron Imaging: Planar and 2π Ring Cameras

4) Ultrasound Imaging
   a) Grey Scale B-scan Principles
   b) Phased Array Transducers: Electronic Steering and Focussing
   c) Real Time B-scans.

HIGH RESOLUTION RADIOLOGY

The technology of x-ray tube construction and performance is obviously one to which the manufacturers place close attention. In this section we discuss one recent approach to high resolution x-ray tubes which provide a focal spot at the anode of the tube smaller than 90 microns. The heat dissipation limits the maximum power that can be used and hence the maximum x-ray exposure.

Figure la shows a conventional medium power x-ray tube using a metal focussing cup around the cathode. Tubes of this type have focal spots ranging from 0.5-3 mm, depending on the power used. A method used to produce a smaller focal spot [1] involves building the tube with a series of focussing grids, as shown in Fig. lb, that can focus the electron beam to spots smaller than 80 microns. Since x-ray pictures are projections of the transmission from the source (x-ray tube) through the object to the film, the overall resolution depends on the geometrical spacing of all three, with the limitation being the effective dimensions of the x-ray source. High resolution x-ray pictures are of use in studying bone structures as shown in Fig. 2a. Another application in which high resolution x-ray imaging has been used recently is in the diagnosis of breast tumors by show-ing the existence of small calcifications ranging in size from 0.3 mm to a few mm in size. Malignant tumors of the breast are often as-sociated with the smaller calcifications [2], a sample of which is shown in Fig. 2b.

COMPUTERIZED TOMOGRAPHY

Tomography is the name given to the process by which a three-dimensional object can be displayed graphically as a series of slices. In medical diagnosis there are presently in use the tomo-graphic procedures defined in Table II and illustrated in Fig. 3.

X-ray C. T. is achieved by detecting electronically the trans-mission of the x-ray beam through an object when viewed from a series of angles. One slice of the object is measured at a time: the x-ray source is moved around 2π angle around the object. Various computational algorithms involving Fourier Transforms are used in reconstructing the image on matrices with size up to 300 x 300 picture elements called Pixels. Fig. 4 shows configurations of various x-ray C. T. machines. The first successful C. T. machine, the EMI head scanner of 1973, accomplished the multidirectional viewing by using a parallel beam geometry with a single sodium iodide crystal detector, moved in a combination of translational and rotational motions.

This slow (~5 min./scan) machine was replaced next by a fan beam geometry in which a series of detectors recorded the attenuation

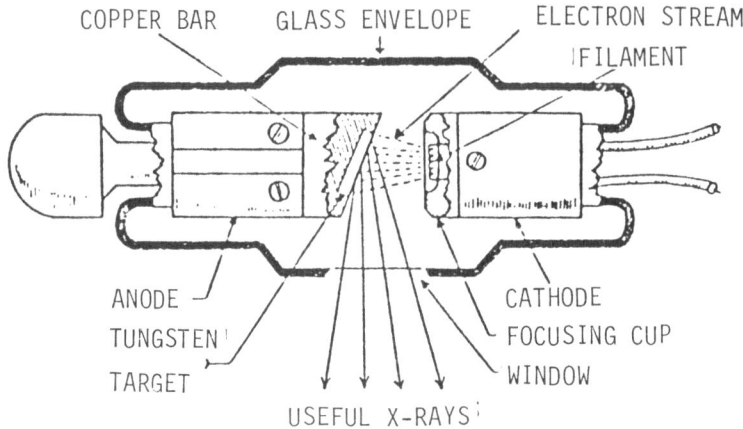

Fig. 1a. Stationary anode x-ray tube with cathode focussing cup.

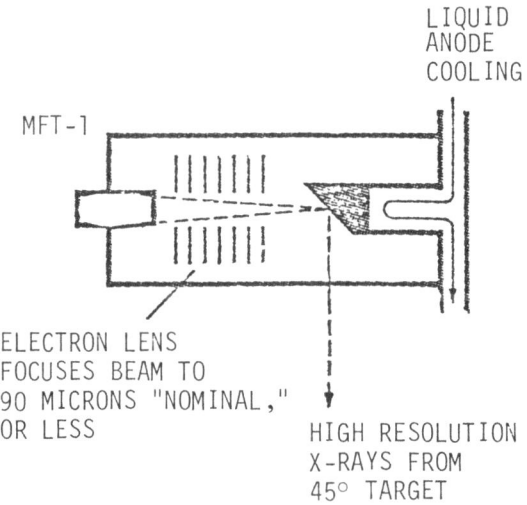

Fig. 1b. Fine focus tube with electron lens focussing.

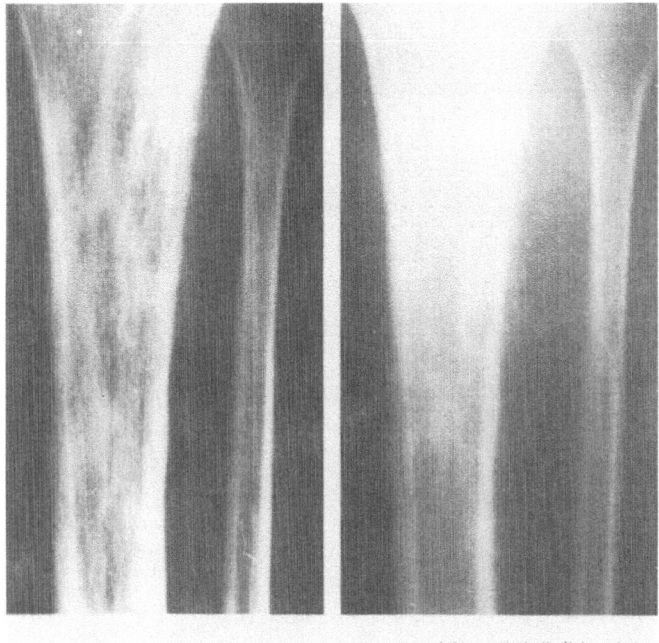

Fig. 2a. Comparison bone structure: conventional and fine focus
tube.

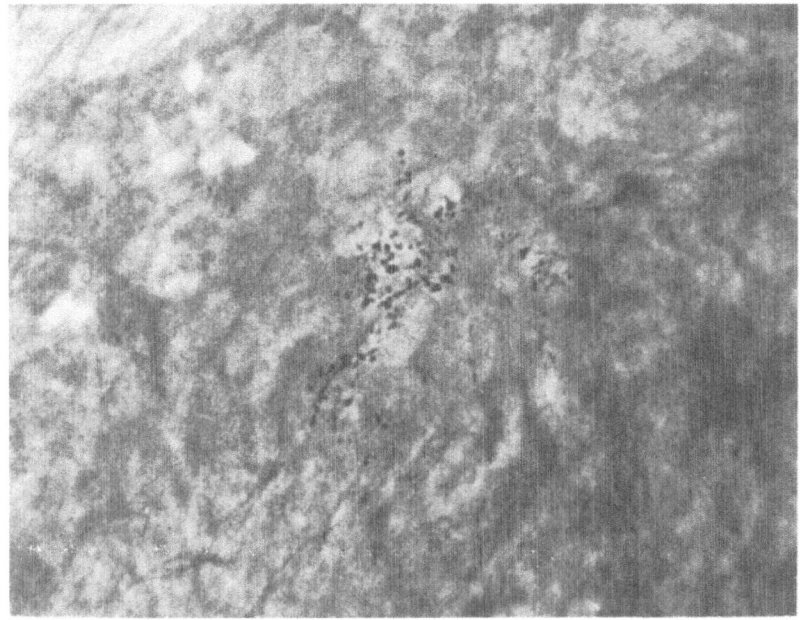

Fig. 2b. Radiogram for breast cancer calcifications: fine focus
tube.

Transmission: X-Ray   Emission: Gamma Rays

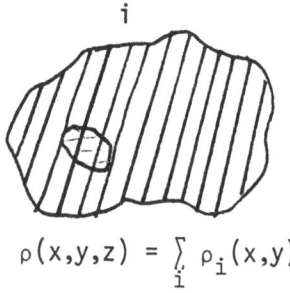   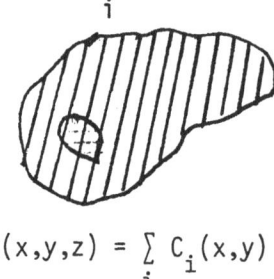

$$\rho(x,y,z) = \sum_i \rho_i(x,y)$$ $$C(x,y,z) = \sum_i C_i(x,y)$$

Anatomy   Physiology

Density altered by tumors,   Concentration of radioisotope
cysts.   altered by t mors, cysts.

Ultrasound: B-Scan

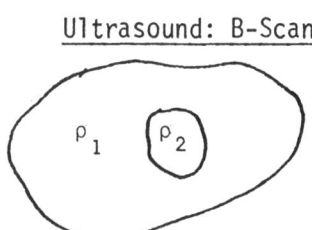

Anatomy and Tissue Characteristics

Cyst or tumor produces density boundary.

Fig. 3.   Comparison of various tomographic imaging modalities.

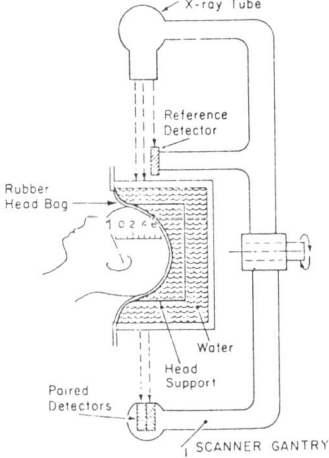

Fig. 4a.   Early version of E.M.I. head scanner.

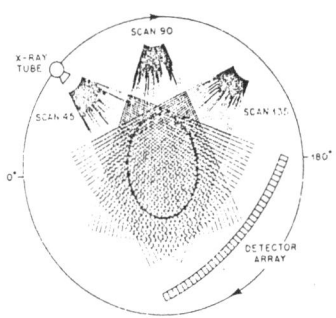

Fig. 4b.   Third generation scanner: rotating x-ray tube coupled
to rotating detector array.

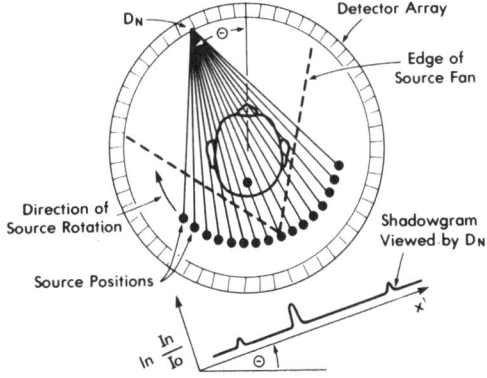

Fig. 4c.   Fourth generation scanner: rotating x-ray tube: fixed
detectors.

Table II. Medical Tomography

Tomography: Greek word meaning "to take a slice," in this case to
     take a 3-D object and display graphically as a series of slices.

Computerized Tomography (X-ray C. T.)L Display density distribution
     of body (organ) section.

Ultrasound B-scan: Display tissue boundaries of body (organ) sec-
     tions.

Emission Tomography: Display concentration distribution of gamma-
     emitting radioisotope in organ section.

---

of the beam from the x-ray source through the object. The x-ray
tube and the detector array moved jointly along the periphery of a
mechanical gantry. This arrangement proved to be unsatisfactory in
that any slight inhomogeneity in the gain of the detectors provided
circular artifacts on the image. This difficulty was eliminated in
the "fourth generation" scanner in which 200-300 individual detectors
are mounted on a circular gantry; the x-ray tube alone moves. The
gain of each detector in the bank is adjusted electronically on each
rotation. This scheme eliminated most circular artifacts.

     In earlier machines, the detectors were xenon gas ionization
chambers; in some of the newer C. T. machines there are now sodium
iodide scintillation counters. Some newer machines will use bismuth germanate scintillation crystals, since their detection efficiency is higher than that of sodium iodide (due to the high Z of
the bismuth). Also, bismuth germanate crystals do not have the
low intensity, long-term fluoresence that sodium iodide has and
therefore favors the main objective of newer C. T. machines, which
is to provide higher resolution images during a shorter period of
time. This is necessary in order to accomplish dynamic imaging of
the heart in motion.

     Fig. 5a shows a picture of the General Electric fast scanner
in research use in the Radiology Department, University of California, San Francisco [3]. This machine, a fourth generation scanner, produces two 180° images during a single 2.4 sec, 360° revolu-
tion. Following a 1 sec pause the scanner reverses direction and
scans back in 2.4 sec. This sequence is repeated until 12 images
at approximately 1.5 sec intervals have been produced. Fig. 5b
shows the kind of star artifact that can appear in a scan if a high
density object such as a metal clip is present.

     Comparable fast scanners have been made by other manufacturers

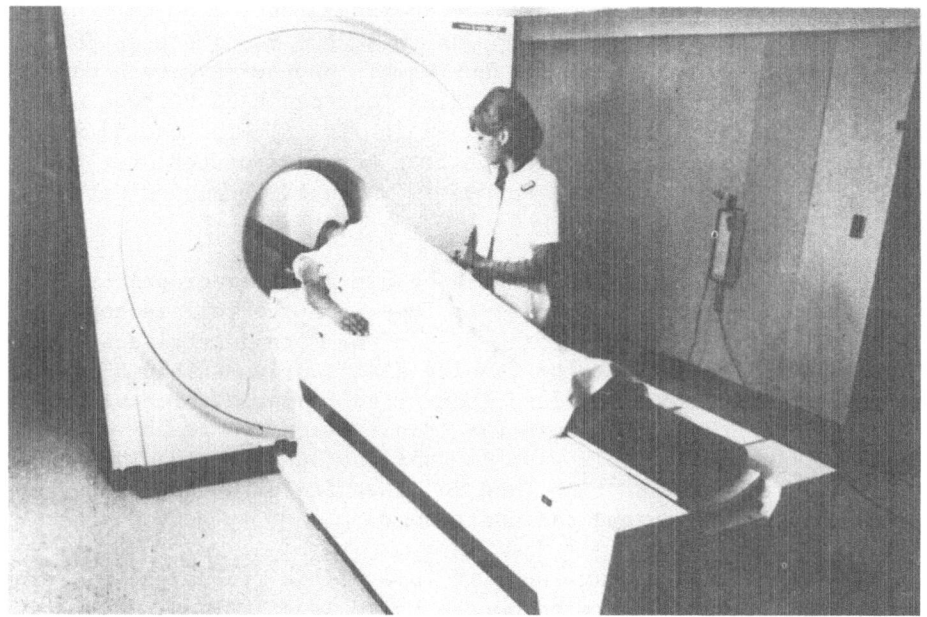

Fig. 5a. General Electric scanner. X-ray tube and detectors in
          circular gantry.

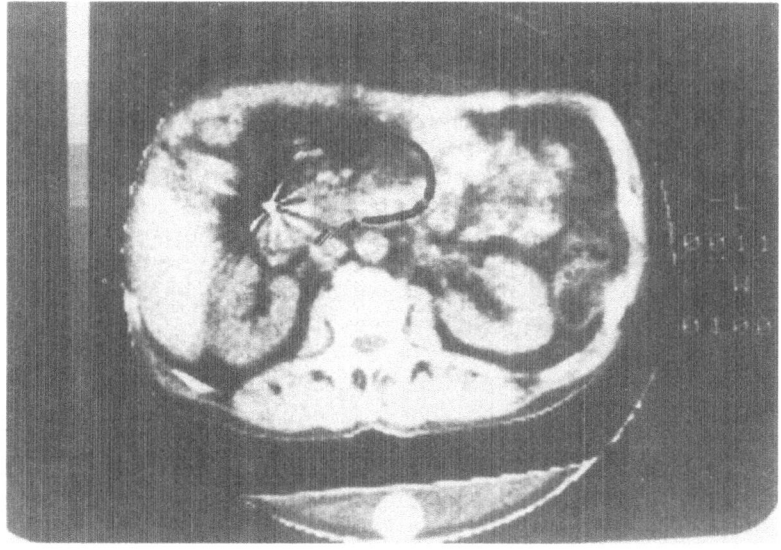

Fig. 5b. Representative tomographic scan showing artifacts produced
          by metal clip in patient's body.

in various countries. The fastest scanner under development is one
at the Mayo Clinic that will provide 60 images/sec, each of 0.01 sec
duration [4]. In order to accomplish this, the machine uses 28
x-ray tubes which are pulsed sequentially. Another approach to fast
C. T. machines is the use of a rotating focussed high voltage electron beam. Two such schemes are shown in Fig. 6; Fig. 6a [5] shows
a proposed machine in which the electron beam is rotated over the
full $360°$ and Fig. 6b shows a scheme in which the beam and detector
cover only $180°$ [6].

The fact that it is possible to obtain good tomographic reconstructions using scanners that cover less than the full 2π angle
in a plane is illustrated in Fig. 7. In the more general case, it
is possible to reconstruct the density distribution within a volume
from projections that cover less than 4π solid angle. For simplicity, we discuss the planar case for transmission tomography. The
mathematical concepts apply also to Radioisotope-Emission Tomography,
discussed in the next section, and to other fields of science such
as electron microscopy and radioastronomy.

The feasibility of high quality tomographic reconstruction from
limited-angle input data rests on the following theorems: (a) The
"Projection Theorem" states that the integral of the absorption of
radiation along a straight line through the object determines the
Fourier components of the density along a line in Fourier space perpendicular to the projection direction [7]. This theorem holds also
for the case of a radiation-emitting object where the concentration
distribution of the emitter is to be obtained from the measurements.
(b) The Fourier Transform of a finite object with a finite maximum
density is an entire function in the Fourier domain, hence knowledge
of the F. T. in any finite region of the Fourier space implies knowledge of the F. T. over all the remaining space.

One way to obtain the complete F. T. over all spatial frequency
space from the directly measured cone, as shown in Fig. 7, is to
perform the reconstruction by an iterative procedure that starts
off with the measured values K_x, K_y in the allowed cone and K_x, K_y
= 0 outside it. By a series of iterations that use the restriction
that the object has a non-negative density within its bounded volume
and is zero outside, it is possible to calculate the missing components with an accuracy that depends on the size of the angle θ and
the statistical quality of the measured data [8]. Other reconstruction methods are described in a special issue of the IEEE Trans.
Nuc. Sci. <u>26</u>, no. 2 (April 1979).

EMISSION TOMOGRAPHY: RADIOISOTOPE IMAGING

Nuclear Medicine is the diagnostic imaging discipline that uses
gamma-emitting isotopes to trace physiological abnormalities within

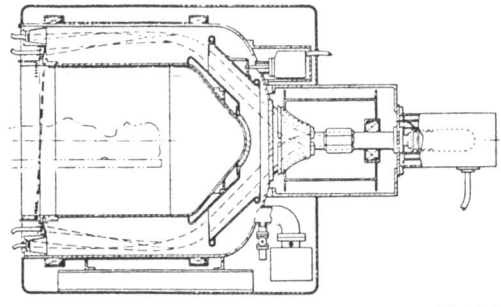

Scanner Showing Beam Transfer from Shielded
Annular Standby Collector to Target.

Fig. 6a. Rotating electron beam scanner. 2π rotation.

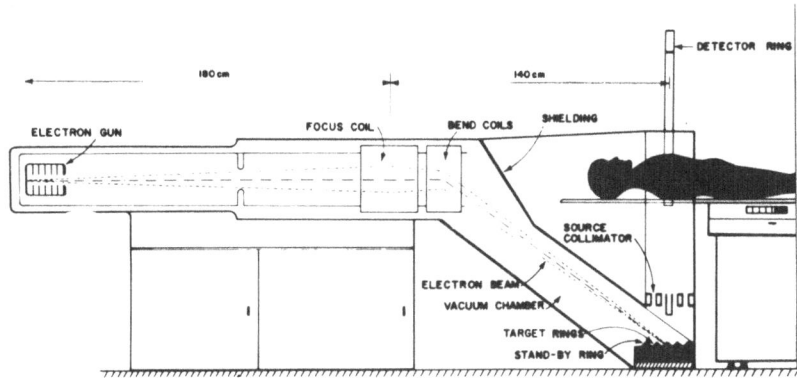

Fig. 6b. Rotating electron beam scanner. π rotation only.

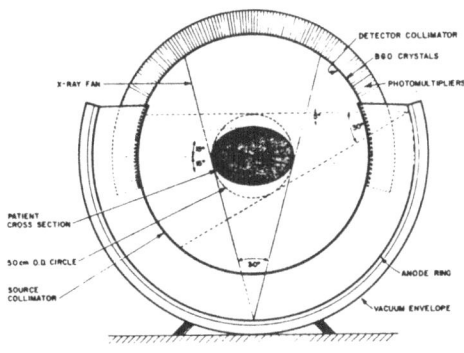

Fig. 6c. End view of π rotation scanner showing overlapping anode
detector ring and scanning x-ray beam fan geometry.

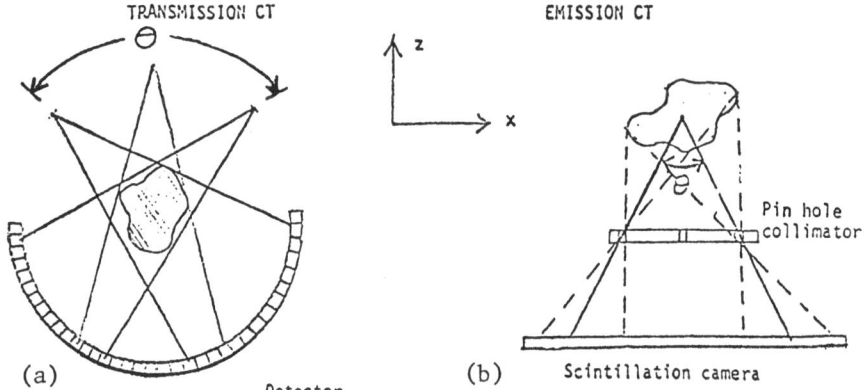

Fig. 7a. Limited-angle transmission C.T. showing fan beam and
          sector of fixed detectors.

Fig. 7b. Limited-angle emission tomography, using discrete pinhole
          collimator.

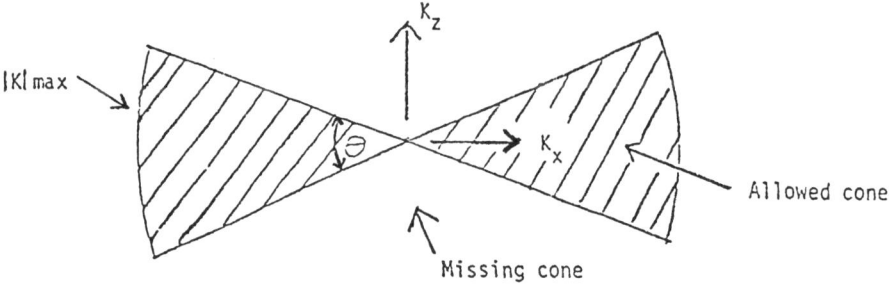

Fig. 7c. Fourier space representation showing components in
          "allowed cone" corresponding to the limited-angle imaging
          in geometry in Figs. 7a and b.

a given organ. Using scintillation cameras with various types of collimators to image a projection of the radioisotope concentration within the organ, it has supplied useful complementary information to the x-ray diagnosis.

Obviously, if the concentration of isotope can be determined in tomographic layers, greater contrast in detecting abnormalities would be achieved. A straight forward method of obtaining tomographic reconstructions with gamma-emitting isotopes is shown in Fig. 8a [9]. The organ in question is viewed from a 2π angular range by taking images at a series of angles with a scintillation camera equipped with a parallel-hole collimator. However, the procedure is time consuming, and the equipment is rather bulky.

A newer approach, for the specific purpose of obtaining tomographic images of the myocardium, was developed at the Denver Veterans Administration Hospital [10]. A seven-pinhole collimator was mounted on a large scintillation camera (LFOV: diameter of NaI crystal = 40 cms) in such a way that seven non-overlapping images of the heart were obtained, as shown in Fig. 8b. The conventional radioisotope for this purpose, ^{201}Thallium (E_γ = 60-80 keV), was used. The information of the separate images was stored on a computer. Since the independent images view the heart from a range of different angles, it is possible to obtain tomographical reconstructions as shown in Fig. 9. This method is limited to tomographical scans of small organs, since the seven separate images have to be recorded with adequate resolution on the detector and without overlapping each other.

Another approach to Radioisotope Emission Tomography is to use the 511 keV gamma-rays from the annihilation of positron-emitting isotopes at rest. This approach was suggested as early as 1953, but it is only during the last few years that it has been used clinically. This became possible primarily because of detector and computer development, and also because of the impetus given to tomographical imaging in general by the success of the C. T. machines in Radiology.

Positron annihilation tomography has the following advantages over single photon tomography:

a) The two 511 keV annihilation gamma-rays come off at 180° when the positrons annihilate at rest; the mean angle is $180^\circ \pm 0.5^\circ$, where the angular spread is due to annhilation by electrons in motion. When the two gammas are detected in coincidence, the position of the emitting atom is known to be on this line. This is shown schematically in Fig. 10a. This fact eliminates the need to use a collimator for defining the direction of the gamma rays, as indicated in Fig. 10b. Since the transmission of a collimator used

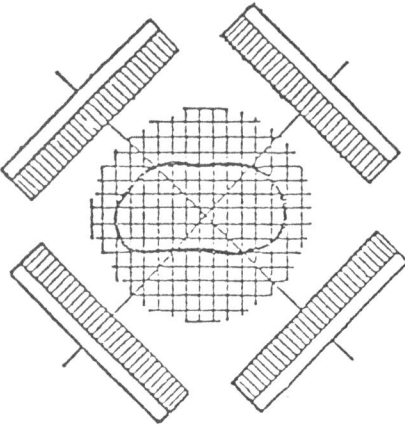

Fig. 8a. Schematic of 2π emission tomography using rotating scin-
          tillation camera with parallel-hole collimator.

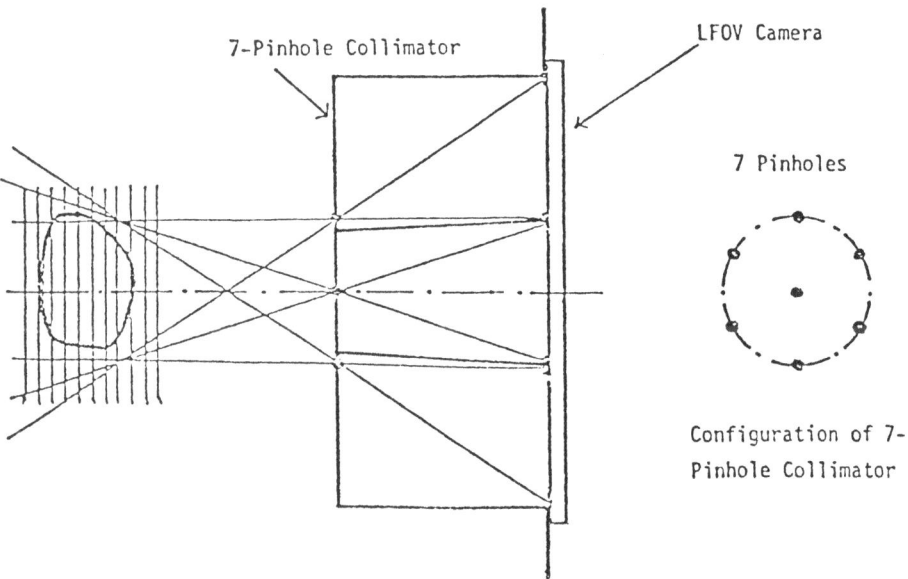

Fig. 8b. Schematic of limited-angle emission tomography using large
          stationary scintillation camera and 7-pinhole collimator.

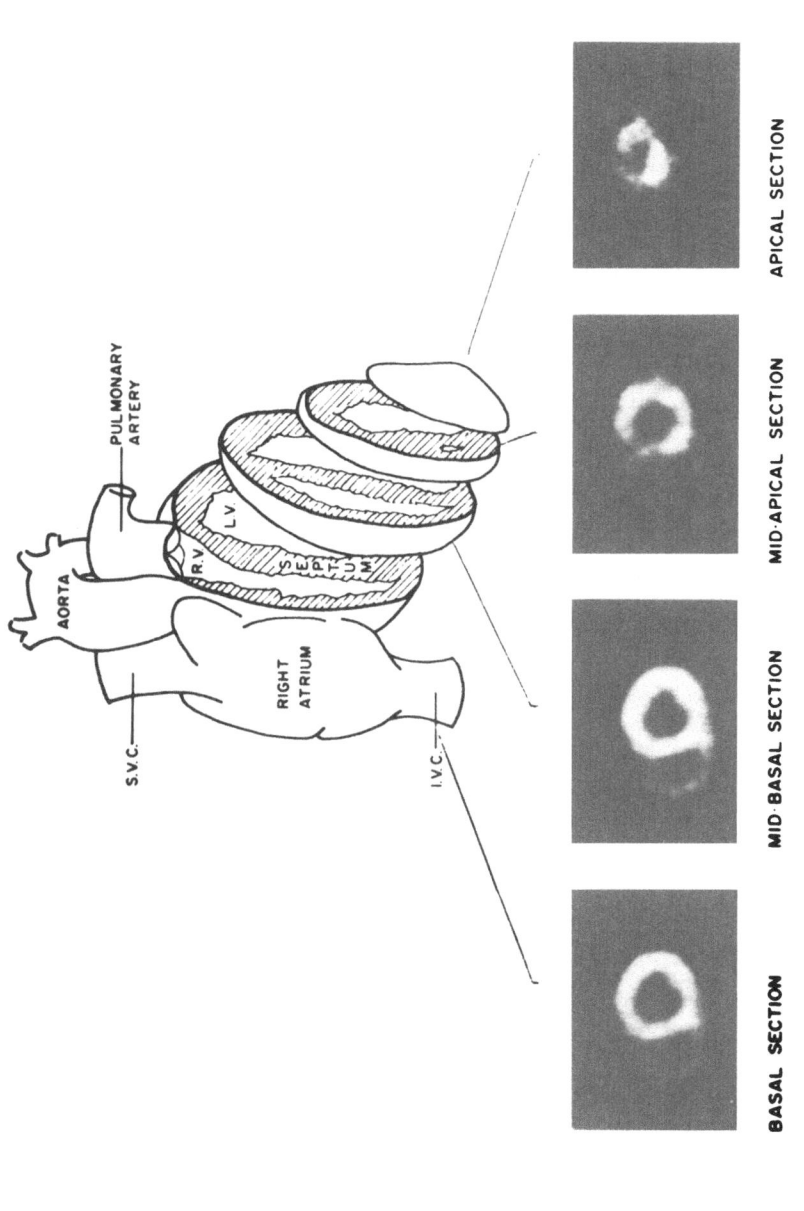

Fig. 9.   Emission tomography using 7-pinhole collimator. Heart imaging with ^{201}Tl. Heart anatomy and four corresponding tomograms are shown. Picture taken by Denver Veterans Administration Hospital Nuclear Medicine group.

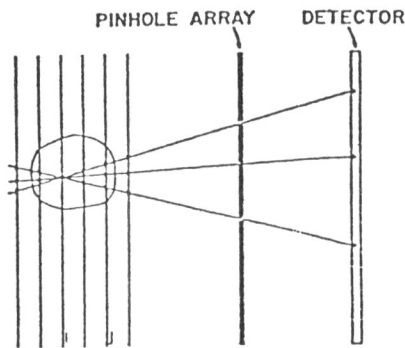

Fig. 10a. Single-photon imaging collimator.

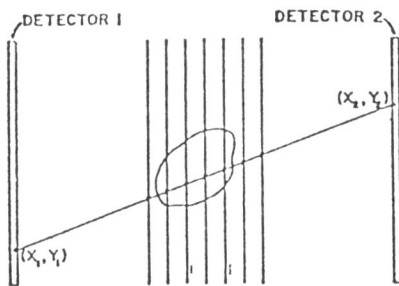

Fig. 10b. Positron imaging. Area detectors. No collimator.

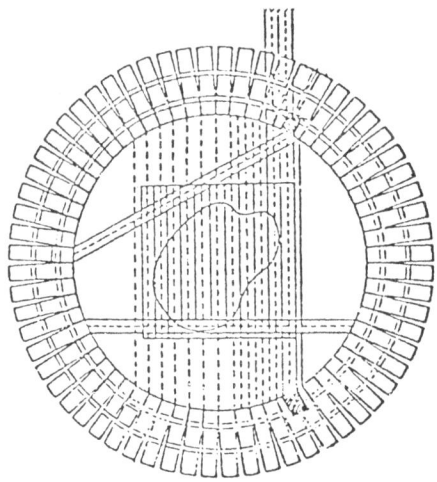

Fig. 10c. Positron imaging. Single slice. NaI ring detector.

in single photon imaging is approximately 10^{-4} its absence allows the use of smaller doses for the positron imaging case.

b) The only reasonably long-lived "gamma-emitting" isotopes of C, O, N are positron emitters. Furthermore, parent-daughter genera-tors of positron-emitting isotopes are available, such as ^{68}Ge (250d) → ^{68}Ga (1.1 hr) and ^{82}Sr (25d) → ^{82}Rb (1.2 m), which can fulfill most imaging requirements.

Positron cameras can be built in two configurations: (a) as two large-area detectors (Fig. 10b) which cover a limited fraction of 4π solid angle; and (b) ring cameras (Fig. 10c), as built by Donner Laboratory, University of California, Berkeley [11], Wash-ington University [12], University of California, Los Angeles [13], McGill [14], and others. Cameras of the first category use two scintillation cameras [15] or two multicrystal cameras [16] or MWPC detectors with converters [17,18]. The ring cameras cover 2π solid angle in one plane and can measure one or more planes at a time.

In Fig. 11a we show some construction details of the Donner Laboratory 280 NaI crystal camera; Fig. 11b shows a representative heart scan taken with ^{82}Rb. The NaI crystals will be replaced by Bismuth Germanate crystals in order to increase the detection ef-ficiency.

In Fig. 12 we show the principles of a multiwire proportional chamber. The MWPC is capable of recording the position of an ionizing event by the signals induced on the orthogonal cathode grids of wires on either side of the anode wire plane. When the ionizing event is a charged particle from an external source, posi-tion accuracy measured on the cathode grids is less than 0.5 mm.

In order to detect 511 keV gamma-rays with reasonable efficiency high Z element converters from which the interaction electrons can be extracted to form avalanches on the anode wires have to be placed within the chamber. Fig. 12a shows a cross-sectional layout of such a chamber as built at the Lawrence Berkeley Laboratory. The effi-ciency of a converter from which electrons can be extracted depends on maximizing the useful-surface area to volume. Fig. 12c shows converters that were used in this camera. They consist of a series of lead platings on plastic backing, corrugated to form holes of approximately 3.5 mm diameter. The insulated strips are necessary so that a suitable electric field can be placed across them to drift the interaction electrons out of the converter region and into the sensitive area of the MWPC between the cathode and anode planes. A more efficient converter can be made by using matrices of high lead content glass (80% PbO) tubes with holes approximately 1.4 mm in diameter. The glass surface is made resistive by hydrogen reduc-tion at high temperatures. Then a voltage of 500 V/cm will produce a suitable electric field for drifting the electrons [19]. Other

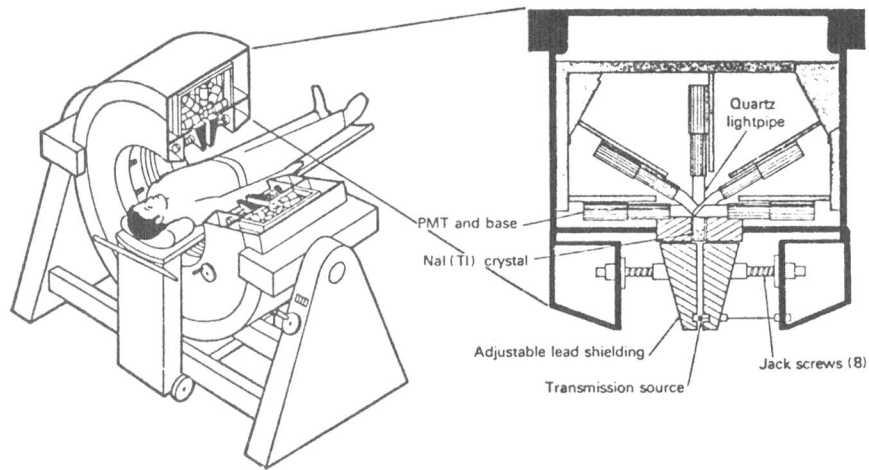

Fig. 11a.   Schematic of Donner Laboratory 280 crystal positron
            camera showing NaI crystal mounts, light pipes and P.M.
            tubes.

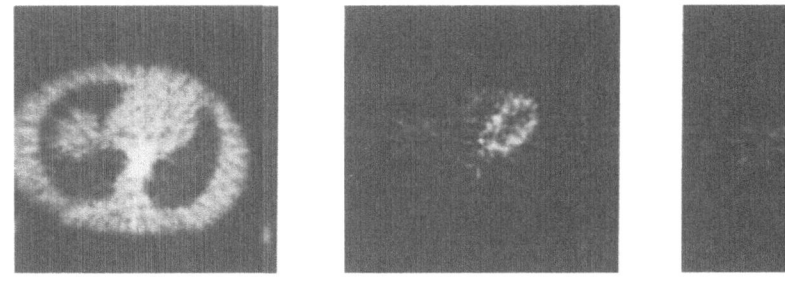

Before exercise   At exercise stress

Fig. 11b. Transmission (left) and emission pictures of heart,
          taken with Donner camera. Images taken with ^{82}Rb
          for a total of 1.5 x 10^6 true counts.

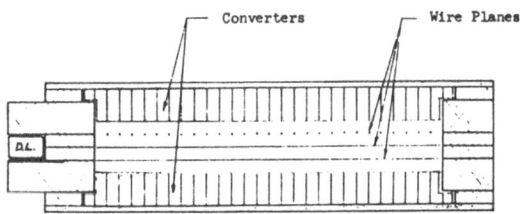

Fig. 12a. Cross section of MWPC gamma detector showing lead converters and wire planes.

Fig. 12b. Photograph of MWPC chamber showing wire planes and lead
converter below.

Fig. 12c. Photograph of honeycomb lead converter showing plated
lead bands on plastic support structure.

schemes for converters are to etch small holes on stacks of thin
lead foils with thin plastic insulated layers between them [20]
or to make arrays of tungsten wires (Prague University, Massachusetts
Institute of Technology).

In Fig. 13 we show schematically the configuration of a MWPC
camera and a head scan of a tumor region imaged with a ^{68}Ga compound
and reconstructed on a series of four planes.

ULTRASOUND IMAGING

In this section we discuss the present methods that are used
in producing the essentially tomographic images of body structures
as a B-scan, and recent developments [21].

B-scans are usually done by sending a succession of short
ultrasonic pulses at a series of overlapping directions into the
patient's body. One planar cross-section of the body is displayed
on an oscilloscope screen at any one time. This is accomplished in
diagnostic scanners by having the transducer mounted on a hinged
arm that permits motion in a single plane, producing a scan such
as that shown in Fig. 14a. The frequency of ultrasound in clinical
imaging is 2.25-3.5 MHz, limited at the high side by the increasing
attenuation of sound propagation at higher frequencies.

The technique is very similar to Radar ranging. A short pulse
of sound is sent along a given direction. Any reflecting surface
in its path will send back a signal at the appropriate angle for
specular reflection. If the surface is perpendicular to the sound
direction, the reflected signal will be detected at a delayed time,
which is proportional to the distance on the send transducer, by
the inverse piezoelectric effect.

The amplitude of the reflected signal, after time-gated amplifier gain correction for the attenuation in the medium, depends on
the coefficient of reflection

$$\mathcal{E} = I_{Ref}/I_{Send} \qquad\qquad \mathcal{E} = (\rho_1 v_1 - \rho_2 v_2)^2/(\rho_1 v_1 + \rho_2 v_2)^2$$

where ρ_1, ρ_2, v_1, and v_2 are the densities and velocities of sound
of the two media at the interface. The reflection coefficient drops
off by a factor of approximately ten within $5°$ of the perpendicular
direction.

Older diagnostic machines produced a bi-stable picture. In
this case, a fixed intensity dot was produced on the oscilloscope
screen whenever the reflected signal exceeded a preset level. Scans
of this type tended to be rather "spotty" with sections of surfaces
missing if their inclination was too far away from the specular

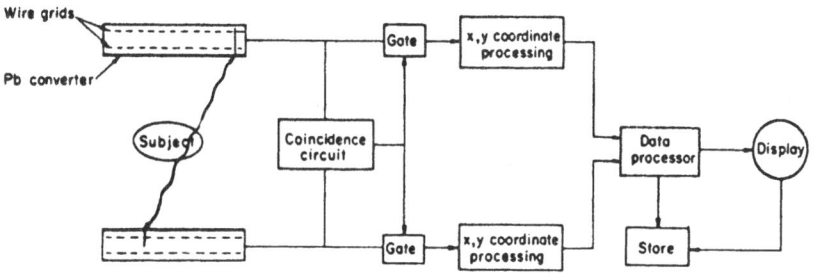

Fig. 13a. Schematic of two area detectors MWPC positron camera
showing electronic logic.

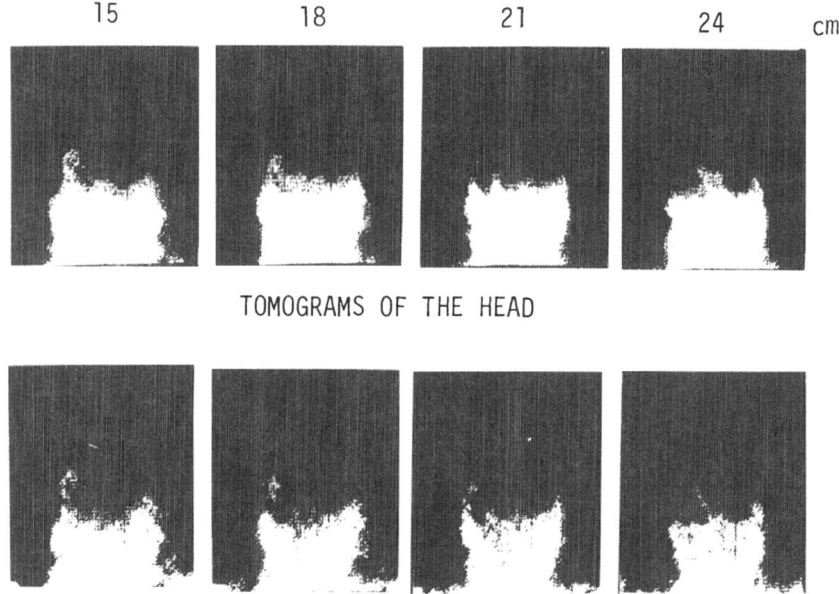

TOMOGRAMS OF THE HEAD

RECONSTRUCTED IMAGES OF THE HEAD

Fig. 13b. Representative tomograms of head showing tumor on left
side taken with MWPC camera. Images are displayed by
two different reconstruction algorithms.

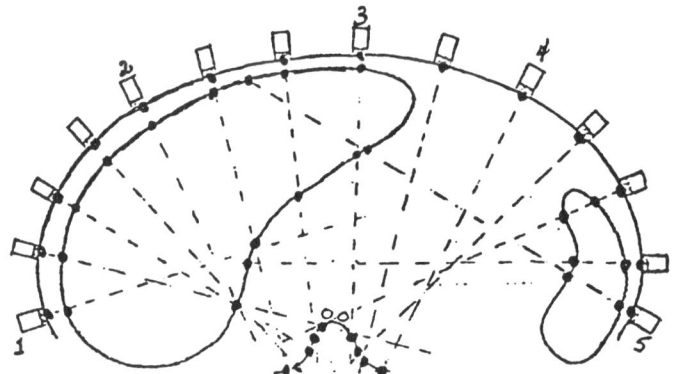

Fig. 14a.   Successive positions of ultrasound transducer during
            transverse B-scan of abdomen.

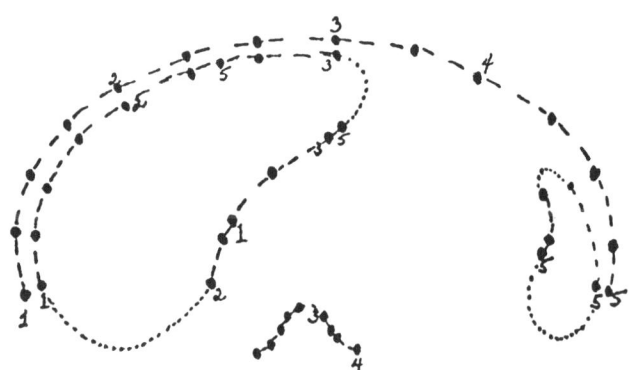

Fig. 14b.   Summation of multiple light spots in transverse B-scan
            showing how image is formed.

backward reflection angle. However, boundaries such as foetal heads, where distance measurements are taken to record growth, were very precisely delineated. Partial filling in of scattering from inclined surfaces can be achieved by using a logarithmic gain amplifier to display dots on the oscilloscope screen, whose intensity is correlated with the reflected amplitude. These "Grey scale" images fill in more of the boundary details and density anomalies within the organs. If the Grey scale gain controls are misused, the resulting scans can either be too full of reflections or omit low-contrast details.

Recent developments in B-scan ultrasound imaging are in the direction of replacing various analog gain compensating devices and display controls by an overall computerized display using digital data throughout the entire system. The present manual methods of positioning the transducer are being replaced by electronically steered and focussed transducers. Fig. 15a shows an early version of a motor-driven transducer, which could map a sector approximately 90° wide in direction. When placed at the proper positions and orientations on the chest wall, it could produce real time images of the various heart valves in motion. Fig. 15b shows a wide multielement transducer that can produce a real time image of the body section parallel to its long axis. The ultrasound beam is shifted in position from one end of the transducer to the other end to provide a linear scan over the entire width of the transducer. This is done electronically by switching the driving pulses across the transducer elements at a rate of a few hundred cycles per second.

Figs. 15c and 15d show how a multielement transducer can have its beam orientation directed and its focal length changed by the sequence of pulses delivered to the various elements. Combination of the electronically controlled deflections, as shown in Figs. 15b, c, and d, under computer control will produce in future years extremely versatile scanning machines with excellent tomographical capability.

In summary, we have reviewed in this paper advances in three clinical diagnostic imaging techniques and shown that their future development is in the direction of providing higher resolution tomography at increased speeds. The main contributor to these advances has been the availability of versatile high-speed small computers.

ACKNOWLEDGMENTS AND REFERENCES

I would like to take this opportunity thank my colleagues Drs. D. F. Boyd, R. Gould, and E. Sickles from the Radiology Department, University of California, San Francisco. The assistance of R. H. Thomas, P. Wiedenbeck, and C. Johnson-Joy of the Lawrence Berkeley Laboratory is deeply appreciated.

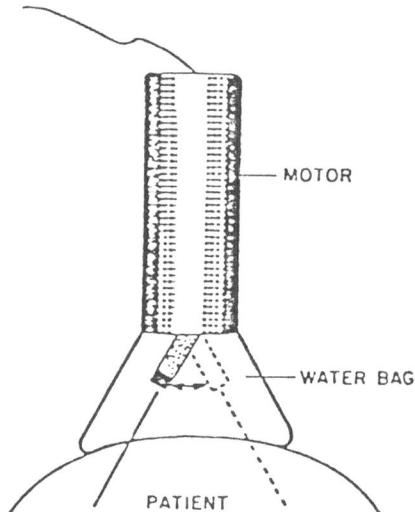

Fig. 15a. Oscillating sector scanner for real-time viewing of
heart valve motion.

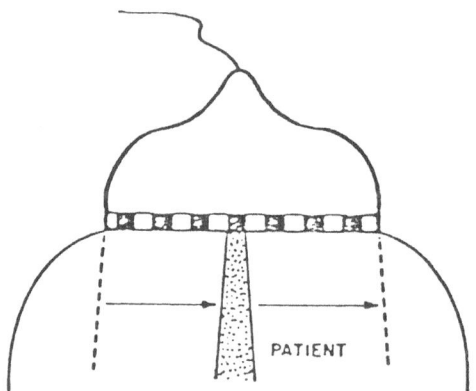

Fig. 15b. Electronically switched linear array. Transducer beam
is programmed to scan entire area without mechanical
motion.

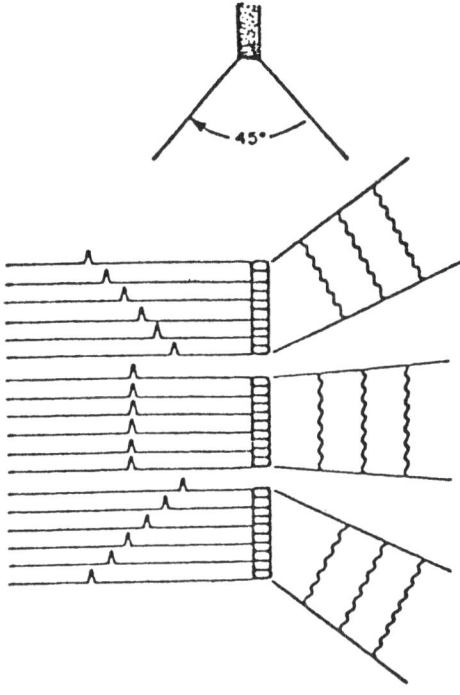

Fig. 15c. Electronically switched linear array to provide different
           scan directions by phase changes in triggering sequence.

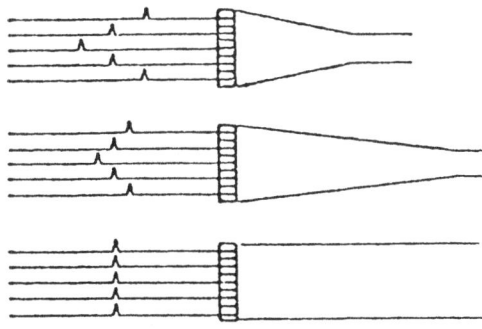

Fig. 15d. Dynamic focussing produced by programmed phasing of
           triggering sequence.

The figures listed below are reprinted with permission of Radiological Sciences, Inc: Figs. 1b and 2a.

The figures listed below are reprinted with permission of Lea and Febiger, Inc. from "Introduction to the Physics of Diagnostic Radiology" (2nd ed.) by Christensen, Curry and Nunally, copyright 1978, Lea and Febiger, Inc: Figs. 4a, 4b and 4c.

Figs. 5a, 5b, 6b and 6c were reprinted with permission of D. F. Boyd and IEEE Trans. Nuc. Sci. from the following two articles: D. F. Boyd, Status of diagnostic computer tomography, IEEE Trans. Nuc. Sci. NS-26:2836 (1979) and D. F. Boyd et al., Proposed dynamic cardiac 3-D densitometer for early detection of heart disease, IEEE Trans. Nuc. Sci. NS-26:2724 (1979).

Fig. 6a is reproduced from J. Haimson, X-ray source without moving parts for ultra-high speed tomography, IEEE Trans. Nuc. Sci. NS-26:2857 (1979).

Fig. 9 is reproduced courtesy of D. Kirsch and R. Vogel, Veterans Hospital, Denver, Colorado.

Figs. 10a and 10b are reproduced courtesy of S. Derenzo and T. F. Budinger, Donner Laboratory, Berkeley, CA.

The figures listed below are reprinted with permission of John Wiley & Sons, Inc. from "Advances in Abdominal Gray Scale Ultra-sonography," B. B. Goldberg (Editor), Ch. 2 by E. N. Carlsen and G. S. Perlmutter, copyright 1977, Wiley Medical Press: Figs. 15a, 15b, 15c and 15d.

This work was supported by the Physics Research Division of the United States Department of Energy under contract no. W-7405-ENG-48.

[1] Tubes of this type manufactured by Radiological Sciences, Inc., Santa Clara, California.
[2] Private communication and picture courtesy of E. Sickles, University of California, San Francisco.
[3] D. F. Boyd, Status of diagnostic computer tomography, IEEE Trans. Nuc. Sci. NS-26:2836 (1979).
[4] R. A. Robb et al., The DSR: A high speed 3-D x-ray computer tomography system for dynamic reconstruction, IEEE Trans. Nuc. Sci. NS-26:2713 (1979).
[5] J. Haimson, X-ray source without moving parts for ultra-high speed tomography, IEEE Trans. Nuc. Sci. NS-26:2857 (1979).
[6] D. F. Boyd et al, Proposed dynamic cardiac 3-D densitometer for early detection of heart disease, IEEE Trans. Nuc. Sci. NS-26:2724 (1979).

[7]   R. N. Bracewell, Strip integration in radioastronomy, Aust. J. Phys. 9:198 (1956).

[8]   K. C. Tam, V. Perez-Mendez, and B. Macdonald, Reconstruction in emission and transmission tomography with limited angular input, IEEE Trans. Nuc. Sci. NS-26:2797 (1979). Also, Lawrence Berkeley Laboratory report LBL8539 (Dec. 1978).

[9]   T. F. Budinger et al., Isotope distribution reconstruction from multiple gamma camera views, J. Nuc. Med. 15:480 (1974).

[10] R. A. Vogel, D. L. Kirsch, M. T. Lefree et al., Thallium-201 myocardial perfusion scintigraphy: Results of standard and multi-pinhole tomography techniques, Am. J. of Cardiology 43: 787 (1979).

[11] S. E. Derenzo, T. F. Budinger et al., The Donner 280 crystal high resolution positron tomograph, IEEE Trans. Nuc. Sci. NS-26: 2790 (1979).

[12] M. TerPogossian, M. Phelps, and E. J. Hoffman, Radiology 114: 89 (1975).

[13] Z. H. Cho, J. K. Chan, and L. Eriksson, Circular ring transverse axial positron camera for 3-D reconstruction, IEEE Trans. Nuc. Sci. NS-23:613 (1976).

[14] C. J. Thompson, E. Meyer, and Y. L. Yamamoto, "Positome II: A high efficiency P.E.T. device for dynamic studies," Proc. First Intl. Symp. on Positron Emission Tomography (Montreal, June 1978).

[15] G. Muehllehner et al., Performance parameters of a positron imaging camera, IEEE Trans. Nuc. Sci. NS-23:528 (1976).

[16] C. A. Burnham and G. L. Brownell, A multicrystal positron camera, IEEE Trans. Nuc. Sci. NS-19:201 (1972).

[17] V. Perez-Mendez, C. B. Lim et al., Characteristics of a MWPC camera for positron imaging, IEEE Trans. Nuc. Sci. NS-21:85 (1974). Also Nuc. Inst. & Methods 156:33 (1978).

[18] A. P. Jeavons and C. Cole, The proportional chamber gamma camera, IEEE Trans. Nuc. Sci. NS-25:164 (1976).

[19] G. K. Lum, V. Perez-Mendez et al., Lead-oxide glass tubing converter for gamma detection in MWPC, Lawrence Berkeley Laboratory report LBL9966 (Oct. 1979). To be published in IEEE Trans. Nuc. Sci. NS-27 (1980).

[20] A. P. Jeavons, G. Charpak, and R. Stubbs, High density lead converters, Nuc. Inst. & Methods 124:491 (1975).

[21] P. N. T. Wells, "Biomedical Ultrasonics," Academic Press (1977).

DOSIMETRY OF HIGH ENERGY PHOTON BEAMS IN NON HOMOGENEOUS MEDIA

Andrée Dutreix

Physics Department
Institut Gustave-Roussy
94800 Villejuif - France

INTRODUCTION

High energy X-ray beams have been used in radiotherapy for about 25 years and many papers have been published on dose measurements as well as the more general considerations on the variation of dose distribution with various parameters. In the present lecture I have chosen to deal with the problems encountered in the dosimetry of high energy photon beams in non homogeneous media, problems which have been too often underestimated.

The need for accurate mathematical models in the use of computers for treatment planning, as well as the greater accuracy on anatomical data resulting from the use of CT scanners in radiotherapy has lead to the necessity to reconsider carefully the problem of non homogeneous media. In order to try to clarify the problem, I shall first remind you of the concept of electronic equilibrium and its consequences on dose distribution.

ELECTRONIC EQUILIBRIUM AND TRANSITION STAGES

Electronic Equilibrium

Electronic equilibrium is achieved at a point in a medium irradiated by X-rays, if, in a sphere of radius R equal to the maximum range of the secondary electrons, two conditions are satisfied.

First condition : the photon radiation field should be uniform in intensity, quality and angular distribution ;

Second condition : the medium should be homogeneous.

Under these conditions a local physical effect such as G_1 produced by the electronic fluence at this point and measured in a small mass Δm of the medium is qualitatively and quantitatively equal to the effect G_d which is produced in the medium along the tracks of the electrons arising from Δm (figure 1) including their δ-rays. The phenomenon the most commonly considered is energy exchange where G_d should be the kerma and G_1 the absorbed dose neglecting the fraction of the electron energy converted into bremsstrahlung which is always less than a few percent in tissue equivalent materials.

The identity $G_d = G_1$ corresponds to an "absolute equilibrium". Such an absolute equilibrium could be achieved only in an infinitely extended γ emitting radioactive medium of uniform concentration. In a high energy photon beam (above a few MeV) the attenuation of the photon beam over a distance R is not negligible since the photon radiation length Λ is not markedly greater than the electron range, e.g. for 10 MeV monoenergetic radiation in water $\Lambda = 25$ cm and R = 5 cm.

Since the emission of the electrons secondary to high energy photons is strongly in the forward direction, therefore at a depth in the medium greater than R, G_1 is higher than G_d and the decrease in the values of G_1 and G_d is governed by the decrease in the photon fluence (figure 2). If this variation is exponential and without any large changes in the spectral distribution, the ratio G_1/G_d remains constant. Such a situation can be considered as a transient equilibrium. When referring to "electronic equilibrium conditions", one

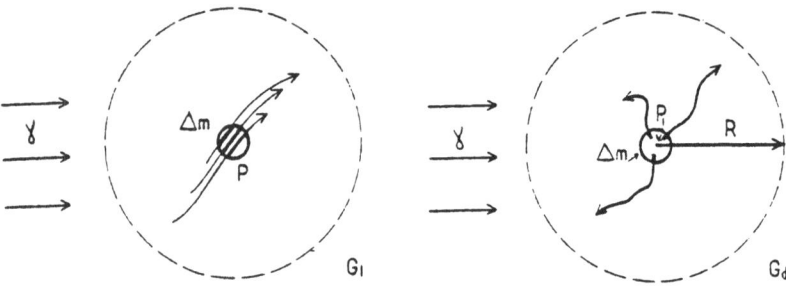

Fig. 1. Every local effect G_1 produced within a mass Δm of material by the electrons crossing this mass is identical to the effect G_d produced in the medium along the entire tracks of the electrons originating from Δm.

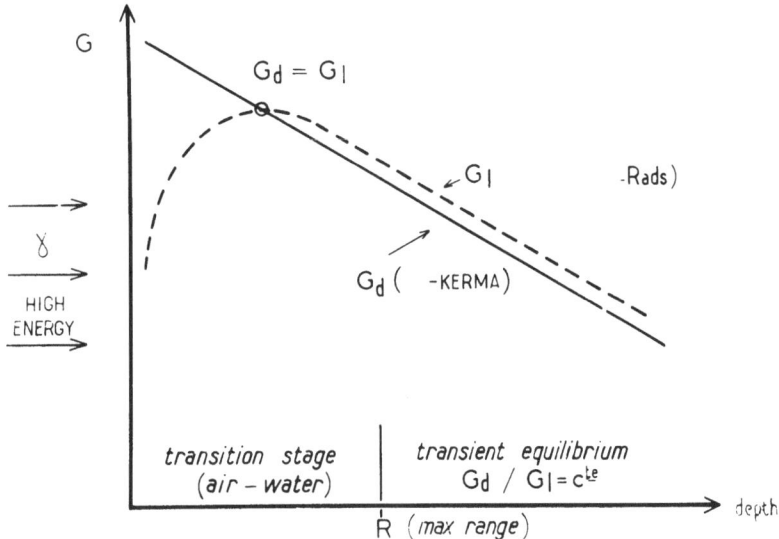

Fig. 2. G_d represents kerma. G_1 represents absorbed dose. If the
units are correctly matched $G_d = G_1$ in absolute electronic
equilibrium. For high energy X-rays, there is only a transient
equilibrium when the depth in the medium is larger than the
maximum electronic range R.

refers usually to the conditions of transient equilibrium.

In practice, photon beams are never monoenergetic and the
effective maximum range of secondary electrons averaged over the
whole spectrum of the primary photons is much smaller than the theoretical maximum range. As a result a practical transient electronic
equilibrium is obtained at a depth smaller than the maximum range
R of the secondary electrons, for instance in a 20 MV X-ray beam
the maximum range of secondary electrons is about 10 cm in water and
the transient electronic equilibrium is in practice obtained at a
depth of about 5 cm which can be considered as the "effective range
of the secondary electrons" R_e in the forward direction.

In clinical conditions there are two kinds of regions where
one of the conditions for electronic equilibrium is not fulfilled :
    - In regions near the edges of the beam the first condition
regarding the uniformity of the radiation fields is not fulfilled.
    - In the vicinity of any interface between two tissues and
in particular near air-medium interfaces ; the second condition

regarding the homogeneity of the medium is not fulfilled.

Lateral Electronic Equilibrium

The first condition for electronic equilibrium concerning the radiation field uniformity implies that the distance between the point under consideration and one of the beam edges be larger than the maximum range of electrons R.

In practice, however, as we have seen, a photon beam possesses a continuous spectrum of energy with a rather small proportion of high energy photons. Furthermore the secondary electrons are not equally distributed in direction and the effective maximum range of electrons decreases continuously from the forward direction to the backward direction. In order to quantify these observations, it is possible to determine "isoinfluence curves" surrounding the volumes which contribute by 25 %, 50 % ... 100 % to the absorbed dose at a point P. Figure 3 shows such isoinfluence curves for 20 MV X-rays in perspex (Dutreix et al, 1965) ; they show that the vicinity of point P has a much greater influence on the absorbed dose at P than the outer parts of the electronic equilibrium volume. Therefore the practical equilibrium volume is considerably smaller than the sphere of radius R and it is possible to calculate the effective range R_e of secondary electrons in a given direction as the distance of the isoinfluence curve surrounding the volume contributing for instance to 98 % of the electronic equilibrium. Such calculations could be

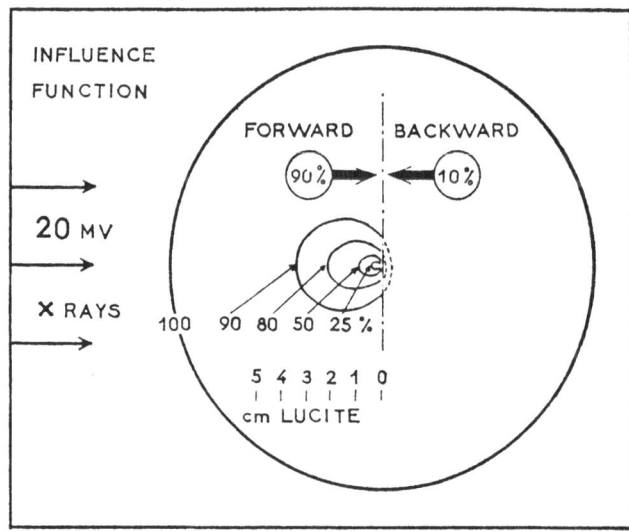

Fig. 3. Isoinfluence curves for 20 MV X-rays in Perspex.

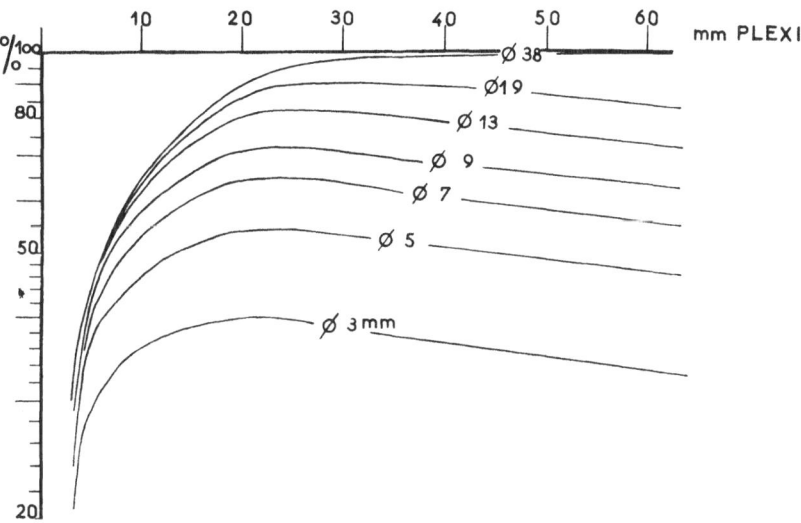

Fig. 4. Transition curves of 20 MV X-rays for beam diameters ranging
        from 3 to 38 mm (Film dosimetry in Perspex).

done by Monte Carlo methods ; they have been approximated by measurement of build-up curves for beam diameters ranging from 3 to 38 mm
(figure 4).

Measurements were performed by film dosimetry in a perspex
phantom at a large distance from the X-ray source. The curves are
corrected for inverse square law and for the exponential attenuation
of the photon beam. They show clearly the problems which should arise
if narrow beams were used for clinical purposes, that is to say
beams, the radius of which would be smaller than the effective range
of secondary electrons in the direction perpendicular to the photon
beam axis R_1 (R_1 = 2 cm for 20 MV X-rays).

Another practical implication is that around the geometrical
edge of an high energy photon beam there exists a region whose width
is equal to 2 R_1 in which G_1 varies from its maximum value G_d to 0.
Within the geometrical edges of the beam there is a lack of secondary electrons, but outside the beam, the absorbed dose is not negligible because some high energy electrons can deliver their energy
outside the geometrical beam. This region which is incorrectly called
a penumbra region by comparison with the geometrical penumbra expe-
rienced in Cobalt beams cannot be reduced by improving the source
size or the collimator design : it is only related to the primary
photon spectrum (figure 5) (Marinello et al, 1973).

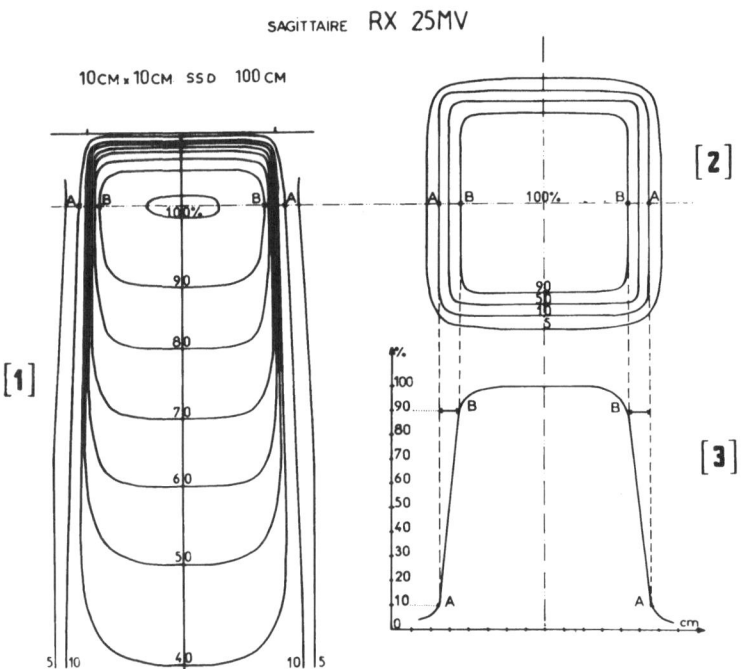

Fig. 5. Isodose curves for a 25 MV X-ray beam. The slow variation
of the dose near the edges of the beam can be seen in the
plane of the beam axis (1) as well as in a plane perpendicular to it (2) at the depth of the maximum absorbed dose.
Curve 3 shows the dose variation along a line perpendicular
to the beam axis at the depth of the maximum.

Air-medium Interfaces

We shall assume for simplicity than the medium differs from
air only by density and that their atomic compositions are approximately the same. For highenergy photon beams, this is the case
for water or soft tissues at a first approximation.

Fano (1954) demonstrated that the electronic equilibrium is
not modified by changes in density. This implies, however, that the
conditions required for electronic equilibrium are fulfilled in both
media. In the case of an air-water interface, this implies that the

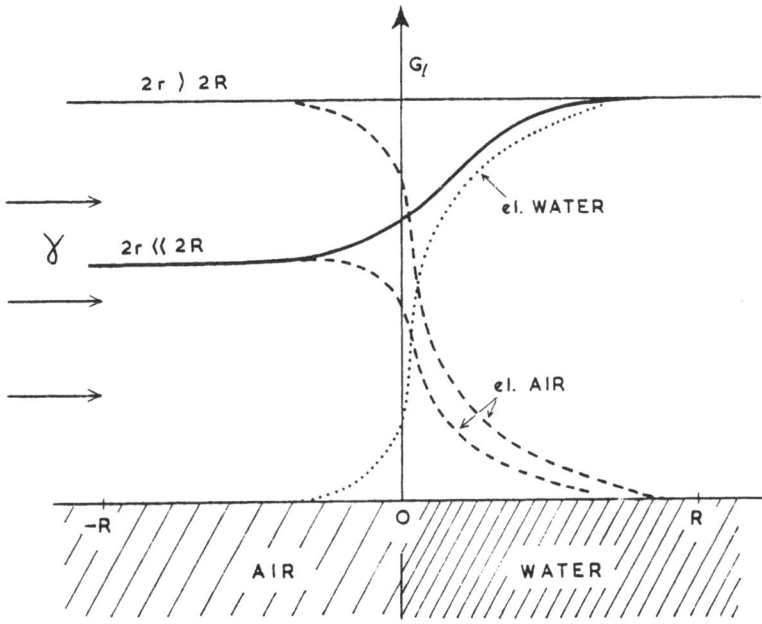

Fig. 6. Dose distribution in the vicinity of an air-water interface.
        If the beam radius r is larger than the maximum range R the
        absorbed dose remains constant. This condition is usually
        not fulfilled in air (r ≪ R) ; there is a transition stage
        over a distance R on both sides of the interface. The dotted
        lines represent schematically the different components of
        the absorbed dose, related to the electrons originating from
        the air and the water.

diameter of the beam is large enough to achieve a practical equilibrium in air. In a 20 MV X-ray beam, for instance, a radius corresponding to 2 g.cm^{-2} would be necessary, that means about 20 metres
of air which is difficult to obtain in practice. The local effect
G_1 (r) corresponding to the beam radius r is smaller than its equilibrium value G_1 (R_e). If for simplicity we neglect the photon beam
attenuation, we can expect the following variation. In air G_1 (r)
remains constant up to a distance R_e from the entrance surface ;
it increases up to the electronic equilibrium value G_1 (R_e) which
is reached at a depth R_e in the medium and remains constant up to
a distance R_e from the exit surface (figure 6). In the air behind
the exit surface, G_1 decreases over a distance R_e back to the value
G_1 (r) (figure 7). This last decrease is very rapid since the electrons originating from the medium are scattered out of the photon
beam at a rather large angle in air.

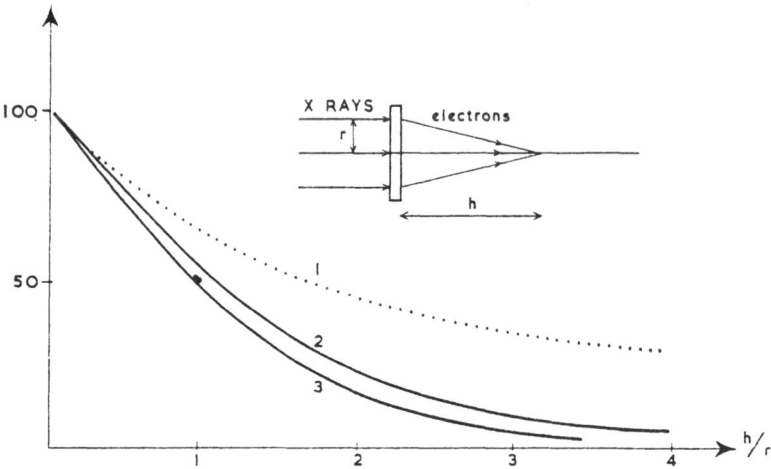

Fig. 7. Electron fluence in air behind a perspex/air interface.
Curve 1 - Theoretical distribution for an infinitely thin
layer of material (Klein-Nishina formula). Curve 2-3 -
Experimental values (ionization measurements) for r = 7.5 cm
and r = 3.75 cm. The distribution is mainly a function of
h/r. The difference between (1) and (2)-(3) shows the influence of the scattering of the electrons.

Figure 8 shows the transition curves obtained in a phantom for
25 MV X-rays from the linac Sagittaire. The entrance dose in a phantom
increases with field size partly because of the increase in the radius
of the air volume, but mainly because of the increase in the electrons
scattered by the collimator walls and by the dense materials through
which the photon beam passes. The origin of the electrons contaminating high energy photon beams has been clearly demonstrated in a
recent paper by Padikal and Deye (1978) and Biggs and Ling (1979).

Figure 9 shows the transition curves obtained in a phantom for
22 MV X-rays when a thickness of 3 cm of perspex is placed in a
photon beam at various distances from the phantom surface (Tubiana
et al, 1956). They show that the transition curve has recovered its
"normal" shape when the distance between the material placed in the
beam and the phantom surface is more than 30 centimeters or so. A
similar variation is observed for metallic filters.

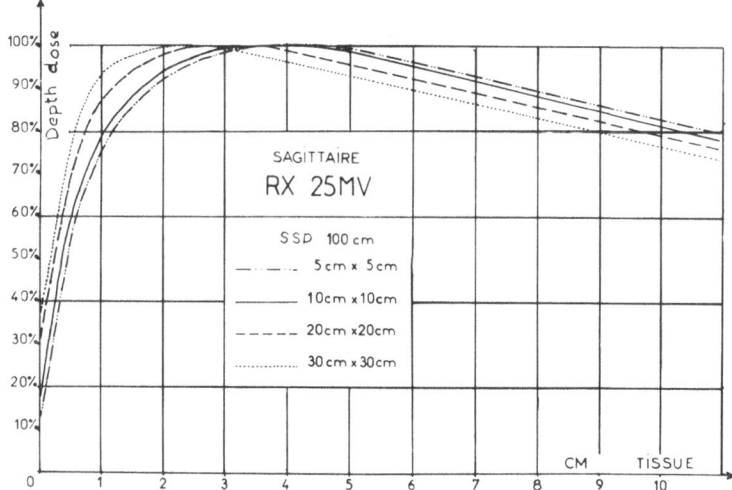

Fig. 8. Variation of the absorbed dose in the first layers of a
tissue equivalent phantom (polystyrene) irradiated by
25 MV X-rays at a source skin distance of one meter.

The important practical fact is that when the radius of the
beam is much smaller than the effective range of electrons in the
low density material, the electrons arising from the dense medium
are spread so much that their intensity in the low density gap is
reduced very rapidly at a short distance from the exit surface of
the dense-medium. We can foresee the application for interfaces
between soft tissues and lung.

Interface high Z - low Z material

When a high energy photon beam passes through interfaces sepa-
rating media of different atomic composition, the discontinuity in
the composition of the media produces a rapid change in the absorbed
dose on the two sides of the interface, in the region of non electronic equilibrium where the flux of secondary electrons is composed
of electrons arising simultaneously from both media.

The practical problem which arises in radiological applications
is the determination of the absorbed dose in water (or soft tissues)
in contact with a medium with higher atomic number. An extensive study

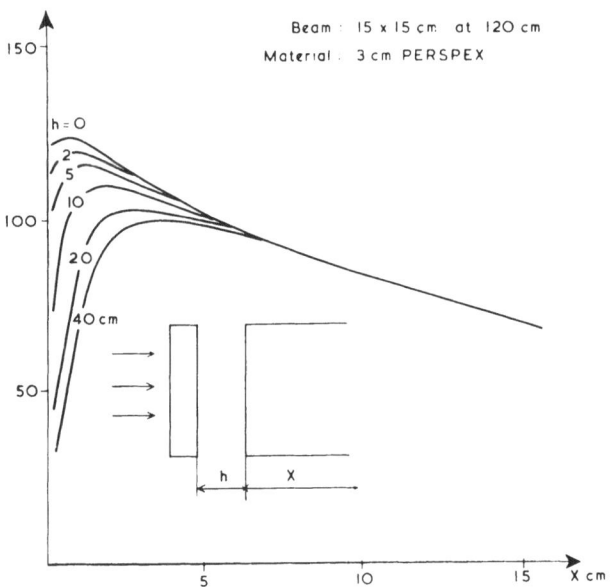

Fig. 9. Depth dose in water behind a slab of perspex (22 MV X-rays)
as a function of the thickness h of the air gap between the
perspex slab (3 cm thick) and the surface of the water
phantom. When h > 40 cm the curve has recovered its normal
shape (without material between source and phantom).

of the problem has been performed several years ago by ionization
measurements (Dutreix et al, 1964). The ionization values do not
strictly represent the variation of the absorbed dose in water but
the correction may be neglected at a first approximation (within a
few percent). The results were checked in water using chemical and
biological dosimetry (Wambersie et al, 1965). Figure 10 shows a set
of typical curves corresponding to the interface between Copper and
Carbon and table 1 shows the dose measured in water at the interface
between glass and water (Dutreix et al, 1966). For high energy X-rays
the soft tissues are always overdosed when in contact with a medium
of higher Z, whatever direction the radiation crosses the interface.
The magnitude of the overdosage increases with Z and the overdosage
zone extends overa thickness which increases with increasing energy
(about 1.5 cm at 10 MV and 3 cm at 20 MV). As the mean Z of bone is
similar to the mean Z of glass, the overdosage which can be expected
in the first centimeters of soft tissues near a bone interface are
similar to those shown for glass in table 1.

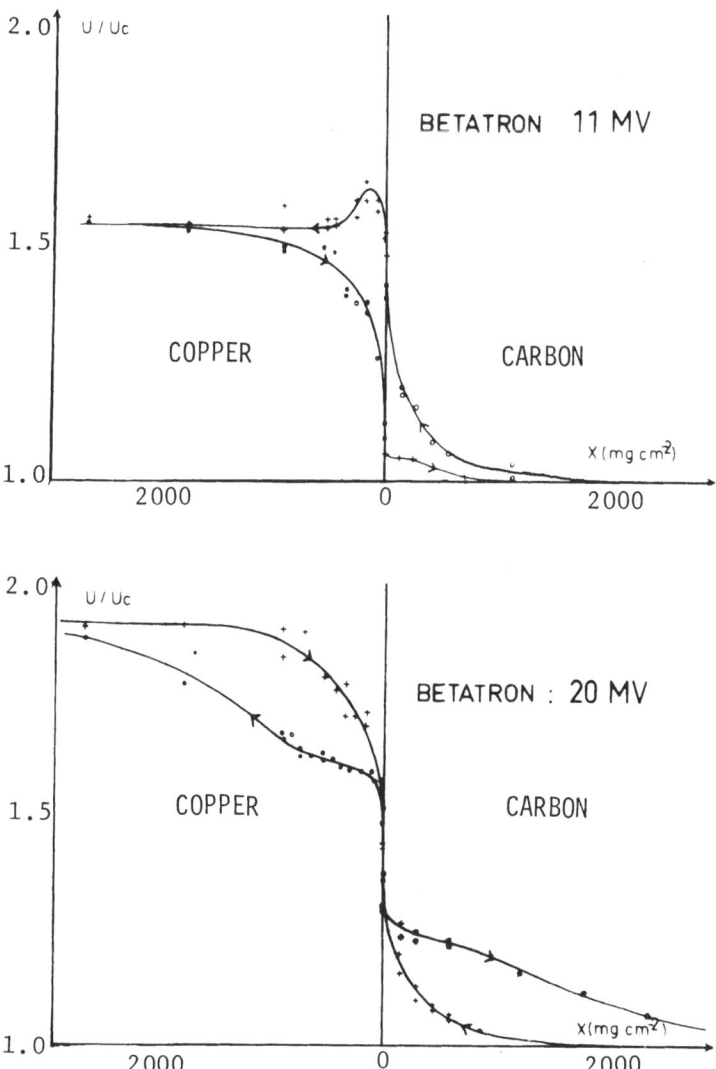

Fig. 10. Variation of the ratio U/U_c of the measured ionisation U
to the reference ionisation U_c in Carbon, with the distance
to the interface X in mg.cm^{-2}. The two curves in each figure
correspond to the photon direction either from Carbon to
Copper or from Copper to Carbon as indicated by the arrows.

Table 1. Dose at the Interface
Between Water and Glass
Relative Values D/D_e
(D_e = dose in water in electronic equilibrium)
Reproducibility of the measurements $\pm$ 2 %

Beam Quality	D/De
	Water–Glass Interface.
11 MV	107
15 MV	107
20 MV	105
	Glass–Water Interface.
11 MV	102
15 MV	104
20 MV	105

DOSE CALCULATION IN NON HOMOGENEOUS BODIES

Three different cases have to be considered in clinical dosimetry :

- Organs and tissues including fatty tissues the atomic composition
of which is not too different from water and the density of which
is near unity (between 0.9 and 1.06). Fat is considered as an important inhomogeneity for neutron beams where the proportion of hydrogen
is of importance, but for high energy photons, where interactions
for low Z materials are mainly Compton interactions the main difference between fat and other soft tissues is the difference in electronic density.

- Lung which differs from soft tissues only by density (Densities
between 0.1 and 1 were measured on patients).

- Bone which differs from soft tissues by density and by atomic
composition. The density of cortical bone is equal to 1.85 $g.cm^{-3}$
but the average density of a bone as a whole varies between 1.15
and 1.65 $g.cm^{-3}$ between the various parts of the skeleton.
The various methods of dose calculation in inhomogeneous media which
have appeared in the litterature use a correction factor applied to
reference dose distributions determined in water phantoms. The
methods recommended by Icru (report 24) are the so-called TAR method
(Tissue Air Ratio) and the power law Tissue-Air-Ratio method proposed
by Batho in 1964 and discussed in 1970 by Young and Gaylord. This
method is often referred to as the Batho method.

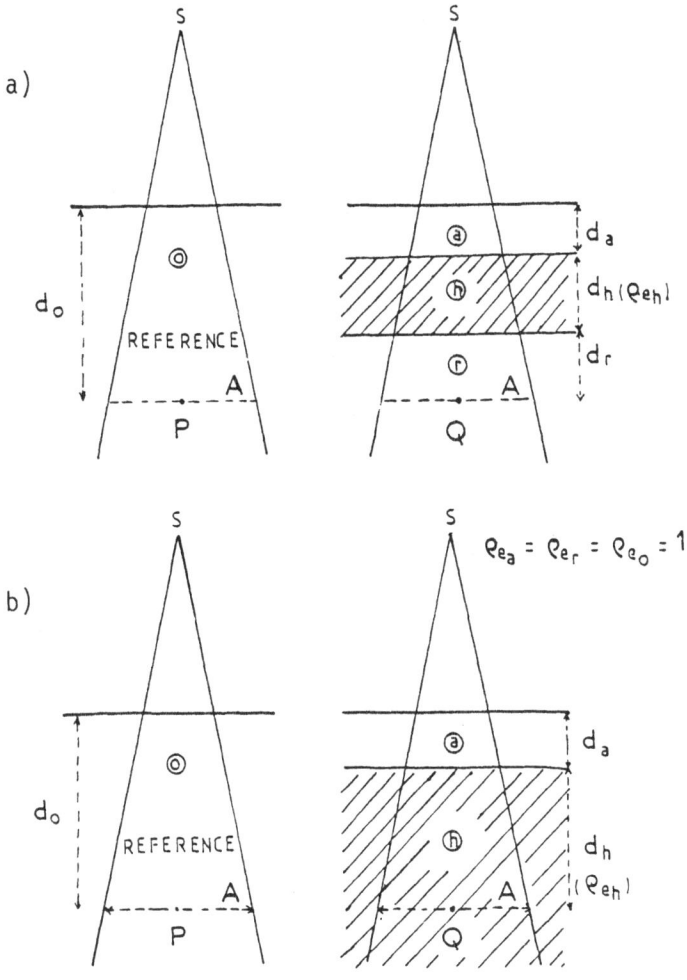

Fig. 11. The absorbed dose at point Q may be estimated by applying
    a correction factor CF to the absorbed dose at point P.

- In the TAR method (figure 11) the correction factor depends only
upon the total thickness of the inhomogeneity expressed in terms
of the number of electrons relative to water $\rho_{eh} \cdot d_h$ where $\rho_{eh} = \frac{\eta_e(h)}{\eta_e(w)}$
is the electronic density (the ratio of the numbers of electrons per
unit of volume for medium h and water) and d_h is the thickness of the

inhomogeneity h. In the TAR method the correction factor CF does not
depend upon the distance between the inhomogeneity and the point of
interest.

$$CF = \frac{TAR \ (d_a + \rho_{eh} \cdot d_h + d_r, \ A)}{TAR \ (d_a + d_h + d_r, \ A)}$$

where A is the field size at the depth of Q.

- In the BATHO method two formulae were proposed taking into account
the position of the point inside or outside the inhomogeneity.

Point Q outside the inhomogeneity (figure 11a)

$$CF = \left[\frac{TAR \ (d_r, \ A)}{TAR \ (d_r + d_h, \ A)} \right]^{(1 - \rho_{eh})}$$

Point Q inside the inhomogeneity (figure 11b)

$$CF = [TAR \ (d_h, \ A)]^{(\rho_{eh} - 1)}$$

- Later on, Sontag and Cunningham (1977) have proposed a generalized
formula based upon the Batho formulae and valid for any point and
any medium. They have introduced the ratio of the mean mass energy
absorption coefficients between the medium where the point lies
and water to get a correct expression of the absorbed dose in the
medium of interest. The generalized Batho formula is to be recommended for Cobalt-60 beams : it agrees, within a few percent, with
experimental data for any point and any biological material.

$$CF = \frac{[TAR \ (d_r, \ A) \ ]^{(\rho_{er} - \rho_{eh})}}{[TAR \ (d_h + d_r, \ A)]^{(1 - \rho_{eh})}} \ \frac{(\bar{\mu}_{en}/\rho)_r}{(\bar{\mu}_{en}/\rho)_o}$$

The formulae, unfortunately cannot be applied directly to very high
energy photon beams (higher than a few MV) where Tissue-Air-Ratios
cannot be measured since they imply the measurement of a tissue
absorbed dose free in air which is the absorbed dose in an elementary
mass of matter in air, the elementary mass of matter being large
enough to ensure electronic equilibrium in the cavity and small
enough for scattering and attenuation of photons to be negligible.
We have seen that for 20 MV X-rays the minimum mass of matter should
be 4 $g.cm^{-2}$ in diameter and 5 $g.cm^{-2}$ in longitudinal direction. The
attenuation of the photon beam can be estimated to be 15 % and the
scattering should contribute to the dose by about 1 % in such an
elementary mass.

TAR is usually replaced in clinical dosimetry by TMR (Tissue Maximum Ratio). TMR is by definition equal to 1 at the depth d_m of the maximum build-up ; d_m is at most, for large distances and small field sizes (figure 8) equal to the maximum effective range R_e of the secondary electrons for the photon beam under consideration. For small depths $d < d_m$ where electronic equilibrium conditions are not fulfilled, TMR is smaller than one. Then if the generalized Batho formula is used by replacing TAR by TMR it will lead to rather big errors everytime the distance d_r between point Q and the interface between the two media is smaller than d_m, including the case where point Q lies within the inhomogeneity ($d_r = o$). Furthermore it is obvious that in clinical practice, in many occasions either the longitudinal or the lateral electronic equilibrium are not achieved and data should only be used with caution. We shall now review the various cases.

Soft tissues. Let us assume that the correction factor CF is calculated in points where the electronic equilibrium is achieved, that is to say at depths larger than R_e or d_m and for field sizes larger than 2 R_1, it is then possible to use the Bathomethod when replacing TAR by TMR and adding d_m to the thicknesses in order to calculate TMR for depths larger than d_m

$$CF = \frac{[TMR \ (d_r + d_m, \ A) \qquad]^{(\rho_{er} - \rho_{eh})}}{[TMR \ (d_h + d_r + d_m, \ A)]^{(1 - \rho_{eh})}}$$

Lung. As the density of lung is usually lower than the density of water the effective range of electrons in lung expressed in centimeters is much larger than it is in water. As a first approximation the range of electrons R_h in a medium h is inversely proportional to the electronic density ρ_{eh} of h, and proportional to the effective range in water R_w.

$$R_h = R_w/\rho_{eh}$$

For instance in a 25 MV X-ray beam ($R_w = 5$ cm), when the lung electronic density is equal to 0.4, the effective range of electrons in lung is $R_h = 12,5$ cm and laterrally $R_e = 2$ cm/0,4 = 5 cm. Electronic equilibrium in lung is then achieved only for a beam size larger than 2 $R_e = 10$ cm and at a depth greater than 12,5 cm. It is then necessary to consider whether longitudinal and lateral electronic equilibrium are achieved before choosing the best method of arriving at a correction factor.

Figure 12 shows build-up curves measured in a lung equivalent material ($\rho_{eh} = 0.32$) as compared to similar curves in polystyrene

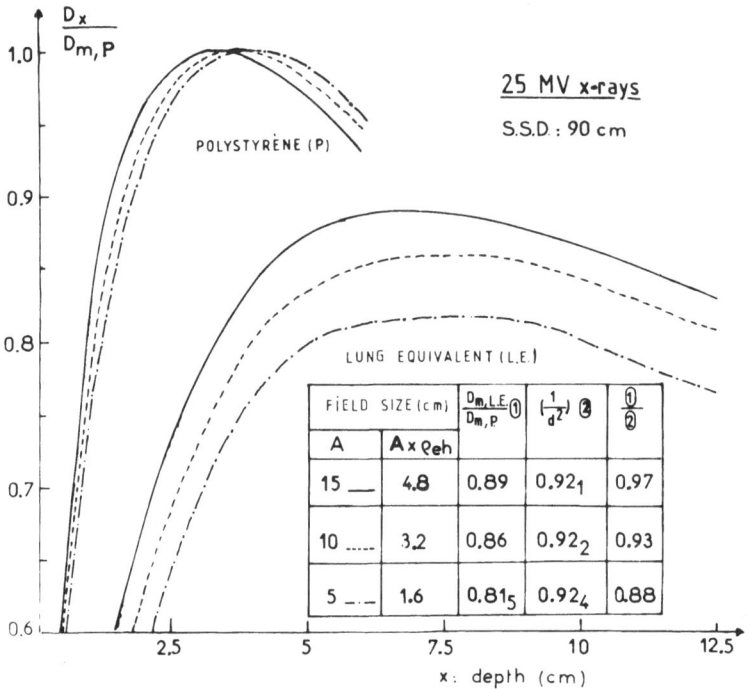

Fig. 12. Build up curves in polystyrene and lung equivalent material
25 MV X-rays SSD 90 cm. Curves are shown for three different
field sizes 15 cm x 15 cm, 10 cm x 10 cm and 5 cm x 5 cm ;
they are normalized to the maximum absorbed dose in polystyrene. For comparison the values have to be corrected for
inverse square law (column 4), the corrected values for the
maximum absorbed dose in lung equivalent tissue are given
in the last column, they show the lack of lateral electronic
equilibrium in lung equivalent tissue, especially for small
field sizes.

for three different field sizes (Hannia, 1979). The last column
in the inserted table shows that the maximum dose (after inverse law
correction) is smaller than the maximum absorbed dose in polystyrene
by 3 to 12 % depending upon the field size because of the lack of
lateral electronic equilibrium.

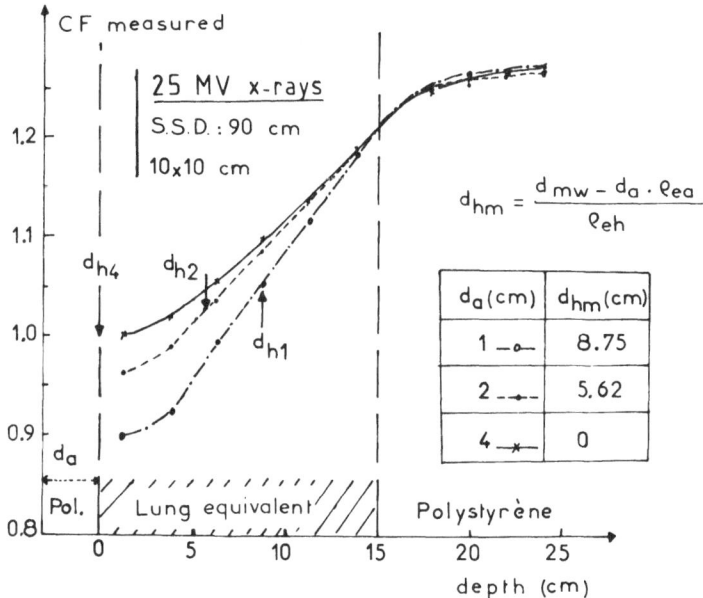

Fig. 13. Variation of the correction factor (CF) for 15 cm of lung
equivalent material placed at a depth (d_a) in polystyrene
of 1,2 or 4 cm. The longitudinal electronic equilibrium
(CF = 1) is restored at depth d_h in lung such as

$$d_h = \frac{d_m - \rho_a \cdot d_a}{\rho_h} \ .$$

Figure 13 shows the variation of the measured correction factor
with depth, when a 15 cm thickness of lung equivalent material is
irradiated in a 10 x 10 cm beam, behind a thickness of soft tissue
d_a varying from 1 to 4 cm. The correction factor is equal to or
larger than 1 at depths d_h in lung, large enough for the longitudinal electronic equilibrium to be achieved.

$$d_h > \frac{d_m - d_a \cdot \rho_{ea}}{\rho_h}$$

At smaller depth, the correction factor CF is smaller than 1,
that is to say that the absorbed dose in lung is lower than it
would be in water.

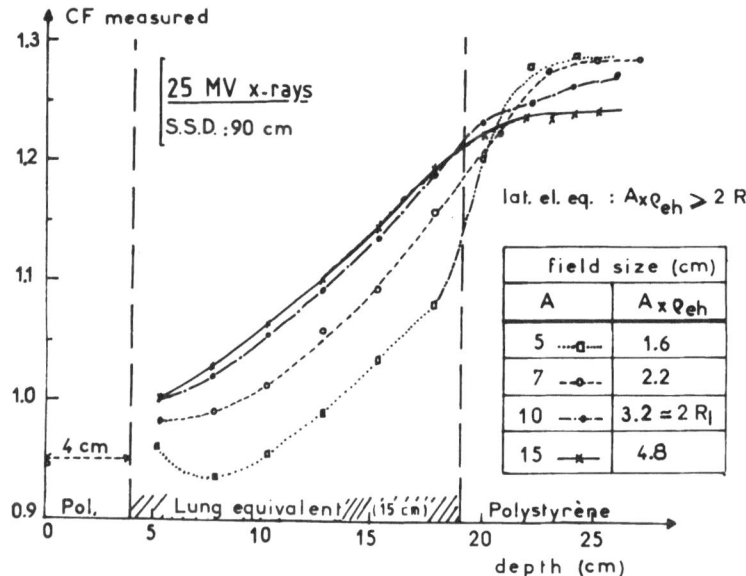

Fig. 14. Variation of the correction factor (CF) for 15 cm of lung
         equivalent material placed at a depth equal to 4 cm of
         polystyrene for four different field sizes 5 cm x 5 cm,
         7 cm x 7 cm, 10 cm x 10 cm and 15 cm x 15 cm. For small
         field sizes FC is lower than 1 in the first ten centimeters
         of lung due to the lack of lateral electronic equilibrium.

Figure 14 shows the variation of CF for various field sizes,
when electronic equilibrium is achieved in soft tissue in front of
the lung. For small field sizes, as the lateral electronic equilibrium is not achieved in lung the correction factor decreases in
the first few centimeters of lung and is larger than one at depths
in lung greater than 9 cm and 14 cm respectively for field sizes
of 7 cm x 7 cm and 5 cm x 5 cm.

Empirical methods for arriving at CF in good agreement with
measured values are proposed (BRIDIER et al, 1979). It is necessary
to consider whether the longitudinal and the lateral electronic
equilibrium are achieved or not at the level of the point Q under
consideration. In clinical practice we shall assume that field sizes
are always large enough to ensure electronic equilibrium in soft
tissues.

- Longitudinal electronic equilibrium is achieved
$(d_x = \Sigma \rho_{ei} \cdot d_i \geqslant d_m)$ and lateral electronic equilibrium is achieved
$(A. \rho_{eh} \geqslant 2 R_l)$: use the corrected Batho method

$$CF = \frac{[TMR \ (d_r + d_m, \ A) \qquad]^{(\rho_{er} - \rho_{eh})}}{[TMR \ (d_h + d_r + d_m, \ A)]^{(1 - \rho_{eh})}}$$

assuming the ratio of the mean mass energy absorption coefficients is equal to 1 between lung and soft tissues.

- Longitudinal electronic equilibrium is achieved
$(d_x = \Sigma \rho_{ei} \cdot d_i \geqslant d_m)$, lateral electronic equilibrium is not achieved in lung $(A. \rho_{eh} < 2 R_l$ and point Q lies in lung $d_r = o)$ as TMR
$(d_r + d_m) = 1$

$$CF = [TMR \ (d_m + d_h, \ A) \ . \ K_e]^{(\rho_{eh} - 1)}$$

Where K_e is a coefficient taking account of the lack of lateral electronic equilibrium and calculated from build up measurements in narrow beams (figure 4).

- Longitudinal electronic equilibrium is not achieved
$(d_x = \Sigma \rho_{ei} \ d_i < d_m)$, point Q lies in lung and lateral electronic equilibrium is achieved $(A \ \rho_{eh} \geqslant 2 R_l)$. We shall assume that longitudinal equilibrium is always achieved for points lying in soft tissues behind lung.

In our experience the Batho method cannot be used but the simple TMR method results in good agreement with measurements.

$$CF = \frac{TMR \ (d_a + \rho_h \cdot d_h, \ A)}{TMR \ (d_a + d_h, \ A)}$$

We should note that TMR $(d_a + \rho_h \cdot d_h, \ A)$ is much lower than 1, since $(d_a + \rho_h \ . d_h) < d_m$ then CF < 1.

- Neither longitudinal nor lateral electronic equilibrium are achieved. Special measurements are then necessary in the irradiation conditions in order to take account of the contamination by electrons and no formula is proposed.

- <u>Bone</u>. As we have seen from Table 1 and figure 10, the absorbed dose in soft tissues near a bone interface is increased

by the increased fluence of the secondary electrons set in motion
in bone. The magnitude of this dose increase cannot by easily predicted and accurate measurements or Monte-Carlo calculations would
be necessary to determine the value of CF for points lying at distances d_h from a bone interface smaller than d_m/ρ_{eh}. For points
at larger distances the corrected Batho method may be used.

$$CF = \frac{[TMR\ (d_r + d_m, A)]^{\rho_{er} - \rho_{eh}}}{[TMR\ (d_h + d_r + d_m, A)]^{(1 - \rho_{eh})}}$$

CONCLUSION

The correction factors to apply to dose distributions in high
energy photon beams (above a few MV) to take account of inhomogeneities are certainly less than for Cobalt beams but they cannot
be neglected. For points at large distances d_h from interfaces
($d_h \geqslant R_e/\rho_{eh}$) simple correction methods similar to those used
for Cobalt beams may be applied. However as the absorbed dose is
governed essentially by the fluence of the secondary electrons
set in motion in a volume of tissue, the radius of which is equal
to the effective range R_e of the secondary electrons, the correction factor CF for points near interfaces ($d_h < R_e/\rho_{eh}$) depends
on both the atomic composition and the electronic density of the
inhomogeneity. A low density may lead to a lack of electronic equilibrium and consequently to a correction factor much lower than the
correction factor calculated by means of usual correction methods.

REFERENCES

Batho H.F., 1964. Lung corrections in Cobalt 60 beam therapy.
    J. Can. Assoc. Radiol. 15, 79.
Biggs P.J., Ling C.C., 1979. Electrons as the cause of the
    observed d_{max} shift with field size. Med. Phys. 6, 291.
Bridier A., Hanna T., Dutreix A., 1979. Problèmes liés à la
    modification de la dose due à la présence d'hétérogénéi-
    tés dans le cas de photons de haute énergie. In procee-
    dings of the 18e Congrès de la Sphee, Nancy.
Dutreix J., Dutreix A., Tubiana M., 1965. Electronic equilibrium
    and transition stages. Phys. Med. Biol. 10, 177.
Fano U., 1954. Note on the Bragg-Gray cavity principle for
    measuring energy dissipation. Rad. Research 1, 237
ICRU, 1976. Determination of absorbed dose in a patient irra-
    diated by beam of X or γ rays in radiotherapy. Report 24.
Marinello G., Dutreix A., 1973. Etude dosimétrique d'un faisceau
    de Rx de 25 MV. J. Radiol. Elect. 54, 951.
Padikal T.N., Deye J.A., 1978. Electron Contamination of a High
    Energy X-ray Beam. Phys. Med. Biol. 23, 1086.

Sontag M.R., Cunningham J.R., 1977. Corrections to absorbed dose calculations for tissue in homogeneities. Med. Phys. 4, 431.

Tubiana M., Dutreix J., Dutreix A., 1956. Dispersion des électrons secondaires mis en mouvement par des Rx de 22 MV. Journ. de Phys. 17, 12A.

Wambersie A., Dutreix J., Bernard M., 1965. Variation de la dose au voisinage de l'interface entre le plexiglas et un métal. Etude par dosimétrie chimique et biologique. Radiabiol. Radiother. 67. 237.

Young M.E.J., Gaylord J.D., 1970. Brit. J. Radiol. 43, 349.

DOSIMETRY AND RADIOBIOLOGY OF PROTONS AS APPLIED TO

CANCER THERAPY AND NEUROSURGERY

Börje Larsson

Department of Physical Biology
Gustaf Werner Institute, University of Uppsala
Box 531, S-751 21 Uppsala, Sweden

INTRODUCTION

In radiology, accelerated light ions, for example the protons considered in this paper, are generally classified as "charged heavy particles". A beam of such particles, accelerated to a kinetic energy of some hundred millions electron volts per atomic mass unit, is able to penetrate thick layers of tissue with only a small amount of scatter, comparable in magnitude with that of the most energetic roentgen rays, at present used in radiotherapy. As with electrons, the charged heavy particles create ionization of practically continuous density along their path of penetration. In contrast, however, to a high energy electron which produces a fairly sparse ionization as it moves nearly at the velocity of light along most of its track, a charged heavy particle induces a marked increase in specific ionization in the last centimeters of its course of penetration, where its velocity decreases gradually with increasing depth. A beam of nearly monoenergetic charged heavy particles has, indeed, a sharp maximum, *"the Bragg peak"* [*], near the depth at which the particles are brought to rest. This effect is accentuated by the fact that the charged heavy particles, due to their large mass, are less influenced by statistical fluctuations in their attenuating collisions, and therefore have little range variation.

By applying the well established theory for the interaction

---

*In honour of W.H. Bragg, who first observed that monoenergetic alpha particles have a well defined range in air and ionize most heavily near the end of their path.

of charged particles with matter Wilson[1] made in 1946 the first
radiological characterization of beams of accelerated protons. He
predicted that such particles should be useful for the precise
irradiation of regions in the depth of the body, as they combine
the most attractive physical properties of therapeutic electron
and roentgen radiation: suitable penetrability, little scattering,
and an almost definite range of penetration in which the dose shows
a marked increase with the depth. This view was soon confirmed experimentally by Tobias, Anger and Lawrence, who were the first to
use ion beams of high energy for radiobiological studies[2]. In 1952,
they estimated the dose distribution in a 2.5 cm wide, 190 MeV deuteron beam, from the synchrocyclotron at Berkeley, and gave results
of measurements on which a calculation of the relative mass stopping power of the various tissues could be based. When exposed to
180 MeV deuterons, wet bone should absorb 20 percent less energy
per gram than soft tissue. This also applies approximately for
protons in the energy interval 20-187 MeV.

The first therapeutic use of ion beams was reported in 1956
by Tobias et al.[3] who tried to destroy the endocrine function of
the pituitary in patients with metastatic carcinoma of the breast.

With the pioneer work at Berkeley as a source of inspiration,
we begun, in 1954, to investigate the potentialities of a 185 MeV
proton beam** in biological and medical experiments. Methods were
developed for the use of high-energy protons for radiotherapy of
large deep-seated tumours[4] as well as for "radiosurgery", i.e. production of small circumscribed lesions in the central nervous system[5]. The present review is mainly based on these developments and
on subsequent biological and clinical studies in Uppsala. Indeed,
protons are also being used for similar purposes, at 160 MeV at
Harvard[6], at 200 MeV in Moscow[7], at 670 MeV at Dubna[8], and at 1000
MeV in Leningrad[9]. In the following, however, reference to publications from foreign laboratories will be made only when originality
has to be emphazised or instructive supplementary information is
called for.

**The synchrocyclotron at the Gustaf Werner Institute completed in
1951-1952 and running to 1976, produced an internal circulating beam
of light hydrogen nuclei, *protons*, which could be used at varying
energies, up to about 200 millions of electron volts (MeV). Since
1956, primarily through the work of Svanheden and Tyrén, it has
been possible to extract an appreciable fraction of the circulating protons at a magnet radius of 100 cm, thus providing an external beam of radiation that can be collimated and conducted to
areas distant from the accelerator. This external collimated radiation is referred to as the 185 MeV proton beam, although its
energy could actually be varied by a few percent (Fig. 1).

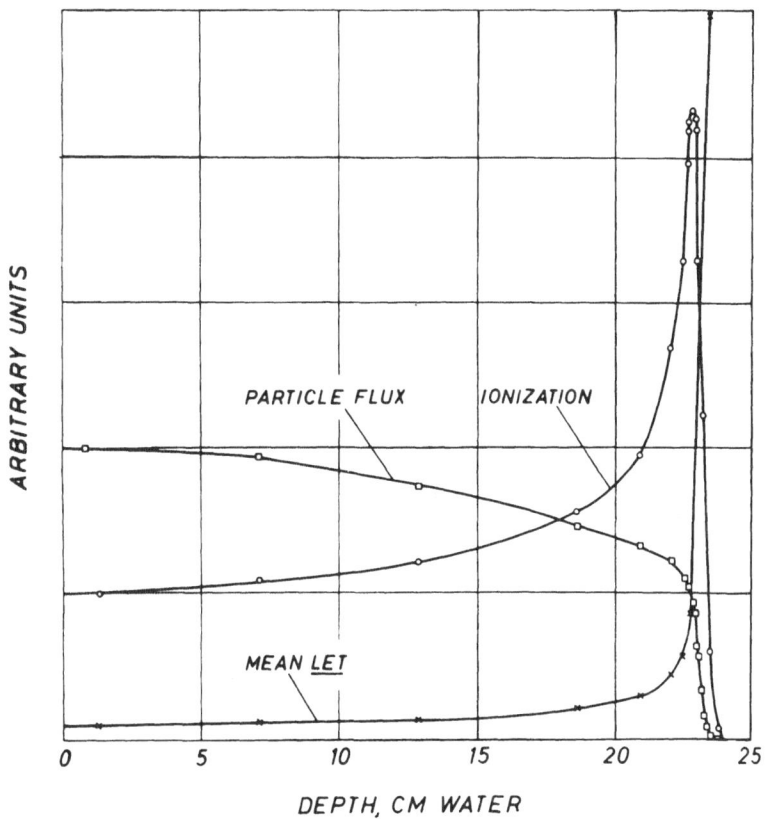

Fig. 1. Relative flux, relative ionization (the Bragg peak), and
mean liner energy transfer (LET$_\infty$) of 187 MeV protons at
different depths of penetration in water. The LET$_\infty$ in the
first part of the range of penetration is 0.5 keV/μ. The
data represent central axis conditions of broad beams,
2 cm or more in diameter. From ref. 10.
(Courtesy Brit.J.Radiol.)

IRRADIATION OF DEEP-LYING TISSUES: A BRIEF REVIEW

Ionizing radiation is the only modality, in cancer therapy,
that permits adequate and, at the same time, more or less uniform
treatment of any chosen target. Ever since 1896, when roentgen radiation was applied successfully for the first time in the treatment of human disease there has been a steady development of methods
for the selective irradiation of various parts of the body.

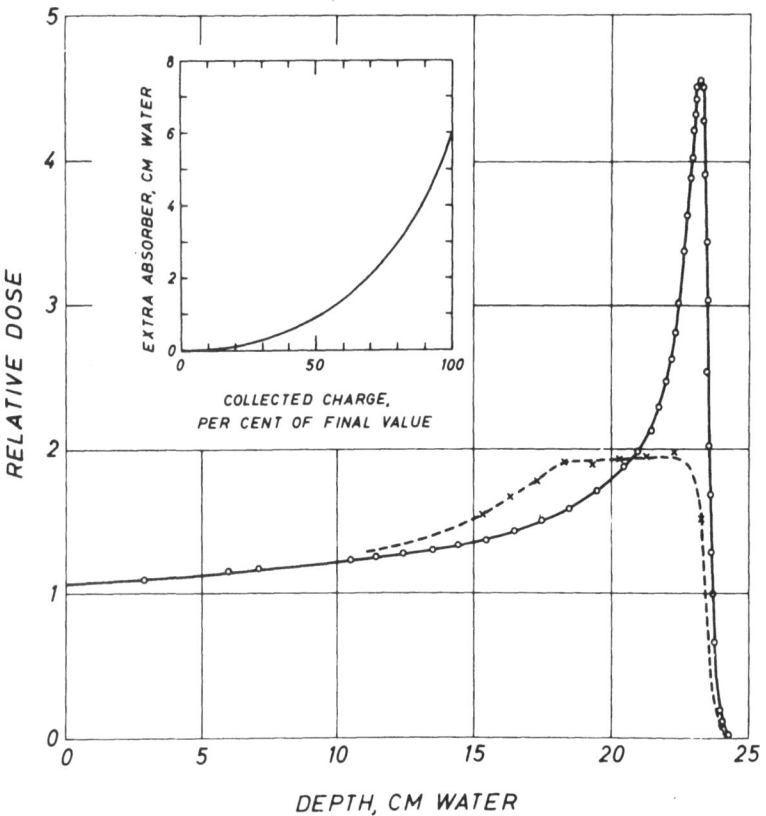

Fig. 2.   Example of transformation of the Bragg peak. Variation of
the thickness of an extra water absorber in front of the
target (cf. Fig. 4) was performed, during irradiation, according to the curve in the inset diagram. The large diagram shows the original (———) and the transformed (---)
depth-dose curves. The crosses give the results of measurement of points on a depth-dose curve obtained by use of
a specially designed "ridge filter". The profile of the
ridges was determined by the shape of the curve in the inset diagram, one cm water being equivalent to 0·18 cm brass.
From ref. 10.
(Courtesy Brit.J.Radiol.)

As regards irradiation of deep-lying structures by external
means, the investigations have progressed along three principally
different lines. The first has been the utilization of new techniques for the production of ionizing radiation of improved penetra-

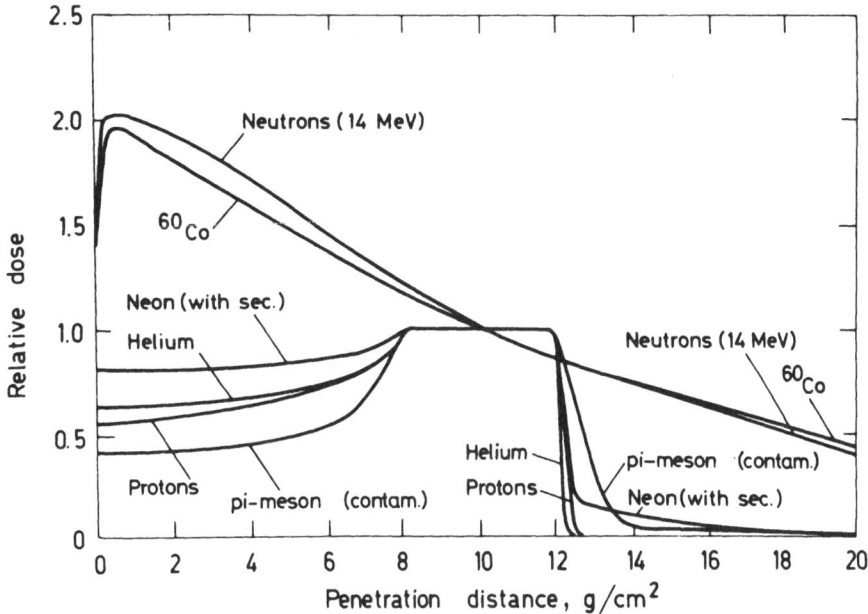

Fig. 3. The dose planning is very much facilitated by radiations
         that permit a free choice of depth of penetration and a-
         daptation of the dose maximum to the area suspected of tu-
         mour growth. Ion beams have, in these respects, particular-
         ly advantageous absorption characteristics. Both protons
         and heavier ions (such as helium or neon) can be made to
         deliver the dose uniformly within a Bragg plateau, that
         can be varied almost at will by the techniques outlined in
         Fig. 2. For comparison, depth-dose curves for ^{60}Co photons,
         fast neutrons, and negative pi-mesons are also shown.
         Redrawn from ref. 11.

bility and lower scatter than those earlier applied. This approach
has, up to recently, represented an obvious way of improving the
distribution of the absorbed dose in the single field. Important
steps towards the present widespread use of supervoltage and tele-
isotope therapy are the successful design of tubes for the generation of hard roentgen rays, the development of the resonant trans-
former for energizing such tubes, and the first therapeutic use of
teleradium and telecobalt equipment. Since World War II we have wit-
nessed the exploitation of radionuclides, as well as the betatron,
the linear accelerator, and the microtron, for the production of
therapeutically useful electron and photon radiation in the multi-
MeV range. The energy range of interest in the application of these "conventional" radiations has now been covered. Protons, and

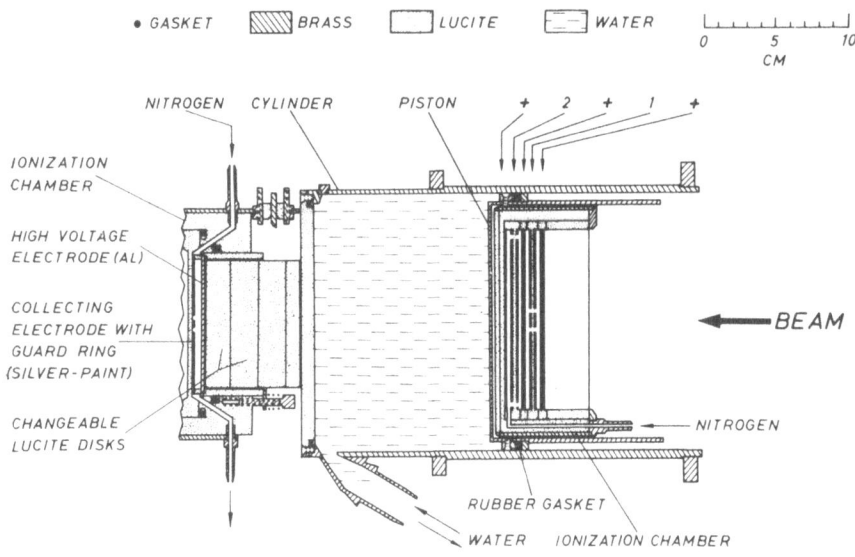

Fig. 4. Section through the ionization chambers and the variable
         cylindrical absorber used for modification of depth-dose
         curves in broad proton beams as outlined in Fig. 2. From
         ref. 10.
         (Courtesy Brit.J.Radiol.)

other accelerated ions, are characterized by near straight-line penetration and by a Bragg peak that may conveniently be transformed
into "Bragg plateaus" of almost any chosen shape at the end of the
well-defined range of penetration (Fig. 2). These radiations seem
to represent the ultimate phase of progress along this line (Fig. 3).

The second important line of development in "deep-therapy" by
external irradiation has been the endeavour to increase the selec-
tivity of the irradiation procedure by moving the radiation field,
continuously or step-wise, around a centre located in or near the
target structure. This method is described by the term "moving field
therapy". The idea of using a multiple field procedure, "cross fire
irradiation", where the radiation fields meet at a point in the
region to be irradiated seems to have been described first by Kohl
in 1906 [12] and has later appeared in many modifications, including
"rotation therapy", "pendulum therapy", and "convergent beam therapy".
This approach has been very fruitful, particularly since simulator
and computer systems were introduced so as to allow dose-planning for
superimposed fields, also in difficult anatomical situations, with
corrections for inhomogeneities and irregular body curvature.

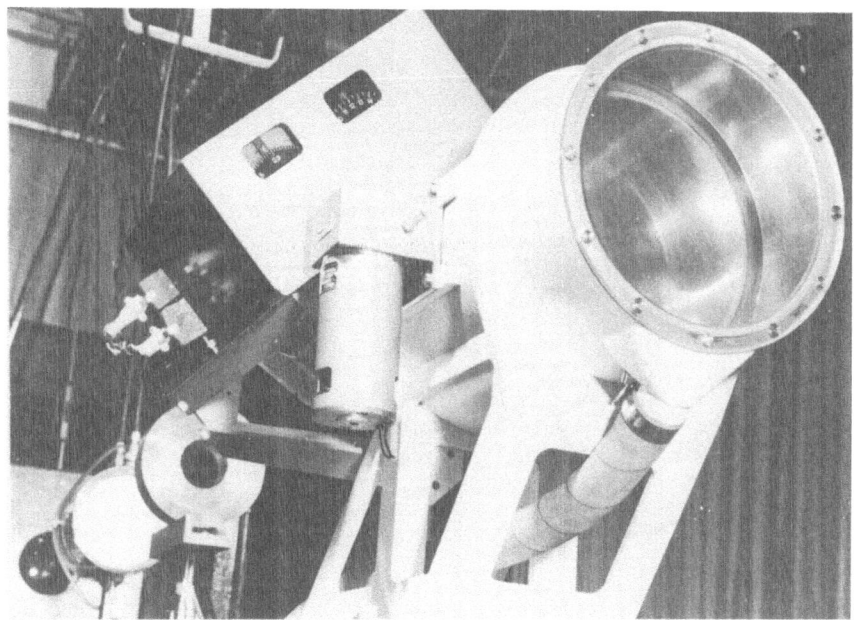

Fig. 5. Photograph of the apparatus used for modification of
         depth-dose curves in the 185 MeV proton beam, as indi-
         cated in Fig. 4. The near-parallel beam of protons enters
         the experimental area through a thin plastic window, at
         the end of the evacuated beam transport system (lower
         left corner).

With conventional radiations, complex multifield arrangements have
to be used to permit treatment of irregular targets. Protons, on
the other hand, often allow sufficiently sophisticated tailoring
of dose distributions with single fields, as exemplified in Figs.
6 and 7. However, there are two important situations when single
proton fields have to be avoided,(i) when compensation for tissue
inhomogeneities cannot easily be made[13], and (ii) in procedures
employing thin beams for irradiation of small deep-lying structures (*vide infra*). These situations have to be taken into consideration in the eventual design of systems intended for routinary clinical application of high-energy protons.

Possible progress along a third, presently much considered line of development depends primarily on the potentialities of high
linear energy transfer (LET), a new radiation quality introduced
in radiotherapy with the use of fast neutrons, negative π-mesons,
and accelerated heavier charged particles, such as neon ions (cf.
Fig. 3). At the necessary energy for clinical use, several hundred

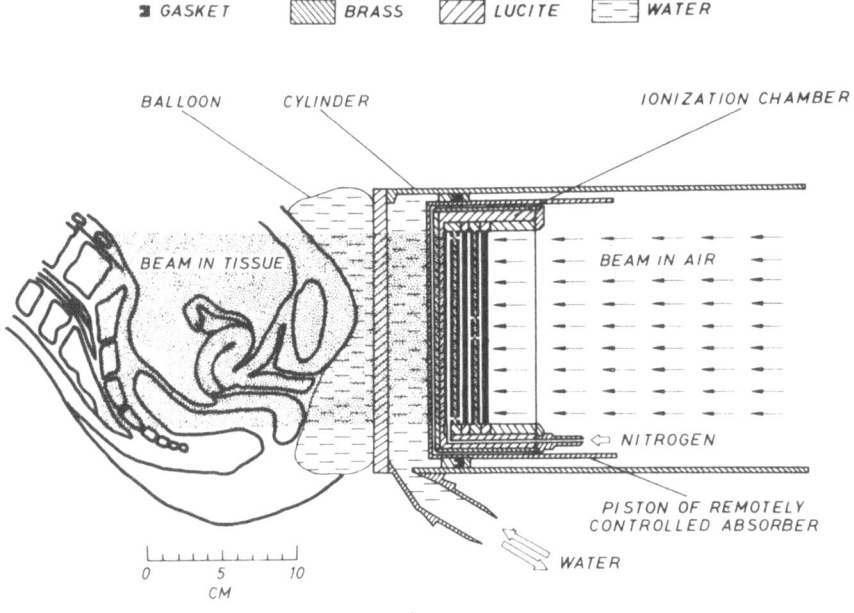

Fig. 6.   Section through the experimental apparatus shown in Fig.5
          arranged for irradiation of tumours in the pelvis by a
          Bragg plateau[14]. The piston controlling the varying thick-
          ness of the absorber is shown at a moment when its position
          gives maximum penetration of the beam.
          (Courtesy Acta radiol.)

MeV/amu, protons deliver most of their dose at $LET_\infty < 10$ keV/μm and
neon ions at $LET_\infty > 100$ keV/μm. These values represent, effectively,
the clinically relevant extremes of a scale of $\overline{LET_\infty}$ on which all radiations, "new" or "conventional", can be placed. With our present
knowledge, this is a relevant and, for practial purpose, suitable
way of presenting the radiobiological quality of the various radiations[15]. From this point of view, that is with regard to *microscopic* distribution of dose, the protons are classified as low-LET radiations together with all conventional kilovoltage or megavoltage
radiations, while heavy ions, together with fast neutrons and negative pions, may be called high-$\overline{LET_\infty}$ radiations. The latter assigna-
tion is, in fact, not entirely correct, fast neutrons[16] and negative pions[17] being both representatives of an intermediate class. This
is easily understood, considering that these latter particles are
mainly acting indirectly through nonuniform mixtures of secondary
protons and other ions of varying $\overline{LET_\infty}$.

Awaiting further radiobiological studies and clinical trials,
we must admit that a reliable judgement as to the place of high $\overline{LET_\infty}$

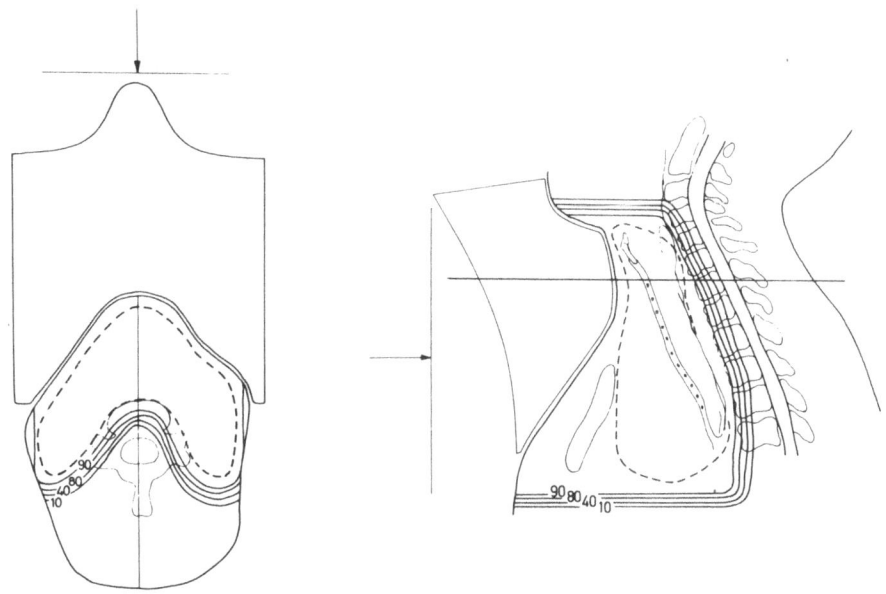

Fig. 7. This case, from a series of therapeutic applications with
185 MeV protons in Uppsala, illustrates the tailor-made
dose distributions which can be obtained with a single
proton beam of suitable energy. The dashed curve indicates the target area, the figures and the full lines represents "isodoses". The area within the isodose 90% is almost uniformly irradiated, while surrounding tissues (*e.g.*,
in the lungs and in the spinal cord) are spared. From ref.
18.
(Courtesy "Atomkernenergie".)

in radiotherapy cannot be given. Neither low $\overline{LET}_\infty$ nor high $\overline{LET}_\infty$ will
probably come out as the "best" type of irradiation, as the choice,
in each case, would depend on the type and extent of tumours and on
various radiobiological factors. At the present state of development,
the possibility of co-ordinated production and use of 200 MeV protons and 50 MeV (possibly 100 MeV) deuterons for fast neutron therapy seems attractive from a clinical-practical point of view[19]. This
concept is being included in the plans for the continued biomedical
use of the Uppsala cyclotron. This accelerator is now under reconstruction, according to modern principles, to permit production of
various ion beams.[20]

In this account, so far, we have been concerned with various
aspects on the use of high-energy protons in general radiotherapy,

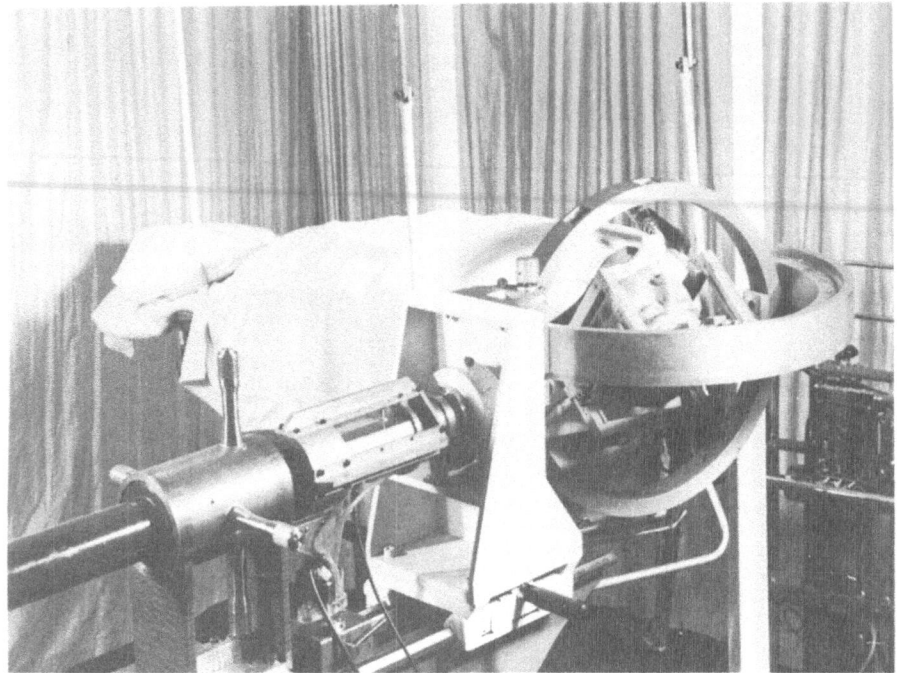

Fig. 8.   Patient with Parkinsonism in position for cross-firing
          of the thalamus with 185 MeV protons. A 3 mm wide beam
          was continuously centered on the target by means of a ro-
          tating stereotactic instrument. From ref. 21.
          (Courtesy Acta Chirurgica Scandinavica.)

for irradiation of comparatively large tumour-infiltrated structures.
res. As shown already in the pioneering studies with the Berkeley
synchrocylotron, however, cross-firing with high-energy beams of
light ions also offers convenient means for precise and effective
destruction of small intracranial structures[3]. In our work[21],well
collimated proton beams have proven themselves very useful, e.g.
for destruction of cerebral tracts or nuclei (Fig. 8). They would
in fact, be similarly applicable to affection of the Gasserian gang-
lion or the trigeminal root, of pituitary tissue, of vascular malformations and small non-infiltrating tumours[22]. Such techniques,
that usually employ very thin beams, 1 - 10 mm in diameter, permit
"surgery" through the intact skull, implying considerable reduction
of risks and discomfort to the patient, as compared to conventional
procedures. Similar experiences have been reported from Harvard where
stereotactic "Bragg peak radiosurgery" with protons has been employed
since 1962. A detailed account of this approach was recently given
by Kjellberg[23].

CONDITIONS FOR RADIOLOGICAL STUDIES WITH THE 185 MEV PROTON BEAM

For the above review of the basic radiological properties of
high-energy protons, in the context of general development in radiotherapy, illustrations were chosen from the discontinued work with
the 230-cm cyclotron, shut down for improvement. It is instructive to consider the conditions for this early work, in some detail.

Control of the beam

The experimental area was situated about 25 m from the synchrocyclotron and it was therefore easy to obtain a well collimated beam,
practically free from contaminating secondary radiation produced in
the extraction apparatus. By the use of a system of magnetic coils,
arranged so as to sweep the beam uniformly over the area to be irradiated, it was possible to reduce effectively the fraction of the
available beam current which was lost in the walls of the beam-de-
fining aperture and at the same time decrease the contribution of
secondary radiation from that source. An analysis of the frequency
of chromosome aberrations, induced in broad bean roots near a brass
block in which protons of a given flux had been stopped, indicated
that the effect of secondary radiation in the therapeutic irradiation procedures was negligible[24].

The mean energy, 187 $\pm$ 2 MeV, and the flux, 10^7 to $5\cdot10^{10}$ pro-
tons/sec, of the protons were controlled before the beam was permit-
ted to enter the experimental area (Fig. 9). The initial energy spread of the particles was of the order $\pm$ 1 MeV. The measuring apparatus was aligned with the axis of the almost parallel beam by means
of an optical bench (Fig. 5). Localization of the biological object
in relation to the beam was often direct, by means of roentgenogra-
phy, or image amplifier techniques[25]. Alternatively, the target region was precalculated in relation to palpable structures and lead
indicators fixed to the skin or, in relation to the coordinate frame
of a stereotaxic instrument[26].

Dosimetry

The ionization chambers used in the biological experiments were of the classical plane-parallel type with large high voltage electrodes and collecting electrodes, surrounded by wide quard-rings.
When the width of the final aperature was more than one centimeter
the electrodes were placed perpendicular to the beam, the sensitive
volume of the chamber being defined by the distance between the electrodes and the size of the collecting electrode (Fig. 10). In the
experiments with beams of 1 cm diameter and less, the electrodes were parallel with the beam, as in a free air "longitudinal" ionization chamber, the sensitive volume being defined by the cross-section of the beam and the length of the collecting electrodes (Fig.
11). This latter arrangement did not permit a direct, accurate cal-

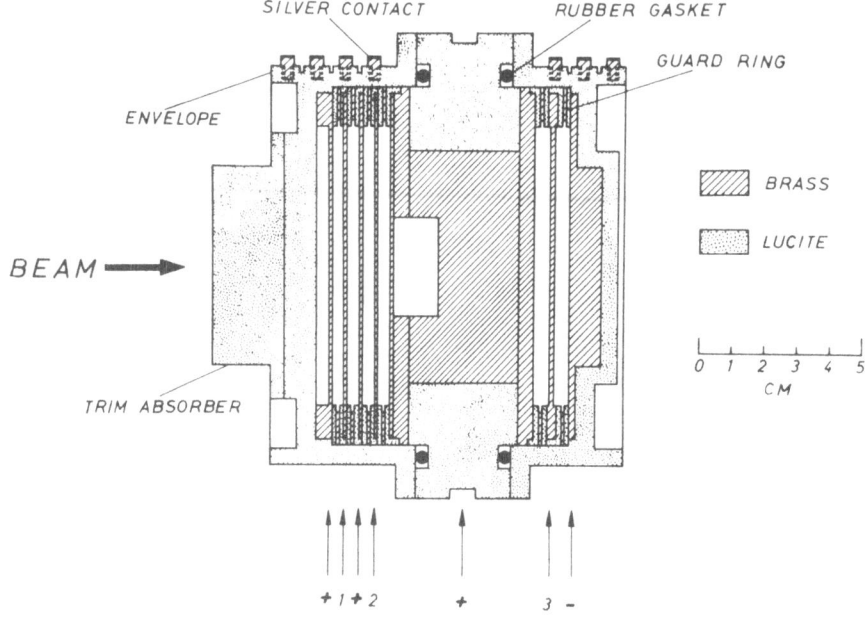

Fig. 9. Section through the "shutter", an arrangement of absorbers
and plane-parallel ionization chambers[5] of cylindrical symmetry used to control the admittance of the proton beam to
the experimental area and to check the energy and flux of
the protons. The polarities of the electrodes carrying high
tension are shown. The collecting electrodes are indicated
by numbers. From ref. 10.
(Courtesy Brit.J.Radiol.)

culation of dose, mainly because of the presence of scattered and
attenuated protons emanating from the wall of the aperture which was
used to define the beam entering the chamber. This chamber was therefore calibrated against the more reliable "transverse" chamber in
Fig. 10 by activation dosimetry, making use of small polystyrene cylinders situated on the central axis of the beam at the point where
the target structure was to be placed during the actual experiment.

    Independent checks on the ionization chamber dosimetry were ma-
de by measurement of the proton current by a "Faraday cup", by activation dosimetry (*vide infra*) and by classical ferrosulphate dosimetry. The Faraday cup was similar to that described by Birge et al.[27]
and their formulas were also used for the calculation of dose from
the measured current and the mass stopping power of nitrogen.

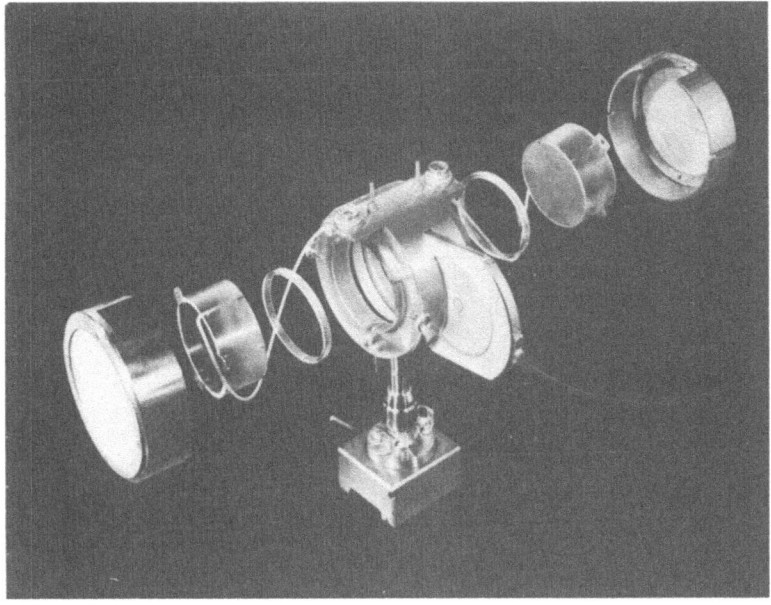

Fig. 10. A "transverse" plane-parallel, nitrogen-flushed ionization chamber, used for dosimetric study and as a reference monitor in the 185 MeV proton beam (cf. Fig. 5).

In the activation dosimetry, the proton-induced activity of ^{11}C
was measured in a well type scintillation counter, calibrated by the
use of standardized activities of ^{22}Na and ^{60}Co. The efficiency of
the counter for the pure positron activity of ^{11}C was calculated statistically from the observed efficiency of the counter for the combined positron and 1.28 MeV gamma radiation of ^{22}Na, and the efficiency for 1.25 MeV gamma radiation, as statistically calculated from
the observed efficiency for the combined 1.17 and 1.33 MeV gamma radiation from ^{60}Co (Fig. 12).

The results of these independent measurements of dose agreed
within 5 per cent with the dose values obtained by ionization dosimetry with the transverse chamber, provided oxidation of the ferrous
ion occurs at the same rate with irradiation by high energy protons
as with irradiation by other ionizing radiation of low linear energy
transfer. In the case of 170 MeV protons, there was a yield of 15.6±
1.0 ferric ions per 100 eV absorbed energy, in excellent agreement
with the values reported for high energy electrons and ^{60}Co gamma
radiation which also fall within this range[28].

The mean range in water was 23 cm for the two thirds of the original number of particles which escaped close nuclear interactions

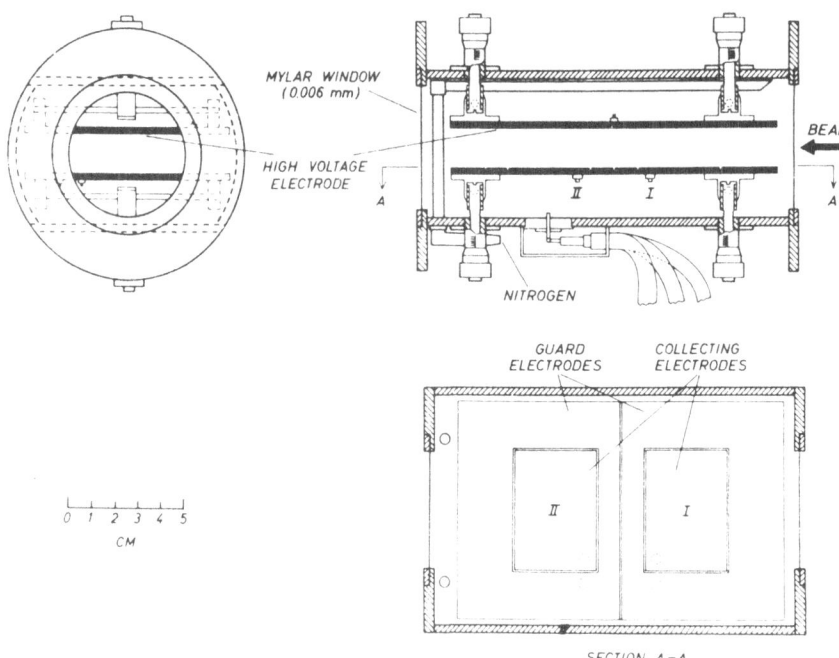

Fig. 11.   The "longitudinal" chamber used for the flux measurements,
and as a dose monitor, in studies with 1 - 10 mm wide proton beams. The electrodes consist of thin layers of silver
paint on Lucite.

during the course through the absorber and were able therefore to
contribute to the dose in the Bragg peak region of the depth-dose
curve. In beams of 3 cm diameter or more, the maximum value of the
depth-dose in soft tissue was found to be 4.3 times higher than the
dose in the region where the protons entered the absorber, the half
value width of the Bragg peak being about 2 cm (Fig. 1). Due to scattering, the Bragg peak was found to become progressively flattened
when the width of the beam was decreased below 3 cm and disappeared
almost entirely when the beam cross-section was less than 1 cm in diameter.

The observation of the significant effect of scatter in the second half of the range was the key to a new method of varying the
depth-dose distribution by means of "ridge filters"[10]. In contrast
to other devices suggested for the controlled transformation of the
energy distribution of the protons the ridge filter introduces an absorbing material whose thickness varies periodically in space rather

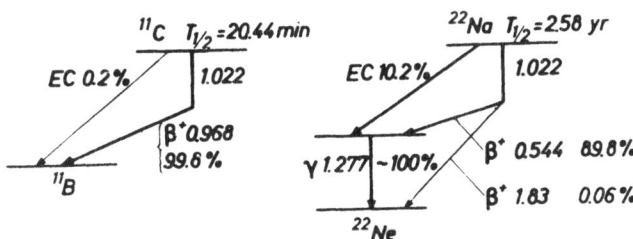

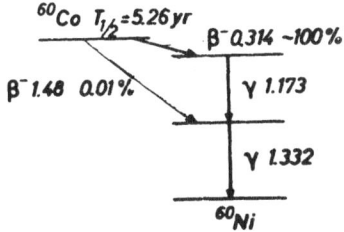

Fig. 12. Determination of ^{11}C in proton-irradiated polystyrene was
carried out by detection of the 0.51 MeV annihilation photons with a NaI(Tl) well crystal. The efficiency of the
counter for the detection of positrons from ^{11}C was statistically calculated from the measured efficiencies for detection of radiation from standard sources of ^{22}Na and ^{60}CO.

than in time. The small variations in the dose distribution introduced
by this filter were found to be smoothed out at depths where the
linear lateral scatter becomes comparable with the space between the
ridges of the filter. The inhomogeneities introduced by the filter
could be entirely avoided, at all depths, by periodic movement of the
filter in its plane, in a direction perpendicular to the ridges[29].

Distributions of dose were measured in Lucite or water, simulating soft tissues (Fig. 4). Complete distributions of doses in narrow
beams, and plane transverse distributions of doses in broad beams were
measured by photographic film dosimetry. Calibration of the density of the film was made by means of a dose-density curve, measured
from films exposed at full energy and developed together with the
films to be analysed. The dose was then calculated from the assumption that the shape of the dose-density curve is independent of the
energy of the protons[28].

THE RELATIVE BIOLOGICAL EFFICIENCY

As mentioned above the variation of radiation quality along the tracks of the protons may be quantitatively characterized in terms of the varying LET_∞. This variation should be of biological significance and therefore an analysis was made of the induction of chromosome aberrations in root tips of onion seedlings at various depths of penetration[24]. The results showed a significant increase in the relative biological efficiency (RBE) for the induction of aberrations at greater depths: the normalised biological depth-effect curve showed in fact a maximum which was at least twice as high as the Bragg peak of the physical depth-dose curve (Fig. 13).

An increase of the LET_∞ of ionizing radiation tends to decrease the enhancing effect of oxygen on the RBE. Attempts to demonstrate a qualitative change in the radiation by comparison of the oxygen-enhancement at high and low energies failed and it was concluded that the proton beam may be regarded, at all depths of penetration, as an ionizing radiation of the low-LET_∞ type when the radiotherapeutic significance of variations in the supply of oxygen to an irradiated tissue is being considered[24].

The protons exert, however, except at energies below 20 MeV, i.e. in the last centimeter of their range in tissue, a $\overline{LET}_\infty$ which does not exceed 5 keV per micron. In view of the fact that the RBE has been found to be little influenced by LET_∞-variations in this interval, it is likely that the RBE for cell damage of protons in the high energy region above 20 MeV is nearly equal to that shown by the types of ionizing radiation which are used routinely in deep radiotherapy and which do not possess a LET_∞ differing by more than a factor of ten from the absolute minimum, 0.2 keV per micron. RBE of 170 MeV protons ($\overline{LET}_\infty$ = 0.5 keV per micron), as compared with 180 kV roentgen rays, for the production of chromosome aberrations in the broad bean root was reported to be 0.70 ± 0.05 and 0.66 ± 0.05 in air and nitrogen, respectively[24]. These figures were in agreement with early reported semi-quantitative findings in experiments on normal rabbit skin[30], and also with simultaneously reported determinations of $LD_{50/30}$ in mice[31], suggesting the RBE of 157 MeV protons as compared with 250 KV roentgen rays to be 0.77 ± 0.1. The approximate equivalence of 180 MeV protons and ^{60}Co photons was demonstrated by comparison of the observed incorporation of iodinated deoxyuridine into intestinal and spleenic DNA in mice[32].

In parallel with these quantitative studies, series of histopathological observations on proton-irradiated healthy tissues and experimental tumours in animals were also made[33-36], prior to the clinical use of the 185 MeV proton beam. A compilation of results from studies performed in the period considered[37], as well as findings in irradiated patients[14,38-40], were in conformity with the view that

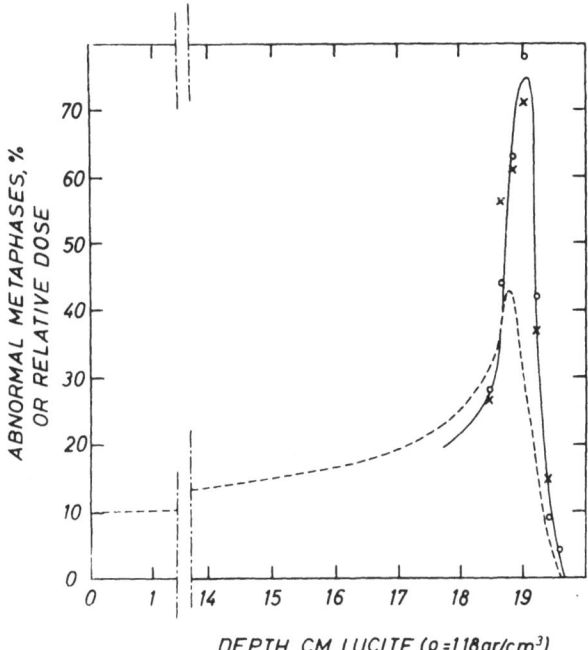

Fig. 13. The efficiency of the 185 MeV proton beam for producing
chromosome aberrations in root-tips of onion seedlings as
a function of the depth in Lucite (——) compared with the
physical dose-depth curve (- - - -). The maximum dose was
1.2 Gy. o and x signify the results of two different experiments. From ref. 24.
(Courtesy Int.J.Radiobiology.)

therapeutically applied, range-modulated proton beams should be
qualitatively and quantitatively similar to conventional therapeutic
radications. An oxygen effect, typical for low-LET_∞ radiation, was
also demonstrated [41].

For therapeutic purpose there seemed to be no reason for extending
the experiments on RBE much further, as the apparently small variations in the "specific" RBE of the protons may, in practice, be less
important than variations due to differences in the macroscopic distribution of the dose in the body of the patient. The significance
of the variation in RBE at the end of the depth-dose curve, should
be diminished when the Bragg peak is flattened and broadened during
irradiation as by a ridge filter. Further applications of the proton

beam to tumour radiotherapy were therefore based provisionally on a
presumed "specific" RBE of approximately "one". To obtain more detailed information of relevance for the clinical use of the radiation,
biological tests of the effect of the protons on tumour tissue and
healthy tissue should be made under conditions closely simulating
those encountered in actual clinical applications.

EFFECTS OF NARROW BEAMS ON CENTRAL NERVOUS TISSUE

At the start of this work, there was little relevant information
on the effects of high dosage irradiation of small volumes of central
nervous tissue. It was therefore necessary to study the neurosurgical use of the proton beam in animals. Observations were first made
on the spinal cord in rabbits. Typically, a dose of 200 Gy (20 000
rad) at 170 MeV given in a 1.5 mm broad beam "transecting"the cord,
produced after 9 days a well defined lesion of approximately the same width as that of the beam. With larger doses the latent period was
shorter. The breadth of the beam also proved to be a factor of great
importance. A 10 mm beam produced at 200 Gy a lesion of corresponding width in 4 - 6 days and this lesion differed histologically from
that produced by a 1.5 mm beam; hemorrhages and vascular changes were very marked in the path of the broad beam, while only slight changes of this type were seen with the narrow beam.

After similar irradiation of the brain of the rabbit with a dose
of 200 Gy and a beam width of 1.5 mm, there was also a well defined
lesion in the brain[43]. It was confined to the path of the beam for up
to 3 months after irradiation and became slightly wider subsequently.
The histological changes following irradiation were recorded at intervals from 2 to 56 weeks. No large hemorrhage was seen and there was
little inflammatory reaction. Two weeks after irradiation, most nerve
cells, myelin sheaths, and axons were destroyed in the obvious path
of the beam. No further changes were observed but,in view of the
possibility that progressive pathologic changes might occur later, longer term experiments were performed on goats (*vide infra*).

On the basis of the qualitative histopathological study in rabbits a semi-quantitative comparison of the effects of local irradiation on the rat brain with different doses and at different times
irradiation was made[44]. As criteria for the evaluation of the relation between the dose and the time after irradiation at which the
characteristic stages of the radiolesion occur, easily demonstrable
effects of irradiation were selected, among them necrosis of brain
tissue. A diagram showing the chronology of events following local
irradiation of the brain was constructed to facilitate the later planning of procedures for the production of circumscribed radiolesions
in the brain (Fig. 14).

A study was made of the effect of cross-fire irradiation on a
target region deep within the goat's brain with a view to the neuro-

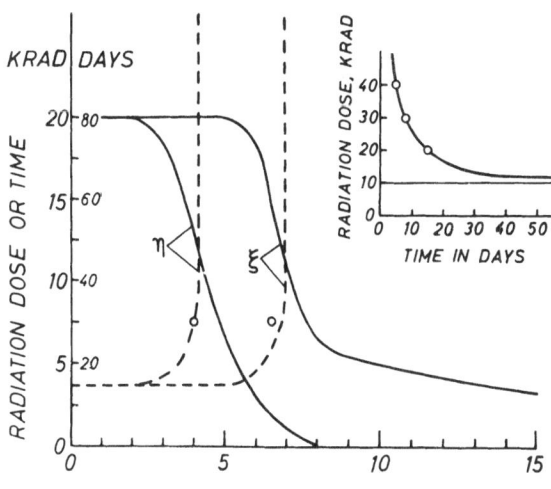

Fig. 14. The calculated dose (———) and the time after irradiation at which necrosis was estimated to appear (- - - -) along to axes (ξ and η) in the target region of the brain. The time curves were constructed from the dose curves and from the data in the small diagram, showing the time at which necrosis has been found to appear in the brain of the rat after irradiation with different doses or protons. The dimensions of a typical lesion in the goat brain, measured 30 days after irradiation, are indicated by circles in the large diagram. From ref. 45.
(Courtesy Acta radiologica.)

surgical use of the proton beam in physiological experiments and in therapy. By use of beams of 2x7 mm^2 or 2x10 mm^2 cross-sections and 20 - 22 fields, directed to an area in the internal capsule by means of Leksell's stereotaxic method[26], it was possible to induce damage to the tissues, with complete destruction of myelin sheaths, axons and glial cells within a sharply demarcated lesion. In conformity with the previous experiments a dose of 20 krad seemed to be suitable for the production of a discrete lesion within a few weeks after irradiation (Fig. 15). When a very high dose, 40 krad, was used, the

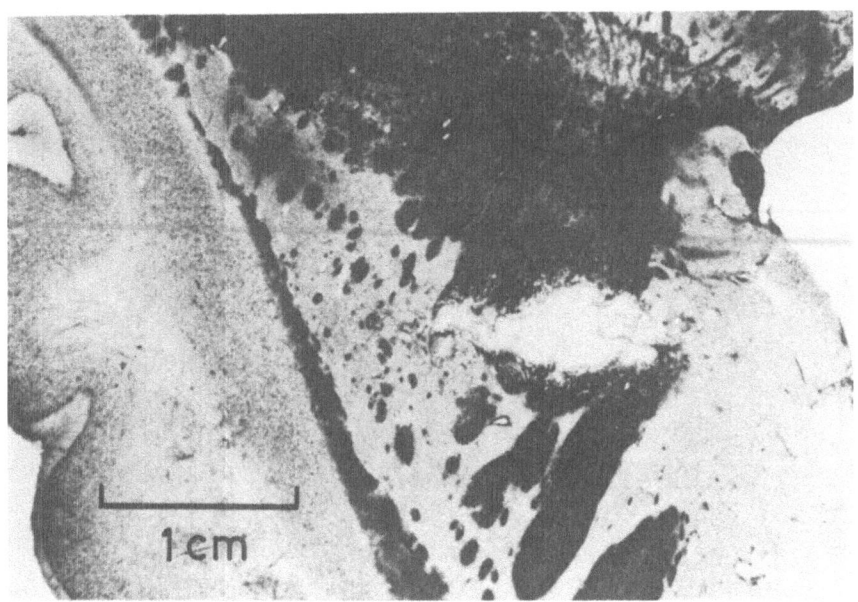

Fig. 15. Histological section from the brain of a goat, one month
          after cross-fire irradiation with 22 proton fields, 10x2
          mm². The absorbed radiation dose, 200 Gy at the center of
          the lesion, has caused a sharply delimited lesion in the
          internal capsule. From a model study[25], performed prior
          to the application of the 185 MeV proton beam to functio-
          nal neurosurgery in man (cf. Figs 8 and 15).

volume of necrotic tissue was large and surrounded by a very intense
tissue reaction. This observation stresses the importance of selecting both the dose of the centre of the lesion and the dose distribution carefully. The size and shape of the lesion should, according
to time-dose relationships given, be dependent on the time after irradiation. It was demonstrated, however, that with a dose of 20 krad
there is reason to believe that the margin of the "final" lesion will
coincide approximately with the surface of the greatest gradient in
the dose distribution.

    It was concluded that the radiosurgical method gave satisfacto-
ry results in the animal experiments so far described, but the extent
to which the procedure might be used in operations on man still depended on the possible late changes following the local administration of high doses of radiation. It was thus decided to test the
operative method in a long term follow-up experiment on goats, irradiated with 200 - 300 Gy. No untoward histopathological side-effects
which could be ascribed to the irradiation have been observed[46].

With the increasing evidence that radiosurgery with the 185 MeV proton beam might be applicable, to operations in man, it was decided to give therapy to a small group of patients with disordered brain fuction, with psychic disorder, with pain due to a metastatizing carcinoma and with Parkinsonism[21]. After stereotaxic irradiation of selected target regions in the brain with doses of 16 - 25 krad, therapeutically effective lesions became evident during the first weeks following irradiation and no side effects due to the radiosurgical procedure were observed. A patient with carcinoma died during the course of investigation and the lesion found at autopsy was studied. The acute radiosurgical lesion in man was then found to be similar to lesions induced in animals by approximately the same dose[47].

PROSPECTS FOR DEDICATED ACCELERATORS FOR USE OF PROTON BEAMS IN CLINICAL RADIOTHERAPY

In general radiotherapy, direct irradiation with proton beams of suitable range of penetration offers outstanding flexibility and precision. The maximum penetration depth of the radiation field can be varied by changing the energy of monoenergetic beams extracted from the accelerator or by suitable absorbers. The energy spectrum of the particles entering the body may be further modulated so that a homogeneous "Bragg plateau" with well delineated borders in both axial and transverse directions is created. The entrance dose can be kept lower than the target dose and the penetration depth can be varied over the cross section of the field by bolus materials. Beams of protons are easily collimated also to irregular cross sections. At the present state of development, however, with a view towards high-LET the possibility of coordinated production and use of 200 MeV protons and 50 MeV (possibly 100 MeV) deuterons for fast neutron (high LET) therapy seems attractive. A convenient design of an accelerator and beam transport system dedicated to these purpose would facilitate the further evaluation of the medical use of cyclotron-produced radiations[48]. The problems concerned with clinical therapy are given consideration in the present program for the reconstructed cyclotron[20].

The concepts out-lined are presently evaluated in the development of apparatus and routines for therapy with charged heavy particles. The state of technical development in Uppsala, from where examples in this paper were chosen, may also serve to illustrate the elementary principles.

A discussion of the usefulness of high energy protons can be based directly on clinical experience with conventional therapy, since high-energy protons deliver almost all dosage by low-LET. In this respect, they are qualitatively similar to photons and electrons. From the point of view of a given irradiated cell, it is, in fact, a minor importance whether a given dose has been delivered with a proton beam or by any of the radiations from conventional apparatus. The conceived merits of high-energy protons, in radiotherapy, would depend almost entirely on their advantageous macroscopic absorption characteristics.

The options thus offered permit unparallelled flexibility and precision, and so, in theory, increased probability of tumour sterilization in any chosen target volume with minimum affection of healthy tissues. The potentialities of effect modification through tempral, chemical or biological means for improving the therapeutic ratio are also the same, in principle, for protons, photons, and electrons. In addition, after treatment with these low-$\overline{\text{LET}}$ radiations, cells are more able to recover from sublethal damage and express more variation in radiosensitivity with cell type and position in the cell cycle than in the case of high-$\overline{\text{LET}}$ radiation.

Proceedings of three international meetings[49,50] are referred to, for conveniency of readers who would prefer recent, detailed accounts of the international situation, rather than a review based on classical examples from early activities in the field of radiotherapeutic research with 185 MeV protons.

The "First International Seminar on the Use of Proton Beams in Radiation Therapy" arranged at Moscow, in 1977[50], is particularly interesting in that context. The papers presented at this meeting show, indeed, that the explorative phase of investigation now is over. Protons, or related particles from medium-energy ion accelerators, are in fact attracting the interest of an increasing number of oncologists, especially in Sweden, U.S.A. and U.S.S.R. The clinical impression can now be based on substantial experience, together over 1000 patients have been treated so far at the various centers. The work is going on in parallel with the exploitation of high-$\overline{\text{LET}}$ radiations in the form of negative pi-mesons, fast neutrons or accelerated heavy ions. We are witnessing a vital development with varying approaches and constructive new ideas. Progress is made simultaneously at several important front-lines, supplementary to clinical radiology: improved dosimetry, computerized dose planning, basic radiobiology of tumours and healthy tissues and techniques for explorative radiography of scintigraphy based on new detector and computer technologies, including the use of accelerator-produced radionuclides.

CONCLUSIONS

The characteristics of the present states of development may be summarized as follows:

1. Proton therapy is being considered for routinary application at various centres, especially in the U.S.A., the U.S.S.R. and Sweden.

2. The choice of beam energy and accelerator parameters is recommended to be based, not only to the requirements given by conceived treatment and protocols, but also on the recognized value of a proton accelerator in radiography and for radionuclide production.

3. In treatment planning, attention is being paid to body inhomogeneties and the potentialities of computer assisted tomography.

4. There is agreement, between centres, about the qualitative equivalence of proton and other low-LET radiations, as far as RBE and chemical or temporal effect modification are being concerned. Observed differences between the RBE of gamma rays and protons call for precise quantitative intercomparison of dosimetry and dose-effect relationships, however.

5. Dose-effect relationships and their dependence on such factors as oxygen tension and other chemical modifications have to be studied in greater detail, quantitatively, in organized tissues and not only in cell culture.

6. As existing and planned proton accelerators differ in their time-intensity characteristics, the effects of varying pulse- and sweep-patterns on dose-effect-relations should be studied in great detail.

7. In addition to the increasing interest in general radiotherapy with protons, there is a renewed interest in narrow-beam techniques for intracranial irradiation of small structures, n.b. arterio-venous malformations, small central tumours and the pituitary.

ACKNOWLEDGEMENT

On behalf of the research group that the author represented at the Erice meeting, he would like to express thanks to collegues in Berkeley, Harvard and Moscow for collaboration and constructive discussions in various phases of the study.

Developments at the Uppsala synchrocyclotron have been mainly supported by the Swedish Academy of Engineering Sciences, the Knut and Alice Wallenberg Foundation, the Swedish Cancer Society, the Swedish Medical Research Council, the Swedish Natural Research Council, and the National Swedish Board for Technical Development.

REFERENCES

1. R. R. Wilson, Radiological use of fast protons, Radiology
   47:487 (1946).
2. C. A. Tobias, H. O. Anger, and J. H. Lawrence, Radiological use of high energy deuterons and alpha particles,
   Am.J.Roentgenol. 67:1 (1952).
3. C. A. Tobias, J. E. Roberts, J. H. Lawrence, B. V. A.
   Low-Beer, H. O. Anger, J. L. Born, R. McCombs, and
   C. Huggins, Irradiation hypophysectomy and related
   studies using 340-MeV protons and 190-MeV deuterons,
   in: "Peaceful Uses of Atomic Energy, Proc. Internat.
   Conf. Geneva", Vol. 10, United Nations (1956).
4. B. Larsson, T. Svedberg, and H. Tyrén, Djupterapi med
   protoner vid Uppsalasynkrocyklotronen, in:"Riksföre-
   ningen för Kräftsjukdomarnas Bekämpande, Årsbok 1952-
   1956," H. Bergstrand, ed., Almqvist & Wiksell, Stockholm (1957).
5. B. Larsson, L. Leksell, B. Rexed, B. Sourander, W. Mair,
   and B. Andersson, The high-energy proton beam as a
   neurosurgical tool, Nature 182:1222 (1958).
6. A. M. Koehler and W. M. Preston, Protons in radiation
   therapy, comparative dose distributions for protons,
   photons, and electrons, Radiology 104:191 (1972).
7. V. S. Choroškov, V. P. Dželepov, L. L. Goldin, M. F.
   Lomanov, O. V. Savčenko, and S. Tesch, Teilchenstrah-
   len in der Medizin, Wissenschaft u. Fortschritt,
   Berlin 23:347 (1973).
8. V. Abasov, B. Astrakhan, N. Blokhin et al., Use of
   proton beams in the USSR for medical and biological
   purposes, Communic.Joint Inst.Nucl.Res. E-5854,
   Dubna (1971).
9. B. A. Kannov, D. L. Karlin, V. B. Nizkovolos, and I.J.
   Senichev, Physical-technical facility for proton
   therapy on LIMP 1000 MeV synchrocyclotron, in:
   "Proceedings of the First International Seminar on
   the Uses of Proton Beams in Radiation Therapy,
   Moscow, December 6-11, 1977," Vol. I, M.I. Lomakin.
   ed., Atomizdat, Moscow (1979).
10. B. Larsson, Pre-therapeutic physical experiments with
    high energy protons. An extended version of the con-
    tribution to the symposium on Therapy with Beams of
    High Energy Particles, at the annual congress of the
    British Institute of Radiology on December 10, 1959,
    Brit.J.Radiol. 34:143 (1961).
11. C. A. Tobias, J.T. Lyman and J. H. Lawrence, Some con-
    siderations of physical and biological factors in
    radiotherapy with high-LET radiations including heavy
    particles, π-mesons and fast neutrons, in:"Progress
    in Atomic Medicine: Recent Advances in Nuclear Medi-

cine," Vol. 3, J. H. Lawrence, ed., Grune and
Stratton Inc., New York (1971).

12. M. Kohl, German patent 192571 (1906), cited by
     F. Wachsmann and G. Barth, in:"Die Bewegungsbestrahl-
     ung," ed., George Thieme, Stuttgart (1959).

13. M. Goitein, The measurement of tissue heterodensity to
     guide charged particle therapy, in "Particles and
     Radiation Therapy, Second International Conference,
     September 14-17, 1976, Berkeley, California,"
     W. E. Powers, ed., Perman Press, New York (1977).

14. S. Falkmer, B. Fors, B. Larsson, A. Lindell, J. Naeslund,
     and S. Sténson, Pilot study on proton irradiation of
     human carcinoma, Acta radiol. 53:33 (1962).

15. G. W. Barendsen, Responses of cultured cells, tumours
     and normal tissues to radiations of different linear
     energy transfer, in:"Current topics in radiation
     research," Vol. IV, M. Ebert and A. Howard, eds,
     North-Holland Publ., Amsterdam (1968).

16. J. F. Fowler, Fast neutron therapy-physical and bio-
     logical considerations, in:"Modern Trends in Radio-
     therapy," Vol. 1, T. J. Deeley and C. A. P. Wood,
     eds, Butterworth , London (1967).

17. M. R. Raju and C. Richman, Negative pion radiotherapy:
     physical and radiobiological aspects, Current Topics
     in Radiation Research Quarterly, 8:159 (1972).

18. S. Graffman and B. Larsson, High energy protons for
     radiotherapy - a review of activities at the 185-MeV
     synchrocyclotron in Uppsala, Atomkernenergie
     27:148 (1976).

19. B. Larsson, Proton and heavy-ion therapy, in:"Health
     and Medical Physics," J. Baarli, ed., Soc.Italiana
     di Fisica, Bologna (1977).

20. S. Dahlgren, A. Ingemarsson, S. Kullander, B. Lundström,
     P. U. Renberg, K. Ståhl, H. Tyrén, and A. Åsberg,
     Conversion studies for the Uppsala synchrocyclotron,
     in:"Seventh International Conference on Cyclotrons
     and their Applications, Zürich 1975," W. Joho, ed.,
     Birkhäuser Verlag, Basel (1975).

21. B. Larsson, L. Leksell, and B. Rexed, The use of high
     energy protons for cerebral surgery in man,
     Acta chir.scand. 125:1 (1963).

22. J. Arndt, E. O. Backlund, B. Larsson, L. Leksell, G.
     Norén, K. Rosander, T. Rähn, B. Sarby, L. Steiner,
     and J. Wennerstrand, Stereotactic irradiation of
     intracranial structures: physical and biological
     considerations, in:"Stereotactic cerebral irradi-
     ation," G. Szikla, ed., Elsevier, Amsterdam (1979).

23. R. N. Kjellberg, Stereotactic Bragg peak proton radio-
     surgery method, Ibid.

24. B. Larsson and B.A. Kihlman, Chromosome aberrations following irradiation with high-energy protons and their secondary radiation: A study of dose distribution and biological efficiency using root-tips of *Vicia faba* and *Allium cepa*, Int.J.Radiat.Biology 2:8 (1960).

25. B. Jung, B. Larsson, B. Rosengren, K. Ståhl, and W. Wretlind, Roentgen stand for field positioning in high-energy radiotherapy, Acta radiol.Ther.Phys.Biol. 7:282 (1968).

26. L. Leksell, "Stereotaxis and radiosurgery - an operative system," Ch. C. Thomas, Springfield (1971).

27. A. C. Birge, H. O. Anger, and C. A. Tobias, Heavy charged-particle beams, in:"Radiation Dosimetry," G. J. Hine and G. L. Brownell, eds, Academic Press, New York (1956).

28. B. Larsson, On the application of a 185 MeV proton beam to experimental cancer therapy and neurosurgery: A biophysical study, Acta Universitatis Upsaliensis. Abstracts of Uppsala Dissertations in Science, No 9 (1962).

29. B. G. Karlsson, Methoden zur Berechnung und Erzielung einiger für die Tiefentherapie mit hoch-energetischen Protonen günstiger Dosisverteilungen, Strahlentherapie 124:491 (1964).

30. S. Falkmer, B. Larsson, and S. Sténson, Effects of single dose proton irradiation of normal skin and Vx2 carcinoma in rabbit ears, a comparative investigation with protons and roentgen rays, Acta radiol. 52:217 (1959).

31. P. Bonet-Maury, A. Deysine, M. Frilley, and C. Stefan, Efficacité biologique relative des protons de 157 MeV, C. R. Acad.Sci., Paris 251:3087 (1960).

32. K. J. Johanson and B. Larsson, Effect of 180 MeV protons and ^{60}Co radiation on the incorporation of ^{125}I-iodo-2'-deoxyuridine into intestinal and spleenic deoxyribonucleic acid in mice, Acta radiol.Ther.Phys.Biol. 11:452 (1972).

33. M. Danielsson, B. Engfeldt, B. Fors, B. Larsson, and J. Naeslund, Effect of high-energy protons on Vx2 carcinoma implanted in the lower abdominal wall of the rabbit, Acta obstet.gynecol.Scand. 47:373 (1968).

34. S. Sténson, Effects of proton and roentgen radiation on the rectum of the rat, Acta radiol.Ther.Phys. Biol. 8:263 (1969).

35. S. Sténson, Weight change and mortality of rats after abdominal proton and roentgen irradiation, a comparative investigation, Acta radiol.Ther.Phys.Biol. 8:423 (1969).

36. B. Engfeldt, B. Larsson, C. Naeslund, J. Naeslund, and B. Tjernberg, Effect of single dose or fractionated proton irradiation on pulmonary tissue and Vx2 carcinoma in the lung of the rabbit Acta radiol.Ther. Phys.Biol. 10:298 (1971).

37. K. J. Johanson, Deoxyribonucleic acid metabolism in the small intestine in vivo and in vitro. A study of the normal and 180-MeV proton or ^{60}Co gamma irradiated mouse, Acta Universitatis Upsaliensis, Abstracts of Uppsala Dissertations in Science, No. 204 (1972).

38. B. Fors, B. Larsson, A. Lindell, J. Naeslund, and S. Sténson, Effect of high energy protons on human genital carcinoma, Acta radiol. 2:384 (1964).

39. S. Graffman, E. Jung, B. A. Nohrman, and R. Bergström, Supplementary treatment of nasopharyngeal tumours with high-energy protons, Acta radiol. 6:361 (1967).

40. S. Graffman, W. Haymaker, R. Hugosson, and B. Jung, High energy protons in the postoperative treatment of malignant glioma, Acta radiol.Ther.Phys.Biol. 14:445 (1975).

41. B. Larsson and S. Sténson, Reduction of radiation damage to the intestinal mucous membrane by local hypoxia, Nature 205:364 (1965).

42. B. Larsson, L. Leksell, B. Rexed, and P. Sourander, Effects of high energy protons on the spinal cord, Acta radiol. 51:52 (1959).

43. B. Rexed, W. Mair, P. Sourander, B. Larsson, and L. Leksell, Effect of high energy protons on the brain of the rabbit, Acta radiol. 53:289 (1960).

44. B. Larsson, Blood vessel changes following loacl irradiation of the brain with high-energy protons, Acta Societatis Medicorum Upsaliensis 65:61 (1960).

45. L. Leksell, B. Larsson, B. Andersson, B. Rexed, P. Sourander, and W. Mair, Lesions in the depth of the brain produced by a beam of high energy protons, Acta radiol. 54:251 (1960).

46. B. Andersson, B. Larsson, L. Leksell, W. Mair, B. Rexed, P. Sourander, and J. Wennerstrand, Histopathology of late local radiolesions in the goat brain, Acta radiol.Ther.Phys.Biol. 9:385 (1970).

47. W. Mair, B. Rexed, and P. Sourander, Histology of the surgical radiolesion in the human brain as produced by high-energy protons, Radiat.Res.Suppl. 7:384 (1967).

48. S. Graffman, B. Jung, and B. Larsson, Design studies for a 200 MeV proton clinic for radiotherapy, in: "Proc. Sixth International Cyclotron Conference, Vancover 1972," American Institute of Physics, (1973).

49. "Particle Radiation Therapy, International Workshop,
     October 1-3, 1975, Key Biscayne, Florida,"
     W. E. Powers, ed., American College of Radiology,
     Philadelphia (1976) and "Particles and Radiation
     Therapy, Second International Conference, September
     14-17, 1976, Berkeley, California," W. E Powers,
     ed., Pergamon Press, New York 1977).
50. "Proceedings of the First International Seminar on
     the Uses of Proton Beams in Radiation Therapy,
     Moscow, December 6-11, 1977," Vol. 1-3, M. I.
     Lomakin, M. F. Lomanov, and T. G. Ratner, eds.,
     Atomizdat, Moscow (1979).

DOSIMETRY OF NEUTRONS

J.J. Broerse

Radiobiological Institute TNO
Rijswijk
The Netherlands

INTRODUCTION

To estimate the risks of mixed n-γ radiation fields and to
predict the responses of irradiated biological systems, it is essential to obtain a quantitative description of the radiation field
or of the energy deposition processes inside an object. For purposes of radiation protection, a rough characterization of the radiation field in terms of type, energy, direction and number of
particles is sufficient in most cases. For medical and biological
applications, the absorbed dose and the radiation quality have to
be determined. The absorbed dose is defined as the quotient of the
mean energy imparted by ionizing radiation to the matter in a
volume element and the mass of the matter in that volume element.

The selection of methods for neutron dosimetry and the precision (i.e., the reproducibility) and accuracy to be achieved will
depend on the specific application. It is now generally recognized
that, for biological and medical applications, the dose in the biological object should be determined with a precision of better than
$\pm$ 2 per cent and an overall uncertainty of less than $\pm$ 5 per cent.
It must be stressed that these requirements are more severe than
generally encountered in radiation protection, where the dose-effect
relationships for radiation-induced late effects are not yet well
known. Although, higher accuracy and precision in dosimetry are
required for medical applications, the dose and dose rates under
these conditions are higher than in the protection situation, which
facilitates the measuring techniques.

The absorbed dose determines only the mean energy imparted
per unit mass in the irradiated medium. For neutron dosimetry, in-

formation on the neutron spectrum is needed, since the neutron interaction cross sections and the kerma to fluence quotients depend
on neutron energy. Specification of absorbed dose does not include
information on the microscopic distribution of the energy deposition. This microscopic distribution, however, can effect the yield
of radiochemical products and it also influences cellular radiation
effects as they depend not only on the total amount but also on the
spatial distribution of energy deposition. Attempts to account for
the microscopic distribution of energy absorption have led first to
the concept of linear energy transfer (LET) and its distributions
and later to the microdosimetric quantities. Although the energy
spectrum of the incident neutrons is an implicit characterization
of radiation quality, the microscopic distribution of energy absorption is decisive for the biological effectiveness of neutrons
and the techniques involved in the determination of LET or of the
microdosimetric quantities are related to procedures in neutron
dosimetry. The probability distributions of y (lineal energy) can
be determined experimentally with proportional counters. Information is available on lineal energy spectra for different neutron
energies (Kellerer and Rossi, 1972; Booz, 1978).

In this contribution, emphasis is placed upon the absorbed
dose distributions inside human phantoms, including perturbations
in charged particle equilibrium which occur at interfaces of materials of different atomic composition. The present adequacy of
neutron dosimetry can be derived from the results of neutron dosimetry intercomparisons. The results of a number of intercomparison
programs will be discussed and recommendations for future research
in neutron dosimetry will be made.

PRINCIPLES OF NEUTRON DOSIMETRY

Neutron fields are always accompanied by gamma rays originating
from the neutron producing target, the primary shielding and field
limiting systems and the biological object or phantom being irradiated. The fraction of the total absorbed dose due to the photon
component of the mixed field increases markedly with increasing
depth of penetration of the neutron beam in a phantom.

Because of the differences in biological effectiveness of these
two radiation components (which will depend on the specific biological endpoint), it is necessary to separately determine the neutron
absorbed dose in tissue, D_N, and the gamma absorbed dose, D_G, of
the radiation field. Generally two instruments are used for the
evaluation of the radiation components: one having approximately
the same sensitivity to neutrons and photons, the second instrument
with a relatively low neutron sensitivity. The dose components in
the mixed field can be derived from:

$$R'_T = k_T D_N + h_T D_G \text{ and } R'_U = k_U D_N + h_U D_G$$

In these equations, the subscript T refers to the tissue equivalent device measuring the total dose and the subscript U refers to the neutron insensitive device. R'_T and R'_U are the quotients of the responses of the two dosimeters in the same mixed beam relative to their sensitivities to the gamma rays used for the photon calibration. Similarly, k_T and k_U are the sensitivities of each dosimeter to neutrons relative to its sensitivity to gamma rays used for calibration and h_T and h_U are the sensitivities of each dosimeter to the photons in the mixed field relative to its sensitivity to the gamma rays used for calibration (ICRU, 1977). It is generally advantageous to use photon dosimeters with small k_U, to minimize overall uncertainties in the neutron and photon absorbed doses.

A common procedure in neutron dosimetry is the use of TE, CH or CH_2 ionization chambers to determine the total dose, in combination with nonhydrogenous ionization chambers (Al/Ar, Mg/Ar, C/CO_2) or GM counters. The principle of the use of ionization chambers in neutron dosimetry does not differ from their use in photon dosimetry. It can be shown that:

$$k_T = \frac{\overline{W}_C}{\overline{W}_N} \cdot \frac{(s_{w,g})_C}{(s_{w,g})_N} \cdot \frac{(K_t/K_m)_C}{(K_t/K_m)_N}$$

where $\overline{W}$ is the average energy expended to create an ion pair; $s_{w,g}$ is the ratio of the average mass stopping power of the wall relative to the gas and K_t/K_m is the ratio of the kerma in tissue to that in the dosimeter material. The subscript c denotes values applicable to the calibration situation, for which gamma rays are commonly used. Although photon spectra in neutron fields are generally not known, h_T and h_U are usually taken to be equal to one. However, this is not necessarily always the case; calibration at more than one photon energy is indicated when there is a possible dependence of h_T and h_U on photon energy.

Ionization chambers are the most commonly used instruments for the determination of absorbed dose and/or kerma in mixed neutron fields. In order to make the ratio of kerma in the dosimeter material to that in soft tissue as close to one as possible, conducting tissue equivalent plastic designated as A-150 is often used for the construction of ionization chambers. The principle compromise in the formulation of A-150 is the substitution of carbon for much of the oxygen required to match muscle tissue. An approximately homogeneous chamber can then be obtained by flushing the chamber with a mixture of gasses of comparable composition (TE-gas). A compilation of characteristics of eight different tissue equivalent ionization chambers can be found elsewhere (Broerse, 1979).

The basic physical parameters which have to be used to convert the dosimeter response to tissue kerma in free air or to absorbed dose in a phantom still have a number of uncertainties. The mean energy required to form an ion pair in a tissue equivalent gas mixture depends on the nature and spectra of the secondary particles and on the composition of the gas. In the European as well as in the American institutes engaged in neutron therapy, values of $\bar{W}_c/\bar{W}_N$ of 0.95 to 0.96 are used for TE gas. However, there are strong indications that this ratio should be taken as 0.94 ± 0.03 at neutron energies greater than 10 MeV. Until more information becomes available, there is no need to deviate from the previous recommendation of 0.95. New compilations and evaluation of W values are in preparation (see e.g. ICRU, 1979). It should be further mentioned that the $\bar{W}$ ratio will increase with decreasing neutron energies.

The difference in the oxygen and carbon content of TE plastic and muscle tissue and differences in the oxygen and carbon neutron cross sections will result in a muscle/TE plastic A-150 kerma ratio which deviates from unity. The ratio is approximately 0.95 for the neutron therapy beams presently used for treating patients. It is recommended that the kerma ratios be calculated on the basis of kerma per unit fluence values published in ICRU Report 26 (ICRU, 1977).

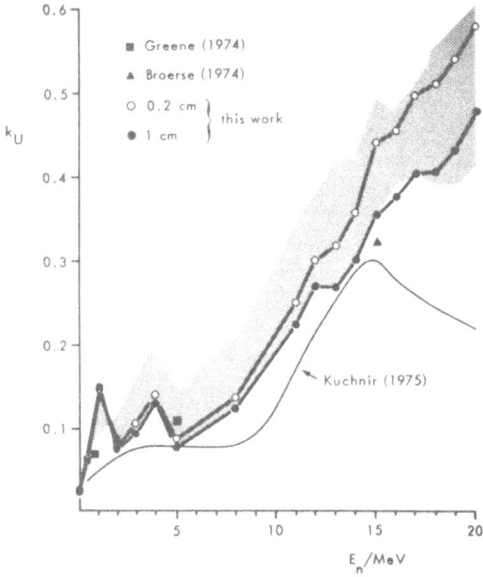

Fig. 1. Relative neutron sensitivity of a C/CO_2 ionization chamber calculated as a function of neutron energy for two spacing distances between electrodes (Makarewicz and Pszona, 1978).

Table 1. Relative neutron sensitivity, k_U of Geiger-Müller counters with different shield design for d+T neutrons. (Mijnheer et al., 1979).

neutron energy (MeV)	types of counter	composition and dimension of shield	k_U (per cent)	
14.1*	ZP 1100	2 mm tin**	2.74 ± 0.40	(AVL)
14.7	ZP 1100	ibid	2.3	comparison at NPL
15	18509	1.35 mm tin + 0.25 mm lead	<0.5	Colvett (1974)
15.5	ZP 1100	2 mm tin**	2.84	comparison at PTB
15.5	ZP 1100	ibid	2.86 ± 0.18	(PTB)
14.1*	18529	1.05 mm tin + 0.55 mm lead**	1.72 ± 0.25	(AVL)
14.7	18529	ibid	1.33	comparison at NPL
14.7	18529	ibid	2.15 ± 0.20	Lewis and Young (1977)
15.5	18529	ibid	1.96	comparison at PTB
15.5	18529	1 mm tin	1.63	comparison at PTB

* collimated
** perforated

If the gas and the wall of an ionization chamber have the same composition, it is generally assumed that they have the same stopping power. There are now some indications that the density effect plays no major role. Until further experimental results become available, there appears to be no severe need to deviate from previous recommendations of using a mass stopping power ratio $(s_{m,g})_C/(s_{m,g})_N$ of one, for homogeneous ion chambers.

The photon component of the absorbed dose can be assessed with a dosimeter which is relatively insensitive to the neutron component. The relative neutron sensitivity for a number of specific neutron-insensitive devices for a range of neutron energies has been derived from calculations and from measurements utilizing various techniques. New data became available on k_U for C/CO_2, Al/Ar and Mg/Ar chambers, especially for energies in excess of 15 MeV (Makarewicz and Pszona, 1978; Ito, 1978a; and Waterman et al., 1979). The information on k_U for the C/CO_2 chamber is given in Fig. 1. Higher k_U values than reported in ICRU Report 26 have been

recently determined by several groups for Geiger-Müller counters
(e.g., Mijnheer et al., 1979; the values are summarized in Table 1).
The new results of k_U for GM counters have clearly demonstrated
that the relative neutron sensitivity will vary for different counter types and different shield design for the same neutron energy.

INTERFACE DOSIMETRY

Experimental conditions are usually arranged to provide approximate secondary charged particle equilibrium for the specimen and the
dosimeter to facilitate accurate measurements of neutron kerma or
absorbed dose. This means that the sensitive volume of the dosimeter
should be surrounded by a layer of material of thickness equal to
the maximum range of the most energetic secondary charged particles
produced. For kerma determinations, the thickness of this equilibrium
layer should not be much greater than required in order to minimize
corrections for attenuation. The situation of incomplete charged
particle equilibrium most commonly encountered in practice is the
exposure of a biological specimen in free air. The number of secondary particles generated in the air, such as electrons in the case
of X- or gamma rays and protons and heavier particles in the case of
neutrons, is usually insufficient to provide charged particle equilibrium in the superficial layers of the biological object. As a
consequence, there will be a build-up of absorbed dose below the surface of tissue for the photons and for neutron beams of different
energies. Another example of interface dosimetry is the irradiation
of cells in monolayer on the bottom of culture dishes of different
atomic composition. In X-irradiations, excess secondary electrons
will be produced in the glass compared with those produced in cells.
Variations in absorbed dose will also occur for neutron irradiations
of cells on materials of different atomic compositions. This is
illustrated by the results shown in Fig. 2 for 15 MeV neutron irradiations of cultured cells adjacent to various materials (Broerse
et al., 1968). For a given tissue kerma in free air, greater survival
is observed for irradiations on nonhydrogenous material, indicating
that a smaller amount of energy is absorbed in the cells in this
situation. These results can be used to estimate the relative absorbed dose in tissue adjacent to different materials. It should be
realized, however, that the energy dissipation at the interface will
change not only quantitatively but also qualitatively. With neutrons
of energies up to a few MeV, charged particle equilibrium for heavy
recoils with relatively high LET is attained at depths less than the
dimensions of the cell, but proton equilibrium is not attained until a tissue depth of a few millimeters is reached. The biological
effects of the "low LET component" of the energy deposition by fast
neutrons due to recoil protons and the "high LET component" due to
interactions with carbon, oxygen and nitrogen are expected to be
different with respect to their relative biological effectiveness
(RBE). The relative contributions of the absorbed dose of each of

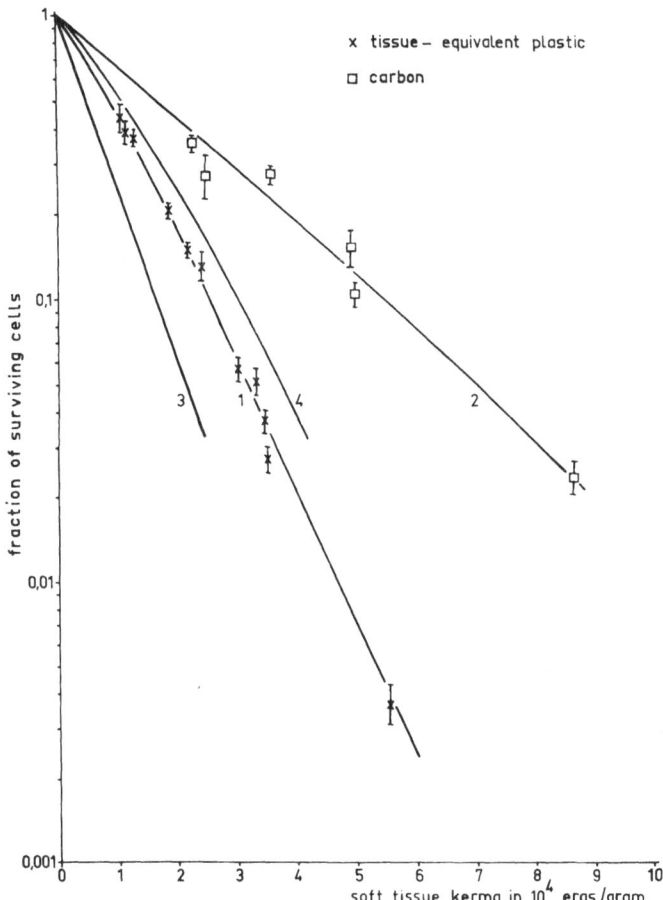

Fig. 2. Survival curves of cultured cells irradiated with 15 MeV
         neutrons whereby discs of tissue-equivalent plastic or
         carbon were mounted under the 6 μ thick Melinex bottoms
         of the dishes (curves 1 and 2, respectively). Curves 3
         and 4 represent survival curves calculated for the high
         LET component and the proton component of the energy dis-
         sipation, respectively (Broerse et al., 1968).

these two components can be derived from the survival curves presented in Fig. 2 under conditions in which proton equilibrium was
provided or deliberately avoided. For the irradiation conditions
with layers of pure carbon mounted in front of the cells, most of
the energy is deposited by densely ionizing α -particles and heavy
recoils. From the survival curves determined under conditions with
and without proton equilibrium, survival curves can be derived for
the absorbed doses of protons and heavier recoils separately and
RBE values for the two radiation components can be calculated.
These studies clearly indicate that changes in both the amount and

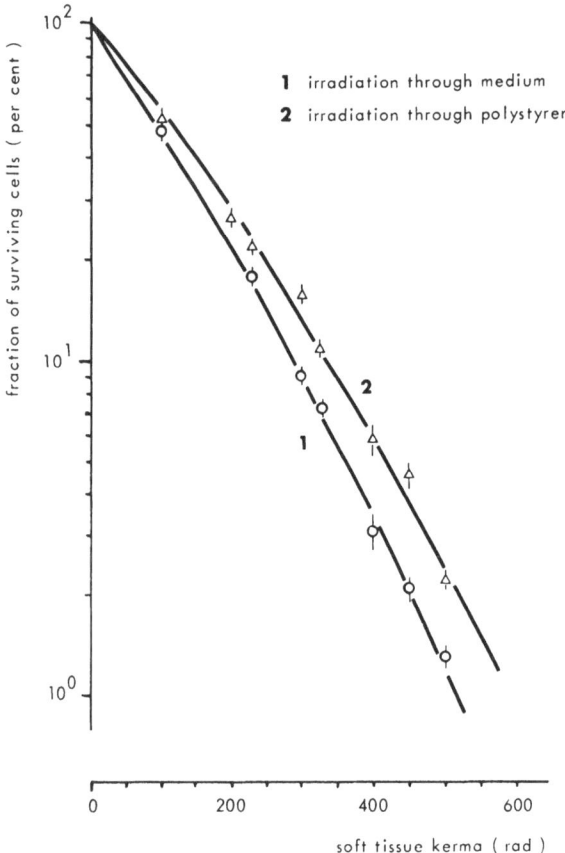

Fig. 3. Survival curves of cultured cells irradiated with d+T
         neutrons through the medium (curve 1) or through the poly-
         styrene bottom (curve 2) of the culture flask (Broerse
         and Zoetelief, 1978).

the quality of the energy dissipated must be considered when evaluating the effects of interface perturbations. A practical implication of these types of studies is shown in Fig. 3, where the survival of cells cultured in monolayer is shown for irradiations performed either through the polystyrene bottom of the culture dishes
or through the medium. It can be seén from the survival curves that,
under these two separate irradiation conditions, there is a difference in absorbed dose of approximately 16%.

DOSIMETRY FOR RADIOTHERAPY

     For radiotherapy applications, three important dosimetric
aspects should be considered:

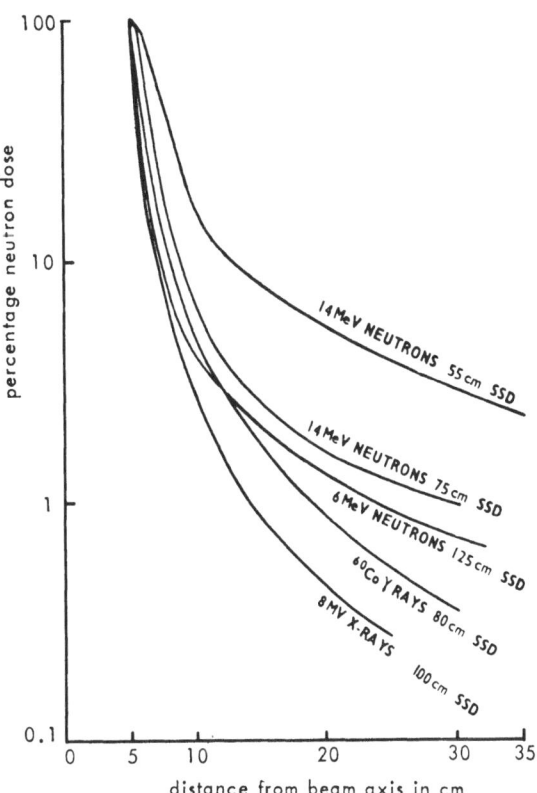

Fig. 4.   The stray radiation from 10 x 10 cm^2 beams of X-rays,
         gamma-rays and fast neutrons at 10 cm depth in a phantom
         (Duncan et al., 1971).

1) The transverse absorbed dose distribution. Optimum beam definition has to be achieved to deliver the required absorbed dose in
   a prescribed volume, while protecting structures outside the
   beam.
2) Central axis absorbed dose distribution. Depth dose data are required to provide an accurate description of the relative absorbed dose distribution throughout the patient.
3) The administration of the prescribed absorbed dose to a specified
   point in the target volume. In this connection, it will be of
   great importance to develop adequate monitoring methods for the
   routine therapy treatments.

Studies of beam profiles of different neutron energy beams have
shown that, at greater distances (in excess of about 20 cm) from
the beam axis, the dose levels for the neutron beams are higher than
those obtained with X- and gamma ray beams (see e.g. Fig. 4). Results
from the transverse absorbed dose distribution for cyclotrons show
that the penumbral width decreases with increasing deuteron energy.

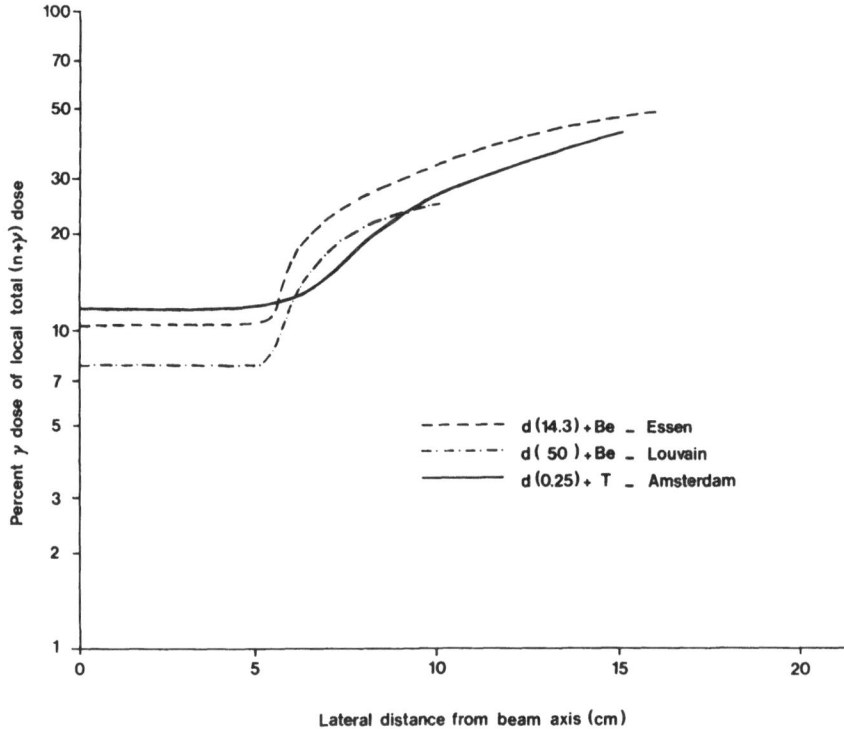

Fig. 5.   Gamma-ray contribution to the total n+γ dose as a function
          of the lateral distance from the beam-axis for three dif-
          ferent neutron beams for field sizes of approximately
          10 x 10 cm² (Mijnheer and Broerse, 1979).

This may be explained by the more forward peaking of the neutrons by
the higher energetic deuteron reactions and the decreasing scattering
cross section of hydrogen at higher neutron energies (Mijnheer and
Broerse, 1979).

     The data for d+T neutron generators show a much larger penumbra
than for the cyclotrons. This is due to the larger target sizes
needed to obtain adequate dose rates, the thinner shield and more
isotropic emission of the neutrons produced in the d+T reaction.
However, the difference in penumbral width becomes smaller if only
the neutron dose is considered.

     It should also be realized, that RBE changes outside the beam
may occur. The RBE will decrease due to the increase in the relative
gamma ray contribution with increasing distance from the central
axis. This is illustrated in Fig. 5. Almost identical curves are

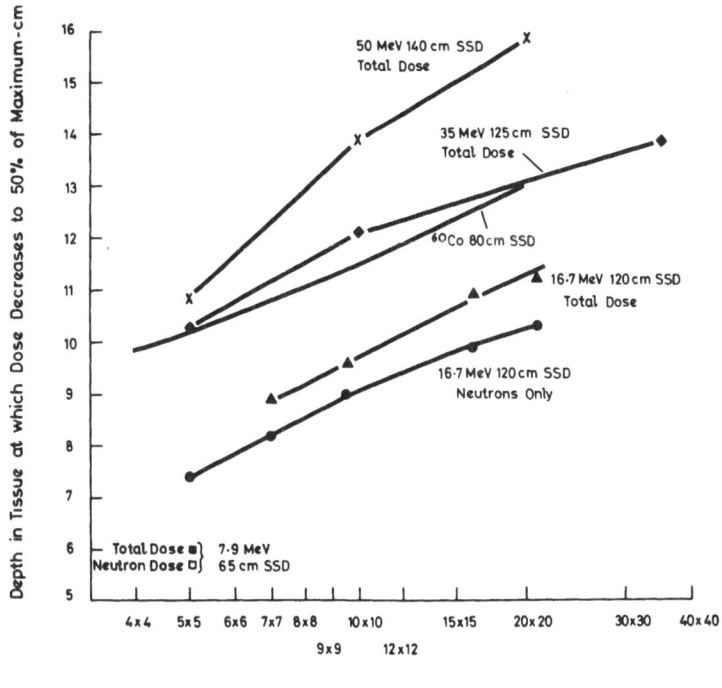

Fig. 6.   Depth dose data of fast neutron beams produced by bombar-
          ding thick beryllium targets with deuterons of 7.9, 16.7,
          35 and 50 MeV energy (Parnell, 1974).

observed for d+T sources as for cyclotrons, whereas changes in the
deuteron energy do not seem to have a great influence on the shape
of this curve, at least not in the penumbra region. On the other
hand, the RBE may increase due to the smaller absorbed dose per
fraction and the lower neutron energy outside the beam as compared
to the centre of the beam. Recently, Bewley and Page (1978) showed
that, when allowance is made for changes in the RBE of the neutron
component, even with perfect shielding the biologically effective
dose outside the useful beam is much higher for d(16)+Be neutrons
than for a ^{60}Co machine. The same statement is probably valid for
other fast neutron therapy beams, indicating that, in neutron therapy, one has to accept a greater hazard to the patient from stray
radiation than with megavoltage X-rays.

     The depth in the phantom at which the absorbed dose is reduced
to half its maximum value, d(50%), can be chosen as a criterion for
comparison of the depth dose data of the different fast neutron
radiotherapy beams. The data for different field sizes are presented
in Fig. 6. A comparison with photon sources shows that a deuteron
energy of 30 to 35 MeV is necessary to achieve a penetration com-

parable to a clinically used ^{60}Co source, whereas the d(50)+Be and
p(66)+Be neutron beams have depth dose characteristics comparable
to a 4 MV X-ray beam. The values of d(50%) are lower for the d+T
sources than for ^{60}Co gamma rays, although an improvement of about
1 cm can be obtained if the SSD is increased from 80 to 100 cm.

The build-up of absorbed dose below the surface of a phantom
can be determined by adding material to a thin-walled ionization
chamber. This can be either a chamber with a fixed distance between
the electrodes or one of the extrapolation type. The ratios of the
ionization measured at the surface of the phantom to the maximum
ionization (which are in first approximation proportional to the
relative entrance dose) as measured by different groups show considerable variations, even for comparable neutron energies. It
should be realized, however, that build-up curves determined by
means of ionization chambers are dependent on several factors such
as the presence of phantom material behind the ionization chamber,
the use of air or tissue equivalent (TE) gas flushing a TE chamber
and the application of an absorber to remove charged particles from
the beam. It has been recently shown that the surface dose of three
different neutron beams, measured under similar experimental conditions, are almost identical (Mijnheer et al., 1978). Therefore, the
differences may partly be attributed to differences in experimental
set-up used by the different groups. For further details on dosimetry for radiotherapy, the reader is referred to the physical
protocols drafted by the European (ECNEU) and American (Task Group
18) physics groups.

RESULTS OF DOSIMETRY INTERCOMPARISONS

The different neutron dosimetry intercomparison programs can
be divided into two classes: intercomparisons of dosimetry systems
performed at a central location and intercomparisons of the neutron
beams actually used for radiotherapy. During the first type of intercomparisons such as the International Neutron Dosimetry Intercomparison, INDI (ICRU, 1978) and the European Neutron Dosimetry Intercomparison Project, ENDIP (Broerse et al., 1978) all participants
brought their systems to central locations (Brookhaven National Laboratory for INDI and Radiobiological Institute and Institut für
Strahlenschutz for ENDIP). In the second type of intercomparisons,
including those carried out under the auspices of Task Group 18,
the former American Neutron Dosimetry Physics Group, NDPG (Almond
and Smathers, 1977) and of the Japan-US Cooperative Cancer Research
Program, CCRP (Ito, 1978b), the dosimetry systems were taken to
each institution and, where possible, reciprocal visits among
institutes were made.

The INDI and ENDIP studies were generally performed with monoenergetic neutron beams produced by the p+T, d+D and d+T reactions
at relatively low kerma rates which varied between 5 and 80 rad/h.

Table 2. Number of evaluated ENDIP results with relative differences, Δx, from the mean (Broerse et al., 1978).

site of inter-comparison	neutron energy	condition		$\Delta x \leqslant 5\%$	$5\% < \Delta x \leqslant 10\%$	$\Delta x > 10\%$
GSF	15.1 MeV	free air	K_N	6/12	3/12	3/12
			K_{tot}	8/12	2/12	2/12
GSF	5.25 MeV	free air	K_N	7/11	3/11	1/11
			K_{tot}	8/11	3/11	0/11
TNO	15 MeV	free air	K_N	11/12	1/12	0/12
			K_{tot}	10/13	3/13	0/13
TNO	15 MeV	5 cm depth	D_N	6/12	5/12	1/12
			D_{tot}	10/13	2/13	1/13
TNO	15 MeV	10 cm depth	D_N	5/12	6/12	1/12
			D_{tot}	9/13	3/13	1/13
TNO	15 MeV	20 cm depth	D_N	3/12	8/12	1/12
			D_{tot}	9/13	3/13	1/13
TNO	5.5 MeV	free air	K_N	8/8	0/8	0/8
			K_{tot}	8/9	1/9	0/9
TNO	5.5 MeV	5 cm depth	D_N	4/8	4/8	0/8
			D_{tot}	6/9	3/9	0/9
TNO	5.5 MeV	10 cm depth	D_N	2/8	5/8	1/8
			D_{tot}	7/9	2/9	0/9
TNO	5.5 MeV	20 cm depth	D_N	4/8	2/8	2/8
			D_{tot}	7/9	2/9	0/9

At each intercomparison site, a ^{137}Cs photon source was available
for the calibration of the participants' dosimeters. The quantities
to be intercompared were the soft tissue kerma free-in-air and the
absorbed dose at three depths in a water phantom for neutrons and
photons separately. The results of the INDI and ENDIP intercomparisons were basically the same. In general, the participants quoted
systematic uncertainties of 7-8% in the neutron and total kerma;
these are mainly attributed to poor knowledge of basic constants,
e.g., $\overline{W}$, kerma ratio and stopping power ratio. Although it is recognized that the calculation of a mean value for the participants'
results has limited relevance, this procedure was used in order to
allow a quantitative comparison of the ENDIP results. In Table 2,
the results of the groups participating in the ENDIP sessions at
GSF and TNO are grouped for relative differences from the meam
below 5%, from 5 to 10% and in excess of 10%. A complete analysis
of the ENDIP results has been published elsewhere (Broerse et al.,
1978); only a few main conclusions will be repeated in this review.

a) The variations in total kerma and total dose are generally
   smaller than those observed for the neutron kerma and absorbed
   dose. For the measurements at TNO, the results show relatively
   small variations for the free air conditions; however, larger
   variations are observed for measurements in the phantom, espe-

cially for the neutron absorbed dose.

b) For the in-phantom conditions, the deviation of the results from the average value for neutron kerma or absorbed dose shows standard deviations in the order of 7-8%. These variations seem to be in accordance with the relatively large systematic uncertainties quoted by the participants. Only for a few specific situations were maximum differences up to 20% observed.

c) The values reported for the gamma ray kerma and absorbed dose show large variations of up to 100% from the mean value. These variations are not acceptable for the measurements in the phantom where relatively large photon contributions have been measured (up to 25% of the total absorbed dose).

d) In the ENDIP and INDI studies, the participants employed hydrogenous ionization chambers of different designs for the determination of the total absorbed dose and they applied divergent basic physical parameters characterizing the detector response for identical experimental conditions.

The NDPG and CCRP measurements were made primarily on cyclotron produced neutrons with fairly high dose rates for tissue kerma in air and absorbed dose at depth. The intercomparisons of neutron dosimetry at the institutes cooperating within the NDPG in the United States show very good agreement: maximum differences of only 2% were observed for the measurements of total absorbed dose (Wootton and Eenmaa, 1978). It should be emphasized, however, that the American groups involved in neutron radiotherapy all used a common set of commercial TE ionization chambers. Typically, the NDPG uses a 1 cm^3 spherical chamber as the principal instrument for measurements of neutron tissue kerma in air and absorbed dose in a phantom and a 0.1 cm^3 cylindrical chamber for spatial dose distribution measurements.

The participants in the INDI and ENDIP dosimetry intercomparisons employed different basic parameters characterizing the detector response. For the conversion from kerma in the TE dosimeter material to kerma in soft tissue (ICRU muscle approximation), the values used by the participants in INDI and ENDIP showed differences of up to 8% for the same neutron energy. Relatively large discrepancies are also observed in the ratio $\overline{W}_N/\overline{W}_C$ as employed for a TE gas consisting of methane, carbon dioxide and nitrogen. The main difference among the various groups is due to the fact that, in some cases, the W ratio was taken to increase with decreasing neutron energy, while other groups employed a constant $\overline{W}$ ratio. These discrepancies led to a maximum difference of 11% for the W ratios used, especially in the case of the lower neutron energies. The various groups also used different values for the relative neutron sensitivity. Where participants employed relatively high or relatively low values of k_U, the reported gamma ray contributions were found to be lower or higher, respectively, than the mean value observed.

Table 3. Variance analysis of ENDIP results at TNO (Broerse et al., 1978).

Values of mean total dose, $\bar{D}_t$, and reduced instrument response, $\overline{R'/N}$, for TE ionization chambers, in rad per 10^5 monitor units with related standard deviations.

measurement condition	$\bar{D}_t$	s	s in per cent	$\overline{R'/N}$	s	s in per cent
5 MeV free in air*	56.1	1.7	3.0	55.0	1.2	2.1
5 MeV 5 cm	60.5	2.3	3.8	59.6	2.1	3.6
5 MeV 10 cm	29.2	1.1	3.7	28.9	1.1	3.9
5 MeV 20 cm	6.9	0.27	3.9	6.8	0.25	3.6
15 MeV free in air*	56.8	2.1	3.7	56.5	1.6	2.9
15 MeV 5 cm	64.8	3.5	5.4	64.5	3.2	5.0
15 MeV 10 cm	37.3	1.9	5.2	37.2	1.8	4.8
15 MeV 20 cm	12.7	0.79	6.2	12.6	0.71	5.6

* under these conditions the values refer to mean total kerma

To exclude the influence of the introduction of differing values for the basic parameters, the relative responses of the participants' dosimeters were also been compared. In Table 3, the variances are given for total kerma and total dose values and for reduced instrument responses for the ENDIP results obtained at TNO by participants using TE chambers flushed with TE gas. It can be seen that the standard deviations for instrument response are of the same magnitude as those calculated for dose and kerma values. This implies that, in addition to the inconsistencies in basic physical parameters, there are also large systematic differences in measurement procedures connected with, for example, the calibration with photons, the gas flow rate, the collecting potential, the polarity, the correction for wall thickness and the choice of effective point of measurement in a phantom. A similar conclusion resulted from an analysis of the INDI results.

The dosimetry intercomparisons performed among the European institutes have shown larger variations among the participants than the results obtained among the American therapy groups. In Europe, the applications of fast neutrons in radiobiology and clinical radiotherapy increased gradually, contrary to the United States where three institutes had to take instantaneous action. For this reason, the American institutes purchased their dosimeters from the same commercial firm and adopted one common ionization chamber. In Europe, the institutes developed and constructed their own dosimeters; these unavoidably show some mutually different characteristics. It must

be concluded, however, that, unless neutron dosimetry groups adopt
one reference ion chamber system, considerable efforts will have to
be expended in performing intercomparisons and expressing concern
about the correction factors to be applied for different chambers.
Although each institute should be free to develop instruments of
special design, it has been recommended that all European groups
use the same type of secondary standard to check their other dosimeters. It is preferable that an international body be involved in
the supervision of the production and use of such a common dosimeter system. It is gratifying that a number of national standards
laboratories are in the process of establishing neutron standard
fields for absorbed dose calibrations. However, the standards laboratories will not be able to offer their neutron fields for intercomparisons before 1982 and for calibration before 1983.

On the initiative of the Commission of the European Communities,
a committee has been formed which will be involved with the collection and evaluation of neutron dosimetry data, CENDOS. One of the
future tasks of the CENDOS committee will be to introduce a set of
reference TE ion chambers for neutron dosimetry in biology and
medicine. In connection with this program, information is presently
being collected on the characteristics of eight different types of
tissue equivalent ionization chambers.

The preliminary results of the CENDOS tests of different
ionization chambers emphasize the need for uniform procedures and
techniques for measuring chamber response and for applying the appropriate corrections (Broerse et al., 1979). In addition to the
introduction of a secondary standard or transfer dosimeter, a consistent set of basic physical parameters should be used and agreement should be reached on the procedures to convert instrument response into absorbed dose values. All of these steps will be of great
importance in the improvement of the consistency of neutron dosimetry for biological and medical applications.

Finally, it has to be stressed that, for an adequate comparison of biological results obtained at different institutes, the
energy deposition processes should be compared in both a quantitative and a qualitative way. The neutron sources employed for biological and medical applications show a wide variety of neutron energy spectra. Even neutron beams produced by the same reaction can
produce different biological results, due to variations in the contribution of scattered radiation. The radiation quality can be related to the neutron energy spectra or microdosimetric spectra and
is an essential concept in view of the dependence of the relative
biological effectiveness (RBE) on these interrelated parameters.
Differences in radiation quality can also be demonstrated by comparing the response of biological dosimeters in the different beams.
If such a standard biological system is taken to various institutes,
a uniform dosimetry system should accompany the radiobiology. The

dosimetry procedures followed for the irradiation of the biological
dosimeter should be standardized. If the dosimetry procedures are
consistent, variations in the response of the biological dosimeter
should reveal true radiobiological differences in the various neutron beams.

CONCLUSIONS AND RECOMMENDATIONS

Interest in the use of fast neutrons for biological and medical
applications has increased considerably in the last two decades.
These applications led to requirements for precision and accuracy
in dosimetry methods which are more demanding than generaly encountered in radiation protection. The following recommendations for
future research in neutron dosimetry for biology and medicine were
made:
1) It is recommended that standards laboratories develop and establish standard instruments and/or radiation fields which will
   be directly applicable to absorbed dose calibrations for the
   neutron energies applied in the life sciences.
2) The results of neutron dosimetry should be reported by giving the
   following values:
   a) The absorbed dose of neutrons in the region of interest and
      its time dependence or fractionation schedule.
   b) The absorbed dose of photons or the magnitude of this compo-
      nent relative to either the neutron or total absorbed dose
      in the region.
   When the spatial variation in absorbed dose in the region of interest can affect the biological response, it is important to
   describe the variation.
   Meaningful comparisons of biological data and therapeutic results
   require that radiation quality be specified, for example, by
   means of neutron energy spectra or by lineal energy spectra.
   These data should be supplied at more than one point when quality
   changes significantly in the region of interest.
3) The data on cross sections and related mass energy transfer coefficients for neutron-produced secondaries in the elements con-
   stituting tissue and for elements employed in dosimeters should
   be expanded to improve the presently available values above 20
   MeV and to provide necessary information above 30 MeV. These
   data are required to compute kerma factors for various tissues,
   dosimeter materials and tissue-substituting materials, particularly for use in radiotherapy.
4) In order to decrease the overall uncertainty of neutron absorbed
   doses determined with ionization chambers, it is essential that
   further determinations of W for the gases used in these chambers
   should be made, especially for protons and heavy charged particles with energies between 0.01 and 10 MeV. This will allow
   effective values of W for neutrons to be calculated more reliably.
5) Decrease in the overall uncertainty attainable with ionization

chambers also requires better data on the relative mass stopping powers with regard to the physical state (solid or gaseous) of the chamber materials, particularly for low energy secondaries.

6) To decrease the systematic uncertainty of the measurement of neutron absorbed doses with calorimeters, further determinations of the thermal defect for the materials used in these instruments should be made.

7) The overall uncertainty of measurements of photon absorbed doses in mixed fields can be decreased by more reliable determinations of relative neutron sensitivities of photon dosimeters as a function of neutron energy than have heretofore been done. In particular, this should include neutron energies above 15 MeV. Evaluation of the effects of dosimeter size and configuration on relative neutron sensitivities should be made.

In addition to improving data presently available, many of these recommendations call for a substantial extension of our knowledge into both higher and lower regions of neutron and charged particle energies than have previously been the focus of study.

REFERENCES

Almond, P.R., and Smathers, J.B., 1977, Physics intercomparisons for neutron radiation therapy. Int.J.Radiat.Onc.Biol.Phys., 3: 169.

Bewley, D.K., and Page, B.C., 1978, On the nature and significance of the radiation outside the beam in neutron therapy. Brit.J. Radiol., 51: 375.

Booz, J., 1978, Mapping of fast neutron radiation quality, in: "Proc. Third Symp. Neutron Dosimetry in Biology and Medicine", G. Burger and H.G. Ebert, eds., Commission of the European Communities, Luxembourg, EUR 5848, p. 499.

Broerse, J.J., 1979, Collection and evaluation of neutron dosimetry data. Compilation of characteristics of tissue-equivalent ionization chambers (second draft). CENDOS 79-1.

Broerse, J.J., Barendsen, G.W., and Van Kersen, G.R., 1968, Survival of cultured human cells after irradiation with fast neutrons of different energies in hypoxic and oxygenated conditions. Int.J.Radiat.Biol., 13: 559.

Broerse, J.J., Burger, G., and Coppola, M., 1978, "A European Neutron Dosimetry Intercomparison Project (ENDIP). Results and Evaluation." EUR 6004, Commission of the European Communities, Luxembourg.

Broerse, J.J., and Zoetelief, J., 1978, Dosimetric aspects of fast neutron irradiations of cells cultured in monolayer. Int.J.Radiat.Biol., 33: 383.

Broerse, J.J., Zoetelief, J., Burger, G., Schraube, H., and Ricourt, A., 1979, "A small scale neutron dosimetry intercomparison", EUR 6567, Commission of the European Communities, Luxembourg.

Duncan, W., Greene, D., and Major, D., 1971, Radiotherapeutic requirements of 14 MeV fast neutron beams with respect to depth-dose and collimation, Eur.J.Cancer, 7: 129.

ICRU, 1977, "Neutron Dosimetry in Biology and Medicine", Report 26. International Commission on Radiation Units and Measurements, Washington, D.C.

ICRU, 1978, "An International Neutron Dosimetry Intercomparison", Report 27. International Commission on Radiation Units and Measurements, Washington, D.C.

ICRU, 1979, "Average Energy Required to Produce an Ion Pair". Report 31, International Commission on Radiation Units and Measurements, Washington, D.C.

Ito, A., 1978a, Neutron sensitivity of C-CO_2 and Mg-Ar ionization chamber, in: "Proc. Third Symp. Neutron Dosimetry in Biology and Medicine", G. Burger and H.G. Ebert, eds., Commission of the European Communities, Luxembourg, EUR 5848, p. 605.

Ito, A., 1978b, Neutron dosimetry intercomparison between Japan (University of Tokyo) and USA, in: "Proc. Third Symp. Neutron Dosimetry in Biology and Medicine", G. Burger and H.G. Ebert, eds., Commission of the European Communities, Luxembourg, EUR 5848, p. 113.

Kellerer, A.M., and Rossi, H.H., 1972, The theory of dual radiation action, Curr.Top.Radiat.Res.Quart., 8: 85.

Makarewicz, M., and Pszona, S., 1978, Theoretical characteristics of a graphite ionization chamber filled with carbon dioxide, Nucl.Inst.Meth., 153: 423.

Mijnheer, B.J. and Broerse, J.J., 1979, Dose distributions of clinical fast neutron beams, Eur.J.Cancer, suppl., p. 109.

Mijnheer, B.J., Visser, P.A., Lewis, V.E., Guldbakke, S., Lesiecki, H., Zoetelief, J., and Broerse, J.J., 1979, The relative neutron sensitivity of Geiger-Müller counters, Eur.J.Cancer, suppl., p. 162.

Mijnheer, B.J., Zoetelief, J., and Broerse, J.J., 1978, Build-up and depth-dose characteristics of different fast neutron beams relevant for radiotherapy, Brit.J.Radiol., 51: 122.

Parnell, C.J., 1974, Depth dose characteristics and beam profile properties of cyclotron-produced neutron beams. Eur.J.Cancer, 10: 335.

Waterman, F.M., Kuchnir, F.T., Skaggs, L.S., Kouzes, R.T., and Moore, W.H., 1979, Energy dependence of the neutron sensitivity of C-CO_2, Mg-Ar, and TE-TE ionization chambers, Phys.Med.Biol., 24: 721.

Wootton, P., and Eenmaa, J., 1978, private communication.

BIOLOGICAL EFFECTS OF NEUTRONS

J.J. Broerse

Radiobiological Institute TNO
Rijswijk
The Netherlands

INTRODUCTION

When a biological system is irradiated with neutrons, the energy is dissipated by fast protons, alpha-particles and heavy recoils
which are produced through interactions with the tissue constituents.
The secondary particles produced by the neutrons generally have a
higher ionization density than electrons produced by X- or gamma-
radiation; consequently, neutrons can be described as high-LET radiation. Due to the complexity of the secondary particle spectra,
neutrons are less suitable than are charged particles for fundamental investigations of the mechanisms by which effects of ionizing
radiation on living cells are initiated. However, fast neutrons are
of great interest for radiobiology, since they offer the practical
possibility of exposing relatively large objects to high-LET radiation with a relatively uniform absorbed dose distribution throughout the subject. In addition, neutrons from certain sources, e.g.,
fission reactors, provide large fields for the irradiation of many
biological objects at a time, an essential feature for the study of
effects in animal populations.

Over the past two decades, the biological effects of fast neutrons have been investigated in a great number of studies with the
following three main objectives. The first is the accumulation of
information on mechanisms through which damage in mammalian tissues
is induced and expressed. The comparison of dose-effect relations
for fast neutrons and high energy X-rays or gamma-rays offers possibilities to analyze the influence of various factors, e.g., repair
of sublethal damage or the oxygenation status of cells. This is

because of the fact that the biological effects of fast neutrons are generally less modified by these factors. Furthermore, investigations of mechanisms of induction of radiobiological damage in organized systems or tissues are frequently of great value for investigations of the properties of unperturbed systems. Studies of dynamic properties of biological systems can be carried out by observing the response of such systems after perturbations in the kinetics caused by toxic agents. Ionizing radiation of different LET can be used as such an agent.

A second objective of studies of neutron-induced normal tissue damage and tumour response is the accumulation of data on the effects on tumours and the tolerance of normal tissues to single and fractionated doses. This information is requested for the application of fast neutrons in radiotherapy. Fast neutrons will provide an advantage over high energy X- or gamma-rays for radiotherapy only if the relative biological effectiveness, RBE, for damage to the tumour is greater than the RBE for damage to those normal tissues which are dose-limiting in the treatment considered.

A third objective of investigations of the biological effects of fast neutrons is the evaluation of the dependence of the RBE on the dose, dose fractionation and dose rate with the aim of extrapolating these data to the low doses relevant to radiation protection problems. It is important to note, however, that the endpoints most relevant for radiation protection are carcinogenesis and genetic damage and it should further be noted that RBE values for cellular damage to various normal tissues and other cell systems are not necessarily equal to the RBE values for other endpoints.

Since this paper is presented within the context of advances in dosimetry and radiobiology for medical applications, the effects of fast neutrons will be discussed for different levels of organization: firstly, at the cellular level; secondly, effects at the level of organs and normal tissues in relation to the occurrence of radiation syndromes; thirdly, effects on tumour systems and the relevance of these results for radiation therapy. Finally, some results on carcinogenesis after fast neutron irradiation will be summarized in this presentation.

EFFECTS AT THE CELLULAR LEVEL

The first studies on the RBE for cell survival as a function of LET were performed by Barendsen and co-workers (1960; 1963) using an established line of cultured cells derived from human kidney. From these studies on cell survival after irradiation with deuterons and alpha-particles, it appeared that the shape of the survival curves for high-LET radiation differed from that obtained after low-LET irradiation, such as X- or gamma-rays. Furthermore, the oxygen enhancement ratio, OER, was lower for radiations with high-LET than for low-LET. The curves on RBE-LET and OER-LET for directly ionizing particles, employing the track segment method, are generally recognized as classical radiobiological data.

In subsequent studies on cell survival after irradiation with fast neutrons of different energies (Broerse et al., 1968; Broerse and Barendsen, 1969; Hall et al., 1973) it was demonstrated that the survival curves for fast neutron irradiation have a much steeper slope than those for X-irradiation, indicating that neutrons are more effective per unit dose than X-rays. Due to these differences in the slopes of the survival curves, the effects of fractionation will be smaller for irradiations with fast neutrons than for X-rays. This reduction in the repair of sublethal damage is demonstrated in fig. 1. In addition, it has been shown that, for neutron beams of

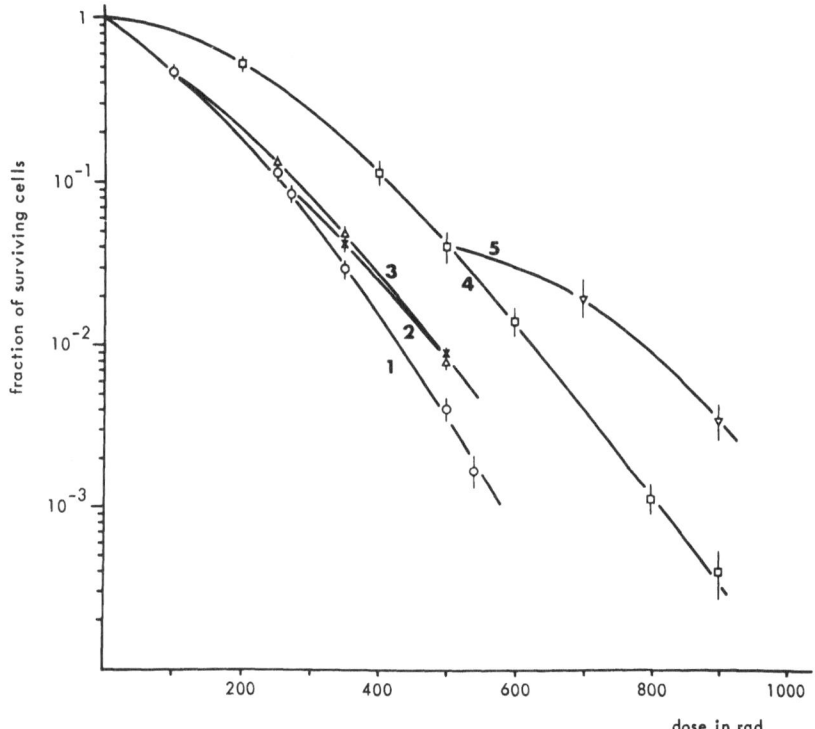

Fig. 1.   Survival of cultured cells after two-dose exposures to 250 kVp X-rays (curve 5) and 15 MeV neutrons (curves 2 and 3). The curves 4 and 1 show the survival after single irradiations with X-rays and neutrons, respectively (Broerse and Barendsen, 1969).

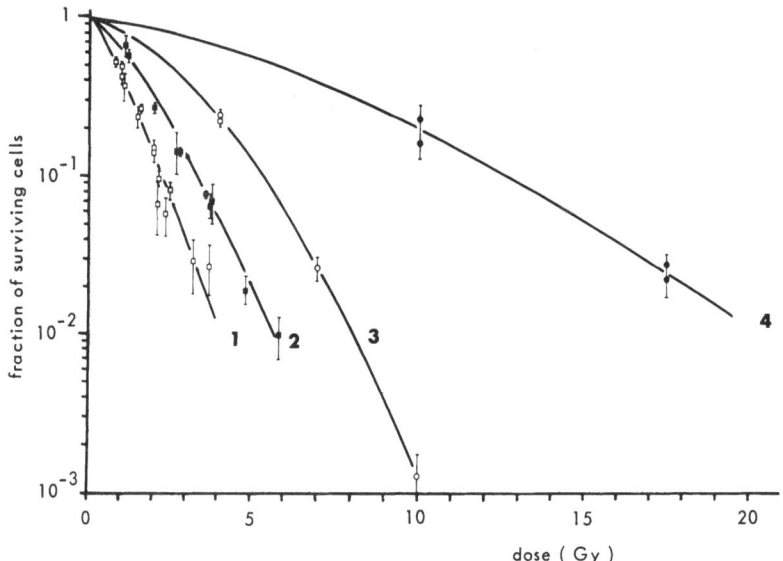

Fig. 2.   Survival curves obtained for well-oxygenated cells
           (curves 1 and 3) and anoxic cells (curves 2 and 4). The
           curves 1 and 2 are obtained after irradiation with 15 MeV
           neutrons, curves 3 and 4 after X-irradiation (Broerse
           et al., 1968).

different energies, the oxygen enhancement ratio is also considerably lower than for X-rays (see fig. 2 for the OER results after 15
MeV neutron irradiation). With a number of biological systems irradiated with neutron beams of different energies, it has now been
demonstrated that the OER varies only between 1.5 and 1.8. This is
lower than that for X-rays, where the OER varies between 2.5 and
3.0. In summary, it can be concluded that the cellular effects of
fast neutrons clearly resemble the survival characteristics of high-
LET radiation. It should be emphasized, however, that a direct comparison of the RBE values of fast neutrons with the RBE-LET relations for directly ionizing particles is not possible, since the
concept of average LET has a limited significance. The quality of
the neutron beams has to be described by LET spectra or, preferably,
y spectra and average values do not correspond with the biological
phenomena observed (Broerse and Barendsen, 1964).

EFFECTS ON NORMAL TISSUES

     Studies of the effects of ionizing radiation on rapidly proli-
ferating tissues such as the bone marrow and the intestinal mucosa
were initially concerned with investigations of the radiation syn-

drome leading to the death of animals. Dose mortality studies have
been carried out with various animal species and neutron beams of
different energies. In previous studies, qualitative differences had
been observed in the pattern of mortality between mice exposed to
fast neutrons and those exposed to X-rays. The older data indicate
that, for whole-body fast neutron irradiation of mice in the LD50
range, the characteristic bone marrow syndrome could not be produced without the concurrent incidence of the intestinal syndrome.
These earlier studies were most probably complicated by bacterial
infections in the animal strains which accelerated the development of
the intestinal syndrome. Later studies have demonstrated the protective effect of autologous bone marrow transplantation in mice
(Davids, 1970) and in monkeys (Broerse et al., 1978a) after irradiation with fast neutron of different energies. Dose mortality
curves for mice are shown in fig. 3, where the 30-day mortality and
5-day mortality are taken as criteria for the hemopoietic and in-

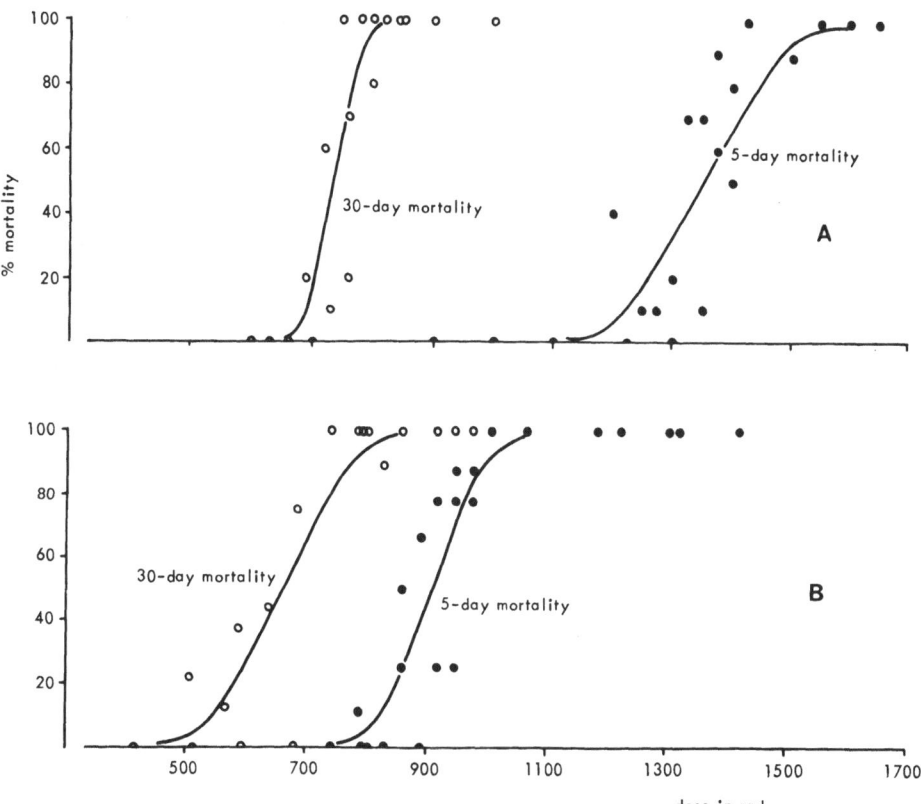

Fig. 3.   Mortality of mice either due to the hemopoietic syndrome
           (30 day mortality) and the intestinal syndrome (5 day
           mortality) after irradiation with X-rays, part A and 15
           MeV neutrons, part B (Broerse, 1969).

testinal syndromes, respectively (Broerse, 1969). Under these exposure conditions, RBE values of 1.1 and 1.4 can be derived for the
bone marrow and intestinal syndromes at neutron doses of 660 and
920 rad, respectively. In addition, RBE values of 0.8 are found for
both syndromes after irradiation with ^{137}Cs gamma-rays. Slightly
different RBE values have been observed for the occurrence of the
intestinal syndrome in other species. However, the finding of the
relatively low RBE value for the hemopoietic syndrome has been
confirmed for other neutron energies and other species.

The introduction of in vivo assay methods such as the spleen
colony technique (Till and McCulloch, 1961) and the microcolony
assay of intestinal crypt cells (Withers and Elkind, 1970) have made
it possible to investigate the cellular changes associated with the
hemopoietic and intestinal syndromes. Survival curves for hemopoietic stem cells and mouse crypt stem cells after irradiation with
300 kV X-rays and 15 MeV neutrons are shown in fig. 4. The survival
curves for the different types of cells after neutron and X-irradiation show relatively large differences in the D_0 values and extrapolation numbers which must be attributed to intrinsic differences
in the radiosensitivity of the stem cells. Also with respect to the
effects of fractionation, different responses are to be expected
for the various cell populations, since some survival curves have
relatively small shoulders even for X-irradiation (mouse bone marrow).

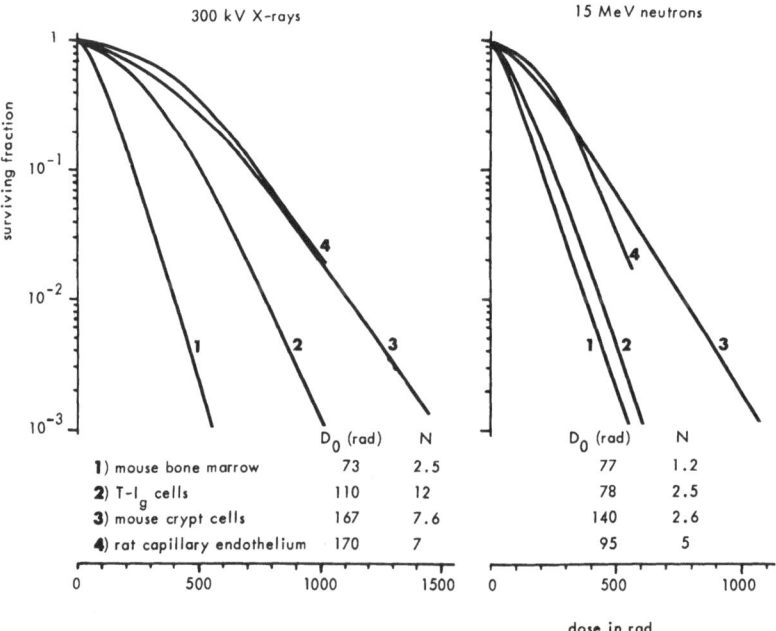

Fig. 4.   Survival curves of clonogenic cells in different types
          of normal tissues (Broerse et al., 1977).

Over the past 5 years, a number of additional scoring methods have been developed to determine the RBE value for both rapidly and slowly proliferating tissues. A summary of RBE values for these different types of tissues as found after irradiation with d+T neutrons is given in fig. 5. The general trend of these results is that the RBE increases with lower doses per fraction. It can also be seen that there is some spread of the RBE data obtained for the different types of tissues by the different groups. All of the information given in fig. 5 pertains to results obtained with irradiation of tissues in experimental animals. It is very important to obtain information on the relevance of the animal data for applications in man. Studies performed at the Hammersmith Hospital have shown that RBE values for skin in experimental animals are not greatly different from those for human skin, which shows the applicability of the animal data for the human situation (Field and Hornsey, 1979).

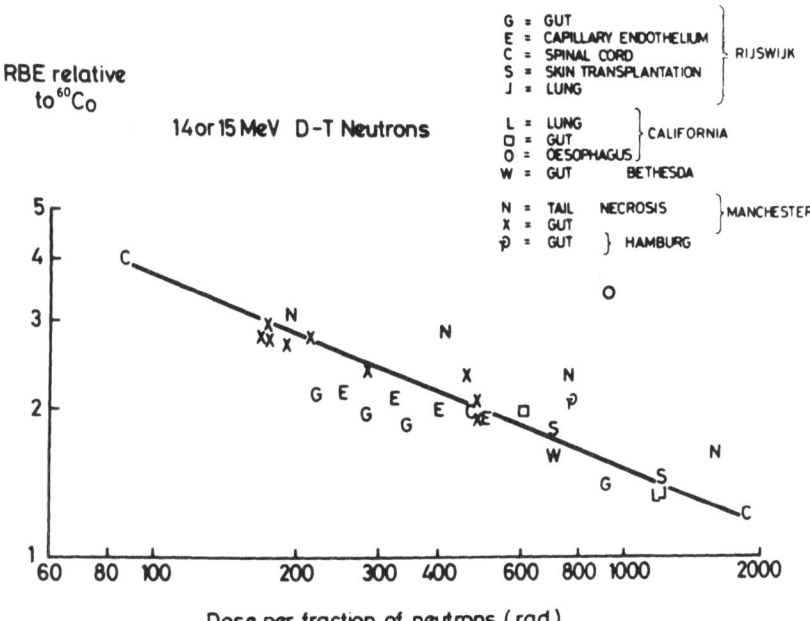

Fig. 5.   RBE of d+T neutrons relative to ^{60}Co gamma-radiation for various normal tissues as a function of the dose per fraction (Field and Hornsey, 1979).

EFFECTS ON TUMOURS

The initial studies on tumour response were mainly restricted to the measurements of the tumour dimensions after irradiation and

the determination of the time interval required for the tumour to
reach its preirradiation volume. Another quantitative assay was the
serial dilution technique developed by Hewitt and Wilson (1959).
The introduction of a cell dissociation technique (Barendsen and
Broerse, 1969) made it possible to evaluate the response of tumours
after irradiation by an in vitro cloning test. The survival curves
for cells from an irradiated rhabdomyosarcoma assayed in vitro
after in vivo irradiation in the living and dead animal are shown
in fig. 6. The same technique has now been applied to a number of
animal tumours, including an osteosarcoma, a ureter carcinoma and

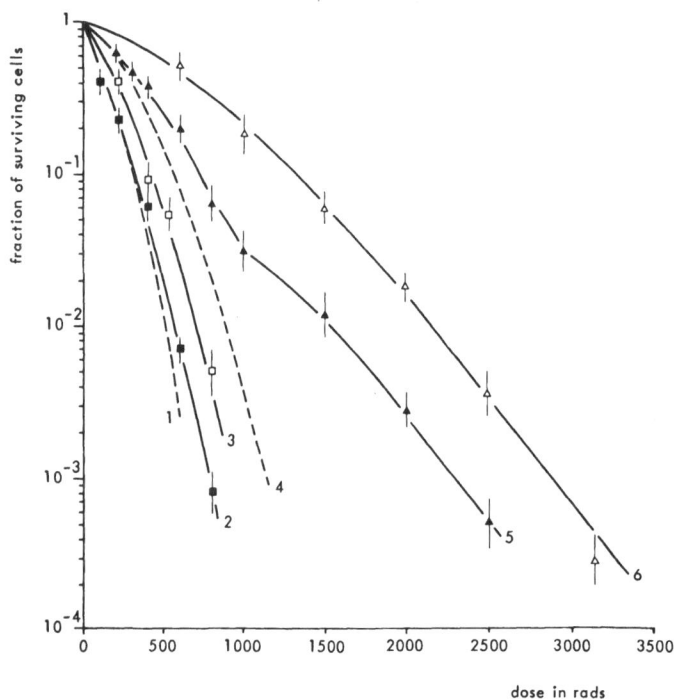

Fig. 6.   Survival curves of R-1 rhabdomyosarcoma cells irradiated
           in the tumour and assayed in vitro; curves 2 and 3: tu-
           mours irradiated with 15 MeV neutrons in living and dead
           rats, respectively; curves 5 and 6: tumours irradiated
           with X-rays in living and dead rats, respectively; curves
           1 and 4: cultured R-1 cells in equilibrium with air, irra-
           diated with 15 MeV neutrons and X-rays, respectively.
           (Barendsen and Broerse, 1969).

a lymphosarcoma (see fig. 7). From the survival curves it can be
concluded that the tumours have distinctly different radiosensitivities. The RBE for the various tumour systems is also considerably
different; e.g., at a dose of about 100 rad, it varies between 1.8

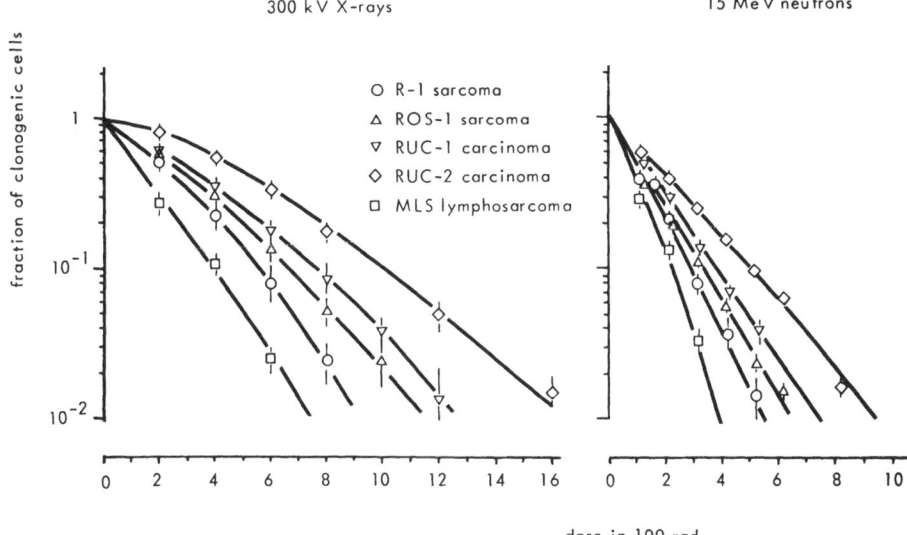

Fig. 7.   Survival curves of cells in culture derived from differ-
          ent animal tumours after irradiation with X-rays and
          15 MeV neutrons (Barendsen and Broerse, 1977).

and 3.0 (Barendsen and Broerse, 1977). The final assessment of the
possible advantages of the use of fast neutrons for radiotherapy has
to be based on a comparison of the RBE values for effects on tumours
with those for effects on normal tissues. For specific combinations
of tumours and dose-limiting normal tissues, an overall gain factor
$RBE_{tumour}/RBE_{normal\ tissue}$ between 1.2 and 1.3 has been observed.
Fast neutron beams are presently applied for radiotherapeutic applications in a number of radiotherapy clinics (see table). The clinical results have shown positive results for the treatment of tumours
of the salivary glands, tumours in the head and neck region, soft
tissue sarcoma and advanced stages of cervix carcinoma (Dutreix and
Tubiana, 1979). It is to be anticipated that the analysis and comparison of all clinical data will give the answer as to which type
of tumours will best respond to this new radiation modality.

CARCINOGENESIS AFTER FAST NEUTRON IRRADIATION

     Extensive information is available on tumour induction in dif-
ferent organs in man and experimental animals after irradiation
(e.g., UNSCEAR, 1977). In view of the risks of mammography, special
attention has been given in the recent years to the induction of
mammary tumours. The initial studies of Vogel and Zaldivar (1972)
and Shellabarger et al. (1974) were performed with the Sprague
Dawley rat strain, in which a relatively high incidence is already
observed spontaneously. Relatively high RBE values have been ob-

Table 1. Fast neutron installations in use for clinical
applications.

reaction employed	location	type of machine	maximum deuteron energy (MeV)	deuteron beam (mA)	start of clinical operation	patients treated until end 1978
d+T	Amsterdam	s. tube*	0.25	18	1975	300
	Glasgow	s. tube*	0.25	30	1978	7
	Hamburg	r. target**	0.5	8	1976	260
	Heidelberg	s. tube***	0.25	500	1978	50
	Manchester	s. tube*	0.25	30	1978	50
d+Be	Chiba-shi	cyclotron	30	0.03	1975	400
	Chiba-shi	v.d. Graaff	2.8	--	1967	--
	Dresden	cyclotron	13.5	0.04	1972	450
	Edinburgh	"	15	0.1	1977	210
	Essen	"	14	0.1	1978	70
	Houston	"	50	0.007	1972	440
	London	"	16	0.1	1969	800
	Louvain	"	50	0.02	1978	--
	Seattle	"	21.5	0.04	1973	240
	Tokyo	"	15	0.1	1976	120
	Washington	"	35	0.01	1973	250
p+Be	Batavia	cyclotron	66	0.008	1976	105

* output : $10^{12}s^{-1}$; ** output : $1\text{-}2.10^{12}s^{-1}$; *** output : $5.10^{12}s^{-1}$

served for mammary tumour induction, especially for neutrons with
energies around 0.5 MeV. In recent studies, a synergistic interaction has also been described between irradiation and hormone administration (Segaloff and Maxfield, 1971). Supplementary studies
(Broerse et al., 1978b; Shellabarger et al., 1978) have shown that
there are considerable differences in the patterns of mammary radiation carcinogenesis in different rat strains. The proportion of animals surviving without a mammary tumour as a function of time is
shown in fig. 8 and fig. 9 for Sprague Dawley rats and Brown
Norway rats. From the studies on mammary radiation carcinogenesis
performed at Rijswijk, a number of preliminary conclusions can be
drawn:
1) The latency period for tumour induction in the 3 rat strains
   studied is considerably longer than that observed by the American groups.
2) Large differences are observed in susceptibility for mammary tumour induction; Sprague Dawley rats have the highest susceptibi-

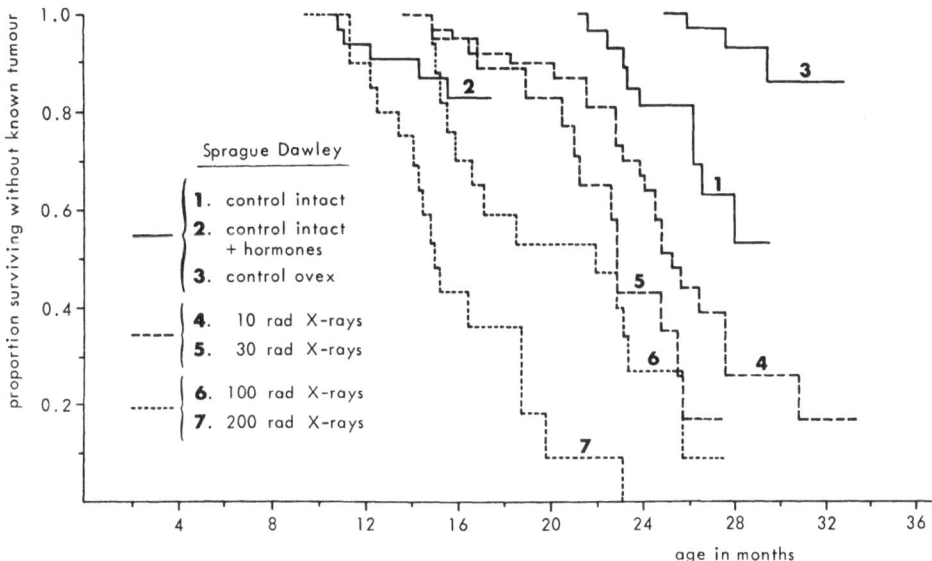

Fig. 8.   Proportion of Sprague Dawley rats surviving after X-irradiation without mammary tumours as a function of age.
(Broerse et al., 1978b).

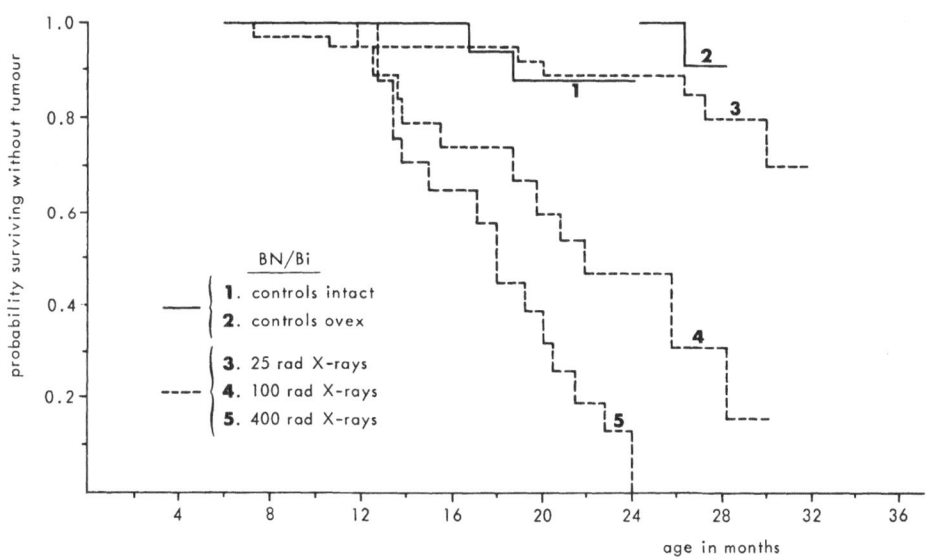

Fig. 9.   Proportion of Brown-Norway rats surviving after X-irradiation without mammary tumours as a function of age.
(Broerse et al., 1978b).

lity and Brown Norway the lowest susceptibility for the neutron
irradiations.
3) The RBE values for 15 MeV neutrons vary between 4 and 2, while,
   for 0.5 MeV neutrons, the values vary between 25 and 8 at a level
   of 30% prevalence. The lowest RBE values for both neutron energies are observed for the Brown Norway rats.

With regard to the RBE values observed for mammary tumour induction after 0.5 MeV neutron irradiation, it is interesting to compare
these data with the observations on long-term surviving monkeys after
irradiation with fission neutrons with a mean energy of 1 MeV (Van
Zwieten et al., 1978). As can be seen from fig. 10, the tumour in-

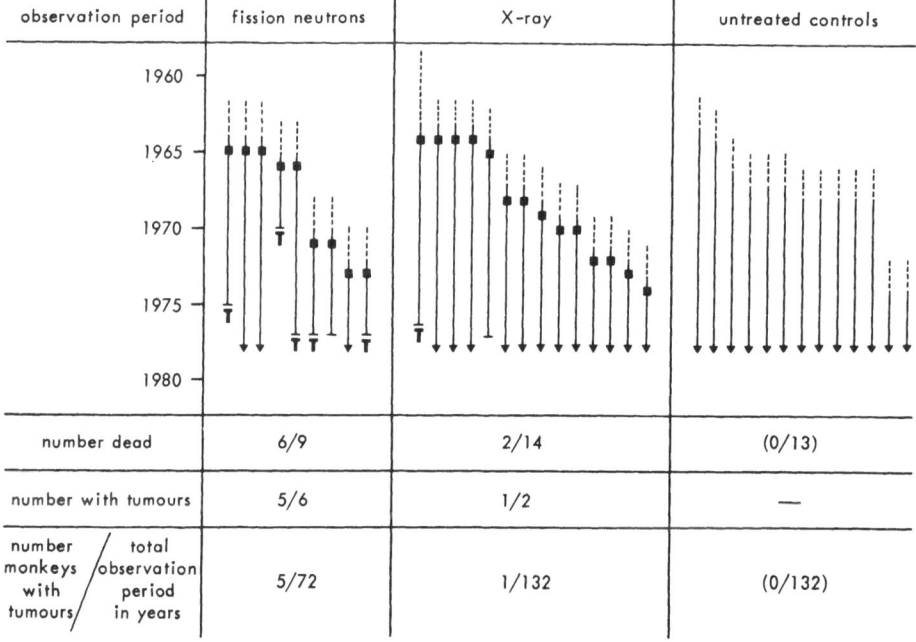

Fig. 10.   Tumour incidence in long term surviving monkeys after
           irradiation and bone marrow transplantation. The solid
           lines indicate the observation period, the squares show
           the time of irradiation, lines ending in cross bars mean
           death of the animal, T indicates tumour, lines ending in
           arrowheads represent monkeys which are still alive.
           (van Zwieten et al., 1978).

cidence in the neutron irradiated monkeys is considerably greater than that in the X-irradiated monkeys and in the untreated controls. On the basis of the number of monkeys dying with neoplasia versus the total observation period in years, it can be concluded that the relative biological effectiveness for tumour induction after neutron irradiation is approximately 10 and this value seems to be in acceptable agreement with the RBE values observed for mammary tumour induction in rats.

CONCLUSION

The fundamental investigations on cell survival characteristics and responses of animal tumours and normal tissues after fast neutron irradiation have resulted in the renewed application of fast neutrons for radiotherapy. At present, there are clear indications that neutrons can offer an advantage for the cure of a selected number of cancers. This development is an encouraging example of the usefulness of radiobiological investigations for radiotherapy applications.

Neutron radiation has been proved to be an effective tool for investigating the differences in biological effects after high-LET and low-LET exposure.

The studies on tumour induction in different rat strains have shown that there are considerable variations in the susceptibility of the different strains. The experimental results obtained in animal models should be handled with care before they are used for extrapolation to the human situation.

In general, it can be stated that the radiobiological studies of neutrons have shown that we still have to increase ouw knowledge in the area of the effects of neutrons on different types of tumours, on different types of normal tissues and the carcinogenic effects of neutrons in different animal species.

REFERENCES

Barendsen, G.W., Beusker, T.L.J., Vergroesen, A.J., and Budke, L., 1960, Effects of different ionizing radiations on human cells in tissue culture. II. Biological experiments, Radiat.Res., 13: 841.

Barendsen, G.W., and Broerse, J.J., 1969, Experimental radiotherapy of a rat rhabdomyosarcoma with 15 MeV neutrons and 300 kV X-rays. I. Effects of single exposure, Eur.J.Cancer, 5: 373.

Barendsen, G.W., and Broerse, J.J., 1977, Differences in radiosensitivity of cells from various types of experimental tumors in relation to the RBE of 15 MeV neutrons, Int.J.Rad.Onc.Biol.Phys., 3: 211.

Barendsen, G.W., Walter, H.M.D., Fowler, J.F., and Bewley, D.K.,
    1963, Effects of different ionizing radiations on human cells
    in tissue culture. III. Experiments with cyclotron-accelerated
    alpha-particles and deuterons, Radiat.Res., 18: 106.

Broerse, J.J., 1969, Dose-mortality studies for mice irradiated with
    X-rays, gamma-rays and 15 MeV neutrons, Int.J.Radiat.Biol.,
    15: 115.

Broerse, J.J. and Barendsen, G.W., 1964, Effects of monoenergetic
    neutron radiation on human cells in tissue culture, p. 309 in:
    "Biological Effects of Neutron and Proton Irradiations,"
    Vol. I, IAEA, Vienna.

Broerse, J.J., and Barendsen, G.W., 1969. Recovery of cultured cells
    after fast neutron irradiation, Int.J.Radiat.Biol., 15: 335

Broerse, J.J., Barendsen, G.W., Gaiser, J.F., and Zoetelief, J.
    1977, The importance of differences in intrinsic cellular radio-
    sensitivity for effectiveness of neutron radiotherapy treat-
    ments, p. 19, in: "Radiobiological Research and Radiotherapy",
    Vol. II, IAEA, Vienna.

Broerse, J.J., Barendsen, G.W., and Van Kersen, G.R., 1968, Survi-
    val of cultured human cells after irradiation with fast neu-
    trons of different energies in hypoxic and oxygenated condi-
    tions, Int.J.Radiat.Biol., 13: 559.

Broerse, J.J., Knaan, S., Van Bekkum, D.W., Hollander, C.F., Noote-
    boom, A.L., and Van Zwieten, M.J., 1978b, Mammary carcinoge-
    nesis in rats after X- and neutron irradiation and hormone
    administration, p. 13, in: "Late Biological Effects of Ionizing
    Radiation," Vol. II, IAEA, Vienna.

Broerse, J.J., Van Bekkum, D.W., Hollander, C.F., and Davids, J.A.G..
    1978a, Mortality of monkeys after exposure to fission neutrons
    and the effect of autologous bone marrow transplantation,
    Int.J.Radiat.Biol., 34: 253.

Davids, J.A.G., 1970, Bone-marrow syndrome in CBA mice exposed to
    fast neutrons of 1.0 MeV mean energy. Effects of syngeneic
    bone-marrow transplantation, Int.J.Radiat.Biol., 17: 173.

Dutreix, J., and Tubiana, M., 1979, Evaluation of clinical ex-
    perience concerning tumor responses to high LET radiation,
    Eur.J.Cancer, Suppl., p. 243.

Field, S.B., and Hornsey, S., 1979, Neutron RBE for normal tissues,
    Eur.J.Cancer, Suppl., p. 181.

Hall, E.J., Rossi, H.H., Kellerer, A.M., Goodman, L., and Marino, S.,
    1973, Radiobiological studies with monoenergetic neutrons,
    Radiat.Res., 54: 431.

Hewitt, H.B., and Wilson, C.W., 1959, A survival curve for mammalian
    leukaemia cells irradiated in vivo, Brit.J.Cancer, 13: 69.

Segaloff, A., and Maxfield, W.S., 1971, The synergism between radia-
    tion and estrogen in the production of mammary cancer in the rat,
    Cancer Res., 31: 166.

Shellabarger, C.J., Brown, R.D., Rao, A.R., Shanley, J.P., Bond, V.P.,
    Kellerer, A.M., Rossi, H.H., Goodman, L.J., and Mills, R.E.
    1974, Rat mammary carcinogenesis following neutron or X-radia-
    tion, p. 391 in: "Biological Effects of Neutron Irradiation",
    IAEA, Vienna.
Shellabarger, C.J., Stone, J.P., and Holtzman, S., 1978, Rat differ-
    ences in mammary tumor induction with estrogen and neutron
    radiation, J.Nat.Cancer Inst., 61: 1505.
Till, J.E., and McCulloch, E.A., 1961, A direct measurement of the
    radiation sensitivity of normal mouse bone marrow cells,
    Radiat.Res., 14: 213.
UNSCEAR, 1977, "Sources and Effects of Ionizing Radiation", United
    Nations, New York,
Van Zwieten, M.J., Zurcher, C., Hollander, C.F., and Broerse, J.J.,
    1978, Longevity studies in Rhesus monkeys after X-ray and neu-
    tron irradiation, p. 165 in: "Late Biological Effects of Ionizing
    Radiation", Vol. II, IAEA, Vienna.
Vogel, H.H., and Zaldivar, R., 1972, Neutron-induced mammary neo-
    plasms in the rat, Cancer Res., 32: 933.
Withers, H.D., and Elkind, M.M., 1970, Microcolony survival assay
    for cells of mouse intestinal mucosa exposed to radiation,
    Int.J.Radiat.Biol., 17: 261.

DOSIMETRY OF PIONS

J. F. Dicello, M. Zaider[*] and D. J. Brenner

University of California
Los Alamos Scientific Laboratory
Los Alamos, New Mexico 87545

INTRODUCTION

Shortly after the discovery of the negative pion, scientists began to consider the possibility of its application to cancer therapy. Fowler and Perkins (1961) were the first to show quantitatively the potential advantages.

Negative pions are hadrons having a rest mass of 139.6 MeV, intermediate to that of the electron (0.511 MeV) and the proton (938.2 MeV). Pions have both electromagnetic and nuclear forces associated with their interaction with other particles as compared with the mu meson which has a comparable mass, but interacts only electromagnetically (weak interactions being disregarded for our purposes). The pion, in fact, is the fundamental particle associated with nuclear forces, playing the same role as the photon in electromagnetic forces. Some of the basic characteristics of pions are given in Tables 1 through 3.

As with all heavy charged particles passing through matter, the probability of a pion losing energy increases as the particle energy approaches a few MeV, reaches a maximum, and then decreases at lower energies. This results in the Bragg-type distribution of dose as a function of depth, as shown for a 180 MeV/c π^+ beam in Fig. 1 (Dicello et al., 1978). The energy deposited by π^- particles

[*] Present address: Radiological Research Laboratory, College of Physicians and Surgeons of Columbia University, New York, NY.

at the end of their range is significantly enhanced by a unique feature of the negative pion. As the π^- reaches electron-volt energies,
it is captured into an outer orbit of an atom or molecule forming
what is called a pionic atom. The pion then cascades down the
atomic energy levels firstly by Auger transitions and then by x-
ray emission producing characteristic x rays. When the wave function
of the negative pion has a large overlap with that of the nucleus,
the pion, being a strongly interacting particle, is absorbed by the
nucleus. The whole process takes about 10^{-10} s (Schneuwly, 1977).
Hence pion decay with a half life at rest of 2.54×10^{-8} s is a rare
process compared with absorption. Because the energy absorbed by
the nucleus (140 MeV, corresponding to the rest mass of the pion)
is considerably larger than the separation energy of particles in
light nuclei (7, 12, and 16 MeV for alpha particles, protons, and
neutrons, respectively, in oxygen), a variety of secondary particles
such as gamma rays, neutrons, protons, and heavier particles are
emitted. This process is called pion star formation because of the
tracks produced in a nuclear emulsion. For carbon, as an example,
about 74% (103 MeV) of the pion rest mass is supplied as kinetic
energy to the various emitted particles, most of the remaining
energy being used to overcome the binding energies of the emitted
particles. Roughly 76 MeV of energy is removed by neutrons alone.

Apart from the neutrons, all of the emitted particles have short
ranges and thus deposit their dose in the immediate vicinity of the
star, increasing the dose in the Bragg region as shown in Fig. 1.
A significant fraction of the energy deposited by these heavier

Table 1.   Fundamental Properties of Pions
           (Leighton, 1959)

Masses:
    $\pi^\pm$ 139.6 MeV
                                 273.2 M_e
    π^o 135.1 MeV
                                 264.4 M_e

Mean Life:
    $\pi^\pm$ 2.54×10^{-8} s
    π^o $\sim 10^{-15}$ s

Decay Modes:
    $\pi^\pm \rightarrow \mu^\pm + \nu$ $\sim$ 100%   Q = 33.9 MeV
           $e^\pm + \nu$ $\sim 10^{-2}$ 139.1

    $\pi^o \rightarrow \gamma + \gamma$ 98.8   135.1
         $e^+ + e^- + \gamma$ 1.2   134.1
         $e^+ + e^- + e^+ + e^-$ 3.6×10^{-3} 133.1

Table 2. Charged Particle Production for Negative
          Pions Stopping in Thin Carbon Target
          (Mechtersheimer, 1978)

Particle	Energy Range	Yield per Stopped π^-	Average Energy per Stopped π^-
p	1.8-99.5	0.452±0.043	10.4±1.8
d	1.9-99.5	0.326±0.032	6.3±1.1
t	2.0-79.5	0.219±0.023	3.0±0.5
^{3}He	8.5-29.5	0.034±0.005	0.5±0.2
^{4}He	2.0-53.5	0.622±0.066	5.5±1.0
Li*	4.5-25.5	0.132±0.019	1.2±0.3

*^{6}Li:^{7}Li $\approx$ 1:1

Table 3. Neutron Production for Negative Pions
          Stopping in a Carbon Target
          (Klein, 1978)

	Yield per Stopped π^-	Average Energy per Stopped π^-
Klein	2.5 ± 0.3	76 ± 9 MeV
Anderson et al.	2.8 ± 0.3	68
Hattersley et al.	2.92 ± 0.36	110 ± 11
Hartmann et al.	2.10	76.04
Guthrie et al.	2.79	64.16

particles, including the neutrons, is at a high stopping power or
LET. These high LET events can produce significantly different biological effects in tissue as compared with events from radiations
conventionally used in therapy, i.e., x rays, gamma rays, or electrons. Specifically, the increased specific energy (energy per unit
mass) along the tracks of the particles can result in an increased
relative biological effectiveness (RBE) and a decreased oxygen enhancement ratio (OER).

From a therapeutical point of view, the fact that one is dealing with a heavy charged particle is also an advantage in shaping
the beam in order to deliver maximum effective dose (absorbed physical dose times RBE) to the treatment volume while minimizing the
effective dose to surrounding tissues and vital structures.

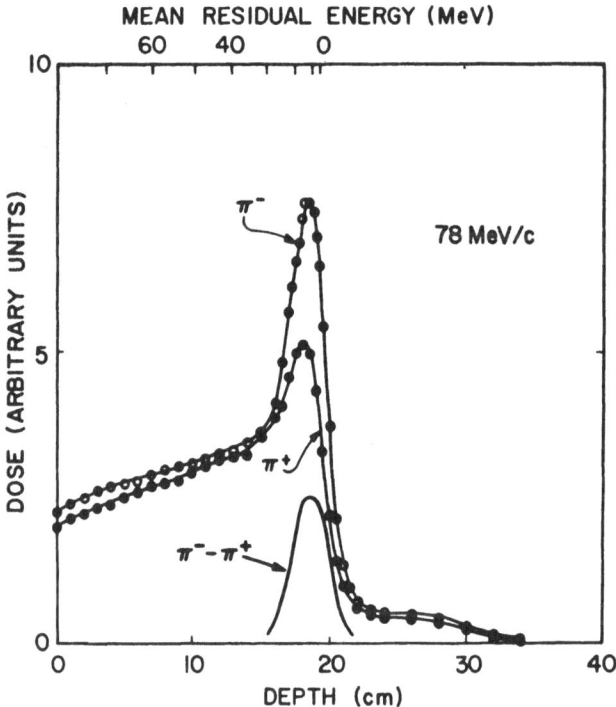

Fig. 1. Depth dose curves for narrow beams ($\Delta p/p$ 0.02) of π^+ and
π^-, and the difference ($\pi^- - \pi^+$). The difference is actually
less than the star dose because of μ^+ decay. See text for
further explanation. (Dicello et al., 1977).

The gamma rays associated with negative pion absorption by a
nucleus can be detected in order to reconstruct the spatial distribution of pions stopping in the treatment volume. That is, it is
possible to monitor and verify a patient treatment during the irradiation.

In summary, some of the potential therapeutic advantages of
negative pions include 1) the increased effective dose delivered to
the treatment volume as compared with the surrounding tissue, 2)
improved beam shaping, 3) a decrease in the resistance of certain
tumors to radiation, and 4) the possibility of monitoring the dose
and the distribution of dose during irradiation.

PION BEAMS

Pions exist virtually in nuclei as the mediators of the nuclear
force thereby holding the nucleons together. They can be produced
by the bombardment of any target material, e.g., carbon or beryllium,
with a beam of particles, such as protons or electrons, with a mean

energy in the center of mass greater than the rest mass of the
pions. The phase space of the secondary pions, along with muons and
electrons produced by other interactions between the primary particles and the target, depends on the initial energy of the primary
beam and the target material and shape. Magnetic quadrupoles are
used to collect and focus the pions. The intensity and composition
are a function of the collection angle and other factors. Bending
magnets are used to select the mean momentum of the beam and, therefore, its average range. Slits and/or wedges are used to establish
the desired distribution in momentum. The entire system is called
a pion transport channel, or simply, a pion channel. Because the
magnetic transport of the particles is only momentum-selective,
muons and electrons produced at the target with the same momentum
as the pions are also transported. In addition, because of the
finite length of the channel, a large number of pions decay in
flight, contributing further to the number of muons and electrons
present. For example, the decay length (36.8% remaining) of a
160 MeV/c beam is about 9 meters.

The presence of muons and electrons is generally undesirable
because they contribute strongly to the entrance and exit doses,
and they reduce the mean LET and, therefore, the biological effectiveness of the beam. Because they have a different spread in
momentum, although the same mean momentum, they will influence the
effectiveness of collimation and beam shaping. Most ways to reduce these contaminants are either not practical (such as reducing
the channel length) or reduce also the pion flux (such as the use
of low Z target materials). Therefore, the amount of contamination
is a compromise with the actual contamination being the maximum
considered acceptable. It is imperative that data on the number
and spectra of the contaminants be known either from measurements
or calculations.

At present, the contamination in an incident pion beam, prior
to entering any absorbing material, is usually measured by time-of-
flight techniques. Results of such measurements for a typical pion
beam at LAMPF are shown in Fig. 2 (Paciotti et al., 1975). Some
typical data from the TRIUMF channel for the amount of contamination for a beryllium target as a function of pion energy are shown
in Fig. 3 (Dicello et al., 1977).

At the present time, there are basically two types of pion
channels in operation. The conventional system, such as that being
used at LAMPF (Paciotti et al., 1975) in Los Alamos, NM and TRIUMF
(Craddock, 1977) in Vancouver, B.C., consists of a single collection quadrupole facing the target followed by a series of quadrupoles for beam shaping and bending magnets and slits for momentum
selection. The Los Alamos biomedical channel is shown in Fig. 4.
The disadvantages of such a setup include the limited solid angle

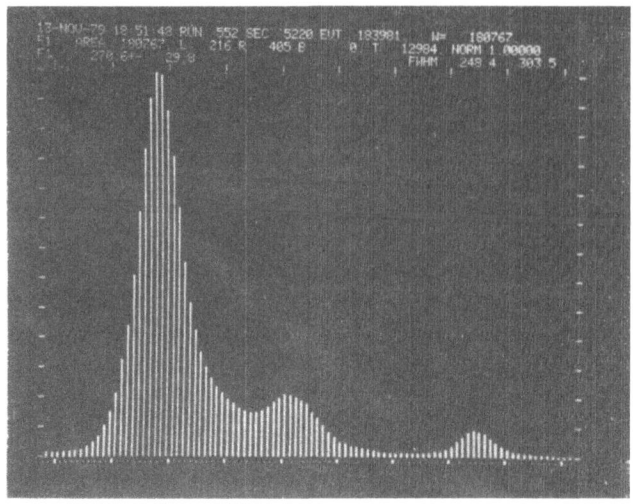

Fig. 2. Number of particles as a function of time to traverse the
         flight path. The peak at the left is for the slowest par-
         ticles, π^-, followed by μ^- and e^-. (Paciotti et al., 1975).

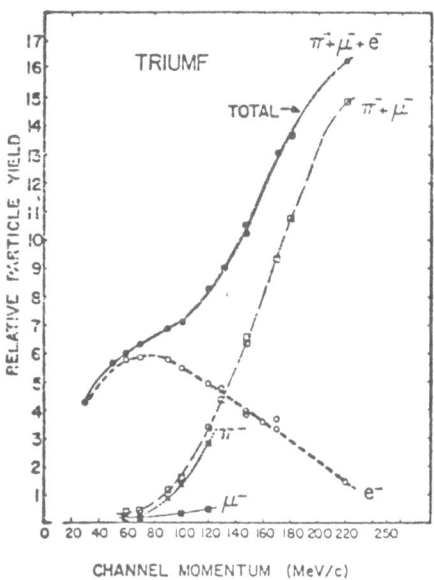

Fig. 3. The relative number of π^-, μ^-, and e^- as a function of
         mean momentum for a beryllium target at the biomedical
         channel at TRIUMF.

of acceptance (thus limiting the beam intensity) and the fixed orientation of the channel, requiring the patient to be rotated for multiport therapy.

A novel system which resolves some of these difficulties was developed at Stanford University (Bagshaw, 1973) and is illustrated in Fig. 5. This pion applicator consists of 60 individual superconducting magnets surrounding a target, all of which are used to collect pions. The pions are then transported to the patient area. The beams are then focused on the treatment volume in the manner illustrated in Fig. 5. The large number of initial magnets increases the pion intensity by about a factor of 50 over conventional techniques. In addition, the availability of 60 simultaneous ports increases the versatility of the treatment planning, although the apparatus does restrict access to the patient. This method of beam transport will be used at the biomedical facility of SIN (Blaser, 1976) in Villigen, Switzerland.

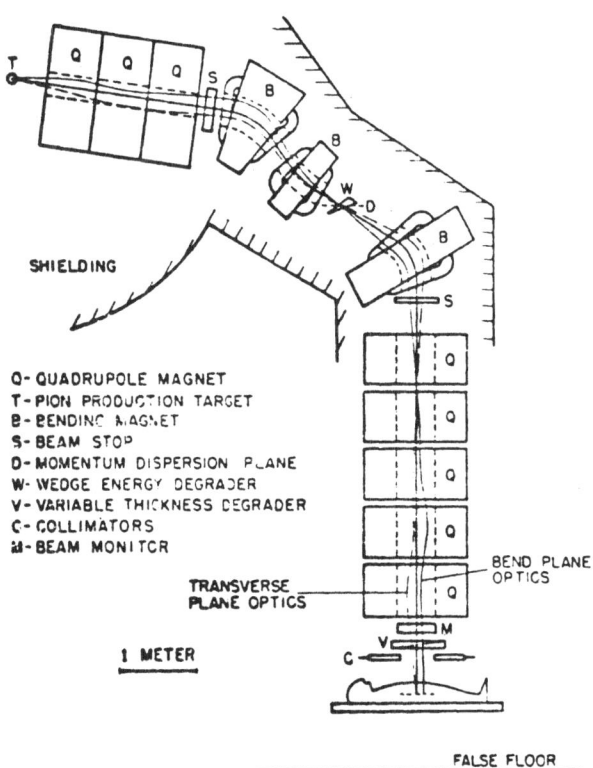

Fig. 4. The biomedical channel at LAMPF.

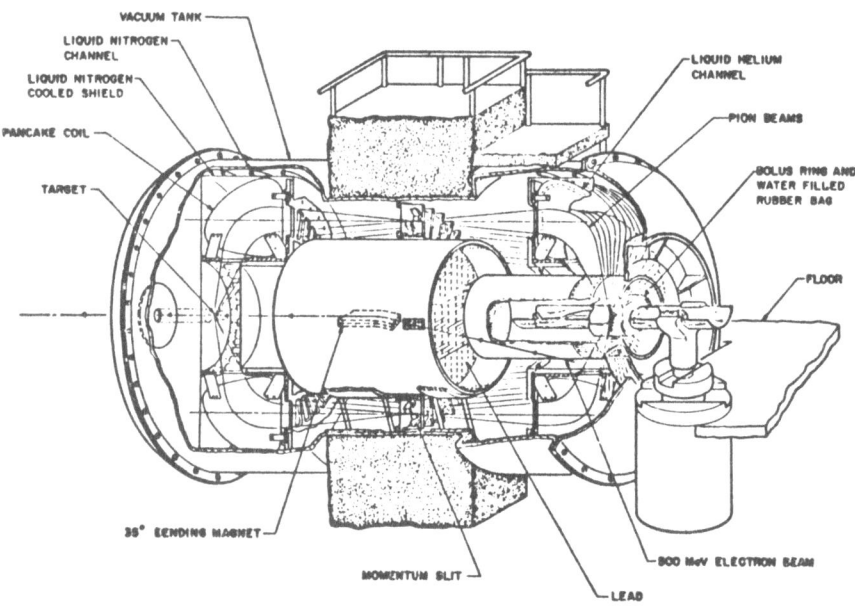

Fig. 5.   Stanford biomedical pion applicator.

DOSIMETRY

     Generally, in conventional therapy, a knowledge of absolute
absorbed dose and its distribution in the patient is sufficient
for treatment. Because of past experience with photons and electrons, the relationship between dose and biological response is
better understood than with other types of radiations. On the
other hand, because of the complex physical character of a pion
field, there is no single physical parameter which directly determines the biological response. Instead, it is necessary to determine several physical quantities in order to characterize the
variation in radiation quality across the stopping pion field and
ultimately to establish a relationship between the physical parameters and biological effect.

     To illustrate the type of information needed and its applica-
tion, results for some typical beams used at Los Alamos are presented.

     The methods used for the measurement of doses and dose distri-
butions for pions are quite similar to those used for photons.
Usually, the measuring instrument is a thimble-type ionization

chamber with an effective volume of between 0.05 and 1.0 cc. The walls of the chamber are constructed of A150 tissue-equivalent plastic and the gas is either air or a methane-based tissue-equivalent gas. The data are obtained in a phantom of water or plastic.

Ionization chambers are used, of course, to measure the charge induced in a gas by ionization of that gas by a radiation. This is not a direct measurement of energy loss or dose, and the conversion of the results to dose may not be simple, particularly in the case of high LET radiations.

The effective volume of the chamber is determined by a measurement of the induced charge in a calibrated photon beam such as cobalt-60. Then a direct application of Bragg-Gray theory gives (Bichsel et al., 1975):

$$V = 3.88 \times 10^6 \frac{A_{eq}}{\rho_{gas}} \frac{W_{gas}}{W_{air}} \frac{(\mu_{en}/\rho)_{air}}{(\mu_{en}/\rho)_{wall}} S_{gas}^{wall} \frac{Q \text{ gR}}{X \text{ C}}$$

where

V = the effective volume of the gas cavity (cm^3);

A_{eq} = a correction factor for the attenuation in the wall, the thickness of which is chosen to establish electron equilibrium,

ρ = the density of the specified medium (g/cm^3),

W_{gas} = average energy necessary to produce an ion pair in the gas,

W_{air} = average energy necessary to produce an ion pair in air,

(μ_{en}/ρ) = mass energy absorption coefficient in the specified medium

S_{gas}^{wall} = ratio of the mean values of the mass stopping power of the wall to gas, properly weighted according to the energy spectrum of electrons crossing the gas volume,

Q = the charge produced during the irradiations in coulombs (C),

X = the exposure in roentgens (R).

Bichsel et al., (1975) have estimated that the standard deviation of the absolute value of the mass (ρV) derived by this method is $\pm 0.025 \, \rho V$.

Given the volume of the cavity, the absorbed dose at a given point in a radiation field from pions can be calculated from the relationship (Dicello, 1975):

$$D = (1/100 \ V\rho) \sum_i A_i Q_i W_i S_i$$

where

D = absorbed dose in rad,

A_i = correction factor for non-tissue-equivalence of detector materials,

Q_i = charge collected in chamber in coulombs (corrected for any inefficiency in collection),

W_i = average energy per unit in ion pair produced in the gas (ergs/coulomb),

S_i = effective mass stopping power ratio for wall to gas,

V = effective gas volume in cm^3,

ρ = gas density in g/cm^3.

The summation is performed over the entire distribution of particle types and their energies at the point of the measurement. For example, in the peak region, the summation includes the pions, muons, and electrons as well as the secondary protons and heavier particles.

The uncertainty associated with the calculation, at the moment, is fairly large. Some conservative estimates of the uncertainties (standard deviation divided by the mean value) for monoenergetic pions in the plateau region disregarding secondaries might be $\Delta A_i / A_i = \pm \ 0.05$; $\Delta Q_i / Q_i = \pm \ 0.01$; $\Delta \omega_i / \omega_i = \pm \ 0.05$; $\Delta S_i / S_i = \pm \ 0.02$; $\Delta V/V = \pm \ 0.025$; and $\Delta \rho / \rho = \pm \ 0.02$. With the assumption of random errors, this gives a total uncertainty to the absolute dose of $\Delta D/D = \pm \ 0.08$. That is, the absolute dose in the plateau, even under these idealized conditions, is still uncertain to about $\pm$ 8%. Moreover, these correction factors and uncertainties will change throughout a given irradiation volume and from beam to beam. A final point that should be made is that in a field as complex as a pion stopping field, the applicability of Bragg-Gray cavity theory is by no means certain and has not yet been subject to direct experimental verification.

Because of the large uncertainty in specifying absolute doses, most researchers either quote the measured charge in the pion beam times the cobalt-60 conversion factor (rads per coulomb) as the dose, or they apply a specified constant conversion factor. The former method has been used throughout this paper.

With an unmodified beam the volume irradiated, and correspondingly the region of stopping pions, is only a fraction of a liter which is too small for most medical applications. To increase the span along the beam axis, either 1) the momentum of the pion beam is varied in order to spread the range of the pions, or 2) an absorber, the thickness of which can be varied, is placed between the primary pion beam and the volume to be irradiated.

The result in either case is to spread the region over which pions are stopping in some prescribed manner.

A sketch of the type of apparatus used to shift the range of the pion beam at Los Alamos (Liska, 1977) is shown in Fig. 6. It consists of a piston filled with oil, the spacing of which can be varied as a function of time by a servomechanism linked to a computer or a microprocessor.

Two typical depth dose curves for broadened beams are shown in Fig. 7 (Amols et al., 1977). One distribution was chosen to have a constant physical dose across the stopping region. The second distribution was chosen to have a decrease in physical dose which would partially compensate for an increasing RBE as a function of depth. The narrow, unmodulated beam is shown for comparison.

There are some characteristics which are common to all pion depth dose curves. First of all, there is a region beginning at the surface of the phantom, commonly called the plateau region, where the dose is relatively constant. This results from the fact that the dose results primarily from the energy losses from energetic pions for which the stopping power is changing slowly. At

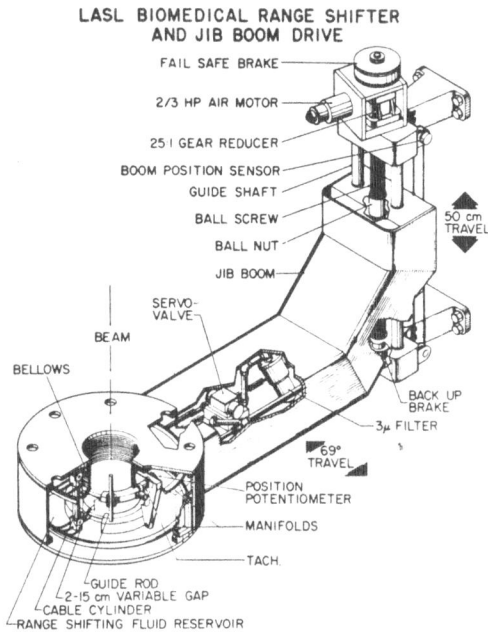

Fig. 6. The type of range shifter used at LAMPF. (Liska, 1977).

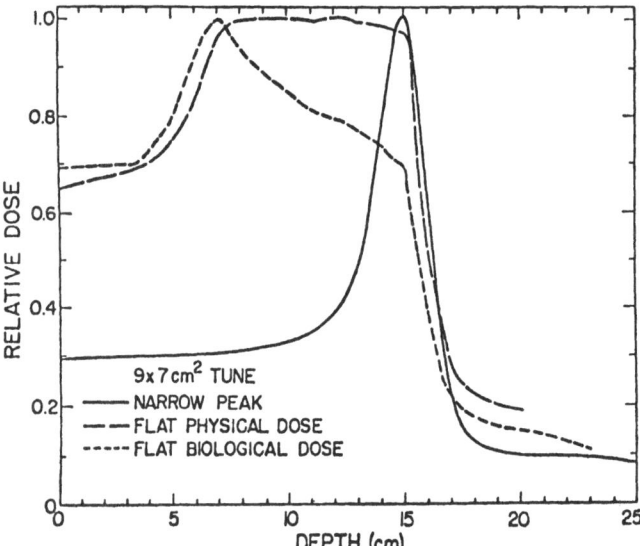

Fig. 7. Depth dose curves for broad beams (Amols et al., 1977).

some depth there is an abrupt increase in the dose because of the
rapid increase in the stopping power as the pions approach the end
of their range. As we have seen, beyond this point there is a large
contribution to the dose from pions which have stopped and produced
stars. Beyond this region, there is a rapid decrease in dose corresponding to the maximum range of the pions. At greater depths,
there is still a significant fraction of dose remaining. This is
primarily a result of muons and electrons produced in the primary
target and by pion decays which have a greater range than the pions.

Finally, it should be noted that the peak-to-plateau dose ratios for the broader beams are smaller than that for the narrow
beam. This can be understood by realizing that the broad beams are
simply superpositions of narrow beams of different mean momenta.
For the same reason, as the peak is widened with a range shifter,
the depth to which the plateau extends decreases, while the magnitude of the plateau dose increases. A consequence of this is that,
although the dose in the distal (downstream) portion of the peak is
a result of pion stars and low energy pions, the dose in the proximal (upstream) portion is the result of more energetic passing pions
with a lower relative contribution from pion stars and low energy
pions. This can be seen in Fig. 8 where the depth dose curve for
a flat physical dose with a π^- beam is compared with the corresponding curve for a π^+ beam (Dicello et al., 1978). Because positive pions do not produce stars, the difference in the two beams
is a result of star dose. In fact, this difference underestimates
the star contributions because the π^+ curve also includes energy
from π^+ decays of the type

$$\pi^+ \to \mu^+ + \nu_\mu; \; \mu^+ \to e^+ + \nu_e + \bar{\nu}_e.$$

These decay products can contribute as much as 25% (Turner et al., 1972) of the dose from the π^+ beam. Nevertheless, qualitatively, one can see that there is more star dose in the distal region than in the proximal portion. Moreover, the proportion of star dose on the average, is much less than at the peak of a narrow beam (Fig. 1) because of the increased contribution from passing pions (Dicello et al., 1977).

Because the higher LET events are associated with star formation, the relative decrease in the average number of stars can reduce the biological effectiveness of the beam, and the nonuniformity in the distribution of stars can result in a variation in biological response over the stopping region. The nonuniformity of the stopping distribution can be reduced by the use of overlapping fields, as can be seen from the calculations of Pistenma et al., 1978, in Fig. 9 for 60 ports. Additionally, the peak-to-plateau ratio is improved dramatically, as is also the case with photons.

As stated in the introduction, one potential advantage of a pion beam (or any heavy charged particle beam) is good beam shaping. An isodose distribution (Amols et al., 1978a) for one of the beams in Fig. 7 after collimation is shown in Fig. 10. A comparison of the dose at the lateral edge of such a beam with 20 MV x rays and

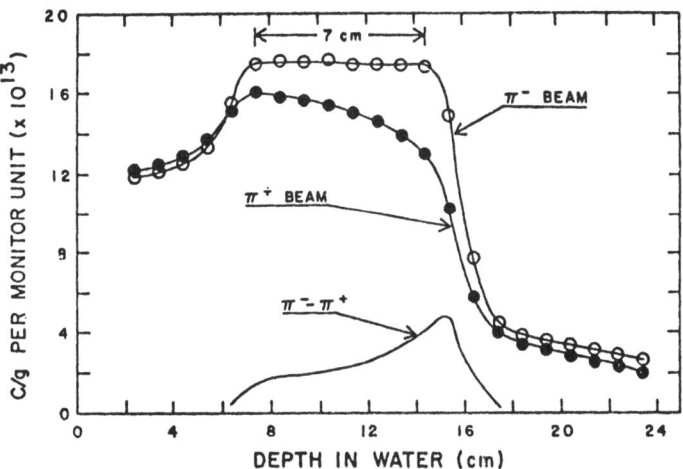

Fig. 8. Depth-dose curves for π^+ and π^- beams and the difference ($\pi^+-\pi^-$) between the two curves. The incident π^- beam contained approximately 62% π^-, 26% μ^- and 12% e^-. The π^+ beam contained approximately 68% π^+, 29% μ^+ and 3% e^+. (Dicello et al., 1978).

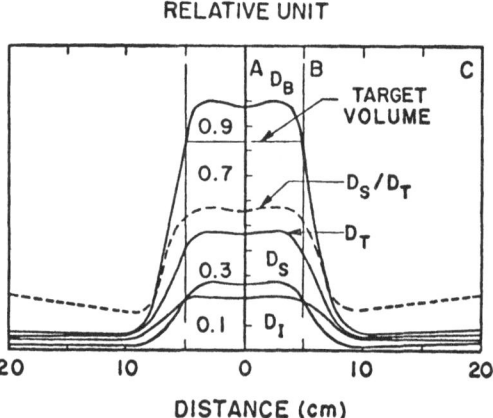

Fig. 9. Calculated profiles of the pion star dose D_S, the ionization dose D_I, the total dose D_T, the biological equivalent
dose D_B and the pion star fraction D_S/D_T for the multiport irradiation of a cylindrical target volume 10 cm in
diameter and 10 cm in length. (Li et al., (1974).

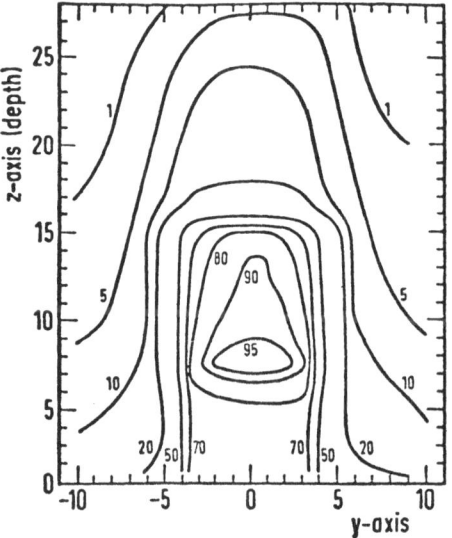

Fig. 10. Isodose profiles in the Z (depth)-Y plane. The collimator
had edges at Y = ± 4.5 cm. (Amols et al., 1978a).

a theoretical calculation for pions including only multiple scattering is shown in Fig. 11. X-ray data are for two opposing fields (Cohen and Martin, 1966). For more details concerning the method refer to Knapp, 1975.

The measured data are poorer than the photon data. The increased spread is a result of the finite phase space of the beam (the distribution of the beam in space, angle, and also momentum) and contamination in the beam (muons, electrons, and neutrons both from stars and from background sources such as the collimator). Hogstrom et al., (1979) have used a semi-empirical approach to calculate beam profiles which are in excellent agreement with the measurements (Fig. 12). However, this agreement was achieved at the expense of assuming a standard deviation for multiple scattering which is between 20 and 40% larger than measured experimentally by Amols et al., (1979), or that derived from Moliére scattering theory. This is at least an indication that improvements in the experimental distributions may still be possible. It should be stressed that

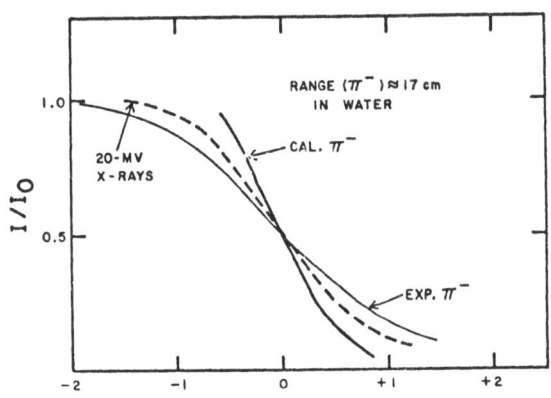

DISTANCE FROM EDGE OF COLLIMATOR (cm)

Fig. 11.   A comparison of the lateral edge of a pion beam with that for 20 MV x rays (2 opposing fields) at a depth of 17 cm. The experimental π^- curve is the dose at the edge of a collimator. The theoretical curve is simply the change in the number of pions from multiple scattering only. It is assumed that the standard deviation, σ, is given by $\sigma = 0.07 \, R_o^{0.02}$ where R_o is the pion range.

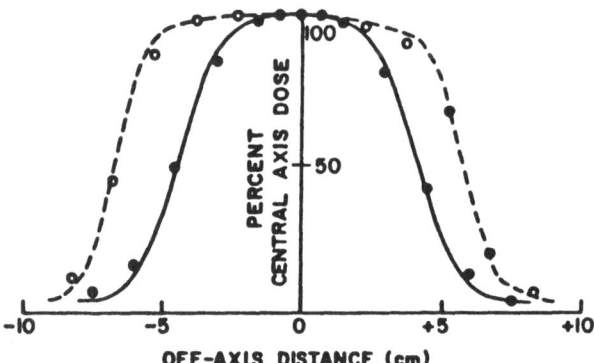

Fig. 12.   Dose profiles in the plane perpendicular to the beam's
           axis at the midpeak position of a broad beam. (Hogstrom
           et al., 1979).

care should be taken not to equate edge sharpness with dose localization. Although edge sharpness is necessary, it is only one
aspect of good dose shaping or localization. The advantages and
disadvantages of improved edge sharpness in therapy have been discussed extensively in the literature (e.g., Kramer and Suntharalingam, 1977; Suit et al., 1977), if the reader wishes to pursue this
topic.

NEUTRONS IN PION FIELDS

     Pion facilities have background radiation, a large fraction of
which is neutrons. More importantly, neutrons are produced by
pions as a result of inflight nuclear interactions and stars. Because of the non-ionizing character of the neutrons and their high
RBEs at low doses, it is worthwhile to consider the consequences of
these particles in pion radiotherapy in some detail. Even at clinical doses the RBE for the biological effects for fast neutrons on
the central nervous tissue of rodents has now been consistently
shown to be around 3.9, as compared with values between 2.4 and 3.0
currently used in neutron therapy calculations (e.g., Geraci et al.,
1978). Much of the dose to normal tissues in patients undergoing
clinical trials with pions is from neutrons from pion stars either
from the treatment volume or from collimators. It was thought that
the potentially higher RBEs could reduce the therapeutic gain factor
for pions.

The experimental data on neutron spectra produced from stars are somewhat inconsistent, but it seems that in carbon, on an average between 2.1 and 2.5 neutrons per star with an average total energy of 76 MeV are produced (with the energy distribution extending up to 120 MeV or more)(Klein, 1978). Therefore, the treatment volume, as well as the collimators, are sources of large neutron fluxes. Because of their long mean free path, they alter the dose distributions, reduce the sharpness of the edges, and change the radiation quality.

Small treatment volumes will have 5% or more of the dose contributed by neutrons (Schillaci and Roeder, 1973; Amols et al., 1978). As the stopping volume increases, the fractional contribution to the dose from neutrons will increase because of the longer range over which the neutrons expend their energy. At the same time, the fractional dose from the heavier charged particles is reduced because of the increased ratio of passing to stopping pions. Without neutrons, if the volume were increased, the high LET component would become continuously diluted. This is partially compensated by an increasing high LET component from neutrons, so that even large treatment volumes may contain a biologically significant component of higher LET events (Dicello et al., 1978).

Much of the available information related to the effects of neutrons is from Monte Carlo calculations, such as those of Wright et al., (1979), Schillaci and Roeder (1973), and Brenner and Smith (1977). These calculations have been limited in the past by the lack of accurate cross sectional data and the unreliability of the measured and calculated spectra of neutrons emitted from stars. Amols et al., (1978) estimated the amount of neutron dose outside of the treatment volume for a narrow pion beam, and these data agree with the results of Schillaci and Roeder (1973) within the specified uncertainties. Dicello et al., (1978) have more recently examined these effects for broad beams with comparable results. Dicello et al., (1978) also have employed a single foil activation technique to determine the neutron distribution inside the stopping volume. Their results are compared with calculations of neutron dose by Berardo and Zink in Fig. 13. The excellent agreement may be fortuitous, first of all, because there may be significant amount of activation resulting from charged-particle induced reactions, and secondly, because it is not obvious that the distribution of induced activity is directly related to the absorbed dose.

Brenner and Reading (1977) used multi-foil activation techniques to map out the energy distribution of neutrons outside of the stopping region and compared it with other available data, obtaining reasonable agreement. Unfortunately, because of the problems mentioned above, i.e., interference from charged particle interactions, no practical method of measuring the neutron spectra

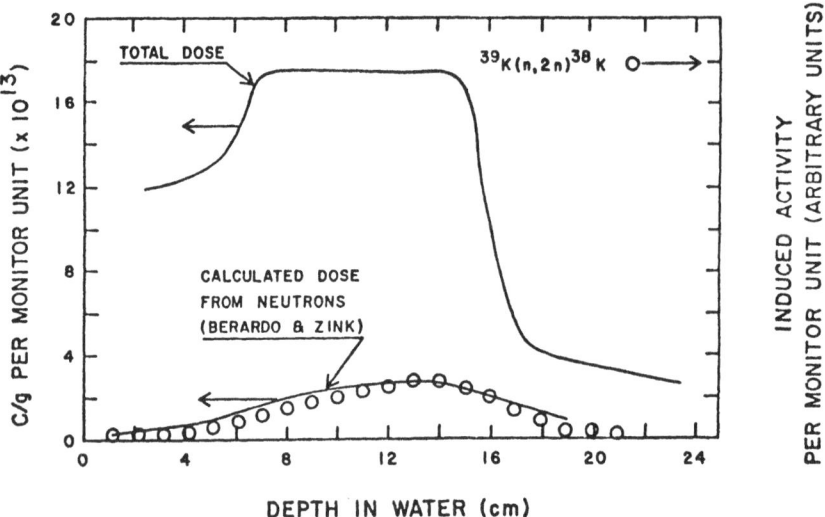

Fig. 13. A calculation of central-axis neutron dose compared with
         the distribution of induced activity from $^{39}K(n,2n)^{38}K$.
         The measured depth-dose curve is shown for comparison.
         (Dicello et al., 1978). The calculation and the experi-
         mental data have been arbitrarily normalized to their
         maximum values.

in the peak region has been developed to date.

DISTRIBUTIONS OF STOPPING PIONS

     A knowledge of the stopping distribution is necessary to cal-
culate, for example, the absorbed dose from passing pions, the
local star dose, and even that from neutrons. Given the phase
space of the incident beam, it is possible to calculate stopping
distributions with some accuracy. However, the status of these
calculations is such that experimental data are still the most
reliable source of information .

     The classical method for measuring the ranges and, therefore,
the stopping distribution of charged particles utilizes three scintillators; the first upstream from the stopping region, the second
in the stopping region, and the third directly downstream of the
second. If a particle is detected in the first and second scintillators, but not the last, it is known to have stopped in the second.
By varying the position of the downstream pair of detectors in a
phantom, the variation in number of stopping particles as a function of position can be determined. Results for a narrow beam in
lucite obtained by this method are shown in Fig. 14. This
classical approach is only practical, however, at low fluxes.

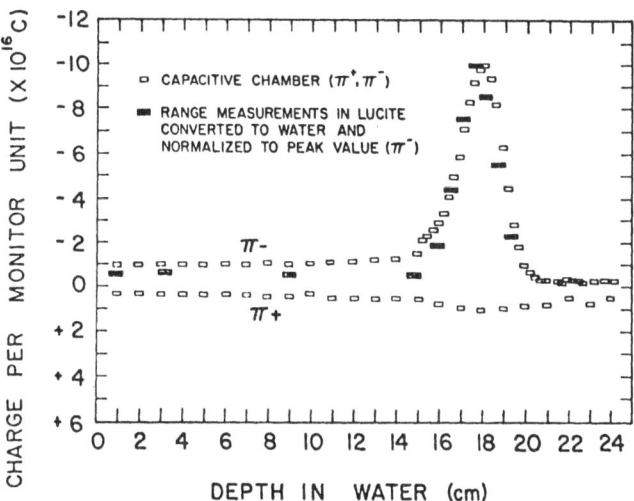

Fig. 14. Stopping distributions for a π^- beam obtained with two techniques described in the text. The π^+ curve for the capacitive chamber is also shown to illustrate the sensitivity of the method to pion stars.

A newer technique, more adaptable to standard dosimetric practices and high fluxes, and sufficiently compact that it could be used for *in vivo* studies was developed by Laughlin (1965), and Van Dyke and MacDonald (1972) for electron beams. This method uses a simple capacitive counter to measure the charge of a particle stopping in it. Shortt and Henkelman (1978) showed that the procedure could be used for negative pions, producing one electronic charge per stopped pion, so that it is an absolute measurement of the number of stopping pions.

Results of the two techniques are compared in Fig. 14. It is seen that the two methods agree well in the stopping region. The slight displacement of the peaks is probably the result of the uncertainty in converting from water to lucite.

Similar data are given in Fig. 15 for the two beams shown in Fig. 5. The distributions obtained by folding the experimental data for the corresponding narrow beam with the range-shifting function is shown for comparison. It is clear that a constant stopping distribution does not produce a flat physical dose because of the contribution from passing pions. By comparing Figs. 13 and 15, one also can see that the dose from neutrons is not strongly correlated to the stopping distribution.

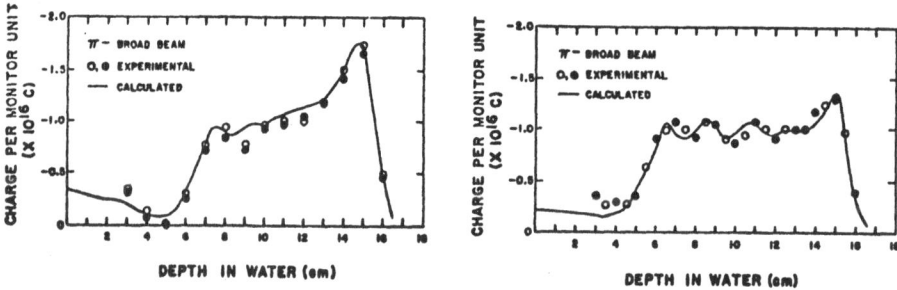

Fig. 15. Stopping distributions as a function of depth for the
broad beams in Fig. 5. The one on the left corresponds
to the flat physical dose. The one on the right corresponds to an inclined dose distribution. (Dicello
et al., 1978).

MICRODOSIMETRY

     Throughout this article, the term "radiation quality" has been
used in reference to those physical characteristics of the radiation other than the absorbed dose which could alter the response of
biological systems. In this context many of the topics discussed
previously are related to the radiation quality. Nevertheless, it
is frequently difficult, if not impossible, to combine them in
order to determine the ultimate biological effects. A more direct
approach is the use of microdosimetric methods. Because Rossi has
reviewed microdosimetry in general and its application to biology
and medicine in a previous paper, let us concentrate on some specific examples and results.

     Amols et al., (1976) obtained the first extensive experimental
data for pions using a narrow beam. Dicello and Zaider (1978)
have more recently examined the two beams shown in Fig. 5. The
microdosimetric distributions at the proximal and distal positions
of the treatment volume are compared in Fig. 16.

     First, consider the flat beam (i.e., flat physical dose). At
first glance, from only the stopping distribution (Fig. 15), one
might expect a large increase in the higher lineal energy portion
of the dose from the proximal to the distal positions. However,
the microdosimetric spectra show a much more subtle change, because the neutrons, contributing significantly to the higher lineal
energies, are not as localized as the stars, and because the dose
from passing pions only decreases about 20% from front to back.
Correspondingly, the dose from stars should only increase about
20% with a similar increase in the dose from higher lineal energies.
This is about what one sees. It is not easy to predict changes

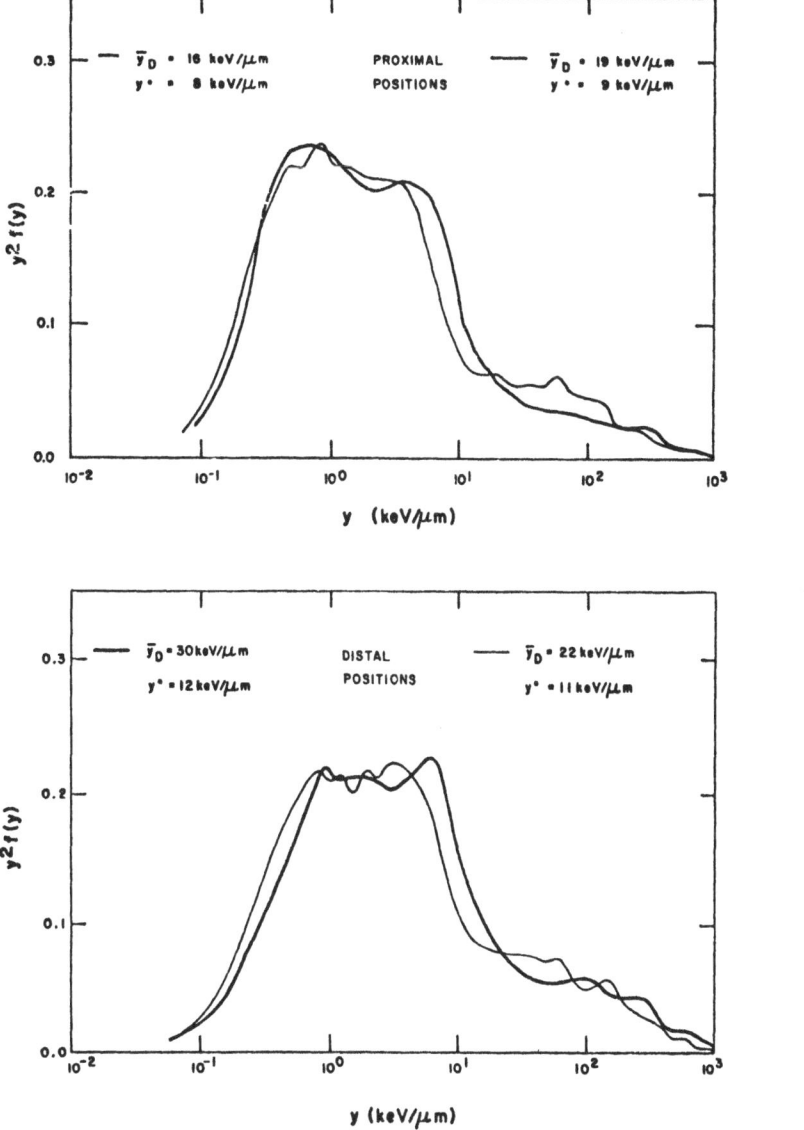

Fig. 16. Microdosimetric spectra for the two beams shown in Fig. 1.
          The heavy lines are for the sloped dose. The light lines
          are for the flat dose. The area under a curve between
          two values of lineal energy, y, is proportional to the
          fraction of dose in that interval of y.

in radiation quality, even with an extensive characterization of
the beam without detailed Monte Carlo calculations.

Now consider the sloped beam. This beam was developed in an
attempt to partially compensate for the increasing RBE as a function of depth in the previous beam by decreasing the physical dose
in the distal position relative to the proximal dose. The microdosimetric spectra show that, in doing this, the mean lineal energy
has decreased in the proximal position, while increasing at the
distal position. Perhaps more important, however, is the fact that
the changes in the mean values result from subtle changes throughout the entire spectra, and it is not apparent what consequences
these will have biologically. Ultimately, it is necessary to relate the spectra to biology through some model. More will be said
about this later.

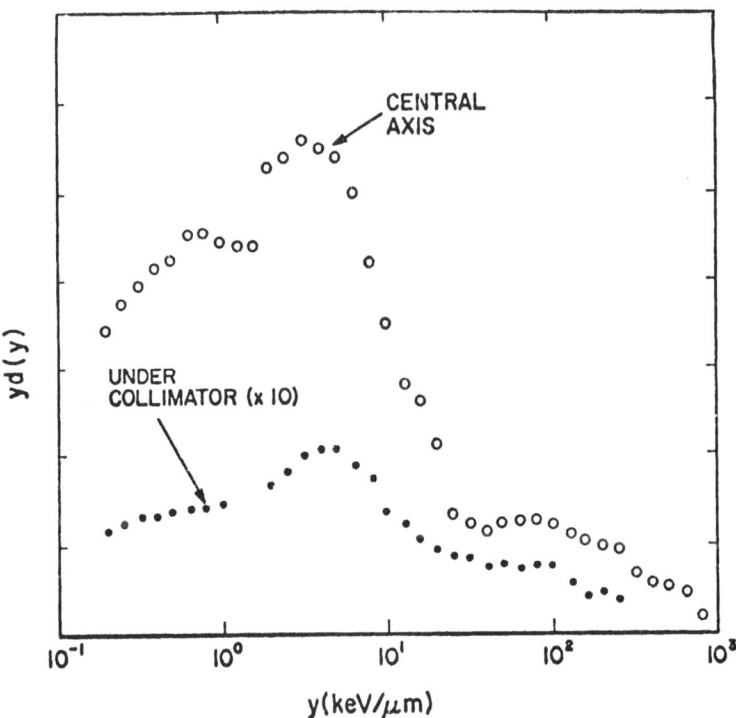

Fig. 17. Microdosimetric spectra for the beam with flat dose in
          Fig. 1. The central-axis data were obtained at the
          center of the stopping distribution. The data under the
          collimator were taken 8 cm off axis in the plateau region
          with the edge of the collimator 4.5 cm off axis.
          (Dicello et al., 1978).

Another interesting application of microdosimetry has been an investigation of the effects of neutrons outside the stopping region. Zaider and Dicello et al. (1978) obtained spectra at the center of the stopping region and under the collimator, as shown in Fig. 17. The two spectra are quite similar in shape. The Theory of Dual Radiation Action (Kellerer and Rossi, 1972 and 1978) predicted that for *in vitro* V79 cells, there would not be a large difference in the RBEs for the two regions. Subsequent biological experiments by Amols et al. (1978a) confirmed these conclusions.

THEORY AND CALCULATIONS

For purposes of clinical trials, one would like to be able to calculate all of the relevant physical parameters from first principles and to relate these parameters to the expected biological response through models. The calculations for treatment planning should be, if not simple, at least fast, so that many repetitions can be accomplished easily.

Much of the earlier calculational work with pions was done with Monte Carlo codes developed at Oak Ridge by Wright et al. (1979), and Armstrong and Chandler (1974). The method of Armstrong and Chandler was very general and capable of obtaining a variety of information in complex situations. However, because of their extensive capabilities, these codes required a great deal of computer time and space, and were very costly to run. The programs of Wright et al. were restricted to pion beams in tissue-equivalent materials. The present versions are much faster than the more generalized approach. Both groups have contributed greatly to our knowledge of pion dosimetry, including the effects of beam composition, beam shaping, and inhomogeneities, especially during that period of time when the beam intensities available were too low for extensive experimentation. However, neither set of programs is fast enough or sufficiently flexible for direct use in treatment planning.

Programs specifically designed for patient treatments have been developed at Los Alamos and SIN. These codes take into account as initial conditions the phase space of the beam, beam composition, collimation, patient boluses, and inhomogeneities. They calculate the absorbed dose, stopping distributions, and the dose from neutrons and heavy charged particles. Some results for a broad beam (Zink and Berardo, 1979) using the Los Alamos code, PIPLAN, are compared in Fig. 18 with experimental data. Because of the necessity of performing a fast calculation for routine patient dosimetry, several physical processes have been approximated, or assigned adjustable physical parameters. For example, the mean neutron energy from the star is adjusted to 55 MeV, whereas the experimental value quoted above is 76 MeV. For detailed calculations, therefore, (for example, to check biological models) more

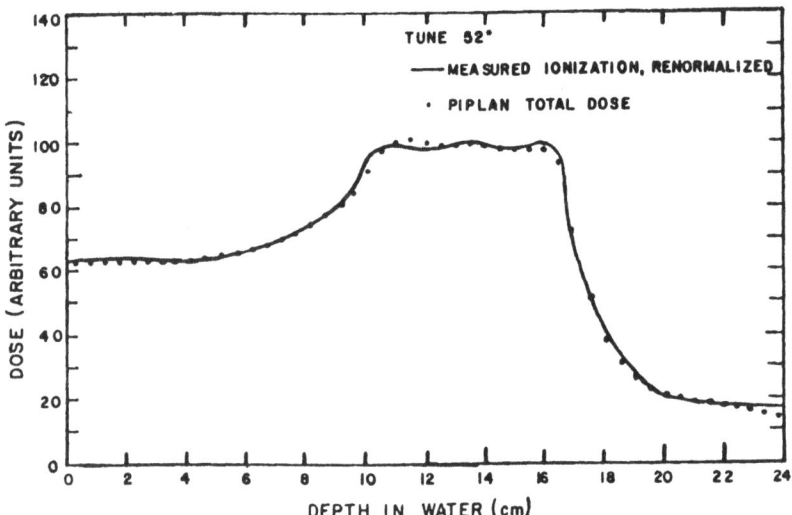

Fig. 18. A comparison of an experimentally measured depth-dose
          curve and that calculated by PIPLAN. The model in PIPLAN
          for range-shifting the peak yields the points indicated
          by the circles. (Zink and Berardo, 1979).

realistic, though slower, procedures may be necessary.

     The semi-empirical approach of Hogstrom et al.(1979) mentioned
earlier has proven to be extremely valuable in calculating dose
distributions for patient dosimetry. The results of this method
are compared with experimental data in Fig. 19. Although more
restrictive than some of the other programs, this program resulted
in a large decrease in the amount of experimental dosimetry necessary prior to a patient treatment.

     Treatment planning codes generally concentrate on the macro-
scopic dose distributions and dose components. Such an approach
does not fully characterize the radiation quality of the beam.
That is, the physical information given is not sufficient to u-
niquely determine the expected response of a particular biological
system. These data are generally supplemented by microdosimetric
data in order to correlate changes in radiation quality with the
corresponding change in biological effects.

     Microdosimetric data for pion beams have been obtained almost
exclusively from experimental results, with only sparse efforts
directed toward calculations. The agreement between calculated LET
distributions and experimental spectra of energy loss was poor and
it was difficult to establish, even qualitatively, the degree to
which the various possible physical processes were contributing to

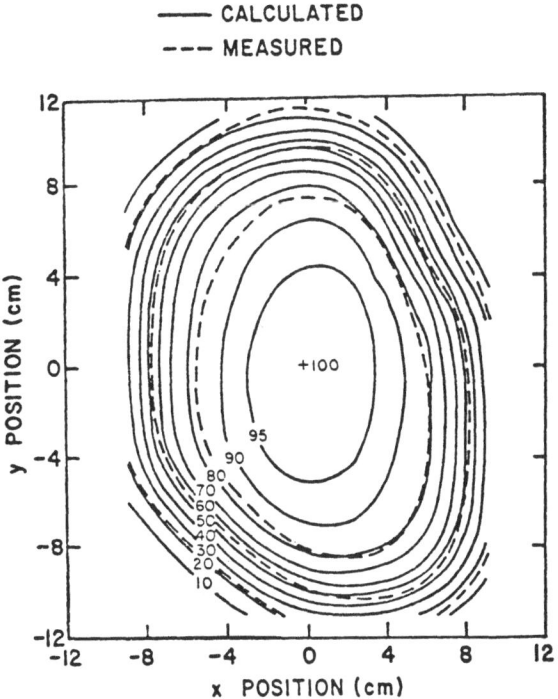

Fig. 19. Calculated and experimental lateral isodose distributions
         under an irregularly-shaped patient collimator.
         (Hogstrom et al., 1979).

energy deposition. For example, it is not possible on the basis of
the experimental data alone to determine the contributions from
energy straggling, delta rays, neutrons, protons, or heavier ions.
Without such detailed knowledge, it is most difficult to interpret
changes in radiation quality except by empirical means. For this
reason, a program has been initiated at Los Alamos to calculate
microdosimetric spectra for heavy charged particles, including
pions. The results have been both encouraging and informative.
Preliminary data of Zaider et al., for 80 MeV pions in the plateau
region are shown in Fig. 20 along with the corresponding experimental results. Calculations in the peak region will be more complex
as the relative contributions from all the different heavy and
light particles must first be known. Also, calculations for regions
distant from the treatment volume are difficult as the high energy
neutrons, which are responsible for most of the energy deposition
in this region, must be transported accurately from the star
region, a calculation made difficult by the lack of reliable

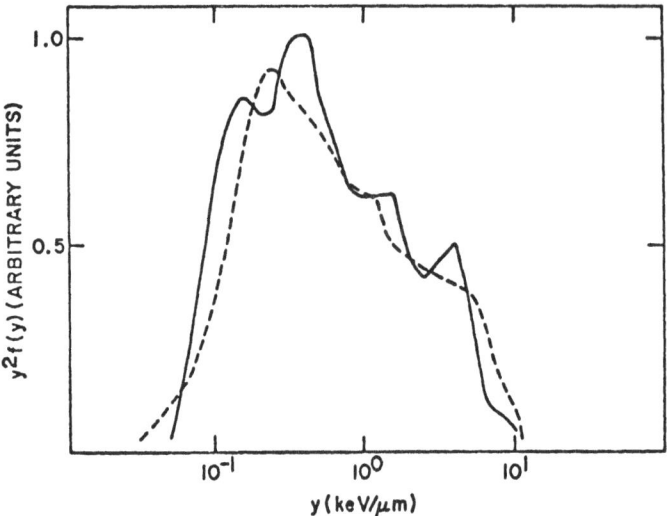

Fig. 20. A comparison of an experimental microdosimetric spectrum
          for a π⁻ beam in the plateau (dashed line) with a prelim-
          inary theoretical calculation (solid line) which includes
          the effects of delta rays. The large fluctuations in the
          calculation are a result of poor statistics.

cross-sectional information for neutrons above 14 MeV.

     As an example of the usefulness of microdosimetric calcula-
tions, consider the data shown in Fig. 20. One unexplained characteristic of pion microdosimetric spectra has always been the
large shoulder in the data extending from about 1 to 10 keV/μm.
This is a characteristic which is also present in spectra from high
energy electrons and protons. Detailed calculations show that it
is a result primarily of energy straggling of the primary particles,
somewhat enhanced by delta rays from the wall of the detector. The
ability to calculate such spectra allows us to interpret the experimental data, and it gives us the capability to extend the data into
regions not accessible by experimental techniques (e.g., for tissue-
equivalent diameters less than 0.2 μm). It allows us to unfold
those characteristics which are inherent in the experiment but not
present in tissue, such as the gas volume of the detector, or the
gas-to-solid interface. It provides for us the capability to correct for such factors as the difference in elemental composition
between tissue-equivalent plastic and gases and tissue.

     The maximum usefulness of all of the dosimetric measurements
and calculations is realized in preclinical and clinical programs
only if they can be related quantitatively to biological effects.
If this can be done, two related objectives are achieved. The
first is a better ability to treat patients. The second is a more

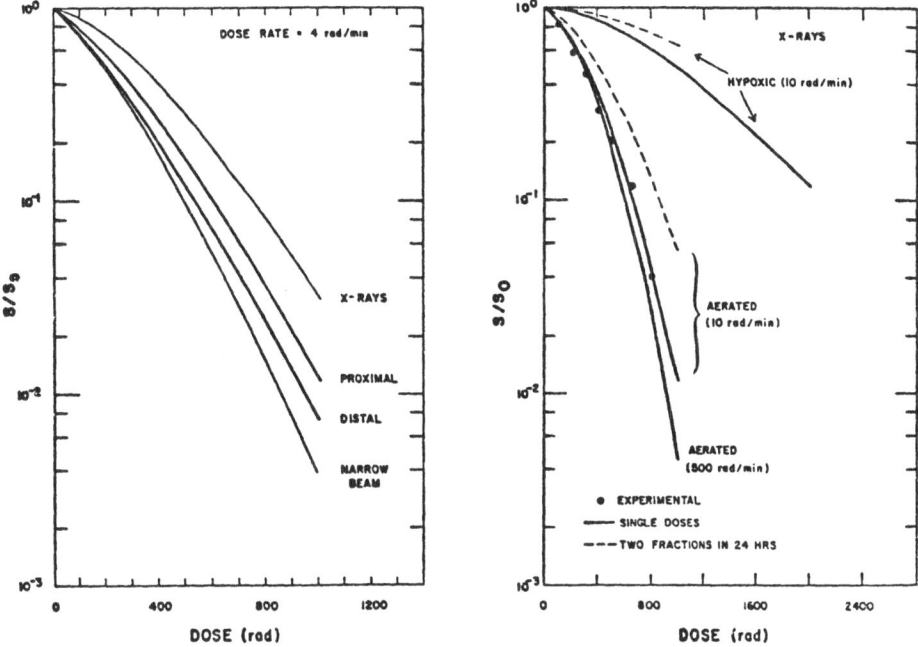

Fig. 21. Survival curves for *in vitro* V79 cells calculated with
the method of Zaider and Dicello (1978).

fundamental understanding of the effects of the radiation. These
two objectives may necessitate different types of theoretical models. The prime necessity for a treatment planning biological
model is that it should predict clinical results accurately, and to
do this, it is reasonable to take a semi-empirical approach with
many parameters that may be fitted to past data. If, however, we
wish to understand the nature of the processes taking place at
some basic level, a more realistic model of the interactions taking
place at this level should be aimed for.

Zaider and Dicello (1978) have applied the Theory of Dual Radiation Action and approaches proposed by Kellerer and Rossi
(1972, 1978) to develop computer codes which calculate biological
responses in simple systems, taking into account such factors as
dose rates, fractionation schedules, radiation quality, recovery
time, differences in repair of sublethal damage, and the degree of
oxygenation. The type of data which can be obtained are illustrated in Fig. 21. This approach has limitations from both a fundamental and practical viewpoint. There are significant discrepancies which exist between the Theory of Dual Radiation Action and
some available biological data. In addition, the present model
does not take into consideration, for example, cell cycle effects,
immune responses, or repopulation. Nevertheless, it is still the

most versatile approach at this time and might act as a foundation
for the development of more comprehensive methods.

One final remark in reference to the calculational efforts.
The accuracy of any calculation cannot be any better than that of
the data it uses as input. In many cases, the lack of available
data is a most severe limitation. Single-collision delta-ray
spectra for energetic pions and for particles heavier than protons
are totally lacking. Reliable experimental production cross sections of charged particles and neutrons for stopped pions in carbon,
oxygen, and nitrogen are only now becoming available, and data for
pion in-flight interactions are still lacking. W-values (average
energy required to produce an ion pair) have been measured only
for energetic pions (with large absolute uncertainties). Even the
accuracy of stopping-power data have proven to be inadequate. It
is clear then that the success of the calculations described above
is highly dependent on good experimental data becoming available,
but once this is obtained, the calculations will, in turn, decrease
the amount of experimentation necessary.

CLINICAL METHODS

Clinical trials with pions began at LAMPF in Los Alamos in
1974 (Kligerman et al., 1976 and 1977). SIN in Villigen, Switzerland and TRIUMF in Vancouver, B.C., will begin treatment shortly.
Procedures for patient dosimetry and treatment planning at Los Alamos have been described in detail by Smith et al. (1977) and
Hogstrom et al. (1978a), and their work will be briefly reviewed
here.

Once an individual has been selected for pion therapy, a rigid
immobilization system is made for proper alignment and positioning
during treatment. An x-ray simulation unit is then used to define
the correct treatment position and to establish the proper orientation and alignment.

CT scans are taken, primarily, to determine the distribution
of inhomogeneities and the treatment volume.

The beam is shaped along the beam axis, and, therefore, in
depth, by a range shifter like that in Fig. 4. The beam is further
modified by the introduction of compensating material upstream from
the patient (bolus) both to shape the beam and to correct for inhomogeneities.

The beam is shaped in the plane perpendicular to the beam axis
by use of a collimator. The designs of the bolus and collimator
are accomplished through computer programs using CT data and the
treatment planning codes described previously. The final arrangement is shown in Fig. 22.

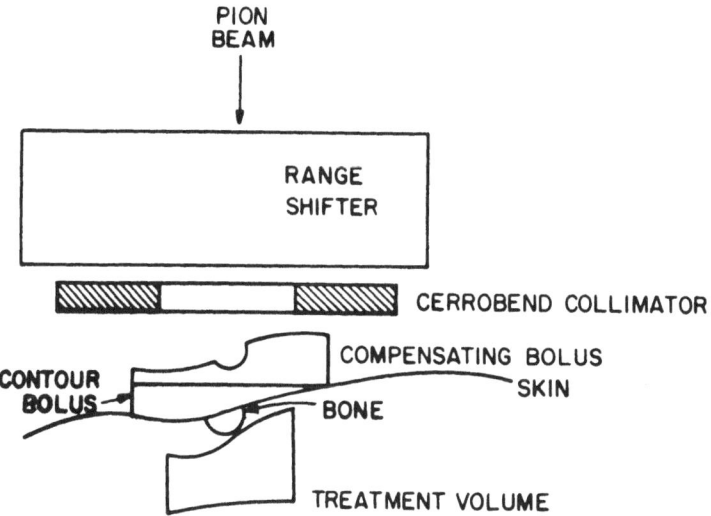

Fig. 22. A schematic of a typical patient treatment. (Courtesy
         of A. R. Smith)

Many of the treatment plans are then checked with dosimetric
measurements in phantoms. Moreover, whenever possible, *in vivo*
dosimetric measurements are also carried out. The dose is measured
with either intracavitary ion chambers or thermoluminescent dosimeters (TLD) (Cooke and Hogstrom, 1979). Other physical properties
of the beam are characterized with silicon detectors (Richman et
al., 1979) and induced radioactivity in aluminum pellets (Hogstrom
and Amols, 1979).

The degree of sophistication of the entire procedure performed
in an efficient and rapid manner, cannot be fully conveyed by this
brief summary. At the same time, the ability to shape the treatment volume will be improved even further by the use of "dynamic
scanning." That is, the patient will be scanned in a prescribed
fashion in the plane perpendicular to the beam while the range
shifter is shaping along the beam axis. This method at Los Alamos,
along with the 60-port method at SIN, will allow maximum versatility in the use of biomedical pion beams.

REFERENCES

Amols, H.I., Bradbury, J.N., Dicello, J.F., Helland, J.A., Kligerman, M.M., Lane, T.F., Paciotti, M.A., Roeder, D.L., and Schillaci, M.E., "Dose Outside the Treatment Volume for Irradiation with Negative Pions," Phys. Med. Biol., 1978, Vol. 23, No. 3, 385-396.

Amols, H.I., Büche, G., Kluge, W., Matthäy, H., Moline, A., Münchmeyer, D., Schmidt, D., Stabl, F., and Walther, H.P., "Pion Multiple Scattering in Thick Targets of Dosimetric Materials," to be published in the Proc. of the Int. Conf. on Med. Phys., Jerusalem, Israel, 1979.

Amols, H.I., Dicello, J.F., and Zaider, M., "The RBE at Various Positions In and Near a Large Negative Pion Beam," Proc. of the 6th Symp. on Microdosimetry, J. Booz and H.G. Ebert, eds. Harwood Academic Publishers, Ltd., 1978a.

Amols, H.I., Liska, D.J., and Halbig, J., "Use of a Dynamic Range-shifter for Modifying the Depth-Dose Distributions of Negative Pions," Med. Phys. 4, 1977, p. 404-407.

Anderson, H.L., Hincks, E.P., Johnson, C.J., Rey, C., and Segar, A.M., "Energy Spectra of Neutrons Emitted Following π^- Capture in C, Al, Cd, Pb, and U," Phys. Rev. 133B, 1964, p. 392-403.

Armstrong, T.W. and Chandler, K.C., "Calculations Related to the Applications of Negatively Charged Pions in Radiotherapy: Absorbed Dose, LET Spectra, and Cell Survival," Radiat. Res. 58, 1974, p. 293-328.

Bagshaw, M.A., Boyd, D.P., Fairbank, W.M., Kaplan, H.S., Li, G.C.C., Schwettman, H.A., and Palos, B.B., "Clinical Dosimetry for Negative Pi Mesons," Radiology 108, 1973, p. 197-202.

Berardo, P.A. and Zink, S., private communication.

Bichsel, H., Eenmaa, Weaver, K., and Wooton, P., "Attainable Accuracy in Fast Neutron Dosimetry Systems," Proc. of an Int. Workshop: Particle Radiation Therapy, American College of Radiology, 1975, p. 71-105.

Blaser, J.P., "First Experiences with the Biomedical Pion Beam at SIN - Introduction and Prospects," Atomkernenergie 27, 1976, p. 146-147.

Brenner, D.J. and Reading, D.H., "A Method for Measuring Neutron Spectra in a Stopping Pion Field," Nucl. Inst. Meth. 153, 1977, p. 137-144.

Brenner, D.J. and Smith, F.A., "Dose and LET Distributions due to Neutrons and Photons Emitted from Stopped Negative Pions," Phys. Med. Biol. 22, 1977, p. 451.

Cohen, M. and Martin, S.J., "Multiple-Field Isodose Charts," Vol. II. Atlas of Radiation Dose Distributions. Int. Atomic Energy Agency, Vienna, 1966.

Cooke, D.W. and Hogstrom, K.R., "Thermoluminescent Response of LiF and $Li_2B_4O_4$:Mn to Pions," submitted to Phys. Med. Biol. (1979).

Craddock, M.K., Erdman, K.L., and Sample, J.T., "Basic and Applied Research at the TRIUMF Meson Factory," Nature 270, 1977, p. 671-676.

Dicello, J.F., "Dosimetry of Pion Beams," Proc. of an Int. Workshop: Particle Radiation Therapy. American College of Radiology, 1975, p. 155-183.

Dicello, J.F., Fessenden, P., and Henkelman, R.M., "Dosimetry of Beams for Negative Pi-Meson Radiation Therapy," Int. J. Radiat. Oncol., Biol., Phys. 3, 1977, p. 299-306.

Dicello, J.F. and Zaider, M., "Investigation of the Microdosimetric Characteristics of Broad, Therapeutic Beams of Negative Pions at LAMPF," p. 469-481, Proc. of the 6th Symp. on Microdosimetry, Brussels, J. Booz and H.G. Ebert, eds., Harwood Academic Publishers, Ltd., London, 1978.

Dicello, J.F., Zaider, M., and Takai, M., "Some Physical Characteristics of Range-Modulated Beams of Pions," Proc. of the 3rd Meeting on Fundamental and Practical Aspects of the Application of Fast Neutrons and Other High LET Particles in Clinical Radiotherapy, The Hague, The Netherlands, 1978. In press.

Fowler, P.H. and Perkins, D.H., "The Possibility of Therapeutic Applications of Beams of Negative π^--Mesons," Nature 189, 1961, p. 524-528.

Geraci, J.P., Jackson, K.L., Christensen, G.M., Thrower, P.D., and Mariano, M., "RBE for Late Spinal Cord Injury Following Multiple Fractions of Neutrons," Radiat. Res. 74, 1978, p. 382-386.

Guthrie, M.P., Alsmiller, R.G., and Bertini, H.W., "Calculation of the Capture of Negative Pions in Light Elements and Comparison with Experiments Pertaining to Cancer Radiotherapy," Nucl. Instr. Meth. 66, 1968, p. 29-36; Nucl. Instr. Meth. 91, 1971, p. 669.

Hartmann, R., private communication to Klein (1978).

Hattersley, P.M., Muirhead, H., and Woulds, J.N., "Neutral Radiations Following Pion Capture in Complex Nuclei," Nucl. Phys. 67, 1965, p. 309-314.

Hogstrom, K.R. and Amols, H.I., "Pion *in vivo* Dosimetry Using Aluminum Activation, submitted to Med. Phys. (1979).

Hogstrom, K.R., Rosen, I.I., Gelfand, E., Paciotti, M., Amols, H.I., and Luckstead, S., "Calculation of Pion Dose Distributions in Water," submitted to Med. Phys. (1979).

Hogstrom, K.R., Smith, A.R., Simon, S.L., Somers, J.W., Lane, R.C., Rosen, I.I., Kelsey, C.A., von Essen, C.F., Kligerman, M.M., Berardo, P.A., and Zink, S., "Static Pion Beam Treatment Planning of Deep-Seated Tumors Using Computerized Tomographic Scans at LAMPF," submitted to Int. J. Radiat.. Oncol., Biol., Phys. (1979a).

Kellerer, A.M. and Rossi, H.H., "A Generalized Formulation of Dual Radiation Action," Radiat. Res. 75, 1978, p. 471-488.

Kellerer, A.M. and Rossi, H.H., "The Theory of Dual Radiation Action," Current Topics Radiat. Res. 8, 1972, p. 85-158.

Klein, U., "Measurement of Neutron Spectra from the Absorption of Stopped Negative Pions in the Biologically Interesting Nuclei ^{12}C, ^{14}N, and ^{16}O. Thesis, University of Karlsruhe, Karlsruhe, West Germany (1978). (In German).

Kligerman, M.M., Smith, A., Yuhas, J.M., Wilson, S., Sternhagen, C.J., Helland, J.A., and Sala, J.M., "The Relative Biological Effectiveness of Pions in the Acute Response of Human Skin," Int. J. Radiat. Oncol., Biol., Phys. 3, 1977, p. 335-339.

Kligerman, M.M., West, G., Dicello, J.F., Sternhagen, C.J., Barnes, J.E., Loeffler, K., Dobrowolski, F., Davis, H.T., Bradbury, J.N., Lane, T.F., Petersen, D.F., and Knapp, E.A., "Initial Comparative Response to Peak Pions and X Rays of Normal Skin and Underlying Tissue Surrounding Superficial Metastatic Nodules," Amer. J. of Roentgenology 126, 1976, p. 261-267.

Knapp, E.A., "Physical Properties of Charged Particle Beams for Use in Radiotherapy," Proc. of an Int. Workshop: Particle Radiation Therapy. American College of Radiology, 1975, p. 107-136.

Kramer, S. and Suntharalingam, N., "Low-LET Alternatives to Particle Irradiations," Int. J. Radiat. Oncol, Biol., Phys. 3, 1977, p. 343-349.

Laughlin, J.S., "Studies of Absorption of High Energy Electron Beams," p. 11-16, Proc. of the Symp. on High-Energy Electrons, A. Zuppinger and G. Porelti, eds., Springer-Verlag, New York, 1965.

Leighton, R.B., Principles of Modern Physics, McGraw-Hill, 1959, p. 637.

Li, G.C., Boyd, D., and Schwettman, H.A., "Pion Dose Calculations Suitable for Treatment Planning," Phys. Med. Biol. 19, 1974, p. 436-447.

Liska, D.J., "Pi Meson Range Shifter for Clinical Therapy," Rev. Sci. Instr. 48, 1977, 52-57.

Mechtersheimer, G., "Measurement of the Energy Spectra of Charged Secondary Particles from the Absorption of Stopped Negative Pions in Carbon Nuclei," Thesis, University of Karlsruhe, Karlsruhe, West Germany, 1978, (In German).

Paciotti, M.A., Bradbury, J.N., Helland, J.A., Hutson, R.L., Knapp, E.A., Rivera, O.M., Knowles, H.B., and Pfeufer, G., "Tuning of the First Section of the Biomedical Channel at LAMPF," IEEE Trans. Nucl. Sci., NS-22, 1975, p. 1784-1789.

Richman, C., Kligerman, M.M., von Essen, C., and Smith, A.R., "High LET Dose Measurements in Patients Undergoing Pion Radiotherapy," submitted to Radiology (1979).

Schillaci, M.E. and Roeder, D.L., "Dose Distributions Due to Neutrons and Photons Resulting from Negative Pion Capture in Tissue," Phys. Med. Biol. 18, 1973, p. 821-829.

Schneuwly, H., "Exotic Atoms," Proc. of the 1st Course of the Int. School of Physics of Exotic Atoms, G. Fiorentini and G. Torelli, eds., Erice, Italy, April 1977.

Shortt, K.R. and Henkelman, R.M., "A Charge Collector to Determine the Stopping Distribution of a Pion Beam," Phys. Med. Biol. 23, 1978, p. 495-498.

Smith, A.R., Rosen, I.I., Hogstrom, K.R., Lane, R.G., Kelsey, C.A., Amols, H.I., Richman, C., Berardo, P.A., Helland, J.A., Kittell, R. S., Paciotti, M.A., and Bradbury, J.N., "Dosimetry of Pion Therapy Beams," Med. Phys. 4, 1977, 408-413.

Suit, H.D., Goitein, M., Tepper, J.E., Verhey, L., Koehler, A.M., Schneider, R., and Gradondas, E., "Clinical Experience and Expectation with Protons and Heavy Ions," Int. J. Radiat. Oncology, Biol., Phys. 3, 1977, p. 115-125.
Turner, J.E., Dutrannois, J., Wright, H.A., Hamm, R.N., Baarli, J., Sullivan, A.H., Berger, M.J., and Seltzer, S.M., "The Computation of Pion Depth-Dose Curves in Water and Comparison with Experiment," Radiat. Res. 52, 1972, p. 229-246.
Wright, H.A., Hamm, R.N., and Turner, J.E., "PION-1. A Monte Carlo Computer Program for Calculations with Negative Pion Beams," Radiat. Res. 79, 1979, p. 1-21.
Van Dyke, J. and MacDonald, J.C.F., "Charge Deposition from High Energy Electron Beams," Radiat. Res. 50, 1972, p. 20-32.
Zaider, M. and Dicello, J.F., "RBEOER: A FORTRAN Program for the Computation of RBEs, OERs, Survival Curves, and the Effects of Fractionation Using the Theory of Dual Radiation Action," Los Alamos Scientific Laboratory Report LA-7196-MS (1978).
Zaider, M., Dicello, J.F., and Bichsel, H., unpublished.
Zink, S. and Perardo, P., "Treatment Planning with Pions: The PIPLAN Approach," Los Alamos Scientific Laboratory Report No. LA-UR 79-2304 (1979).

GLOSSARY

The authors have included, for the benefit of those unfamiliar with the jargon of this field, explanations of some of the more common terms. Formal definitions can be found in the appropriate reports of the International Commission on Radiation Units and Measurements (ICRU).

delta rays - high energy secondary electrons produced along the track of a directly ionizing particle, which are themselves capable of producing ionizing events.

dose - the energy per unit mass absorbed at a point in a material Note that it is the energy absorbed by the material, not the energy lost by a particle passing through the material.

lineal energy (y) - the energy deposited by an individual event in a microscopic volume of material (usually tissue) divided by the mean path length through that volume.

linear energy transfer (LET) - the energy absorbed in a medium as a result of a charged particle passing through the material per unit path length of the particle.

microdosimetry - the study of the physical mechanisms involved in energy deposition in matter at the microscopic scale from radiation. (Microdosimetry is not the development of small dosimeters.)

oxygen enhancement ratio (OER) - the ratio of the doses which produce equal effects between hypoxic and aerated cells. For example, $\text{OER(PIONS)} = \text{DOSE}_\pi(\text{HYPOXIC})/\text{DOSE}_\pi(\text{AERATED})$.

radiation quality – those physical characteristics of the radiation
    other than absorbed dose which can alter the response of
    biological systems.

relative biological effectivensss (RBE) – the ratio of the dose for
    a reference radiation, such as 250 kVp x rays or cobalt-60, to
    the dose for a specified radiation which produces the same
    biological response. For example,
    RBE(PIONS) = DOSE(250 kVp x rays)/DOSE(PIONS)

stopping power (dE/dx) – the energy <u>lost</u> by a charged particle per
    unit path length of the particle as it passes through a
    specified material.

RADIOBIOLOGICAL EFFECTS OF PIONS

J. F. Dicello

University of California
Los Alamos Scientific Laboratory
Los Alamos, New Mexico 87545

INTRODUCTION

Biological studies with negative pions have been motivated almost exclusively by their therapeutic potential. There are basically two biological characteristics of negative pions which could prove useful for cancer management. 1) Negative pions have a lower oxygen enhancement ratio (OER) in the Bragg region as compared with conventional radiations, and the variation in sensitivity throughout the cell cycle is reduced with pions. In certain cases, this could produce an increased effect on malignant cells relative to normal cells. 2) The relative biological effectiveness (RBE) of pions in the Bragg region can be greater than that in the entrance region. This coupled with the good localization properties of pions could minimize the damage to normal tissues outside of the treatment volume.

The physical mechanisms by which negative pions deposit energy near the end of their range are unique. Nevertheless, there is no evidence that this results in any biological effects not capable of being induced by other heavy particles.

Although the therapeutic capabilities of negative pions can be ascertained ultimately only with clinical data, the interpretation of the clinical results and comparison with other high LET radiations will depend on our understanding of those fundamental parameters responsible for the observed responses. Therein lies the necessity for basic research with simple biological systems.

This article will review some of the more general biological characteristics of pions. Conforming to the format of the previous

465

article, we will systematically examine the properties of one or
two beams, rather than attempt to review the literature.

SURVIVAL AS A FUNCTION OF DOSE

Simple cellular systems, with cell death as the endpoint, are
powerful tools for the comparison of radiations. In many respects,
they are the dosimeters of biology, and much of what we know about
pions has been derived from such *in vitro* systems.

Three beams examined by Raju et al. (1977, 1978c) are shown
in Fig. 1. These central-axis depth dose curves are quite similar
to the beams discussed in the previous paper. The lateral dimensions are narrower, however, in order to improve the dose rate.
These beams are approximately 2 cm wide at the 80% isodose level
as compared with about 10 cm previously. Survival of human kidney
cells (T_1) as a function of dose along the central axis has been
measured (Raju et al., 1977) and some results are shown in Fig. 2.

Plateau pions are about as effective as x rays, although the
cell killing at the beam entrance of the 10-cm beam in Fig. 1 is
slightly greater. This is probably because the minimum depth
available with this beam was close to the proximal edge of the peak.
Survival decreases with depth in the broad beam because of the increasing percentage contribution to the dose from stars. However,
the maximum killing, achieved at the distal portion, is less than
that at the center of the narrow beam. This is most likely a result of the higher fractional contribution to the dose from muons
and electrons in the broader beam.

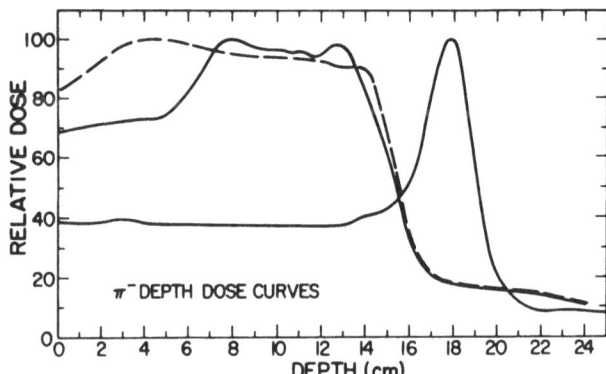

Fig. 1. Depth-dose distribution of negative pions. The 5- and 10-
         cm wide peaks were obtained by using a dynamic range
         shifter (Raju et al., 1977).

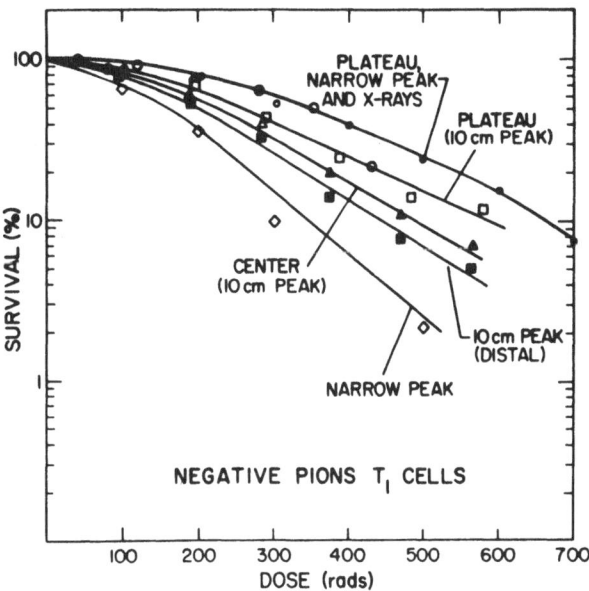

Fig. 2. Percent cell survival vs dose at the plateau and peak for
a narrow peak beam, and at the plateau, peak center, and
peak distal for a 10-cm wide peak (Raju et al., 1977).

REPAIR OF SUBLETHAL DAMAGE

     Note that, in Fig. 2, all of the survival curves have a signi-
ficant shoulder at lower doses, i.e., the same incremental change
in the dose delivered produces less killing at lower doses. If
this shoulder is a result of repairable sublethal damage, the same
total dose delivered in fractions should be less effective because
the cells will have time to repair. The data in Fig. 3 (Raju, et
al., 1977) show that this is the case. Realize that if pions were
totally high LET radiation, there would be no shoulder from sublethal damage and fractionated doses would result in the same survival curve. In fact, most of the dose from pions is a result of
lower LET events so that the effectiveness of fractionation is not
surprising. Typically, 75% or more of the dose in the peak region
is from events with a LET less than 10 keV/μm, which encompasses
the region over which the LET is distributed for cobalt-60 or x
rays.

CELL-CYCLE SENSITIVITY

     If, as the data suggest, there is sublethal damage associated
with pions, it is probable that the response of cells will depend
strongly on their mitotic stage. In order to determine the age
response of cells to pions, Raju et al. (1978b) synchronized CHO

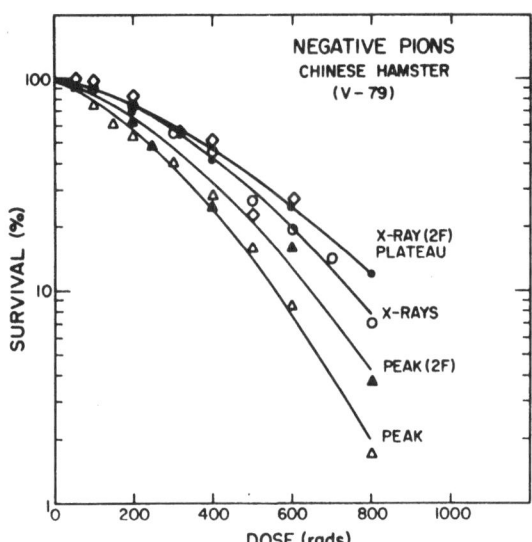

Fig. 3.   Percent cell survival (V79) vs dose for single and two
          fractionated exposures at the pion peak (narrow) and for
          x rays (Raju et al., 1977).

cells at the G/S boundary with hydroxyurea. When the medium with
hydroxyurea is replaced with drug-free medium, the cells begin
again to progress through their cycle in phase. Their response
can then be measured as a function of time after synchronization.

     The results, presented in Fig. 4, clearly show that sensitivity
as a function of cell age is reduced but not completely eliminated.

SURVIVAL AS A FUNCTION OF DEPTH

     The data presented in Fig. 2 show that survival is not only a
function of dose, but also of position in the beam. This is because of changes occurring in radiation quality. Such effects can
be examined in more detail with a technique developed by Skarsgard
and Palcic (1974) with which they suspend cells in gelatin contained
in a long plastic tube. The tube is irradiated, after which it is
sliced into thin sections. The sections are dissolved in medium
and then assayed for survival according to standard procedures.

     Some results of Raju et al. (1977) are shown in Fig. 5 where
the survival is plotted as a function of depth and dose. Cell
killing in the peak region is greater than in either the entrance
or exit regions with the relative differences increasing with increasing dose. Correspondingly, at equal doses there is greater
killing in the peak than in the plateau as we saw in Fig. 2.

Although the dose is approximately constant in the peak region, there is a definite decrease in survival with depth apparently because of the change in the radiation quality. This is exemplified in Fig. 6 where Raju et al. (1978c) have compared survival for a beam having constant dose in the stopping region with that for a beam with a constant number of stopping pions and a beam sloped with depth to produce a uniform biological response. The constant

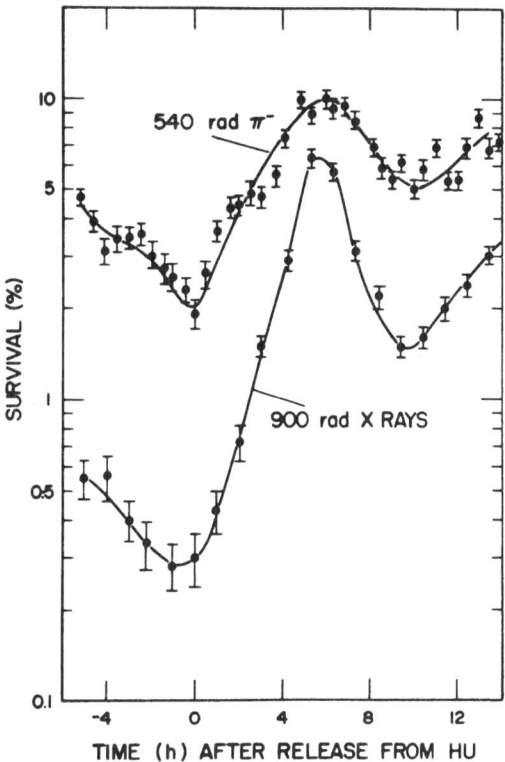

Fig. 4. Percent cell survival after exposure to 900 rad of x rays or 540 rad of peak pions as a function of time after release from hydroxyurea (Raju et al., 1978b).

dose with a single port requires a greater number of stars at greater depths resulting in a decrease in survival. The constant star dose approximates a constant high LET dose but requires greater total dose proximally because those pions stopping downstream will also deposit energy upstream as they pass through. Therefore, there would be greater killing in the proximal region.

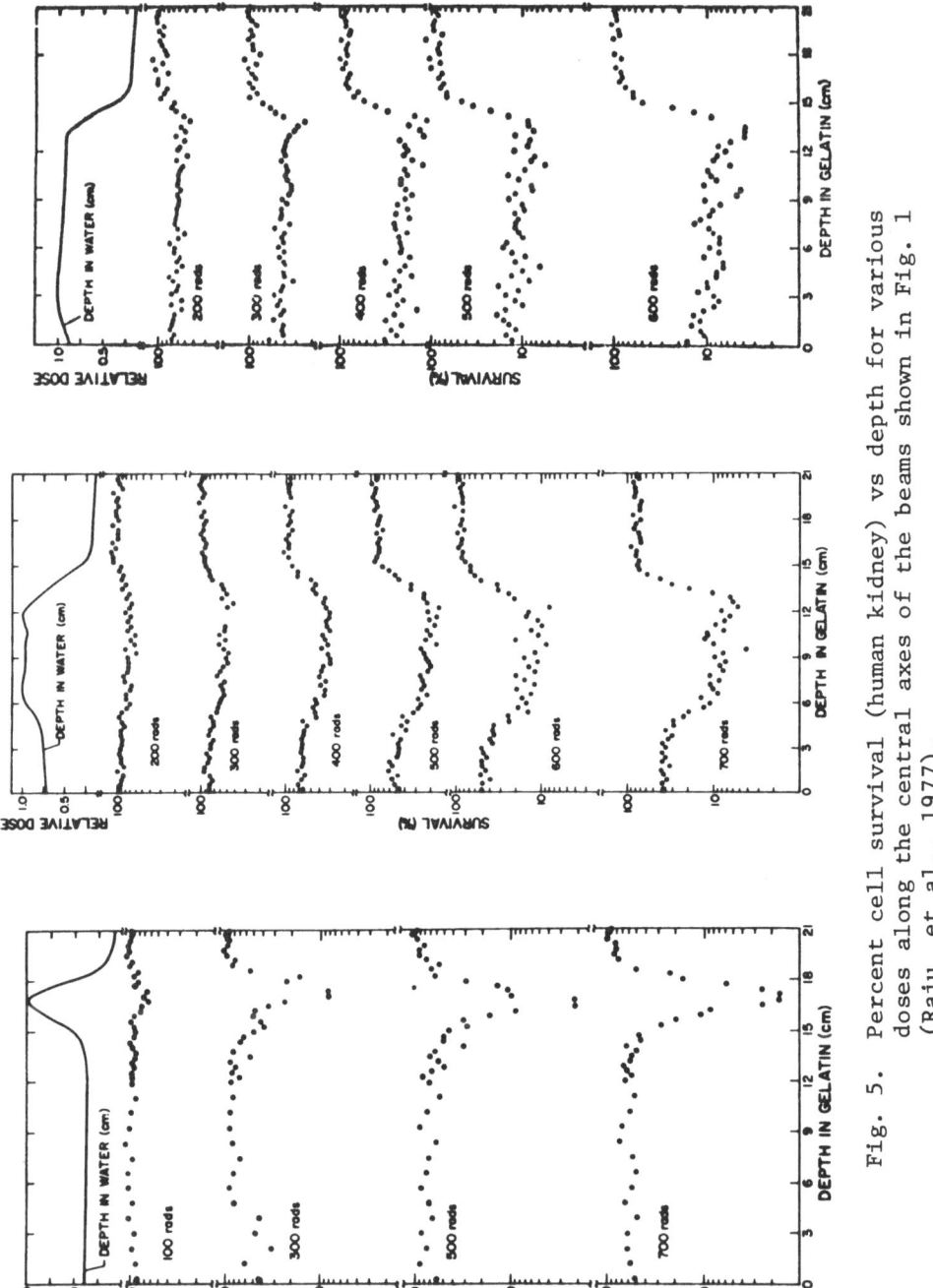

Fig. 5.   Percent cell survival (human kidney) vs depth for various
          doses along the central axes of the beams shown in Fig. 1
          (Raju et al., 1977).

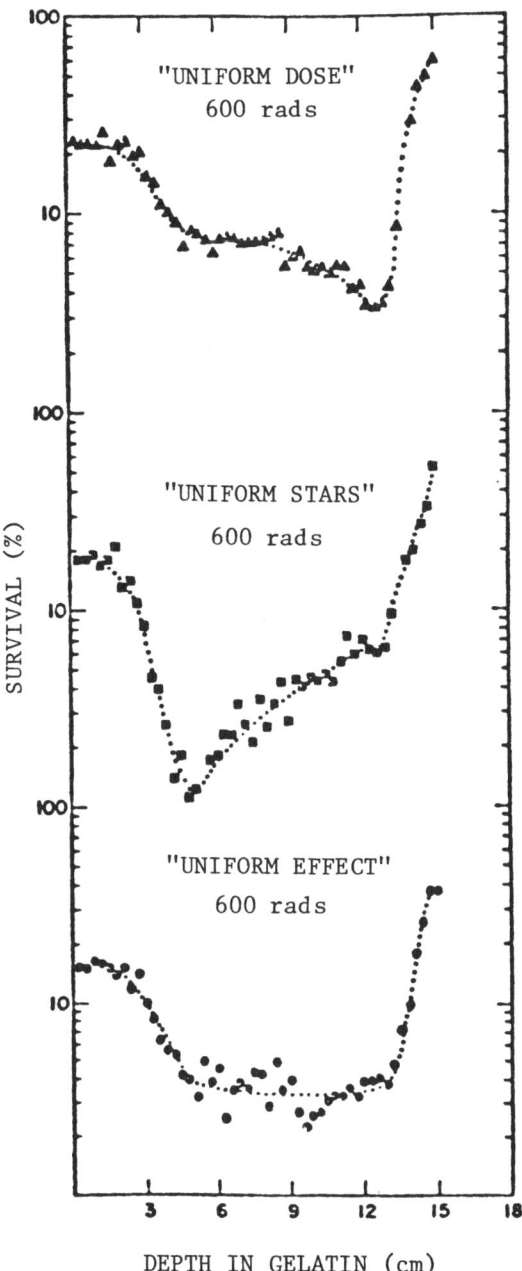

Fig. 6. Cell survival as a function of depth for the three beams
described in the text (Raju et al., 1978c).

The third case represents the situation in which the total
dose and the star dose are adjusted to compensate for the changing
biological response. But we see in Fig. 5 that the relative changes
in biological response are a function of the dose level. Therefore,
what has been achieved is a uniform effect at a given survival
level. It may not be uniform at another survival level. This
problem can be overcome by using the beam of "uniform effect" with
opposed, overlapping fields. Then survival would be more uniform
at any dose. Obviously, the use of opposed, overlapping field
would improve the uniformity of any beam. However, unless the
single port field is shaped for uniform response itself, there
could be an asymmetry in survival from the center of the stopping
region to the edges.

RELATIVE BIOLOGICAL EFFECTIVENESS (RBE)

The change in response as a function of dose and position can
influence the choice of maximum tumor dose, if it is assumed that
normal tissue response is a limiting factor. In many cases, data
are not presented in terms of survival curves, but rather in terms
of RBEs, i.e., the ratio of the dose of a reference radiation,
such as x rays or cobalt-60, to the dose of pions which produces
the same biological effect.

The advantages of an analysis in terms of RBE are two-fold.
First of all, the use of dose ratios eliminates some absolute uncertainties associated with the experiments themselves. Thus interpretation and comparison of data become more precise. Secondly,
much of our knowledge concerning the interaction of radiations in
tissue is empirically derived from experience with photons. It
behooves us to use RBEs with a certain degree of caution, however,
particularly in the case of high LET radiations, because changes
may be a result of the radiation under investigation, the reference
radiation, or a combination of both. In many cases, the RBE is
changing at least partially from changes caused by the reference
radiation. A classical example of this situation is in a comparison
of an exponential survival curve, such as that obtained with low-
energy neutrons, with the corresponding curves for x rays. For a
given increment in dose, the change in the logarithm of survival
is a constant, irrespective of the absolute value of the doses
delivered. Yet the RBE for neutrons is increasing dramatically at
low doses because of the shoulder on the survival curve for x rays.
With this *caveat* in mind, let us consider RBEs for pions.

The RBE for passing pions (plateau) is close to unity for most systems. The biological effects are comparable to x rays. There are some data which are at least suggestive of both higher or lower values. The higher values may be caused by the small component of high LET radiation from inflight nuclear interactions or neutrons. Lower values may result from the extremely low LET_∞ (stopping power) for energetic pions. For example, 80-MeV pions have a stopping power in water of only 270 eV/μm.

The RBE in the stopping region is generally greater than one, the actual magnitude depending on the volume and shape of the stopping region (see Fig. 2 and 6). As the depth over which the pions stop is increased (i.e., as the peak region is broadened), the fractional contribution of charged particle dose from stars decreases. This in itself would generally cause the RBE to decrease. However, as the stopping volume increases from changes in either the depth or the lateral dimensions, the fractional dose from neutrons becomes larger (Schillaci and Roeder, 1973, Dicello, et al., 1978). The greater high LET dose caused by more neutrons can at least partially compensate for the reduced high LET dose from charged particles. Moreover, the longer mean free path of the more energetic neutrons will tend to reduce variations in the RBE.

The RBEs beyond the range of the pions tend to be close to one because most of the dose comes from the long-ranged muons and electrons, although there are some neutrons, too.

Finally, the RBEs tend to increase with decreasing dose. This, of course, is not a result of increased absolute effectiveness of low doses of pions. Quite the contrary. The shoulder on the survival curves show that pions are less effective in producing lethal damage at lower doses. This simply means that at lower doses pions are increasingly more effective than x rays with correspondingly less sublethal damage.

There are no published data of RBEs for fractionated doses. If it is assumed that there is greater killing and less sublethal damage for pions because of the high LET dose, everything else assumed to be equal, we would expect the RBE in the peak to increase with increasing number of fractions. Preliminary data for mouse skin by Raju and his associates, Withers, et al., and Goldstein, et al. do indicate that the RBE is indeed increasing with increasing number of fractions.

OXYGEN ENHANCEMENT RATIO (OER)

Tumor response may depend on the degree of hypoxia in that tumor, particularly in the case of large tumors because cells usually are more resistant to radiation when they lack sufficient oxygen. High LET radiations are usually more effective in killing hypoxic cells because of the greater amount of energy deposited locally in the particular cell or sensitive site traversed per unit macroscopic dose absorbed. The ratio of the doses to produce equal effects between hypoxic and aerated cells is the oxygen enhancement ratio (OER). (It is generally assumed that hypoxic cells will be, at best, equally sensitive to radiation.)

From Fig. 7 we see that, indeed, hypoxic cells are less susceptible to pions than are aerated cells. But pions are better than x rays for inducing lethal damage in hypoxic cells. Raju et al. (1979) reported OERs between 2.4 and 2.1 for peak pions as compared with an OER of 2.9 for x rays. At the same time, the RBE varied between 1.2 and 1.7. That is, a 40% change in the RBE is correlated with only a 15% change in the OER. Raju et al. concluded that the increased neutron dose seemed to compensate for the decreased number of stars, noting that the OER is more sensitive to small amounts of high LET radiation than is the RBE.

The OER in the plateau was reported to be 2.4, significantly lower than that for x rays. An interesting observation is that the smaller OER in the plateau was because of less sensitivity of aerated cells to pions, with hypoxic cells being about equal in sensitivity to plateau pions and x rays. Perhaps this is because of nuclear interactions in the plateau or dose-rate effects. Regardless, most of the cells in the plateau region will usually be well oxygenated so this OER may be of less significance.

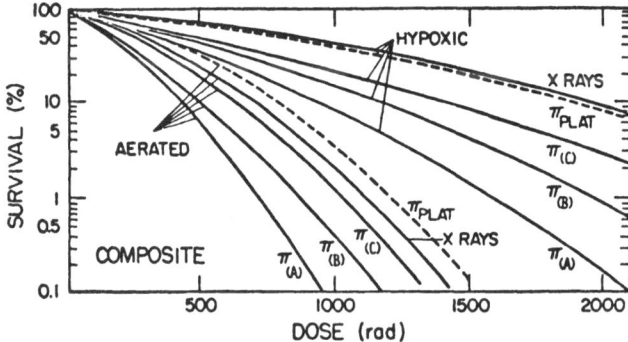

Fig. 7. Survival curves for aerated and hypoxic V79 cells. The notation refers to depth dose curves similar to those in Fig. 1 (Raju et al., 1979).

All of the published data on OERs for pions are for single doses. It is known that, as the number of fractions increases, conventional radiations become more effective in killing cells which were hypoxic, presumably because of progressively more available oxygen available to surviving cells as the total number of cells diminishes (a phenomenon known as reoxygenation). It is usually assumed that there will be less of a therapeutic gain with high LET radiations for fractionated doses because the oxygen effect is diminished already. The gain, however, may still be significant.

IN VIVO ACUTE RESPONSE

In vitro systems measure an effective dose under relatively simple circumstances which may not extrapolate to *in vivo* cases. Ultimately, the goal is to project our understanding of these results to *in vivo* mammalian systems including man. Until recently, the dose rates with pion beams were too low to be used with most *in vivo* systems, and data are just beginning to become available.

Normal Tissue Response

Kligerman et al. (1976, 1977) investigated the acute response of human skin with a narrow beam to establish a base for normal tissue. Data were obtained for 13 fractions in 14 days. An RBE of about 1.4 was determined by calculating from the slopes of the dose response curves the doses of pions and x rays which produced equivalent skin reactions 43 days after the first treatment.

Tumor Response

Kligerman et al. (1975) irradiated spontaneous mammary adenocarcinomas in C_3H mice in the peak region of a narrow pion beam ($\Delta p/p \sim 0.02$). They obtained an RBE of 1.7 ± 0.2 (compared with 250 kVp x rays) for five daily fractions.

In another series of experiments with human metastatic skin nodules, Kligerman et al. (1978) concluded that peak pions were 37% more effective in delaying regrowth of the nodules as compared with 100 kVp x rays for comparable skin injuries. The doses were delivered in 13 fractions over 14 days.

LATE EFFECTS

There is not necessarily a correlation between early and late effects. Severe late effects, particularly when there are only mild acute reactions, can be a dose-limiting factor. Therefore, thorough studies in this area are needed. Several groups have initiated such projects but no data have yet been reported.

AUXILIARY PROBLEMS

There have been many ancillary studies complementing those discusssed in the previous sections.

Because interactions of pions, including star formation, is a function of the atomic number, Z, it is possible that there could be strong dose-interface effects at the surfaces of inhomogeneities. For example, calculations of Brenner and Smith (1977) show that the dose in a bone phantom from neutrons will be 24% lower than in a tissue phantom. However, the neutron dose in a 2.5 x 2.5 cm^2 bone inserted in a tissue phantom to simulate a spinal column is only about 5% less than the dose in the surrounding tissue indicating that small inhomogeneities may not significantly perturb the total dose. Nevertheless, one might expect different LET distributions for bone or bone-tissue interfaces because of the decreased proportion of hydrogen in bone. Raju et al. (1978a) reported that there was no difference observed in killing of V79 cells plated on either plastic or glass. Glass has elements of high atomic number (Z) as are present in bone while plastics simulate soft-tissue composition. These data should be interpreted only to mean that interface problems may be small for systems with the same radioresistivity.

Amols et al. (1978) compared V79 cell killing in and outside of the treatment region in order to determine if there would be a high RBE outside of the stopping volume caused by neutrons. They obtained RBEs of 1.1 downstream from the stopping region and 1.5 under the collimator as compared with 1.3 in the middle of the peak. Two points are worth noting here. First, the maximum RBE was observed, not in the middle of the stopping region, but rather under the collimator, which is an abundant source of neutrons. Second, this experiment was motivated by prior microdosimetric measurements which showed only slight differences in radiation quality. RBEs for V79 cells calculated on the basis of the microdosimetry predicted the subsequently observed biological responses.

Raju et al. (1978) have combined pion beams with misonidazole, a sensitizer in order to determine if the combination was more effective in reducing the radioresistivity of hypoxic cells. The results indicated that relatively low concentrations of misonidazole (0.4 mM) in conjunction with negative pions was more effective than with x rays or pions alone. Again, this is probably a result of the interaction of the drug and sublethal damage by the low LET component of the pion beam.

CLINICAL TRIALS

In addition to skin and nodule experiments, phase I and II trials have been in progress at Los Alamos with tumors in a variety

of sites having been treated. Preliminary results and conclusions
have been presented by Kligerman et al. (1979) and phase III randomized trials should begin shortly.

ACKNOWLEDGEMENTS

     Much of the radiobiological data in this paper was the result
of work by M. R. Raju and his colleagues. The author is grateful
to Dr. Raju for his cooperation and support in the preparation of
this manuscript. I also would like to acknowledge the help and
suggestions of Drs. James N. Bradbury, D. J. Brenner, and Marco
Zaider.

REFERENCES

H. I. Amols, J. F. Dicello, and M. Zaider, "The RBE at Various
Positions In and Near a Large Negative Pion Beam," Proc. of the
6th Symposium on Microdosimetry, J. Booz and H. G. Ebert, eds.
(Harwood Academic Publishers, Ltd., 1978).

D. J. Brenner and F. A. Smith, "Dose and LET Distributions due to
Neutrons and Photons Emitted from Stopping Negative Pions," Phys.
Med. Biol. 22: 451 (1977).

J. F. Dicello, M. Zaider, and M. Takai, "Some Physical Characteristics of Range-Modulated Beams of Pions," Proc. of the 3rd Meeting
on Fundamental and Practical Aspects of the Application of Fast
Neutrons and Other High LET Particles in Clinical Radiotherapy,
The Hague, The Netherlands, 1978. In press.

L. S. Goldstein, private communication.

M. M. Kligerman, J. F. Dicello, H. T. Davis, R. A. Thomas, C. J.
Sternhagen, L. Gomez, and D. F. Petersen, "Initial Comparative
Response of Experimental Tumors to Peak Pions and X Rays," Radiology
116: 181-182 (1975).

M. M. Kligerman, C. F. von Essen, M. K. Khan, A. R. Smith, C. J.
Sternhagen, and J. M. Sala, Cancer 43: 1043-1051 (1979).

M. M. Kligerman, J. M. Sala, S. Wilson, and J. M. Yuhas, "Investigation of Pion Treated Human Skin Nodules for Therapeutic Gain," Int.
J. of Radiat. Oncology, Biol., Phys. 4: 263-265 (1977).

M. M. Kligerman, A. Smith, J. M. Yuhas, S. Wilson, C. J. Sternhagen,
J. A. Helland, and J. M. Sala, "The Relative Biological Effective-
ness of Pions in the Acute Response of Human Skin," Int. J. Radiat.
Oncology, Biol., Phys. 3: 335-339 (1977).

M. M. Kligerman, G. West, J. F. Dicello, C. J. Sternhagen, J. E.
Barnes, K. Loeffler, F. Dobrowolski, H. T. Davis, J. N. Bradbury,
T. F. Lane, D. F. Petersen, and E. A. Knapp, "Initial Comparative
Response to Peak Pions and X Rays of Normal Skin and Underlying
Tissue Surrounding Superficial Metastatic Nodules," Amer. J. Roent.
126: 261-267 (1976).

M. R. Raju, H. I. Amols, E. Bain, S. G. Carpenter, J. F. Dicello,
J. P. Frank, R. A. Tobey, and R. A. Walters, "Biological Effects of
Negative Pions," Int. J. of Radiat. Oncology, Biol., Phys. 3: 327-
334 (1977).

M. R. Raju, H. I. Amols, and S. G. Carpenter, "A Combination of
Sensitizers with High LET Radiations," Brit. J. Cancer 37: 189-193
(1978).

M. R. Raju, H. I. Amols, J. B. Robertson, "Effect of Negative Pions
on Cells Plated on Glass and Plastic Surfaces," Radiat. Res. 75:
439-442 (1978a).

M. R. Raju, H. I. Amols, S. G. Carpenter, R. A. Tobey, and R. A.
Walters, "Age Response of CHO Cells Exposed to Negative Pions,"
Radiat. Res. 76: 219-223 (1978b).

M. R. Raju, H. I. Amols, E. Bain, S. G. Carpenter, R. A. Cox, and
J. A. Robertson, "Cell Survival as a Function of Depth for Modulated
Negative Pion Beams," Int. J. Radiat. Oncology, Biol., Phys. 4:
841-844 (1978c).

M. R. Raju, H. I. Amols, E. Bain, S. G. Carpenter, R. A. Cox, and
J. E. Robertson, "OER and RBE for Negative Pion Beams of Different
Peak Widths," Brit. J. Radio. 52: 494-498 (1979).

M. E. Schillaci and D. L. Roeder, "Dose Distributions Due to Neutrons and Photons Resulting from Negative Pion Capture in Tissue,"
Phys. Med. Biol. 18:821-829 (1973).

L. D. Skarsgard and B. Palcic, "Pretherapeutic Research Programs
at π‾ Meson Facilities," Proc. of the XIIIth Int. Congress of
Radiology (Madrid), Int. Congress Series No. 339, Radiology 2:
447-454 (1974).

R. H. Withers, private communication.

HEAVY-ION DOSIMETRY

Walter Schimmerling

Division of Biology and Medicine
Lawrence Berkeley Laboratory
Berkeley, CA 94720

INTRODUCTION

A comprehensive and accurate definition of the physical characteristics of a radiation field is a necessary, but not sufficient, condition for understanding (and predicting) the probability of observing a biological effect. Part of the difficulty of doing this stems from the fact that very often the biological final state accessible to observation is the end result of a long sequence of events following the irradiation, when most of the information has been distributed among many degrees of freedom. Demography is not a very good starting point for the study of photosynthesis, yet that is, in essence, what we are often forced to do.

In some cases, the observed effect only depends on average properties of the radiation field, e.g., the absorbed dose. Thus, for example, we are able to predict that 50 percent of the human population exposed to approximately 500 rads of low LET radiation will die. However, even here there are effects that cannot be attributed to dose, and nobody could say for sure what the LD50 of neon ions for humans might be. It is one of our objectives to ensure that experimental data on this topic never come into existence.

Even within its limitations, the concept of absorbed dose represents significant progress, especially if one takes into account an understanding of the actual averaging procedures used in arriving at it, to which Harald Rossi and his colleagues have contributed so much. The measurement of absorbed dose alone is also sufficiently demanding that an entire subfield has sprung

up around it--the science and art of radiation dosimetry. However, none of us who have an interest in this field are content
with simply measuring one quantity when we know, as stated
above, that a deeper understanding is required.

Accordingly, this lecture will deal with some of the more
important physical characteristics of relativistic heavy ions
and their measurement, with beam delivery and beam monitoring,
and with conventional radiation dosimetry as used in the operation of the BEVALAC biomedical facility for high energy heavy
ions (Lyman and Howard, 1977; BEVALAC, 1977). Even so, we shall
not be able to discuss many fundamental aspects of the interaction of relativistic heavy ions with matter, including
important atomic physics and radiation chemical considerations,
beyond the reminder that such additional understanding is
required before an adequate perspective of the problem can be
attained.

Heavy ion beams were first accelerated to relativistic
energies in the Princeton Particle Accelerator in 1971, following suggestions made by Cornelius Tobias thirty year earlier
(Vosburgh, 1973). At present, the only operating facility is
the Berkeley BEVALAC, a schematic diagram of which is shown in
Fig. 1. The SuperHILAC is a heavy ion linear accelerator, which
is operated in a time-shared mode as the injector of the
Bevatron, a weak-focusing synchrotron. The BEVALAC has accelerated beams with masses up to iron. Beams are first accelerated
to 8.5 MeV/nucleon in the SuperHILAC, then bent into an 800-ft
transfer line, where they are transported into the Bevatron for
acceleration up to energies of 2 GeV/nucleon. The beams used
for biomedical research are carbon, neon, and argon, with
energies determined by their desired range in tissue (400 to
500 MeV/nucleon). Dose rates for 20-cm diameter field sizes are
typically between 100 and 300 rad/min, but focused beams can be
obtained with dose rates of the order of 5000 rad/min. The beam
is pulsed, with a frequency dependent on the maximum field
strength required for each beam energy, typically at a rate of
15 pulses per minute. The beam has a microscopic time structure
(that can be smoothed out upon extraction) of typically
40-70 nsec "buckets" every 400-700 nsec.

The BEVALAC Biomedical Facility is an area specially
designated for tumor, tissue, cellular, molecular, neuro-
developmental and space radiobiology, radiography, and radiological physics. The Biomedical Facility has its own control
room, three irradiation caves, and two preparation rooms.
Figure 2 is a floor plan of the facility, showing two of the
irradiation areas. The third irradiation area, known as the
"minibeam" cave (because it is intended for use with narrowly
focused beams), is on top of the shielding, where the beam is

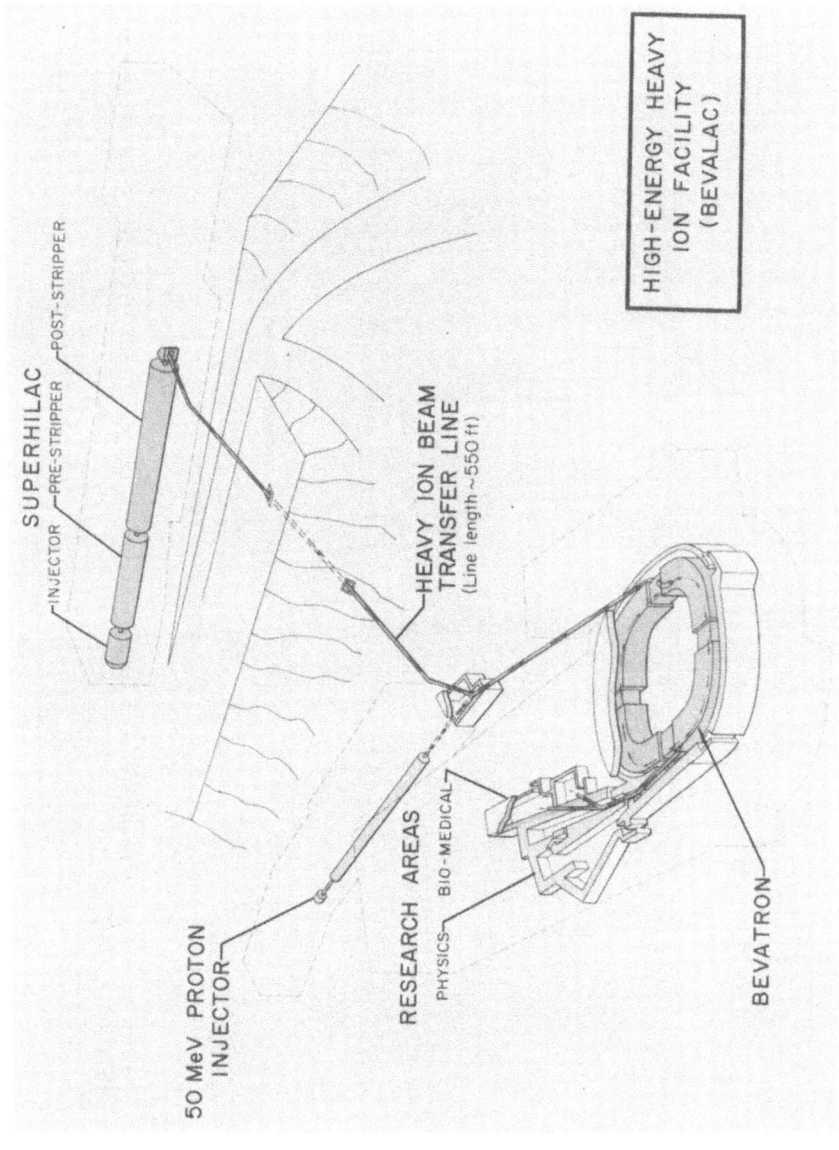

Fig. 1. Schematic diagram of the BEVALAC high-energy heavy-ion facility. Reprinted, by permission, from BEVALAC, 1977.

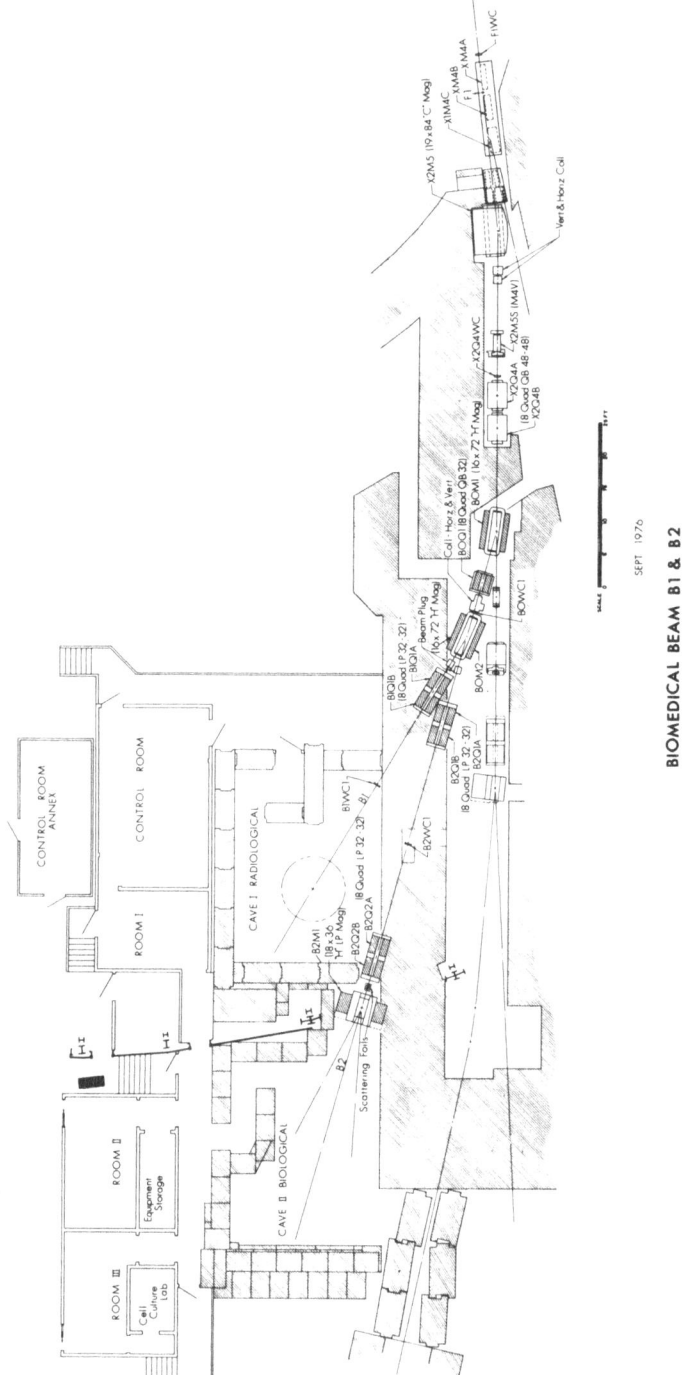

BIOMEDICAL BEAM B1 & B2

Fig. 2.   Floor plan of the BEVALAC Biomedical Facility. Reprinted, by permission, from BEVALAC, 1977.

incident at an angle of 30⁰ from below. Cave I is intended for therapy and diagnosis, and has a patient positioner and related equipment (Alonso et al., 1979a). Cave II is intended for physical, chemical, and biological research. The minibeam is intended for neurophysiological studies with radiation, and consists of an electrostatically shielded room.

SOME CHARACTERISTICS OF HEAVY ION BEAMS

Physical characteristics of high energy heavy ions that are of interest in their application to biomedical research are summarized in Table 1 (Tobias, 1973). In practice, these properties are influenced by the following factors:

1. The mean free path for nuclear interactions of heavy ions with energies of therapeutic interest is comparable to their range, as shown in Fig. 3. The calculated curves are based on geometrical cross sections and range-energy tables. For the energies of interest here, it may be seen that approximately two-thirds of the incident primary beam will interact with nuclei in the absorbing material to produced secondary fragments. An experimental Bragg peak, illustrating the geometric dose distribution, is shown in Fig. 4, including calculated contributions due to secondary particles (J. Lyman, private communication).

Table 1.   Some Important Characteristics of High Energy Heavy Ion Beams

---

1. Stable particles, i.e., intensity, energy, and beam purity are controlled by accelerator.

2. Well-defined range in matter, dependent only on initial energy, i.e., minimal straggling.

3. Reduced multiple scattering which allows the design of sharply defined beams.

4. Well-defined geometrical pattern of energy loss in matter. For monoenergetic beam, leads to increasing dose as a function of depth, up to a peak region located at a depth dependent on the initial energy, with a sharp fall-off at the distal end of the beam.

5. High LET, leading to reduced oxygen enchancement ratio and higher biological effectiveness.

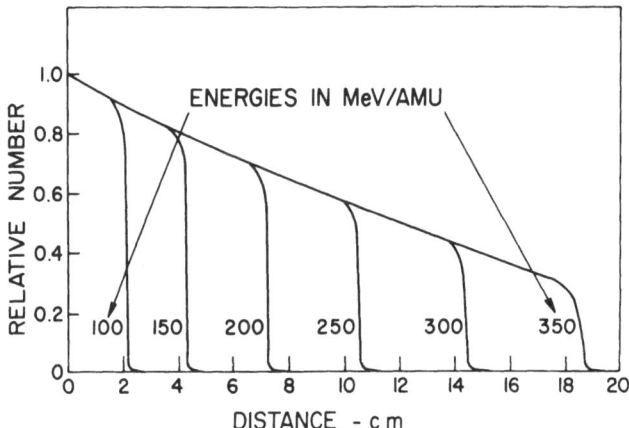

Fig. 3. Mean free path for nuclear interactions of heavy ion
         with energies of therapeutic interest compared to
         their range. Reprinted, by permission, from Schim-
         merling et al., 1973.

2.   Multiple scattering, mainly due to many independent
small-angle deflections by Coulomb scattering with the nuclei in
the medium, results in beam broadening. This can be used to
advantage in the beam flattening schemes discussed below, but
constrains the design of collimators and air paths in the beam.
Secondary fragments, in addition, will contribute a penumbra due
to the combination of their own multiple scattering distribution
with the angular distribution resulting from nuclear interactions.

3.   The beams produced at accelerators are not monoenergetic. When the beam range is controlled by a variable
absorber, the initial energy spread is broadened by straggling.
The fact that slower particles lose energy at a higher rate
results in a slightly asymmetric energy distribution, especially
near the Bragg peak, and a broadening of the peak that depends
on the amount of absorber interposed.

4.   The finite thickness of ionization chambers commonly
used to measure Bragg curves results in a measured peak width
determined by the average ionization in the chamber.

5. The relative biological effectiveness (RBE) is related to the LET_∞ as shown in Fig. 5, showing a compilation of results obtained for cells (E. Blakely, private communication). However, many sources of variability still exist, and LET by itself is insufficient to predict biological effects completely.

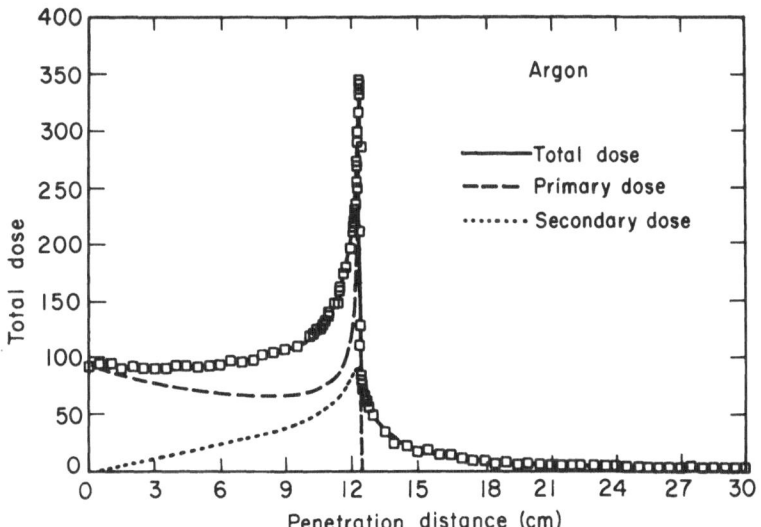

Fig. 4. Bragg curve for an unmodified argon beam. Reprinted, by permission, from Lyman and Howard, 1977a.

The inadequacy of LET_∞ as a predictor of biological effect may be seen more clearly in Fig. 6, showing the inactivation cross sections for human kidney cells irradiated with carbon, neon, and argon beams (Blakely et al., 1979). According to model predictions, the curves should be a function of Z^4/β^4, where Z is the charge of the ion and β its velocity (in units of the speeds of light, c). In the case of hypoxic cells, the data are, rather, a function of $Z^4/\beta^{4.6}$, indicating a velocity dependence of biological effects not described by LET_∞ alone. Blakely et al. conclude that the description of the biological effects of heavy ions requires at least three variables, e.g., fluence, velocity, and charge.

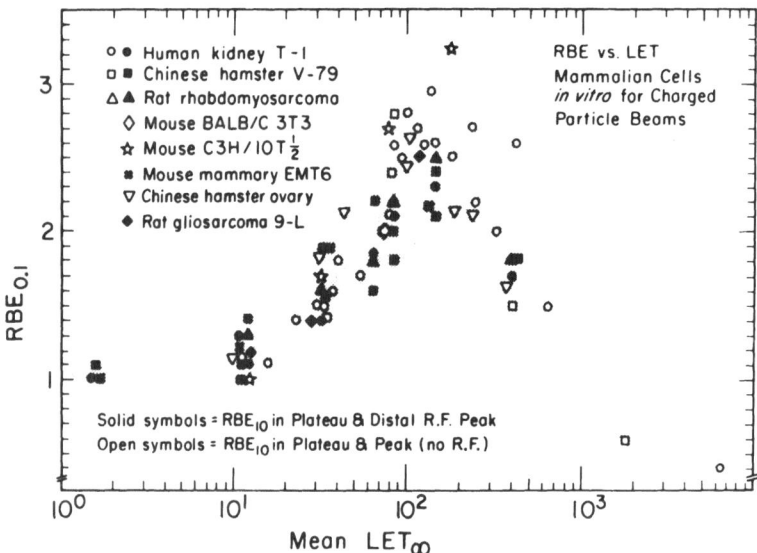

Fig. 5. Relative biological effectiveness as a function of LET
for a variety of mammalian cells. Courtesy of E. A.
Blakely.

The radiation dosimetry of heavy ions thus deals with the
situation depicted schematically in Fig. 7 (with apologies to R.
Magritte): the beam is delivered to a target site, and must be
adequately characterized and monitored in order to provide a
well-defined physical input for biomedical evaluation. The
primary beam, shown coming from the left, interacts with the
components of the beam line used to measure it, and becomes
modified by them. The beam and its reaction products then
interact with the target and its supporting structures, both
directly and indirectly via subsequent scattering.

The incoming beam properties are: the atomic mass, A, and
the charge, Z, of its constituents, the fluence, $\phi_0(x,y)$ in the
plane perpendicular to the beam, the velocity, β (or energy per
nucleon, ϵ), the velocity spread, $\Delta\beta/\beta$ (or energy spread, $\Delta\epsilon/\epsilon$;
sometimes a more useful quantity is the momentum spread, $\Delta p/p$),
the spatial location of the beam, and the angle the beam makes
with the propagation axis, z ($x' = dx/dz$, $y' = dy/dz$).

The material in the beam will consist of beam transport
elements (magnets), vacuum windows and air gaps, beam delivery

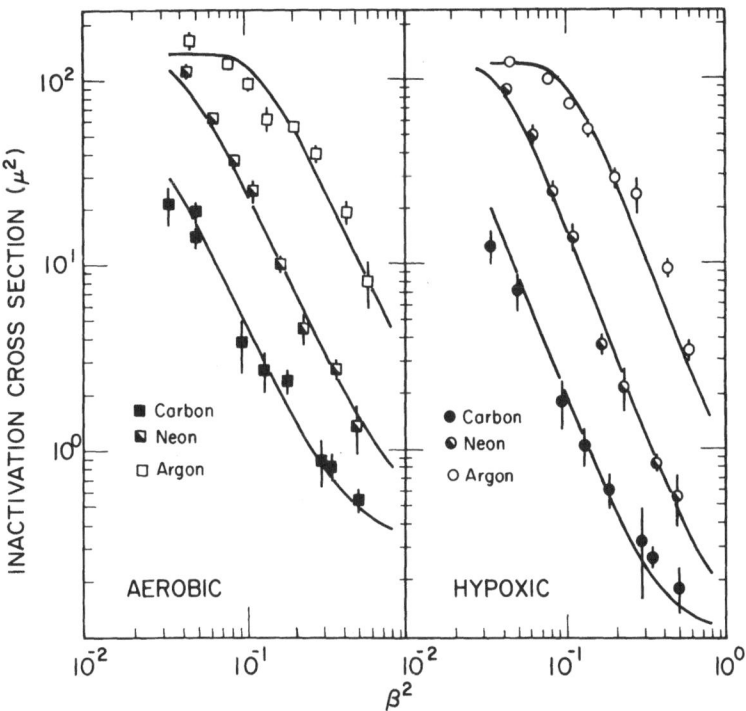

Fig. 6.   Inactivation cross sections for human kidney cells
irradiated with carbon, neon, or argon beams.
Reprinted, by permission, from Blakely et al., 1979.

devices (collimators, beam flatteners, ridge filters, variable
thickness absorbers), and beam monitoring and measurement devices
(ionization chambers, wire chambers, scintillation counters).
These materials are characterized, in turn, by their density, ρ,
their total cross section for nuclear interaction, σ_T (or
attenuation length $\lambda = 1/n\sigma_T$, where $\underline{n}$ is the number of
nuclei/cm^3), various differential cross sections describing the
angular and energy distributions of nuclear interaction
secondaries, denoted by $d^2\sigma/d\theta dE$, their radiation length L_{rad}
(a thickness of material relating to multiple scattering), the
average ionization potential, I_{ave}, that enters into the
stopping power, and their own atomic mass, A_T, and number, Z_T.

Quantities that relate to the interaction of the beam with
the material in the beam line are the range, R, which can be
scaled to a good approximation as $(AZ^2)R_p$ (R_p = range of a

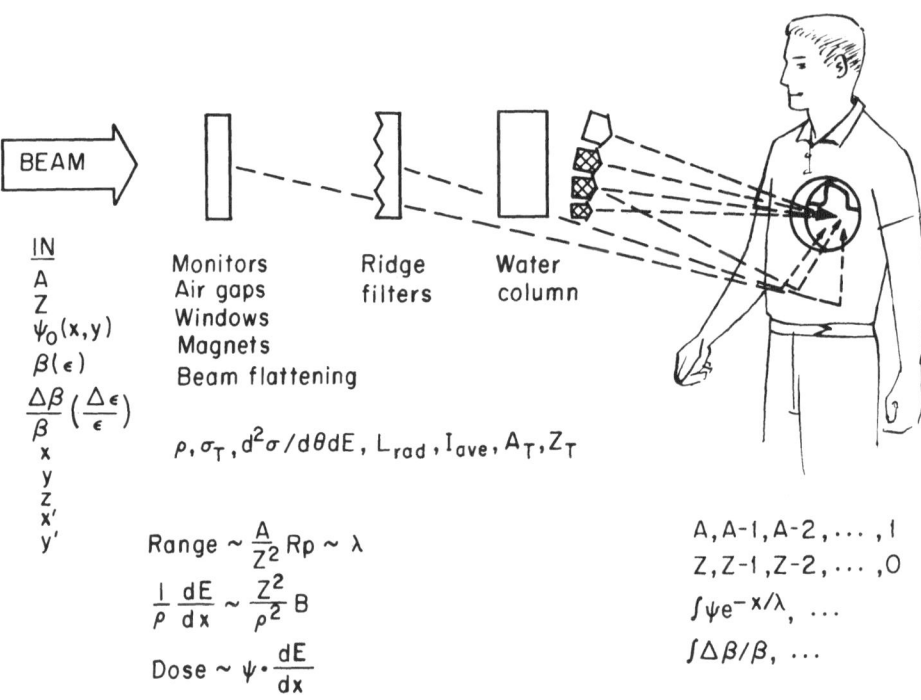

Fig. 7.   Schematic of heavy-ion irradiation arrangement showing
the beam and material properties relevant for radiation
dosimetry. The integral signs are only used as a
mnemonic to denote the thick target attenuation and
velocity degradation.

proton with the same velocity) (Schimmerling et al., 1973), and
is, as noted, generally also of the order of magnitude of the
attenuation length for the energies considered here, the mass
stopping power (since thicknesses are more conveniently measured
in g/cm^2), $(1/\rho)(dE/dx)$, which also can be scaled to a good
approximation as (Z^2/β^2) with respect to the stopping power
of heavy ions with different charge (the logarithmic term in the
stopping power formula varies slowly), and the "dose,"
$D \simeq \phi \cdot dE/dx$. The latter is an approximation to the absorbed
dose, since $dE/dx \equiv LET_\infty$, and the existence of charged particle
equilibrium cannot always be established.

Finally, the actual beam incident on the target may consist
of varying contributions of secondaries, with atomic number Z,
Z-1, Z-2, etc., (experimentally, pickup reactions leading to
fragments with Z+1, or A+1, etc., are not observed, and are not
favored kinematically). For each of these species, isotopes with
mass A, A-1, A-2, etc., may also be assumed to exist, and the
observed flux and velocity distributions will be the integrated

result of all the interactions. In practice, these contribute a small, but not negligible, amount of the actual delivered dose.

BEAM DELIVERY AND SPATIAL DISTRIBUTION

Beam energies at the BEVALAC Biomedical Facility are selected by specifying certain fixed maximum ranges adequate for research and clinical work (Lyman and Howard, 1977; Alonso et al., 1979a, 1979b). The values used most often are listed in Table 2 (BEVALAC, 1977). The restriction to a fixed set of energies is of considerable practical importance in order to obtain reproducible beam line tunes easily, limiting the number of necessary ridge filters, and allowing intercomparison of data obtained in standard beam configurations.

Table 2.   Commonly Used Beams at the BEVALAC Biomedical Facility.

Ion	Nominal Energy (MeV/amu)	Rate (particle/pulse; rad/min)	Range to Bragg Peak (cm)
$^{4}_{2}$Helium (Bevatron)	225	3×10^{8} to 1×10^{9} p/pulse 10 to 50 rad/min	30.4
$^{12}_{5}$Carbon (Bevatron)	400	5×10^{7} to 2×10^{8} p/pulse 20 to 80 rad/min	25.9
$^{12}_{6}$Carbon (BEVALAC)	400	5×10^{8} to 2×10^{9} p/pulse 350 to 1500 rad/min	25.9
$^{20}_{10}$Neon (BEVALAC)	400	2×10^{8} to 1×10^{9} p/pulse 250 to 1200 rad/min	14.6
$^{40}_{18}$Argon (BEVALAC)	500	2×10^{7} to 5×10^{7} p/pulse 200 to 500 rad/min	12.3

Circulating beam in the Bevatron is extracted by means of a resonant extraction system, and provides a "spill" length of up to 1500 msec per Bevatron pulse. Shorter spill can be provided without RF structure. The beam size at the object focus F1 is about 3 cm high, and may be reduced by collimation. The energy spread of the extracted beam is approximately 0.3 percent (BEVALAC, 1977).

The residual range of the primary beam energy can be reduced, in order to accommodate the requirements of different irradiations, by inserting an absorber in the beam. A variable absorber is used to meet these varying requirements. Figure 8 shows a schematic diagram of the 15-cm diameter variable water column used at the BEVALAC Biomedical Facility. The maximum depth is 30 cm of water, variable by computer or manual control. The minimum thickness is the 0.6 cm of the lucite windows. The window is moved by a piston driven by a stepping motor, and is read out to 0.1 mm ± 1 digit.

Another possible design, used for the minibeam area (Schimmerling et al., 1977a), is shown in Fig. 9. This design is based on a double wedge, with some advantages due to its simplicity, lower cost, faster response, and easy exchange of

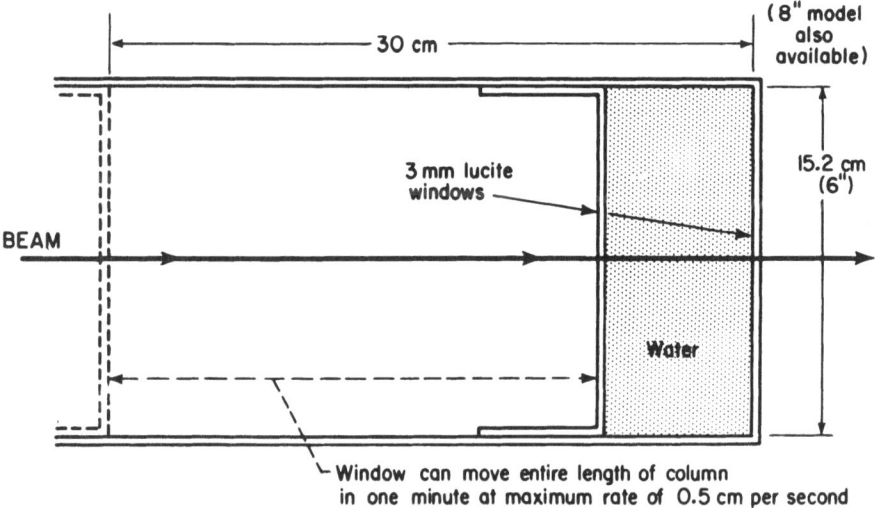

Fig. 8.   Diagram of variable water absorber. Reprinted, by permission, from BEVALAC, 1977.

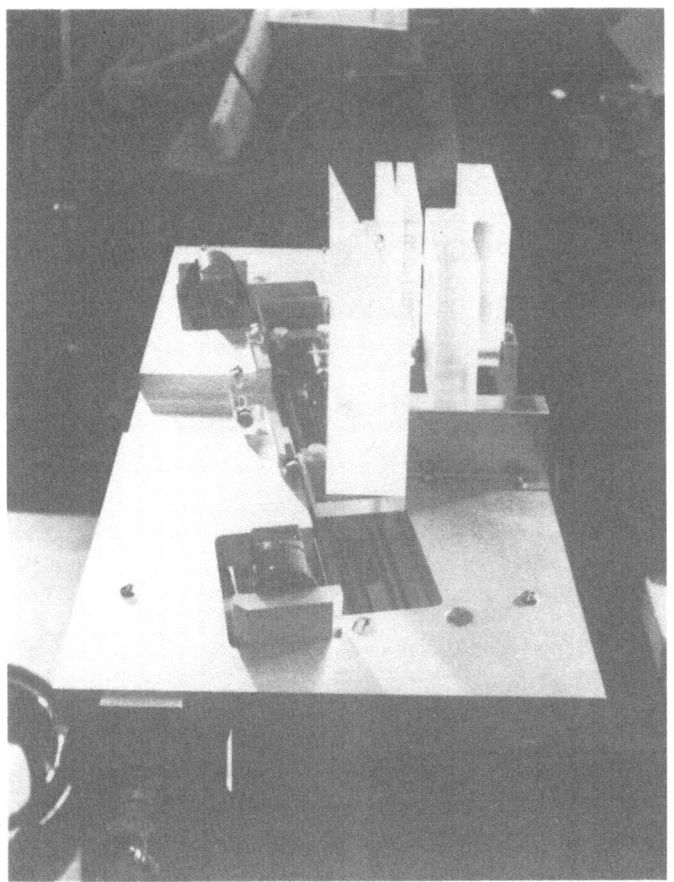

Fig. 9.   A double-wedge variable absorber. Reprinted,
          by permission, from Schimmerling et al., 1977a.

absorber material. Its disadvantages are the limited dynamic
range, due to the minimum thickness required in the thinner
wedge, and the maximum practical dimensions of the larger wedge.
A double wedge is, of course, necessary in order to present a
constant thickness surface to the finite beam spot. It is
possible, and often necessary, to obtain a dose distribution in
the target volume (e.g., a tumor region) that differs from that
shown in Fig. 4. Such a distribution may be synthesized by
shifting the range of the particles in a time short compared with
that of the irradiation. The desired dose distribution in the
target volume will then be the time-averaged superposition of
Bragg curves of the type shown in Fig. 4.

There are several means of accomplishing this goal. For example, Tobias (private communication) has suggested the use of a wedge, performing a programmed motion under computer control, to obtain the desired dose distribution curve, since the faster response of this type of variable absorber is required for implementing the concept. At the present time, prescribed dose distributions are achieved with the use of spiral ridge filters, a photograph of which is shown in Fig. 10. These consist of a brass plate with a precisely machined spiral groove. The exact shape and depth of the groove determine the form of the spread Bragg curve.

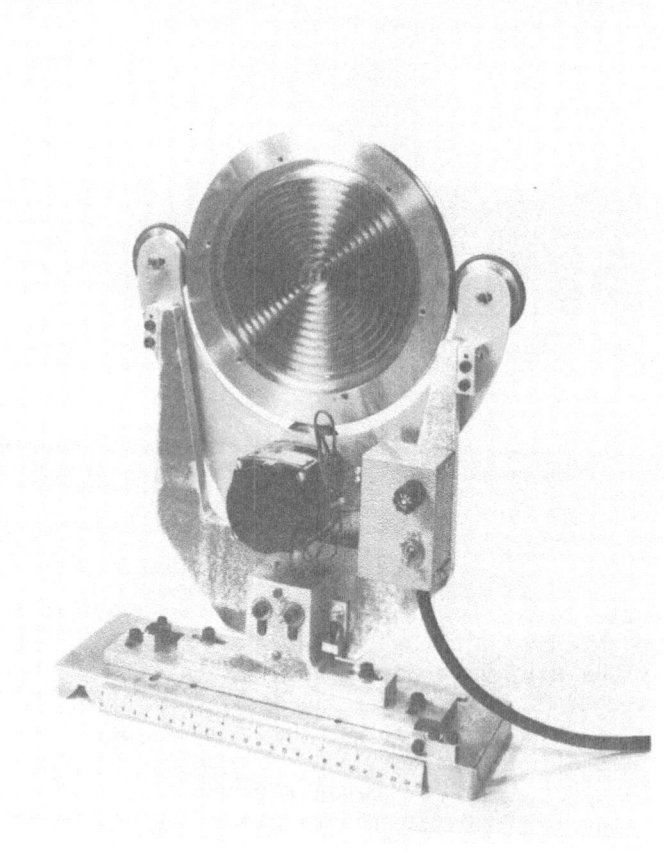

Fig. 10. A spiral ridge filter. Reprinted by
permission, from BEVALAC, 1977.

Examples of such spread Bragg curves are shown in Fig. 11 for carbon, neon, and argon beams with "short" ($\sim$15 g/cm^2) and "long" ($\sim$25 g/cm^2) ranges, and for "narrow" (4 g/cm^2) and "broad" (10 g/cm^2) spreads. The shape of these curves is not flat (isodose), but falls off at the distal end to compensate for the greater RBE, and to yield a constant biological effect (isoeffect criterion). It should be noted that RBE is a function of the biological end point, and therefore the ridge filter will not result in isoeffect curves for different end points (i.e., cell-survival level). As may be evident from Fig. 6, this is true a fortiori for aerated and hypoxic cells--an isoeffect curve for one will not in general be an isoeffect curve for the other.

The spiral ridge has a "dead spot" at the center, where the groove cut ends. For radiobiology experiments, which in general do not require very large beam spots, this problem is circumvented by positioning the filter off-axis from the beam. This is not possible for the large beam spots used in therapy, and the dead region is blocked out by the second scatterer of the beam flattening system described below.

In addition to obtaining a prescribed dose distribution in depth, it is also necessary to define the lateral distribution of the beam. In practice, the simplest way of accomplishing this is by generating a uniform transverse distribution and shaping it to the final contours with collimators. The lateral spreading of the beam into a uniform field is done with the scattering-foil occluding ring system developed at Harvard (Koehler et al., 1977), and at the Berkeley 184-inch Synchrocyclotron (Crowe, 1975). A lead foil, typically 0.6-cm thick, located 10 m upstream of the target volume, scatters the beam into a symmetric Gaussian distribution. A second scatterer, consisting of the ridge filter, an additional 4-mm brass plate, a central brass cylinder, and a concentric brass ring, is positioned approximately 4 m downstream of the first scatterer. The central brass cylinder and the concentric ring block out the central region and an excentric portion of the Gaussian distribution. The brass plate on the filter acts as a diffuser, filling in the blocked-out regions, so that, after traversing the path to the target, the beam profile is completely flat with an intensity variation of no more than 2 percent over the entire 20-cm diameter field. Approximately 40 percent of the original beam intensity lies within this field, thus providing better beam utilization than would be possible by blowing up the beam size and selecting a uniform central region with collimators. A photograph of the assembled ridge filter with occluding rings is shown in Fig. 12.

Operation of this system requires great alignment accuracy, and several alternatives are under study and development. One

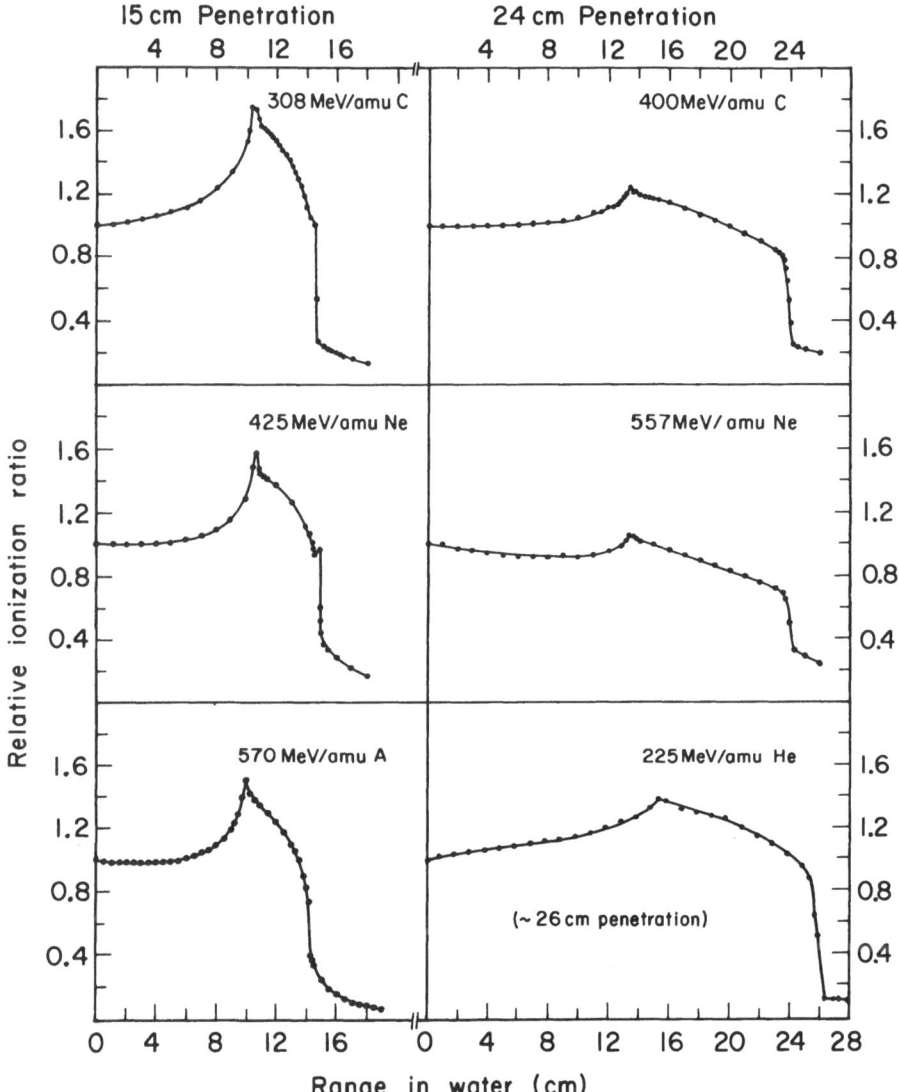

Fig. 11.   Bragg curves spread with a ridge filter for different
          incident beam energies and filter range. Reprinted, by
          permission, from, Lyman and Howard, 1977a.

Fig. 12. A spiral ridge filter covered with an
          occluding ring. Reprinted, by permis-
          sion, from Lyman and Howard, 1977a.

such an alternative is the use of a beam "wobbler," shown
schematically in Fig. 13 (Alonso, private communication). The
device consists of a rotating dipole field, provided by a
stationary sextupole magnet, which sweeps the beam in a circular
pattern to fill in a ring-shaped region, and is turned off to
fill in the central region. Further development will ultimately
lead to a full three-dimensional, computer-controlled scanning
system, receiving its input from treatment-planning and CT

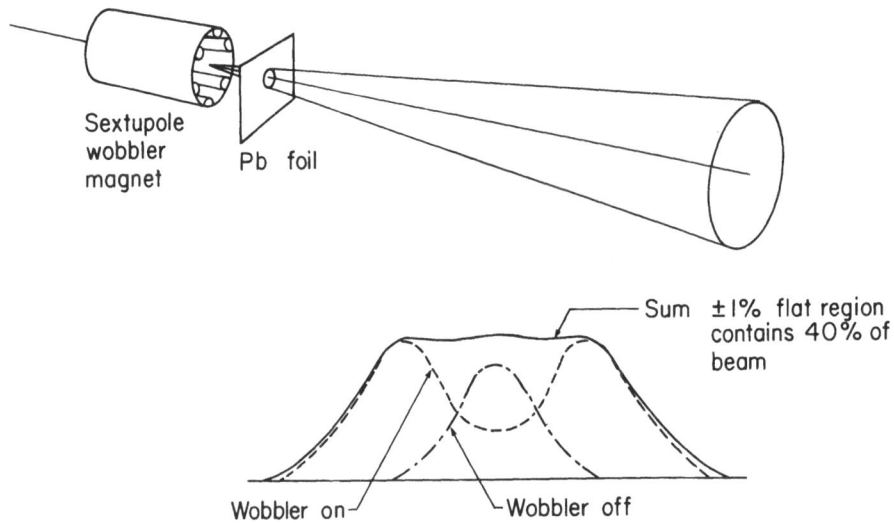

Fig. 13. Beam delivery wobbler system. Courtesy of J. Alonso.

scanning computers. A prototype scanning magnet and power supply
have already been constructed. A schematic diagram of the
concept is shown in Fig. 14.

Verification of the beam flatness can be accomplished
approximately with the concentric-ring ionization chambers
described below. For greater accuracy, exposed photographic
emulsions and densitometry scans had to be used until recently.
This involved six to eight exposures covering a dose range of
1:100, so that a dose vs. film density calibration was obtained
simultaneously.

This procedure is time-consuming and cumbersome, and has led
to the development of a multiwire, multiplane proportional
chamber for reconstruction of beam profiles. The instrument,
called MEDUSA (Alonso et al., 1979b), integrates the charge
collected from a single beam pulse on capacitors connected to
each of the 64 signal wires in each of the 16 planes. The planes
are rotated at equal intervals around 180°, perpendicular to
the beam. The integrated charge profile from each plane constitutes a projection of the beam intensity (but not of dose!).
The beam profile is reconstructed from the 16 different projections, using algorithms developed for CT scanning. Ultimately,
reconstruction is expected to take 2 seconds, sufficient to

enable interactive beam tuning. Accuracy of the device is expected to be better than 0.5 percent intensity resolution. A photograph of the assembled chamber is shown in Fig. 15, and a reconstructed beam profile is shown in Fig. 16. The beam profile was obtained with a copper bar and lead brick placed in front of the chamber. The sharpness of the edges is a good indication of the spatial resolution of the instrument (10 percent to 90 percent in 3 pixels) and its ability to reconstruct asymmetric patterns.

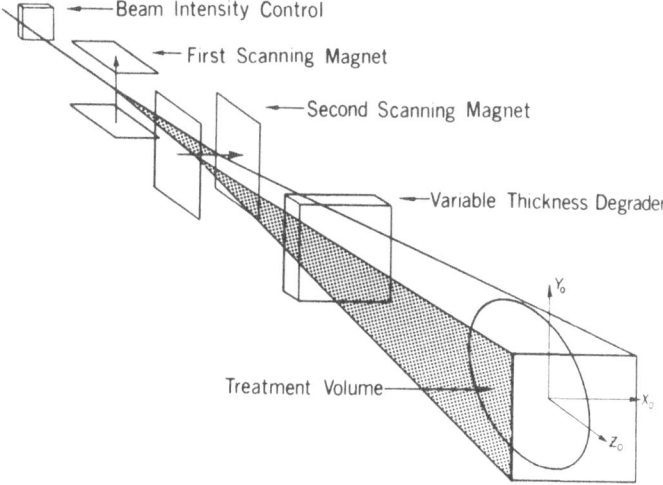

Fig. 14.   Schematic depiction of a three-dimensional scanning method. Reprinted, by permission, from Lawrence Berkeley Laboratory Report LBL-7230.

DOSIMETRY

It must be emphasized that dosimetry serves both a predictive and an archival function. It is as important to know the physical characteristics of beams that will be delivered as it is important to know the characteristics of beams that have been delivered. Most of the methods described in this section are

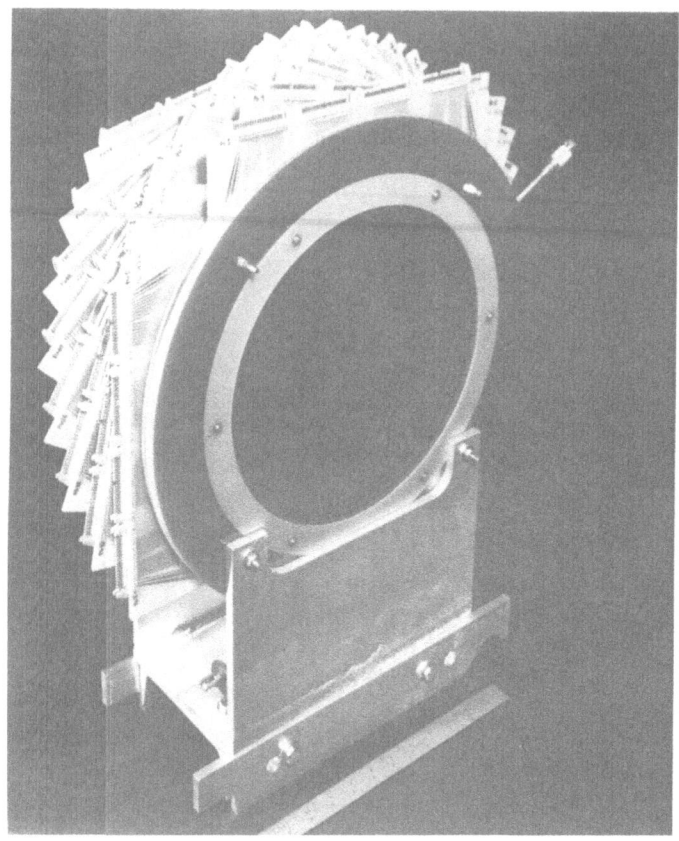

Fig. 15.   MEDUSA chamber.   Reprinted, by permission,
          from Alonso et al., 1977b.

used for beam verification.   Instruments used for routine operation at the BEVALAC presently measure dose intensity distributions.   Microdosimetric measurements have been made, and are
described in other lectures of this course.   Other studies of
beam quality will be discussed below.

The most useful monitor for the heavy ion beams is a parallel
plate ionization chamber.   The basic ionization chamber used in
biomedical dosimetry at the BEVALAC (1977) consists of three
foils mounted on insulating frames in a gas-tight aluminum

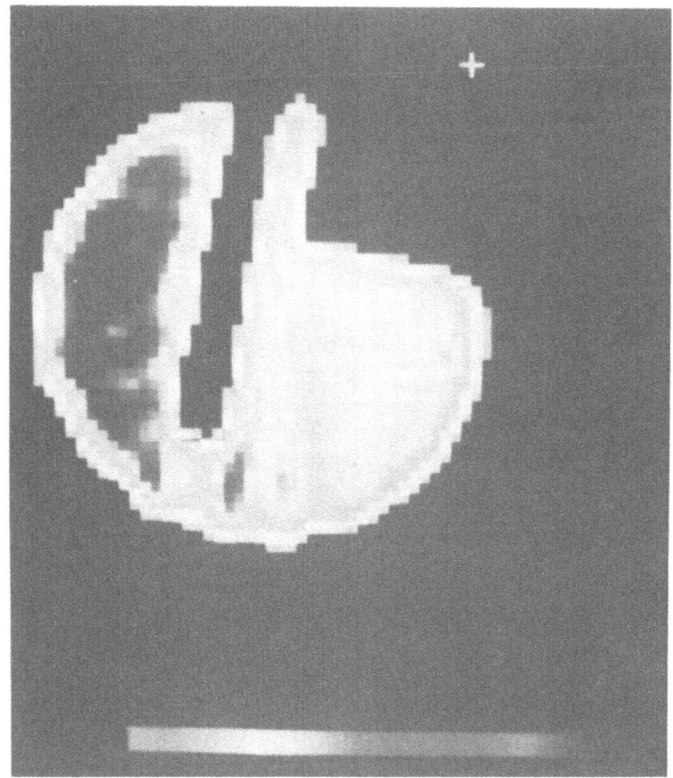

Fig. 16. Reconstruction of a radiation
          field obstructed by a rectangular
          metallic absorber after smoothing
          the raw data. Reprinted, by per-
          mission, from Alonso et al., 1977b.

housing, as shown in Fig. 17. The three foils are a quadrant
foil, a high voltage foil, and a concentric, ringed foil. The
foils are constructed of thin plastic membranes and gold plated
on both sides in the appropriate pattern. The charge collected
by the various electrodes is integrated and measured by linear,
seven-decade voltage-to-frequency converters.

The signals from diammetrically opposed quadrants in the
quadrant foil can be added and modulated with a sine-wave
generator. When the resultant signals are applied to the x- and
y-sweeps of an oscilloscope, the resultant Lissajous pattern is a
circle or ellipse with semi-axes or radius proportional to the
beam intensity, centered on a location that can be referred to
the beam position. In this way, a continuous indication of
relative intensity and position is obtained.

The ringed foil allows monitoring of the dose as a function of radial distance from the beam axis (Lyman et al., 1975) as well as a measurement of total dose (when the beam spot is smaller than the active area of the chamber). A photograph of the disassembled chamber is shown in Fig. 18.

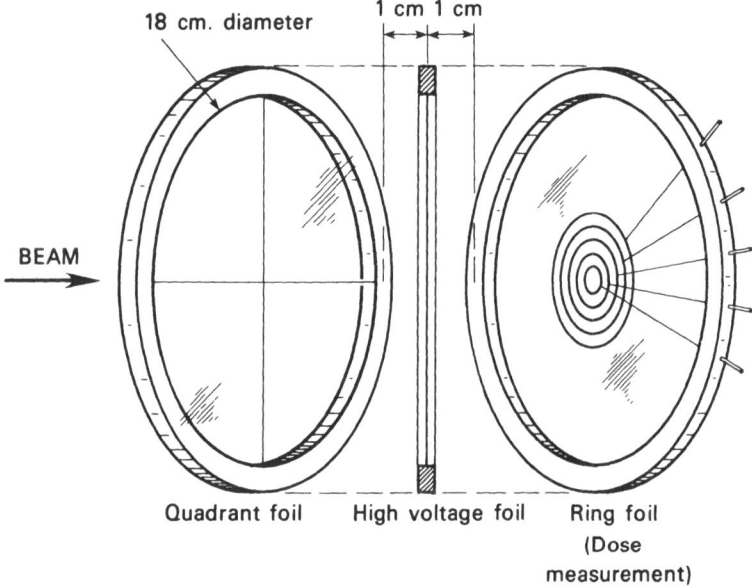

Fig. 17. Ionization chamber foils. Reprinted, by permission, from BEVALAC, 1977.

Figure 19 shows a scale drawing of the ionization chamber used in Princeton (Schimmerling et al., 1976). This chamber is similar in geometry, and does not feature multiple electrodes. Emphasis in this design was placed on eliminating the use of adhesives that could contaminate the filling gas with organic vapors. Also noteworthy in this design are the provisions for vacuum baking for initial cleanup, and the pierced foils at each end of the chamber, used for damping microphonic noise, to which this type of instrument is extremely susceptible.

Fig. 18.   Disassembled components of the ioniza-
            tion chamber used in the LBL Biomedical
            facility (drawn schematically in Fig. 17.
            Reprinted, by permission, from BEVALAC,
            1977.

The Berkeley chamber is flushed with dry nitrogen gas, while
the Princeton chamber used argon. The choice of filling gas is
determined by the desire to have a well-defined material as well
as by the advisability of using the chambers in the electron
collection mode. The latter, especially, is important in order
to deal with the high ionization of heavy ions, and to collect
the produced charges in a time which is short relative to recombination lifetimes. Nitrogen and argon have low electron

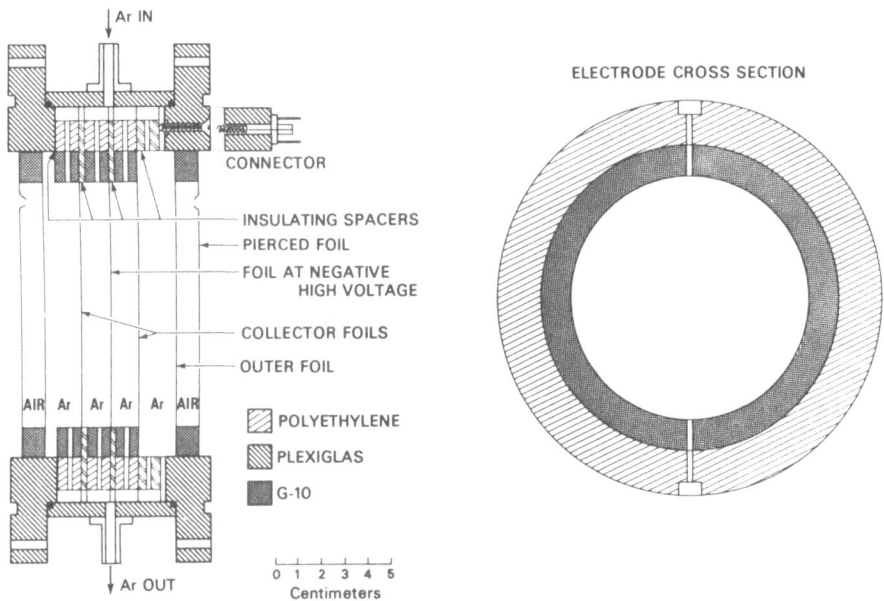

Fig. 19.   Scale drawing of the ionization chamber used at
           Princeton. Reprinted, by permission, from
           Schimmerling et al., 1976.

affinities, and are useful for this reason. When ionization
chambers, such as the one shown in Fig. 19, are used in pulse
mode, it is further important to minimize electron collection
times to ensure proper linearity (collection of all the produced
charge) and maximize counting rates. In that case, propane and
methane give the highest electron mobilities and are the filling
gas of choice. Tissue-equivalent gases that have the same
electron density as muscle tissue are also used in order to avoid
errors introduced by the calculation of absorbed dose in tissue
from the measured ionization.

The measured charge, or ionization, in the chamber can be
related to the energy loss in the chamber and, hence, to the
absorbed dose, if the average energy required to produce one ion
pair, W, is known. This quantity is not easily calculated in an
accurate form (Myers, 1968), and has become the object of revived
theoretical and experimental interest (Dennis, 1972; ICRU, 1979)
in the light of its importance for charged particle dosimetry.

The experimental determination of W requires an independent measurement of beam intensity in order to determine the average ionization per incident beam particle. W in argon was determined (Schimmerling et al., 1976) for nitrogen ions of 520 MeV/u by using carbon activation detectors to monitor the beam passing through the ionization chamber, based on the measured $^{12}C(^{14}N,x)^{11}C$ cross section (Skoski et al., 1973). The value of W obtained, 26.2 ± 1.8 eV/i.p., was in good agreement with the value of 26.3 ± 0.1 for alpha particles in argon (Myers, 1968). Stephens et al. (1976) have used thermoluminescent dosimeters to measure the fluence of 250 MeV/A carbon ions, and obtained a value of 36.6 ± 0.7 eV/i.p. for W in the nitrogen gas used at the BEVALAC. Subsequent measurements report values of 35.3 ± 1.5 eV/i.p. for neon at 375 MeV/A and 34.6 ± 1.4 eV/i.p. for 429 MeV/A argon ions in nitrogen (Thomas et al., 1978). The difference between these values is not statistically significant, and the authors of this work conclude that the average value of W = 35.2 ± 0.9 eV/i.p. in nitrogen may be taken as the best value presently available, independently of particle charge or velocity. This is to be compared with the value of W = 34.9 eV/i.p. used for routine BEVALAC operations (1977).

All these measurements rely upon a two-step procedure to establish the absolute beam fluence. In a first step, the method is calibrated at low intensities, where the signal-to-noise ratio in ionization chambers operated in the current mode is too small for reliable comparison. In this mode scintillation counters (Schimmerling et al., 1976) or nuclear emulsions (Skoski et al., 1973; Stephens et al., 1976), can be used for absolute counting. The second step, at intensities within the operational range of the ionization chamber, compares the secondary standard (activation foils, thermoluminescent dosimeters, etc) with the ionization current. Figure 20 shows the typical efficiency of thermoluminescent detectors relative to ^{60}Co gamma rays and comparison with some theoretical models (Patrick et al., 1976).

It is possible to circumvent this two-step procedure by using the ionization chamber in a pulse mode (Epstein et al., 1971) and counting beam particles directly. This is done in the fragment dose experiments described below. Preliminary measurements (Schimmerling et al., unpublished results) show agreement between the value obtained for 525 MeV/A argon ions in propane and the value calculated using Bragg's rule, W = 29.6 eV/i.p. Similarly, measurements by Goodman and Colvett (1977) have shown that the value of W for argon ions in air and TE gas is not significantly different from that obtained with a ^{137}Cs source. It seems safe to conclude that, except perhaps for a narrow region beyond the Bragg peak where the velocity of the heavy ions becomes comparable with that of orbital electrons, W values of heavy

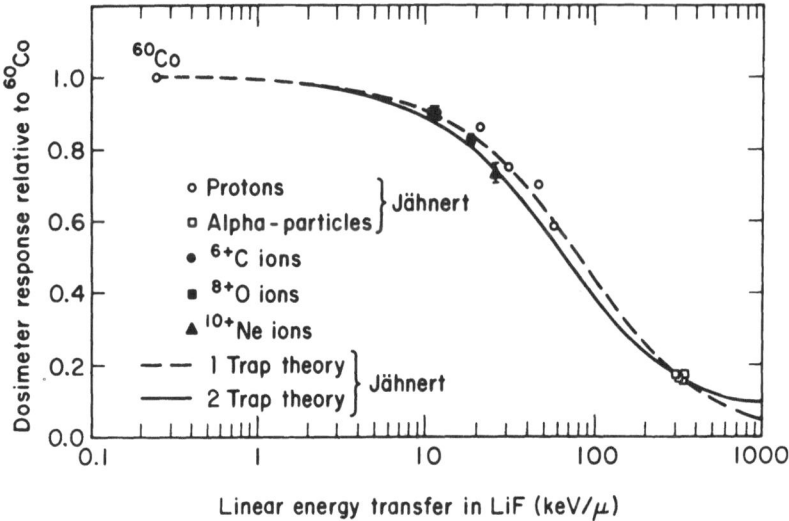

Fig. 20.   Efficiency of thermoluminescent detectors relative
           to ^{60}Co. Reprinted, by permission, from Lawrence
           Berkeley Laboratory Report LBL 6710.

ions differ by less than 5 percent from those of lighter charged
particles.

The accuracy of absolute dosimetry is dependent upon W, a
knowledge of the stopping power, the dimensions of the ionization chamber, and the method of charge collection. The accuracy
of W measurements is, as we have seen, not much better than
5 percent. Stopping power cannot be calculated to an accuracy
better than a few percent at the present time, and experimental
measurements are also not more accurate, for the energies and
ions considered here. Improvements in these accuracies would
require quite sophisticated and fastidious experiments. Thus,
the accuracy of absolute physical dose determinations cannot be
assumed to be better than 5 to 10 percent.

Fortunately, evaluations of the merits of radiation therapy
rely on relative observations and reproducibility of dose
measurements can be attained at the few percent level or better,
as evidenced by preliminary results of recent intercomparisons.
These comparisons utilized 1 cm^3 tissue-equivalent ionization
chambers to eliminate sensitivity to field shapes, and referred
the measurements to ^{60}Co doses. This work is being performed

under the auspices of the Charged Particle Dosimetry Group (CPPDG) of the American Association of Physicists in Medicine.

Figure 21 shows the complete panoply of instrumentation used for routine high-energy heavy ion dosimetry at the BEVALAC, mounted on a standard alignment bench. The beam is incident from the left, and the instruments in the beam path, from left to right, are: collimator, upstream ionization chamber (for intensity normalization), scintillators, collimator, spiral ridge filter, variable water absorber, aperture holder, downstream ionization chamber, remote control sample translator, EG and G TE ionization chamber, alignment pointer, cross hairs and x-ray tube

Fig. 21. The instrumentation used for radiation dosimetry at the BEVALAC Biomedical Facility. Shown from left to right: brass collimator, upstream ionization chamber, collimator, spiral ridge filter, variable thickness water absorber, aperture holder, downstream ionization chamber, remote control sample translator, EG and G TE Ionization chamber, alignment pointer, cross hairs and x-ray tube, and a utility fixture with a custom-poured collimator. Reprinted, by permission, from BEVALAC, 1977.

also used for alignment, and a utility fixture with a custom-poured collimator. The alignment bench for the minibeam is shown in Fig. 22, showing the double wedge, ionization chambers, and a microscope system to be used for on-line studies of cell function. The microscope is equipped with remote handling facilities, as well as a movie camera and a videotape unit for time-lapse photography. The television monitor in the figure displays a culture of neuroblastoma cells being used by T.C.H. Yang of LBL for preliminary studies.

The dose measured by the ionization chamber behind the water column is related to the absorbed dose in samples behind the ionization chamber by a displacement correction to account for the material between the sample and the chamber (measured with

Fig. 22. Photograph of alignment bench used in BEVALAC mini-beam. Shown from left to right are: upstream ioniza-tion chamber of the type shown in Fig. 19, double wedge, downstream ionization chamber, beam spot film holder, and microscope system. Reprinted, by permis-sion from Schimmerling et al., 1977a.

the water column) and by the ratio of stopping powers in the material and the chamber gas. This procedure has been validated by Goodman and Colvett (1977) who concluded that the depth-dose distributions measured for high-energy argon ions in a water phantom and in a tissue equivalent liquid were essentially the same, after taking into account the difference in electron densities between the two liquids, at the 5 percent level. Similarly, the depth-dose distribution measured behind the water absorber was equivalent to that measured inside a water phantom. Finally, the results showed no significant difference between dosimetry in air and in tissue-equivalent gas in the cavity of a tissue-equivalent ionization chamber.

NUCLEAR INTERACTIONS AND BEAM QUALITY

As discussed in the Introduction, the mean free path of the heavy ion beams used for biomedical applications is comparable to their range. Thus, secondary and, possibly, subsequent generations of nuclear interaction products can be expected to make a significant contribution to the dose, especially in the vicinity of the Bragg peak and in the spread depth dose distribution. For the purposes of therapy planning, of course, it must be remembered that nuclear interaction products will constitute 100 percent of the exit dose. In addition, it has been shown that biological effect depends not only on the LET, but possibly on the charge and velocity in different ways. Thus, one approach to characterizing high-energy heavy-ion beams in a manner that would eliminate uncertainties about their physical characteristics from the radiobiological evaluation, is the measurement of velocity and ionization spectra for identified fragments as a function of depth.

Preliminary measurements, using only velocity spectra, showed that the present state of knowledge about fragmentation cross sections and their dependence on the incident particle energy is insufficient to permit accurate predictions (Schimmerling et al., 1977b). Estimated cross sections for fragments produced in the interaction of 251 MeV/u nitrogen ions on carbon, based on the semi-empirical model of Silberberg and Tsao (1973) and calculated as described by Schimmerling et al. (1977b), are shown in Fig. 23. These cross sections are generally in agreement with experimental data at the 20-30 percent level down to energies of 150 MeV/u. Data between 20 MeV/u and 100 to 200 MeV/u are unavailable at present. Such data are required for every element in tissue, or must be calculated by a scaling procedure. In addition, the angular distributions and energy spectra, as well

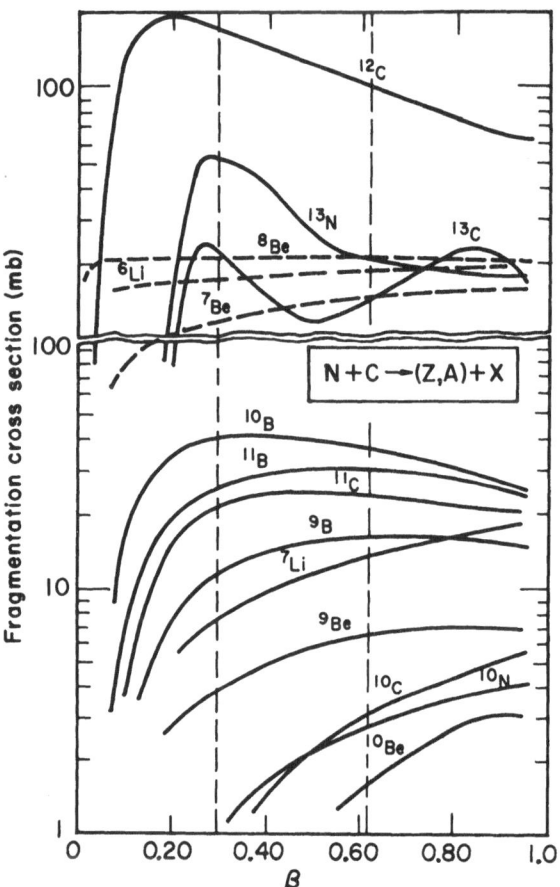

Fig. 23.   Cross sections for producing fragments
           from the interaction of nitrogen ions
           with a carbon target, as a function of
           velocity, estimated from the semi-
           empirical model of Silberberg and Tsao
           (1973). Reprinted, by permission from
           Schimmerling et al., 1977b.

as the multiplicities of light particles (neutrons, protons), must be known for an accurate calculation. Few such data are at present available.

Nuclear physics experiments provide data that are useful for biomedical purposes only after incorporation into cumbersome transport calculations of limited accuracy. Therefore, the approach of obtaining such data directly is currently being pursued at the BEVALAC. Preliminary results (Schimmerling et al., unpublished data), were obtained with the arrangement shown in Fig. 24. R is a variable thickness absorber. The detector consists of a multiwire proportional chamber to measure particle position, a channel plate time-of-flight telescope with a resolution of about 100 picoseconds (Gabor et al., 1975) (corresponding to an energy resolution of about 0.1 percent at low energies), and a pulse ionization chamber for single particle ionization measurements. S_0, S_1, and S_2 are scintillation detectors used to monitor the beam intensity and to impose a range requirement: a signal in all detectors except S_2 indicates that the particle stopped somewhere between S_1 and S_2, and gives a measurement of its range. A photograph of the experimental setup is shown in Fig. 25.

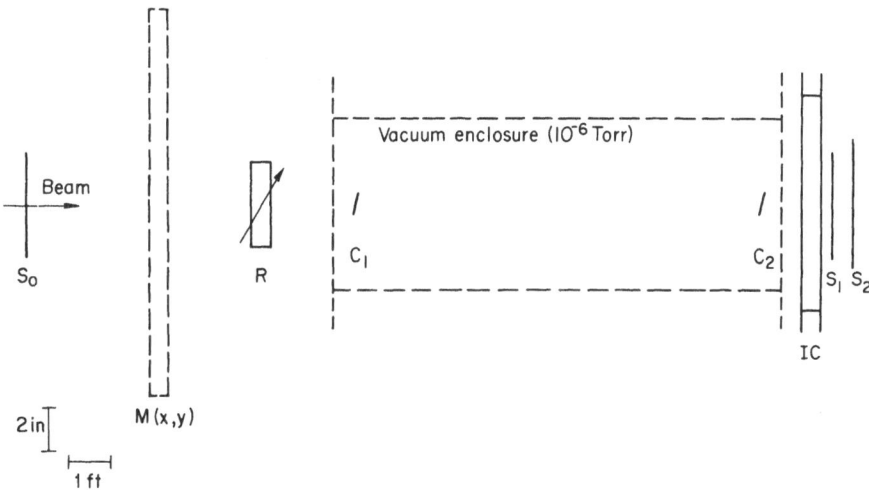

Fig. 24.   Schematic diagram of instrumental arrangement used
           for element identification of fragments emerging
           from a thick absorber R. S_0, scintillator; M(x,y),
           wire chamber; C_1 and C_2 foils of a channel plate
           time-of-flight spectrometer; IC, pulse ionization
           chamber; S_1 and S_2, range discrimination scintil-
           lators.

Fig. 25.   Photograph of experimental arrangement shown in
           Fig. 24. The long pipe contains the time-of-
           flight spectrometer and is connected to a high-
           vacuum pumping system. The beam is incident from
           the left.

     The purpose of this arrangement was to build an ionization-
sensitive particle identifier with minimum mass, in order to
separate the secondary particles close to the Bragg curve. The
charge resolution of the ionization chamber is limited by the
statistical distribution of energy losses. This "Landau tail"
(Rossi, 1952) is, however, limited in a thin ionization chamber
because very energetic electrons (delta rays) can only deposit a
small fraction of their energy in the chamber (Epstein et al.,
1971). In this sense, the ionization chamber is a "restricted
energy loss detector," and its charge resolution, approximately
7 percent, is much better than would be predicted without this
consideration. The counting rate of the chamber is limited by
the collection time for the electrons (~3 μsec). Even though
this is much slower than, e.g., thin plastic scintillators, the
pronounced saturation effects of these place them at a disadvantage with respect to ionization chambers in applications to

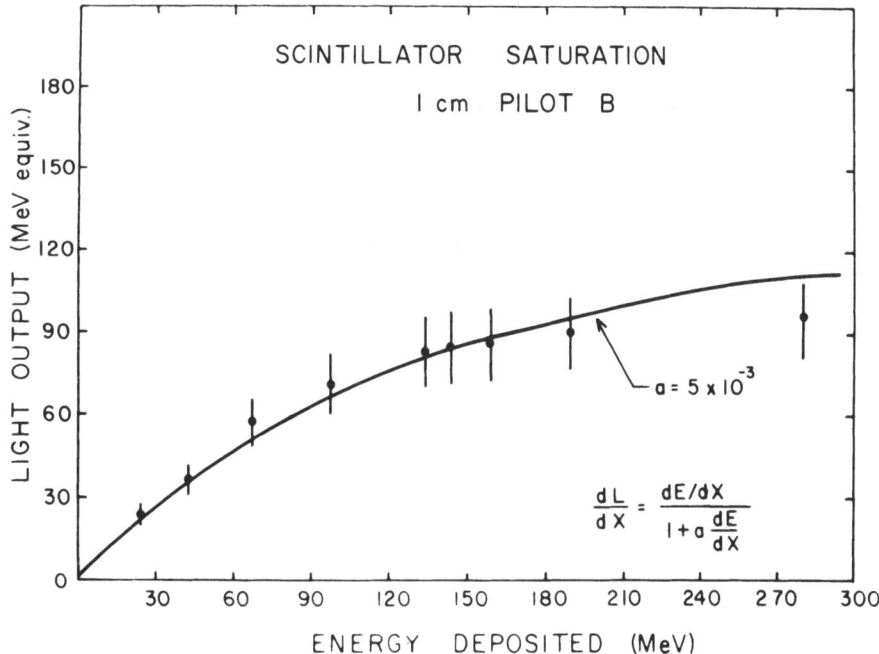

Fig. 26.   Example of the saturation in scintillator output.
Reprinted, by permission, from Kidd et al., 1979.

heavy-ion dosimetry. Figure 26 shows an example of the saturation in scintillator output (Kidd et al., 1979). In experiments
being set up at this time, a solid-state detector telescope
(Greiner, 1972), with isotopic resolution of the order of
0.2 amu, will be used in conjunction with this apparatus.

A typical time-of-flight spectrum of 525 MeV/A argon beam,
near the end of the argon range, is shown in Fig. 27. Velocity
increases from right to left, with the slowest particles having
the longest flight time (corresponding to the higher channels).
The residual peak of argon can still be seen, as well as the
substantial flux of secondary particles, mixed in the distribution to the left of argon.

Ionization pulse height spectra at a depth of 12.5 g/cm^2 of
Lucite for an incident 525 MeV/A argon beam are shown in
Fig. 28. Cuts on the data were made using the known charge
resolution of the ionization chamber and the calculated
velocity-ionization profile of argon in the detector. The lower
left panel shows the ionization pulse height spectrum without any

cuts. The upper right hand panel shows the distribution of all
secondary particles (corresponding to the region left of the
argon peak in Fig. 27). The charge resolution is rather poor,

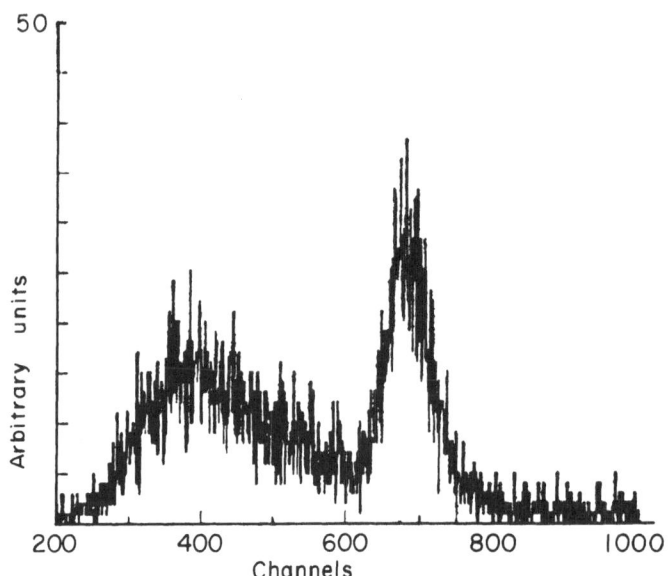

Fig. 27. Time-of-flight spectrum of 525 MeV/nucleon argon
          ions after traversing 12.5 g/cm^2 of Lucite. Time
          increases from left to right, and velocity corres-
          pondingly decreases from left to right.

but some structure seems evident and should be resolved in future
work. Finally, the ionization spectrum for particles identified
as argon is shown in the lower right hand panel of Fig. 28. The
average ionization can be obtained from spectra such as this, and
plotted as a function of absorber thickness to obtain the
separated depth-dose curve for argon shown in Fig. 29.

Near the peak (at 14.6 g/cm^2) in Fig. 29, the data are
adequate to provide a crude estimate of the effect of secondaries on the dose distribution of argon. Assuming that lower-Z
particles have a mean charge of Z = 15, and using the measured
mean velocity, the estimated contribution of lower-Z secondaries
to the dose is 23 percent, at an average LET_∞ of 130 keV/µm. The

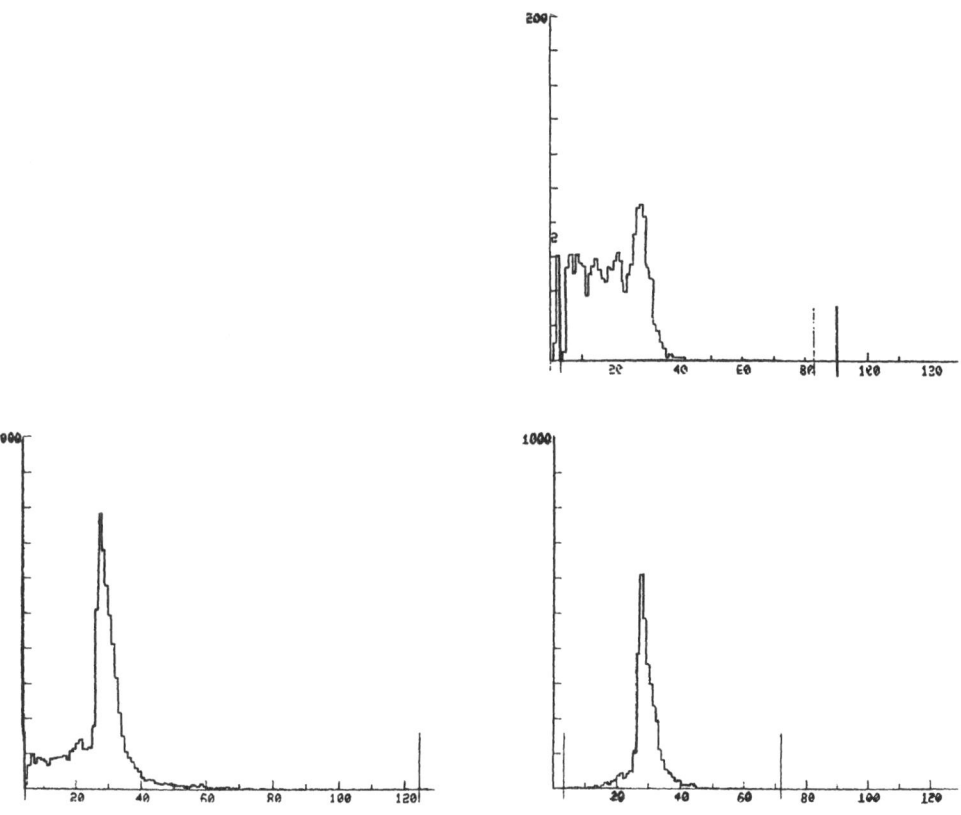

Fig. 28. Ionization pulse height spectra at the exit of 12.5 g/cm^2 of Lucite for a 525 MeV/nucleon incident argon beam.

primary argon is estimated to contribute 39 percent of the dose at this point, at an LET$_\infty$ of 323 keV/μm. Lighter isotopes of argon (e.g., ^{38,39}Ar) and chlorine (e.g., ^{35}Cl) begin to stop in this region, and can be identified with the range requirement. Their contribution to the dose can be estimated as being 37Δ, and their average LET$_\infty$ can be estimated to be approximately 490 keV/μm. The apparatus used in these experiments did not allow clear identification of the particles in the peak region itself, where they are considerable smeared in both ionization and velocity. The relative contributions of fragments can be expected to vary rapidly in this region, and it is expected that work currently in progress will yield important information about these phenomena, especially in the spread, depth- dose curve. With these data, calculation of any desired measure of beam quality, for any beam configuration, should become practical. Some of the problems that become accessible to study are discussed by Curtis (1977).

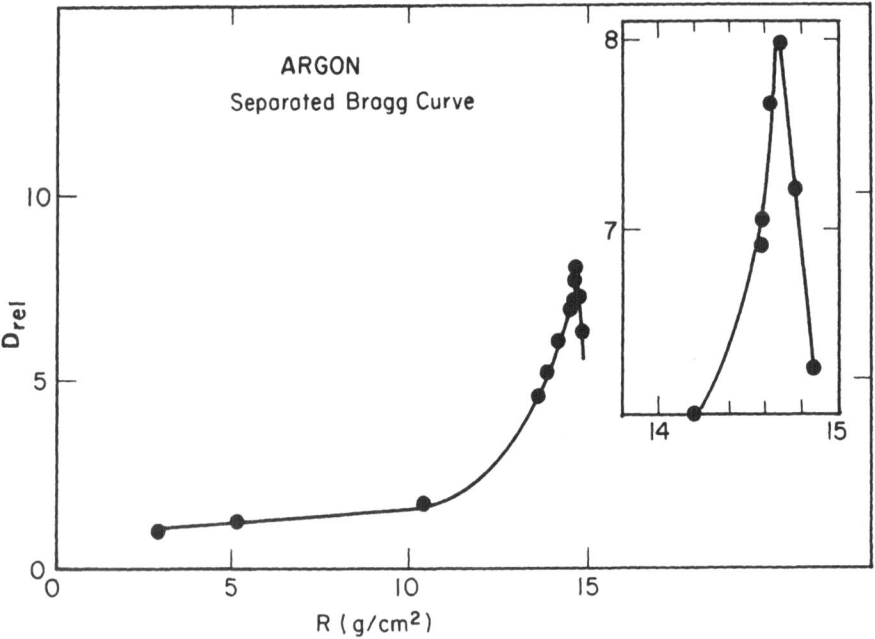

Fig. 29.   Separated Bragg curve for 525 MeV/nucleon argon ions.

ACKNOWLEDGEMENTS

     The author is greatly indebted to W. T. Chu and C. Weber for
the loan of graphical material. Many enlightening conversations
with J. Alonso, E. Blakely, J. Howard, J. Lyman, R. Thomas and
C. A. Tobias are also gratefully acknowledged, as well as their
loan of graphical material. My colleague and collaborator,
S. Curtis, has greatly contributed to my education in this
field. D. Ortendahl, G. Gabor, S. Kaplan, and V. Perez-Mendez
have collaborated with the author in some of the unpublished
experimental work reported here. Finally, the fine effort of
G. Walpole has been of immeasurable help in producing these
lecture notes. This work was supported by the Office of Health
and Environmental Research of the U. S. Department of Energy
under Contract No. W-7405-ENG-48.

REFERENCES

Alonso, J. R., Howard, J., and Criswell, T., 1979a, IEEE Trans.
     Nucl. Sci., NS-26: 3074.

Alonso, J. R. Tobias, C. A., and Chiu, W., 1979b, IEEE Trans. Nucl. Sci., NS-26: 3077.

BEVALAC, 1977, Bevatron/BEVALAC User's Handbook, Lawrence Berkeley Laboratory Publication 101.

Blakely, E. A., Tobias, C. A., Yang, T. C. Smith, K. C., and Lyman, J. T., 1979, Radiat. Res., 80:122.

Crowe, K., Kanstein, L., Lyman, J. T., and Yeater, F., 1975, Lawrence Berkeley Laboratory Report LBL-4235 (unpublished).

Curtis, S. B., 1977, Int. J. Radiat. Oncol. Biol. Phys., 3: 87.

Dennis, J. A., 1972, in: "Proceedings, Third Symposium on Micro-dosimetry," H. G. Ebert, ed., Euratom, Luxembourg.

Epstein, J. W., Fernandez, J. I., Israel, M. H., Klarmann, J., and Mewaldt, R. A., 1971, Nucl. Instr. Methods, 95: 77.

Gabor, G., Schimmerling, W., Greiner, D., Bieser, F., and Lindstrom, P., 1975, Nucl. Instr. Methods, 130: 65.

Goodman, L. J., and Colvett, R. D., 1977, Radiat. Res., 70: 455.

Greiner, D., 1972, Nucl. Instr. Methods, 103: 308.

ICRU, 1979, International Commission on Radiological Units and Measurements, "Average energy required to produce an ion pair," Report No. 31.

Kidd, J. M., Wefel, J. P., Schimmerling, W., and Vosburgh, K., 1979, Phys. Rev. C, 19: 1380.

Koehler, A. M., Schneider, R. J., and Sisterson, J. M., 1977, Med. Phys., 4: 297.

Lyman, J. T., and Howard, J., 1977a, in: "Biological and Medical Research with accelerated Heavy Ions at the BEVALAC, 1974-1977," Lawrence Berkeley Laboratory Report No. 5610.

Lyman, J. T., and Howard, J., 1977b, Int. J. Radiat. Oncol. Biol. Phys., 3: 81.

Lyman, J. T., Howard, J., and Windsor, A. A., 1975, Med. Phys., 2: 163 (abstract).

Myers, I. T., 1968, Ionization, in: "Radiation Dosimetry," F. H. Attix and W. C. Roesch, eds., Academic Press, New York.

Patrick, J. W., Stephens, L. D., Thomas, R. H., and Kelly, L. S., 1976, Health Phys., 30: 295.

Rossi, B., 1952, "High Energy Particles," Prentice Hall, Englewood Cliffs, N. J.

Schimmerling, W., Vosburgh, K. G., and Todd, P. W., 1973, Phys. Rev. B, 7: 2895.

Schimmerling, W., Vosburgh, K. G., Todd, P. W., and Appleby, A., 1976, Radiat. Res., 65: 389.

Schimmerling, W., Alonso, J., Morgado, R., Tobias, C. A., Grunder, H., Upham, F. T., Windsor, A., Amer. R. A., Yang, T. C. H., and Gunn, J. T., 1977a, IEEE Trans Nucl. Sci., NS-24: 1049.

Schimmerling, W., Curtis, S. B., and Vosburgh, K. G., 1977b, Radiat. Res., 72: 1.

Silberberg, R., and Tsao, C. H., 1973, Astrophys. J. Suppl. Ser., 25: 315.

Skoski, L., Merker, M., and Shen, B. S. P., 1973, Phys. Rev. Lett., 30: 51.

Stephens, L. D., Thomas, R. H., and Kelly, L. S., 1976, Phys. Med. Biol. 21: 570.

Thomas, R. H., Lyman, J. T., and deCastro, T., 1978, Lawrence Berkeley Laboratory Report LBL-6710 (submitted to Radiat. Res).

Tobias, C. A., 1973, Radiology, 108: 145.

Vosburgh, K. G., 1973, Enciclopedia Della Scienza E Della Technica Mondadori, Annuario della EST, S and T, 17: 270.

THE BIOLOGICAL PROPERTIES OF HIGH-ENERGY HEAVY CHARGED PARTICLES

Stanley B. Curtis

Biology and Medicine Division
Lawrence Berkeley Laboratory
Berkeley, CA, 94720, U.S.A.

INTRODUCTION

A number of studies are presently underway at Lawrence Berkeley Laboratory to evaluate the potential of high-energy charged-particle beams for use in radiotherapy. The scope of the ongoing studies is such that only a portion can be reviewed in the present paper. Emphasis is placed here on the results of those studies in which comparisons have been made between the beams of the available ions, i.e., carbon, neon and argon. Also, studies are stressed which have gathered data on the broad beams spread in depth, which are most interesting to the radiotherapist. After a survey is presented of the properties of relative biological effectiveness (RBE) and oxygen enhancement ratio (OER), some rather interesting and not altogether expected characteristics of these beams are discussed.

CELL SURVIVAL PROPERTIES OF UNMODIFIED BEAMS

In order to understand the biological characteristics of the modified or spread beams with extended peak regions, it is useful first to present results from unmodified beams with sharp peaks of ionization as a function of depth. An extensive series of experiments has been performed by Blakely et al. (1) to measure the RBE and OER as a function of depth in water for the carbon, neon, and argon beams. The variation of RBE and OER calculated at 10% cell survival for T-1 human kidney cells is shown in Fig. 1. The dose as a function of depth is shown in the lower portion of the figure. The LET of the carbon beam reaches a value high enough to affect significantly the values of RBE and OER only

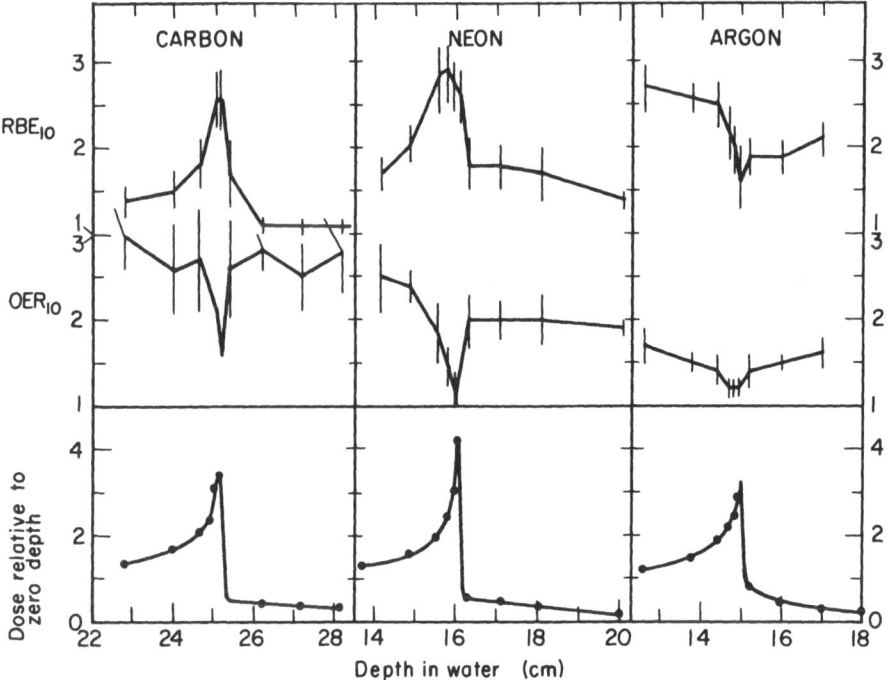

Fig. 1. RBE and OER at 10% survival levels for
        T-1 human kidney cells as a function
        of water depth for unmodified carbon,
        neon, and argon beams. Relative dose
        vs. depth curves are also given indi-
        cating the positions of the peak regions.
        Data from Blakely et al. (1).

at the very end of the range of the carbon ions. The situation
is similar for the neon ion beam, but since the LET is higher
for neon ions than for carbon ions with the same residual range,
the region in the peak where the RBE is high and the OER is low
has become broadened significantly. For the argon beam, the
LET is so high even at the shallower depths that the RBE is
initially quite high and decreases through the peak region of
ionization. This is due to the well-known saturation or "overkill" effect at very high LET. The OER, on the other hand,
decreases slowly and reaches a limiting value of 1.2 throughout
a considerable portion of the peak region. The fact that it
does not reach unity is probably due to the presence of a considerable contribution of lower LET nuclear secondaries produced
by the argon ions as they undergo nuclear interactions and fragment into lighter ions before reaching the stopping region.
Although these nuclear secondaries are present in all three beams

to some extent, the higher probability for nuclear interaction of the argon ions makes the nuclear secondaries more important in the argon beam than in the other two beams.

CELL SURVIVAL PROPERTIES IN SPREAD BEAMS

Because the sharp peaks created in the dose vs. depth curves shown in Fig. 1 are on the order of only two millimeters in width, these beams are not useful to the therapist who must treat tumors often extending over many centimeters. Therefore, the beams must be spread to produce extended peak regions. This can be done by introducing an absorber into the beam with variable thickness (called a "ridge filter"). This absorber is moved across the beam so that any one point in the peak region experiences the passage of particles having a broad spectrum of energies and LETs. This effectively spreads or broadens the peak region, and at the same time unavoidably decreases the peak height. Fig. 2 shows dose vs. depth curves for carbon, neon, and argon beams obtained using ridge filters designed to spread the peak region to 10 cm in depth. The three curves shown with a decreasing dose through the peak region were obtained with the same ridge filter. The decreasing dose was purposely obtained because the RBE was expected to increase through the peak region for the carbon and neon beams. Thus, if equal cell survival is desired through the peak region, the dose in the proximal or shallower end must be greater than the dose at the distal or deeper end to compensate for the lower RBE expected in the proximal region.

Survival of T-1 human kidney cells immobilized in gelatin was measured by Raju et al. (2) throughout these beams. Typical results of survival vs. depth are shown as data points in Fig. 2. These data were selected from an extensive series obtained by this group at various dose levels. The results here compare dose levels producing roughly 15% cell survival through the peak region. Two points of interest are: (a) the carbon and neon beams produce approximate isosurvival at this dose level with the ridge filter used, while there is a distinct increase in survival through the peak region for the argon beam; and (b) there is a well-delineated sharp increase in survival corresponding to the distal edge of the dose curve in each beam. The latter point illustrates an important property of all these charged particle beams--sharp demarcation of areas of high and low cell survival in the beam exit region of single port beams. This property also is true laterally since beam scatter is very small due to the high mass of the ions. The lack of isosurvival through the spread region of the argon beam is due to the fact that the RBE is at a maximum value in the proximal region of the peak. Other experiments, mentioned below, have shown this

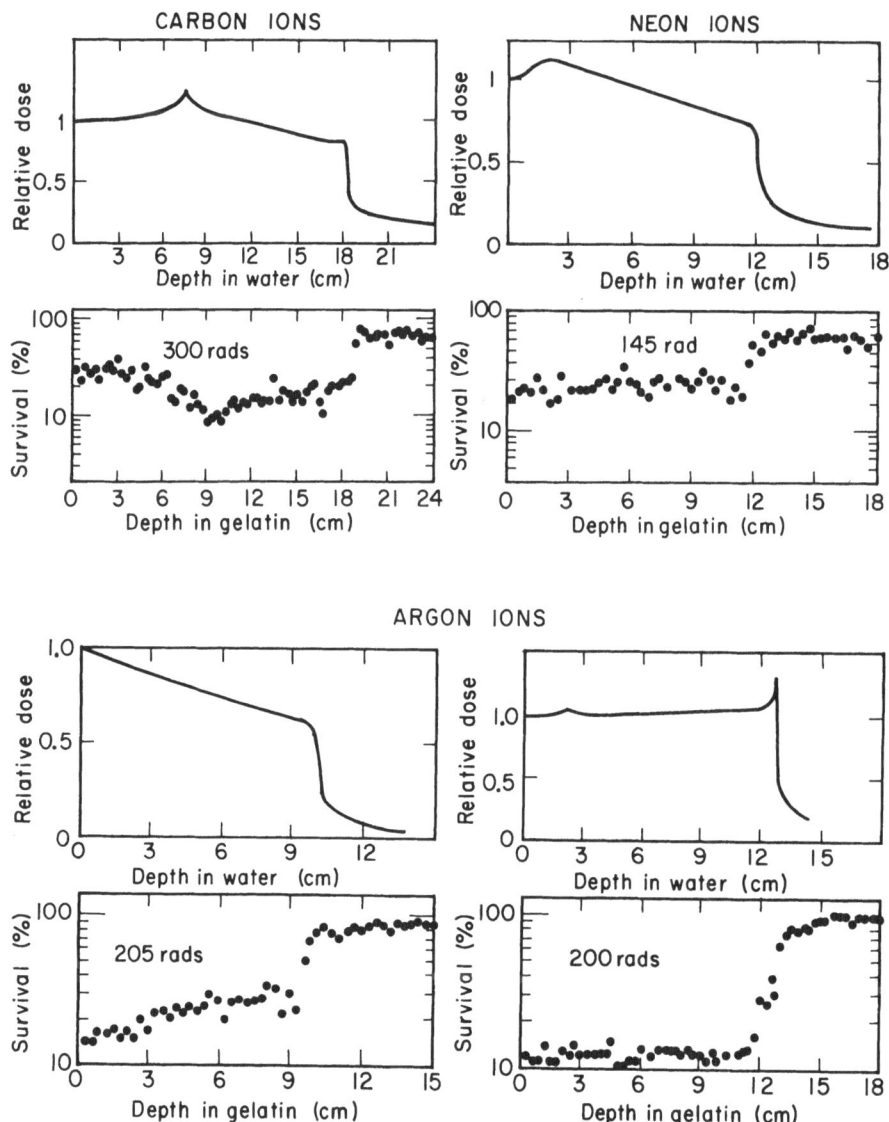

Fig. 2. Relative dose and cell survival of V79
Chinese hamster cells as a function of
water or gelatin, respectively, for 10-
cm spread peak regions of carbon, neon,
and argon beams. Data from Raju et al.
(2).

to be the case, and some systems have shown a significant decrease in RBE through the spread peak region. A ridge filter was designed to produce a flat depth-dose distribution with a small peak at the distal portion of the argon beam. The dose vs. depth and resulting cell survival vs. depth are shown in the lower right hand portion of Fig. 2. It is clear that this ridge filter design effectively results in isosurvival for this T-1 human kidney cell line.

The variations of RBE and OER through the spread regions of carbon, neon and argon beams are shown in Fig. 3 for the R-1 rhabdomyosarcoma tumor cell line irradiated in suspension (3). The positions and extent of the cell suspensions are shown on the dose vs. depth in the lower portion of the figure. Thus, the values plotted in the upper portion of the figure are average values over the 1.5-cm depth occupied by the glass vessels containing the cells.

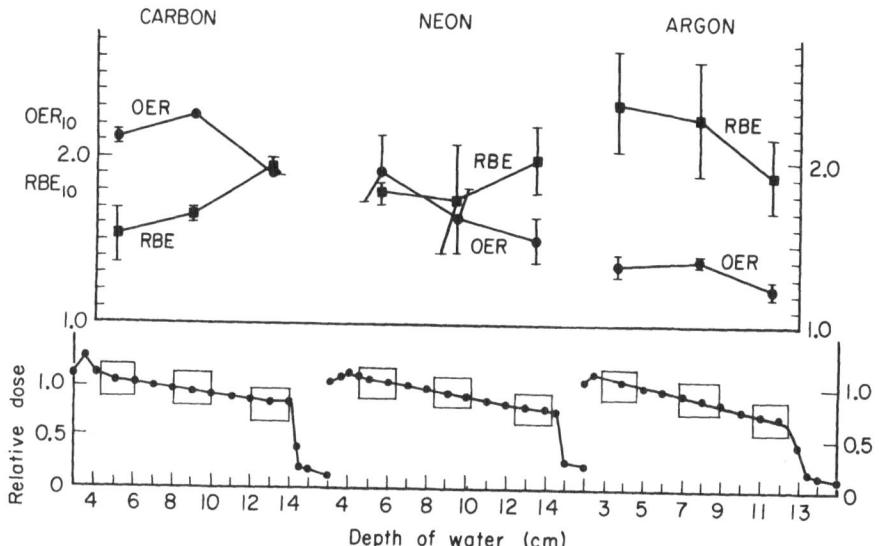

Fig. 3. RBE and OER at 10% survival level for
        R-1 rhabdomyosarcoma cells in suspension
        as a function of the water absorber for
        carbon, neon, and argon beams. The
        relative dose vs. depth curves indicate
        the positions and extent of the glass
        vessels containing the cells. Data from
        Schilling et al. (3).

In general, the results from this cell system indicate that the RBE varies more than the OER in the carbon beam peak region, with the OER remaining close to 2.0. In contrast, the OER varies more than the RBE through the neon-beam peak region, decreasing from around 2.0 to 1.5. Finally, the RBE decreases through the argon-beam peak region, while the OER is low (< 1.4) and remains so, although it does not reach unity at any point. Again, the failure of the OER for the argon beam to reach unity is probably due to the presence of contaminating low-LET nuclear secondaries in the peak region.

The results of the above studies lead to a consistant picture of the variation of the important radiobiological quantities, RBE and OER, through the peak regions of charged particle beams. In order to compare quantitatively results obtained with different cellular systems, proper normalization must be made so that intrinsic differences between cell lines or in experimental technique do not affect the comparisons. One way to obtain this in studying the reduction of the oxygen effect by these beams is to calculate the oxygen gain factor (OGF) defined as the ratio of the OER for x-rays to the OER for the heavy-ion beam in question. Such a comparison is made in Table 1 for R-1 rhabdomyosarcoma cells and T-1 human kidney cells at the midpoints of 4-cm spread beams with a range of 14 cm and 10-cm spread beams with a range of 24 cm (3,4). Another comparison is made for V79 Chinese hamster cells and T-1 human kidney cells in Table 2 at four positions in the unmodified argon beam (1,5). The agreement is remarkably good between cell lines.

Earlier data obtained by Chapman et al. (6) and by Raju et al. (7) substantiate the general features of the beams as outlined above.

Table 1

OXYGEN GAIN FACTORS ($OER_{x-ray}/OER_{H.I.}$)

	Midpoints			
	14cm range — 4cm R.F.		24cm range — 10cm R.F.	
	R-1 cells Schilling et al., 1979	T-1 cells Blakely et al., 1979	R-1 cells Schilling et al., 1979	T-1 cells Blakely et al., 1979
Carbon	—	1.3	1.3	1.1
Neon	1.6*	1.3	1.4	1.2
Argon	2.1*	2.2	—	—

*Average value between proximal and distal positions.

Table 2
Oxygen Gain Factors: Unmodified Argon Beam

	V79 Cells Hall et al. (5)	T-1 Cells Blakely et al. (1)
Plateau	1.4	1.3
Ascending	2.1	1.9--2.0
Peak	2.3	2.3
Tail	2.0	1.9

RESPONSE OF ORGANIZED TISSUES

Results have been accumulating on several organized tissues which indicate that for some tissues RBEs can peak at considerably lower LET than for the in vitro cell lines mentioned above. Alpen et al. (8) have shown for the intestinal crypt cell system that the lowest D_Q and therefore the highest RBE in the neon 4-cm spread beam is found in the proximal and not in the distal position of the beam. Also, in experiments in which mouse testis weight is measured, this group has again found that the highest RBE for the neon beam occurs in the proximal, not in the distal position of the beam (9).

EVIDENCE FOR A NET POTENTIATION OF DAMAGE AFTER SPLIT AND FRACTIONATED DOSES

Evidence is being obtained from two sources to indicate that there is, for the neon peak region and for the argon peak, a net potentiation of damage after a split or fractionated radiation schedule. Ngo et al. (10) have obtained survival curves for V-79 cells that show more cell killing after two doses split by a 3-hour interval than if the same dose is given all at one time. The survival curves for plateau and peak carbon, neon, and argon ions are shown in Fig. 4 for single and split-dose experiments. The two-dose split course is more effective than a single dose in the neon and argon peak regions. Reassortment through the cell cycle is ruled out because other studies have shown that synchronous and asynchronous cells kept at room temperature during the 3-hour interval to inhibit progression through the cell cycle also show this effect.

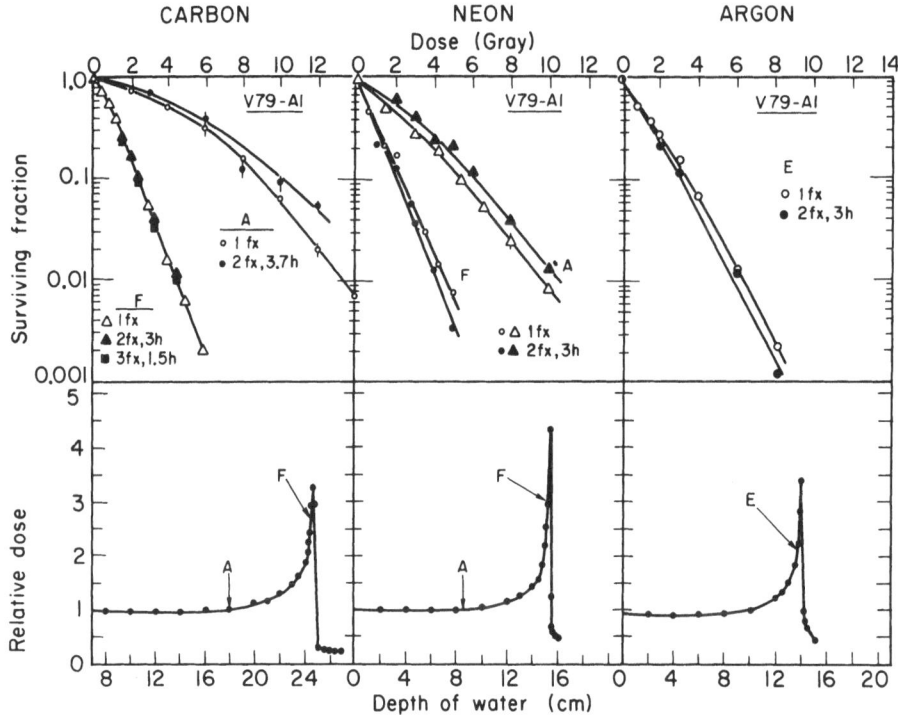

Fig. 4. Survival curves of V79 Chinese hamster
cells for single (open symbols) and split
(solid symbols) doses of carbon, neon,
and argon beams. The lower relative
dose vs. depth curves indicate the
positions in the beams (A, the plateau
region; E or F, the peak region) where
the survival was measured. Data from
Ngo et al. (10).

The second piece of evidence for a net potentiation of
damage comes from the Phillips group (11,12). Intestinal crypt
cell survival was assayed after 5 or 10 fractions (3-hour interval) of the various heavy-ion irradiations. The "recovered dose"
per interval can be calculated by the usual formula:

$$D_r = (D_n - D_s)/(n - 1),$$

where D_n is the total dose in the fractionated experiment that
will produce a given crypt cell survival level (here taken to
be 10 surviving cells/circumference), D_s is the total single
dose to produce the same survival level, and n is the number
of fractions. Another useful parameter is the fractional dose

recovered per interval per fraction:

$$f_r = n \, D_r/D_n.$$

These quantities are shown in Table 3 for the three ions in the plateau and 4-cm spead peak regions. Negative values for D_r in the table indicate the opposite of recovery, that is, potentiation of damage, i.e., more net cell killing if a dose is given fractionated than at one time. Potentiation is clearly evident in the 5- and 10-fraction spread-peak neon results. Because of the many processes occurring in this rapidly dividing

Table 3

INTESTINAL CRYPT CELL SURVIVAL [a]

Fractionation Data (3 hour intervals)

End point: 10 surviving crypt cells / circumference   4 cm ridge filter

		RBE (relative to ^{137}Cs γ's)	Recovered dose per interval (rads)	Percent dose recovered per interval per fraction (%)
^{137}Cs γ-rays	5 fractions	—	200	45
	10 fractions	—	100	43
Carbon ions	1 fraction Plateau	1.2	—	—
	Peak	1.1	—	—
	5 fractions Plateau	1.3	148	42
	Peak	1.8	~0	~0
	10 fractions Plateau	1.3	70	39
	Peak	1.9	~0	~0
Neon ions	1 fraction Plateau	1.2	—	—
	Peak	1.4	—	—
	5 fractions Plateau	1.9	~0	~0
	Peak	2.6	−38	−22
	10 fractions peak	2.6	−13	−15
Argon ions	1 fraction Plateau	2.5	—	—
	Peak	1.6	—	—

a. 1– and 5– fraction data from (11) , 10– fraction data from (12)

cell system, such as reassortment of cells in the cell cycle and possibly even recruitment of cells into the proliferative compartment, it is difficult to single out a specific cause for this enhanced damage. It is not unreasonable that it is caused, at least partially, by the same mechanism that caused the damage enhancement seen in vitro by Ngo et al. (10). There is also some evidence for potentiation in the mouse esophagus with $LD_{50/28}$ as an end point, but there is no evidence for this phenomenon in the mouse lung with $LD_{50/160}$ as an end point (12). Both these studies used one-day intervals between fractions.

SUMMARY

The heavy charged-particle beams are undergoing intensive experimentation to measure their radiobiological characteristics. The following conclusions can be drawn from cellular and animal studies to date:

1. The carbon OER does not drop below 2.0 in the spread peak region.

2. The neon OER is somewhat lower and reaches values similar to those obtained with various neutron beams presently being used for therapy.

3. The argon OER drops to values significantly below 1.5 even for beams spread to 10 cm in depth and, therefore, remains of therapeutic interest if an OER close to unity is desired. The OER does not, however, reach unity, even for the unmodified sharp peak argon beam.

4. The RBEs of the in vitro cell systems are understandable in terms of the LETs involved. The RBEs of two organized tissue systems, the intestinal crypt cell and testis, appear to peak at a lower LET than those for the in vitro cell systems.

5. There is clear evidence for a potentiation of damage in the peak regions of the heavier ion beams during fractionation both from in vitro cell studies and from studies in vivo of rapidly proliferating tissue, such as the intestinal crypt cell system. Such potentiation, however, is not a general property of organized tissue, as it is not evident in the mouse lung system.

ACKNOWLEDGEMENTS

This paper originally appeared in the Proceedings of the Sixth International Congress of Radiation Research, May 1979. It is reprinted here with the permission of S. Okada, M. Imamura, T. Terashima, and H. Yamaguchi, editors of that volume, and the Japanese Association of Radiation Research.

Special thanks go to Drs. E. L. Alpen, E. A. Blakely, F.G.H. Ngo, T. L. Phillips, and M. R. Raju for allowing inclusion here of their data before publication. These studies were supported by the National Cancer Institute, U. S. Department of Health, Education, and Welfare, and the Office of Health and Environmental Research of the U. S. Department of Energy under Contract No. W-7405-ENG-48.

REFERENCES

1. Blakely, E. A., Tobias, C. A., Yang, T.C.H., Smith, K. C., and Lyman, J. T. Inactivation of human kidney cells by high-energy monoenergetic heavy-ion beams. Radiat. Res. 80, 122 (1979).

2. Raju, M. R., Bain, E., Carpenter, E. A., Cox, B. S., and Robertson, J. B. A heavy particle comparative study. Part II: Cell survival versus depth. Brit. J. Radiol. 51, 704 (1978); M. R. Raju, private communication.

3. Schilling, W. A., Curtis, S. B., Tenforde, T. S., Crabtree, K. C., and Daniels, S., private communication.

4. Blakely, E. A., Tobias, C. A., Yang, T.C.H., and Smith, K. C. Cell survival parameters for extended Bragg peaks. Radiat. Res. 74, 588 (1978).

5. Hall, E.J., Bird, R. P., Rossi, H. H., Coffey, R., Varga, J., and Lam, Y. M. Biophysical studies with high-energy argon ions. 2. Determinations of the relative biological effectiveness, the oxygen enhancement ratio, and the cell cycle response. Radiat. Res. 70, 469 (1977).

6. Chapman, J. D., Blakely, E. A., Smith, K. C., and Urtasun, R. C. Radiobiological characterization of the inactivating events produced in mammalian cells by helium and heavy ions. J. Radiat. Oncol. Biol. Phys. 3, 97 (1977); and Chapman, J. D., Blakely, E. A., Smith, K. C., Urtasun, R. C., Lyman, J. T., and Tobias, C. A. Radiation biophysical studies with mammalian cells and a modulated carbon-ion beam. Radiat. Res. 74, 101 (1978).

7. Raju, M. R., Amols, H. I., Bain, E., Carpenter, S. G., Cox, R. A., and Robertson, J. B. A heavy particle comparative study. Part III: OER and RBE. Brit. J. Radiol. 51, 712 (1978).

8. Alpen, E. L., Powers-Risius, P., and McDonald, M. Survival of intestinal crypt cells after exposure to high Z, high energy charged particles. Submitted to Radiat. Res.

9. Powers-Risius, P., Alpen, E. L., Connell, G. M., and McDonald, M. The relative biological effectiveness of helium, neon, and carbon ions on mouse testes. Radiat. Res. 74, 588 (1978).

10. Ngo, F.Q.H., Blakely, E. A., Tobias, C. A., Chang, P. Y., and Smith, K. C., private communication.

11. Goldstein, L. S., Phillips, T. L., and Ross, G. Y. Enhancement by fractionation of biological peak-to-plateau relative biological effectiveness ratios for heavy ions. Int. J. Radiat. Oncol. Biol. Phys. 4, 1033 (1978).

12. Phillips, T. L., Goldstein, T. L., and Ross, G. Y., private communication.

TREATMENT PLANNING FOR CHARGED PARTICLE RADIOTHERAPY

George T. Y. Chen

Biology and Medicine Division
Lawrence Berkeley Laboratory
Berkeley, CA 94720

INTRODUCTION

This lecture will cover the general principles involved in treatment planning for charged particle radiotherapy. Development of treatment planning codes for charged particle radiotherapy is being actively pursued at the Massachusetts General Hospital/Harvard Cyclotron (protons), at the Los Alamos Scientific Laboratory (LASL) and at SIN (pions), and at the Lawrence Berkeley laboratory (LBL) (heavy ions). The general outline of the presentation will include (1) a review of beam properties and depth dose distributions used in treatment planning, (2) the use of CT as a quantitative method for evaluating the relative stopping powers of inhomogeneities. (3) treatment planning algorithms currently in use, (4) representative treatment plans and distribution of inhomogeneities, and (5) factors which affect compensator design (such as possible motions, compensator misregistrations, and multiple Coulomb scattering).

THERAPY BEAMS

The narrow Bragg peaks of charged particles are broadened to therapeutically useful dimensions by either ridge filters (Raju et al., 1972) or by the use of a dynamic range shifter. In either case, the function is the same. A fixed energy beam of charged particles enters the filter and a variable amount of material is introduced as a function of time. This has the effect of sweeping the Bragg peak back and forth along the beam axis in time. The resulting physical dose distribution is the superposition of a number of narrow Bragg peaks with different ranges, and is shaped to produce a region of uniform biological effect. With the fixed ridge filter system, it is practical only to have a finite number of spread Bragg peaks for use in therapy. At LBL, a family of spiral ridge filters from 4 cm to 14 cm in 2-cm steps is used for radiotherapy. Figure 1 shows the physical and "isoeffect" dose distributions for carbon ions. The physical dose distribution for each ridge filter is decreased with increasing depth to compensate for an increasing RBE, with the net result of producing a flat isoeffect region. The spiral or propeller ridge filter is used at LBL and Harvard, while a dynamic range shifter is used at LASL for modifying the depth dose distributions of negative pions. The dynamic range shifter (Amols et al. 1977) consists of a fluid filled piston which has a computer controlled time-dependent thickness. The fluid alters the residual range of the pions such that predetermined depth dose distributions may be obtained. Figure 2 shows the range shifter schematically. This device has the advantage of producing continuous spread Bragg peaks, rather than being quantized as at Berkeley.

The depth dose distributions most commonly used at LBL are the isoeffect or CoRE dose distributions (Castro et al., 1977). This acronym stands for cobalt rad equivalent, and represents the distribution of low LET dose equivalent to the biological effect produced by the heavy charged particle beams.

A depth dose distribution comparison of the various ions reflects some of the physical interactions these different particles undergo. Figure 3 shows proton, carbon, and pion spread beams. Protons deposit their energy primarily through ionization processes with few nuclear interactions, and their depth dose distributions show the primary beam coming to rest sharply with very little tail. The CoRE distribution of carbon, on the other hand, shows a lower entrance dose than protons, due to the RBE differential between spread peak and plateau, but exhibits a tail due to nuclear fragmentation which produces lighter daughter nuclei that travel past the primary beam. Pions have a low entrance dose and a tail due to neutrons and star products when the pion comes to rest and interacts with the medium. A description of other beam properties of the various particles is shown in Table 1.

References which describe the beam layout and dosimetry for these therapy facilities are found elsewhere (Koehler et al., 1977; Crowe et al., 1975; Smith et al., 1977; Alonso et al., 1979).

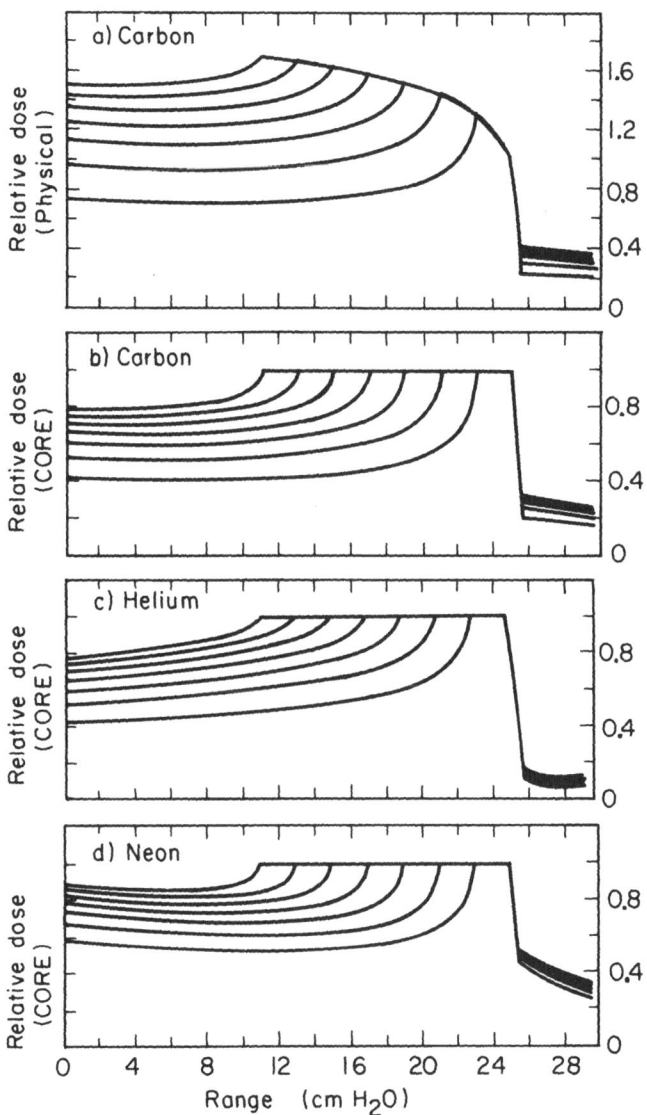

Fig. 1. (a) Physical dose distribution for carbon; (b) isoeffect distribution for carbon; (c) isoeffect for helium; (d) isoeffect distribution for neon. Originally published in Chen et al. (1979).

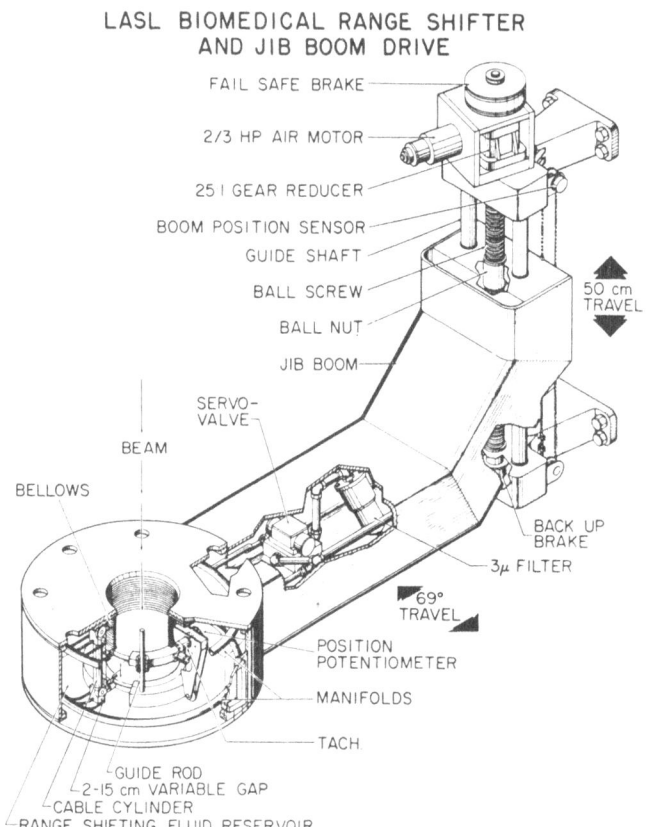

LASL BIOMEDICAL RANGE SHIFTER AND JIB BOOM DRIVE

FAIL SAFE BRAKE
2/3 HP AIR MOTOR
25 I GEAR REDUCER
BOOM POSITION SENSOR
GUIDE SHAFT
BALL SCREW
BALL NUT
JIB BOOM
50 cm TRAVEL
SERVO-VALVE
BEAM
BELLOWS
BACK UP BRAKE
3μ FILTER
69° TRAVEL
POSITION POTENTIOMETER
MANIFOLDS
TACH.
GUIDE ROD
2-15 cm VARIABLE GAP
CABLE CYLINDER
RANGE SHIFTING FLUID RESERVOIR

Fig. 2.   LASL range shifter.   Reprinted, by permission, from
         Amols et al. (1977).

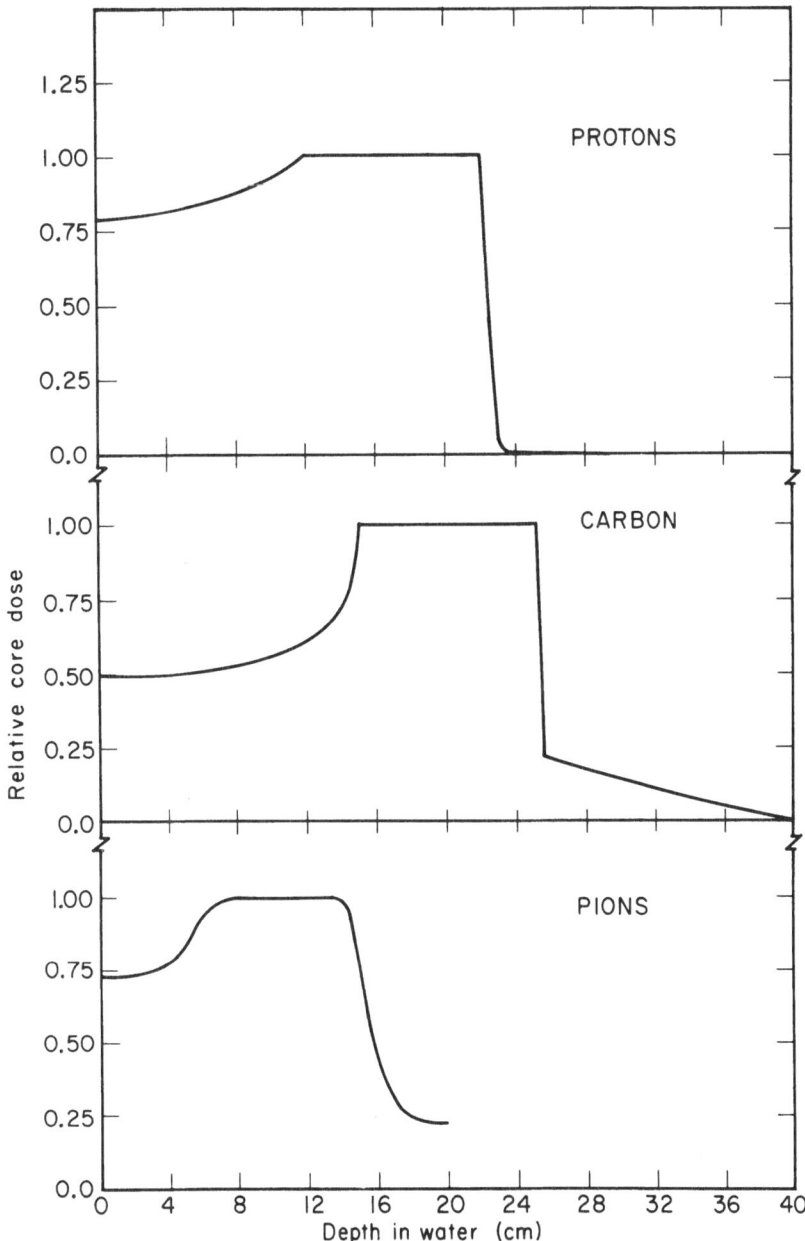

Fig. 3. Spread Bragg peak beams for protons, carbon,
and pions.

Table 1. Beam Properties

Institute	Ions	Range (cm)	Max Field Size	Dose Rate (rad/min)	Penumbra (relative)
Harvard	Protons	16	30	100	1.0
LBL	Carbon	26	30	200	0.25
LASL	Pions	28	20	30	2.0

QUANTITATIVE USE OF CT IN PLANNING

Because sharp dose gradients exist at the end of the beam range, it is critical in charged particle therapy to accurately assess the range required to reach the distal end of the target. Since the presence of inhomogeneities in the beam path may affect the beam stopping point, their effects must be quantitatively estimated. In addition to being used to define the position of the treatment volume relative to critical organs, CT data are used to quantify the inhomogeneities which perturb the particle beams. The CT value of each pixel must be converted to a water equivalent path length in order to provide the expected change in residual range of the beam. As they are calculated, CT numbers are a measure of the linear attenuation coefficient of each pixel at diagnostic photon energies, which run about 70 to 80 keV. At these energies, the linear attenuation coefficient is dependent on the effective atomic number, which contributes to the photoelectric effect, and the electron density, which is the dominant factor in charged particle transport equations.

The equation relating these quantities is:

$$\mu = \rho No \frac{Z}{A} [\sigma^{KN} + \sigma^{PE}] \quad ,$$

where ρ = density
No = Avogadro's Number
Z = atomic number
A = atomic weight
σ^{KN} = Klein Nishina cross section per electron
σ^{PE} = photoelectric cross section per electron.

The quantity $\rho NoZ/A$ is the electron density of the material in electrons per cubic cm. At 70 keV, the photoelectric effect contributes about 5 percent of the interactions in muscle and about 26 percent of the interactions in compact bone.

In contrast, the equation governing the slow down of charged particles in a medium is:

$$\frac{dE}{dx} \propto k_1 \rho No \frac{Z}{A} \left[\ln \frac{k_2}{I} - \beta^2 \right] \quad ,$$

where $k_1(\beta)$ = term including projectile charge and velocity.
     $k_2(\beta)$ = term including velocity.
     I = adjusted ionization potential of the medium.
     β = projectile velocity.

The effective dynamic range of atomic numbers found in the human body is not large (about 6 to 12) and the logarithmic dependence of energy loss on ionization potential results in only small variations due to Z.

The task at hand is to extract the electron density from CT scans. This could be done exactly with the use of dual energy scanning, where sequential scans of the same slice are performed at two different potentials. This technique (Rutherford et al., 1976) would allow for the exact extraction of both Z and electron density. However, at the present time, problems with the effective energy varying throughout the body and pixel misregistration tend to limit its usefulness. It may be possible in the future to produce dual energy scans of the body simultaneously, which would eliminate these problems.

With approximations, it is possible to correlate the water equivalent path length of each pixel with the CT number (Chen et al., 1979) from a single energy CT scan with sufficient accuracy for clinical use in most cases. Specifically, at LBL we have been using a method to define a calibration curve relating CT numbers of tissue analogues with the observed water equivalent pathlength. The model involves conceptualizing tissues of various densities as mixtures of two appropriate materials (Genant and Boyd, 1977; Kijewski and Bjarngard, 1978; Witt, 1973).

In order to simplify the process, the CT interval is divided into two subregions. For CT numbers less than 0, tissues producing the observed CT number are assumed to be a mixture of air and water. Since both of these materials have very nearly the same Z and Z/A, the CT number is proportional to the water equivalent pathlength. Fat, which has a Z of 6.2, is the only exception, and calculations show that for this material the lower Z is balanced somewhat by a higher Z/A, resulting in a net error of the calculated water equivalent pathlength from CT of about 1.5 percent.

For CT numbers greater than 0, the mixture is assumed to have a well-defined correlation between density and atomic number. This assumption allows for the water equivalent range of bone to be extracted from a single energy CT scan. The rationale for this model rests on the assumption that all intermediate density bone may be thought of as a mixture of water and compact bone. A realistic bone analogue (Witt, 1973) (dibasic potassium phosphate) is used in the calibration. This material may be dissolved in water to produce a number of solutions with varying density and Z.

A calibration curve relating CT number and water equivalent length is shown in Fig. 4a. These data were obtained by scanning the phantom materials in a CT scanner and then placing the sample solutions in a helium-ion beam to measure the relative stopping power. The tissue phantom materials used in the generation of the calibration curve are shown in Table 2. The data points were fit to a polynomial for CT < 55 and to a straight line for CT > 55.

An important consideration in pixel-by-pixel treatment planning where the CT numbers are use quantatively is the stability of the CT numbers as a function of time. Abols et al. (1978) have measured the CT numbers of reference solutions over a four month period. Figure 4b shows their results for a set of potassium phosphate solutions. Over the 100 day period, the CT number of the most concentrated bone analogue solution decreased by over 20 percent. Short term drifts in CT numbers have also been observed. C. Cann (private communication) found that CT values of test solutions changed by 5 percent when 12 CT slices were scanned over a 40 sec period. These variations in CT number must be monitored and corrected when treatment plans are calculated. One method of monitoring the drift is to build standard calibration solutions into the patient couch assembly. This produces images of the calibration solutions along with each slice of the patient anatomy.

It is also likely that CT data editing will be required in the treatment planning process. The CT scans performed at the time of the patient workup may contain transient inhomogeneities such as gas in the gastrointestinal tract. Transient inhomo-geneities could result in overshoot or undershoot of the beam relative to the tumor position. Since the CT matrix is used in the design of compensators (to be discussed later), methods for handling these transient inhomogeneities must be developed. Several possible solutions must be explored. First, the problem of bowel gas in the charged particle beam path may possibly be

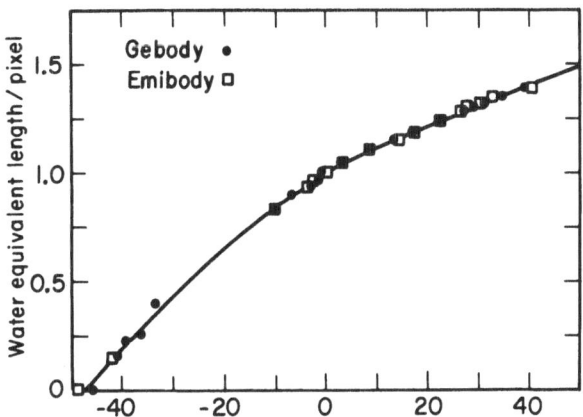

Fig. 4a. Calibration curve for converting CT numbers to
water equivalent range. Originally published in
Chen et al. (1979).

Table 2. Tissue Phantom Materials

Tissue	Analogue
soft tissue	water
fat	water and alcohol solutions
bone	water and potassium phosphate
lung	rando lung, cork, urethane

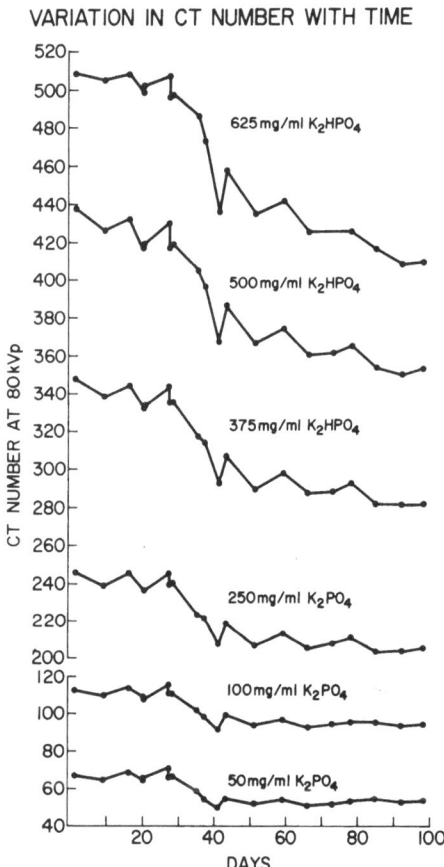

VARIATION IN CT NUMBER WITH TIME

Fig. 4b. CT number drift as a function of time. Reprinted,
by permission, from Abols et al., (1978).

minimized by careful diet management to reduce the production of gas. If gas is present in the scan, but it is expected to be of transient nature, it should be edited out of the CT data. This may be done by altering the CT numbers of the gas bubble to water equivalent density. While such a procedure may result in the possible beam overshoot if the bubble is present during therapy, it minimizes the possibility of tumor underdose if the bubble is not present during therapy but is taken into account during compensation design.

TREATMENT PLANNING ALGORITHMS

The approach to the actual calculational procedure is strongly dependent upon beam characteristic. At LBL, the heavy-ion beams are essentially parallel with about a 10 meter source to isocenter distance. This allows us to plan treatments on separate axial planes through the target volume. The algorithm used at Berkeley is a straightforward water equivalent pathlength technique. The radiotherapist marks the target volume on each CT slice using a cursor and CT graphics display unit. The parameters selected by the treatment planner only include the ion type, number of fields, and beam entry angles. The ridge filter and field size are determined from the target contours. The program evaluates the water equivalent pathlength from the distal end of the contour to the surface of the body along the beam direction, and then calculates the required compensation to stop the beam at the appropriate point. The plans are then calculated plane by plane. The resulting isodose or isoeffect distributions superimposed on the CT image may be displayed on the graphics unit or as a hard copy generated by an electrostatic printer plotter.

The problem at Los Alamos is somewhat more complex due to the nature of the beam line and particle characteristics. At LASL, the beam is converging to a focus or "tune," and rays traverse axial planes other than the plane in which the particle finally comes to rest. In addition, the dose at a plane is affected more strongly with pions than with heavy ions by the star and neutron components of adjacent planes, and by the substantial amounts of multiple Coulomb scattering pions undergo. All of these factors necessitate that the dose calculation for pion treatment planning be performed from the beginning in three dimensions (P. Bernardo, private communication). Measured particle trajectories are traced through a three-dimensional cube of CT-derived density data to establish the stopping point of the pion, and energy deposition from ionization and stars are calculated. The complexity of this problem requires that the calculation be done on a large frame computer, and the one used at LASL for treatment planning is the CDC7600.

REPRESENTATIVE TREATMENT PLANS

In photon treatment planning, the physicist uses weighting and various beam angles in order to achieve a uniform dose to the target and yet spare adjacent critical organs. Planning with charged particle beams is substantially easier since one can consider that the beams may be stopped at the distal contour of the target with an appropriately designed compensator. Therefore, because of the finite range property and the ability to shape dose, planning with particles is conceptually simpler and requires fewer iterations to avoid overdosing critical structures. However, the steep dose gradients impose more stringent conditions on the technical aspects of dose delivery, such as correct range determination and compensation alignment. As this type of treatment planning uses pixel-by-pixel data literally, one should examine the shape of the designed compensator and choose that beam entry angle which both avoids critical organs and minimizes the complexity of the compensator shape. Given these initial remarks, we now examine some representative treatment plans using charged particle beams.

Figure 5a shows an isoeffect treatment plan for the irradiation of a pancreatic lesion using carbon ions. Here, equally weighted anterior and lateral ports have been used. In addition to a surface flattening bolus, a compensator (not visible here) has been calculated and correctly registered to stop the beam at the distal contour. In the abdomen, the major types of tissues and inhomogeneities found are soft tissue, fat, ribs, and bowel gas. At LBL an attempt is made not to traverse the spinal cord when possible, since this is a complex inhomogeneity with range perturbations of about 1.5 cm. The pixels within the radiation field may be histogrammed to illustrate the quantity and type of the tissue traversed (Fig. 6). It can be seen that the major components in the abdomen are primarily fat and soft tissue. The ribs, although present, perturb the beam minimally, on the order of 3 mm. This is consistent with an expected 1 cm pathlength through the rib with an average density of about 1.3 g/cc. Note that the target volume is uniformly irradiated to 100 percent CoRE and the spinal cord and kidneys are essentially spared. The overlying tissue proximal to the tumor receives about 45 percent of the tumor dose. The distribution is characterized by sharp lateral penumbras and a step dose gradient at the end of the primary beam. The 10 percent isodose lines extending past the primary beam stopping point are due to the fragmentation dose component of the heavy-ion beam.

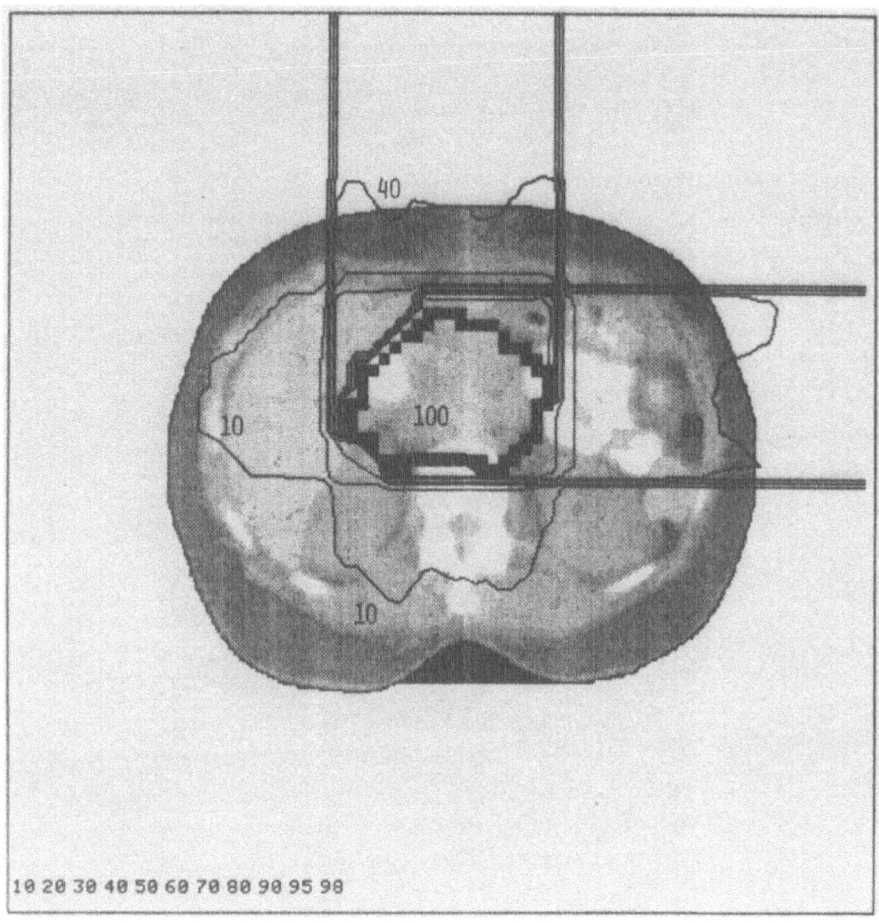

Fig. 5a. Isoeffect plan for irradiation of the pancreas with carbon beams.

Figure 5b shows the physical dose distribution for the isoeffect plan in Fig. 5a. This distribution shows a highly inhomogeneous physical dose area covering the target volume in contrast to the uniform biological effect region of the isoeffect plan. Although physical dose distributions are of use in the complete documentation of new radiotherapy modalities, without a complete map of the corresponding RBE at each point, the physical dose distribution is of limited value.

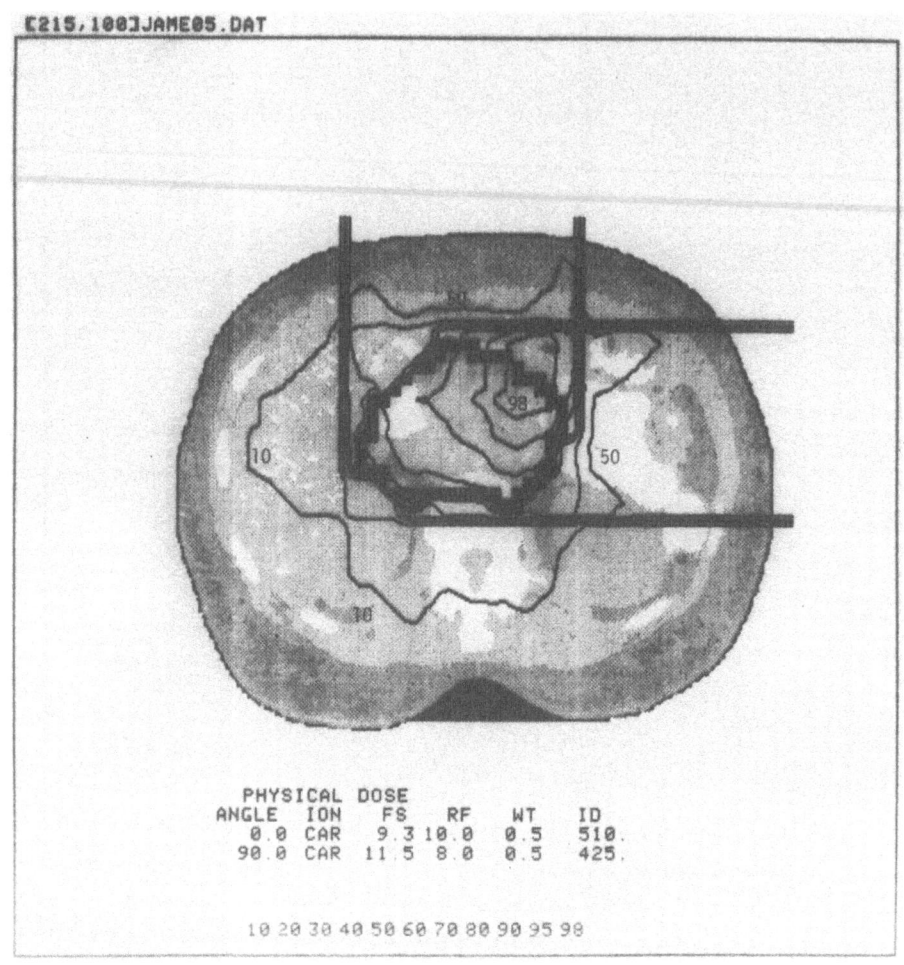

Fig. 5b. Physical dose distribution for treatment plan 5a.

A pion treatment plan for the irradiation of a different abdominal lesion is shown in Fig. 7, where an AP-PA field arrangement was used. The use of parallel opposed fields is preferred by the LASL group because it tends to produce more uniform physical and biologically effective dose distributions,

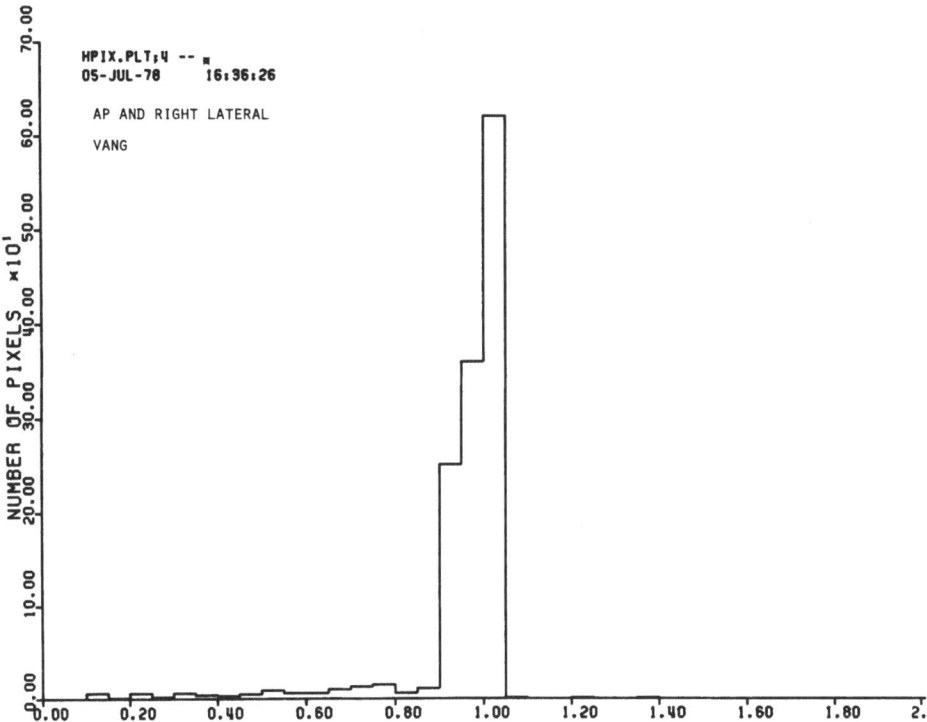

Fig. 6. Histogram of pixels within the radiation field of
Fig. 5. Originally published in Chen et al. (1979).

given the uncertainties in RBE for pions. In this plan, the
tumor is irradiated to the 80 percent level (minimum) with an
inhomogeneity in dose of 20 percent. This plan was generated
from a CT density matrix (Hogstrom et al., 1979). Note that, in
contrast to heavy ions, pions have a substantial lateral
penumbra. The dose fall off from 80 percent to 20 percent
extends over several cm. In this particular plan, the spinal
cord receives about 50 percent of the tumor dose, which leaves
this critical structure well below tolerance. This physical pion
dose distribution may be transformed to an isoeffect distribution
by scaling the 80 percent line by a factor of 1.4 and
renormalizing.

If a four-field photon plan were used to irradiate this
target volume, the isodose distribution in Fig. 8 might have been

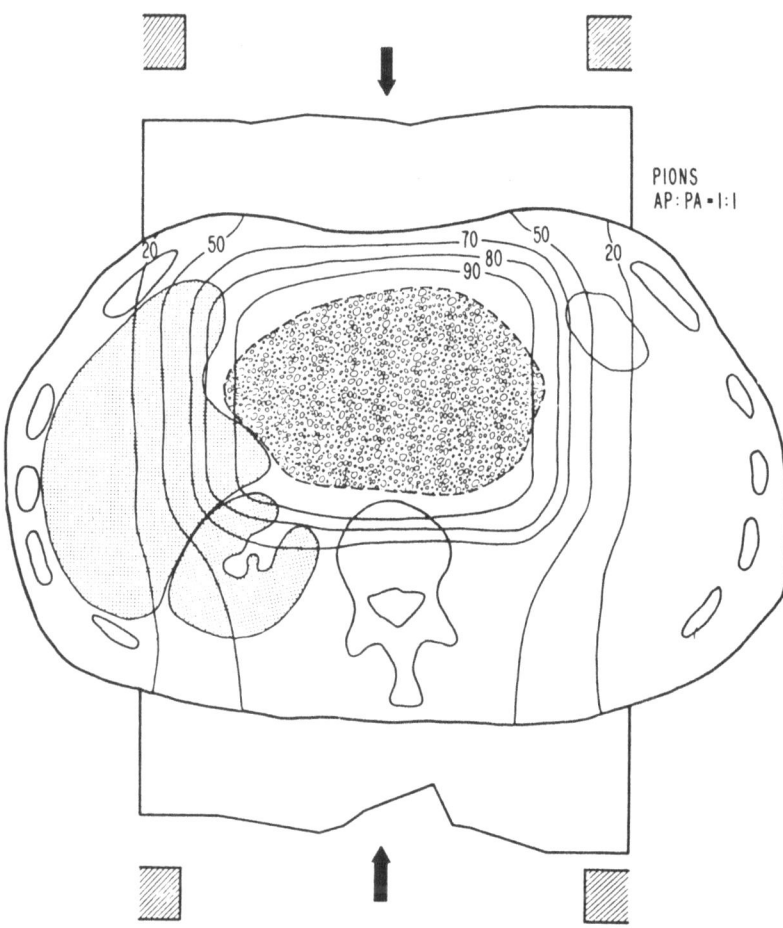

Fig. 7. Treatment plan for irradiation of abdomen with pions.
Reprinted, by permission, from Hogstrom et al. (1979).

developed. Here, 10 MeV photons are used to irradiate the same
target volume. The target volume is enclosed by the 90 percent
isodose line. However, because lateral photon beams were used,
the liver and a large amount of GI tract to the left of the tumor

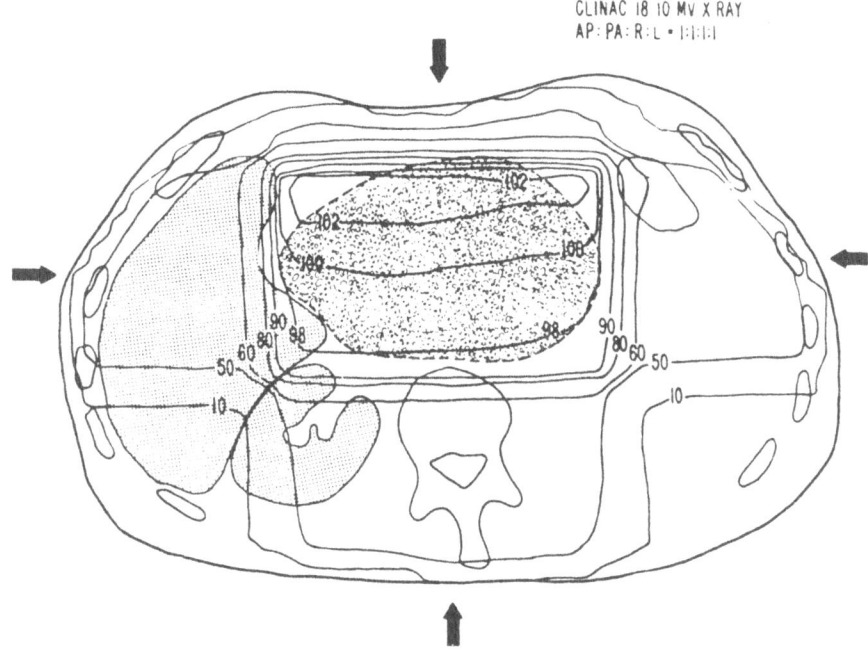

Fig. 8. Treatment plan for irradiation of abdomen with 10 MeV photons. Reprinted, by permission, from Hostrom et al. (1979).

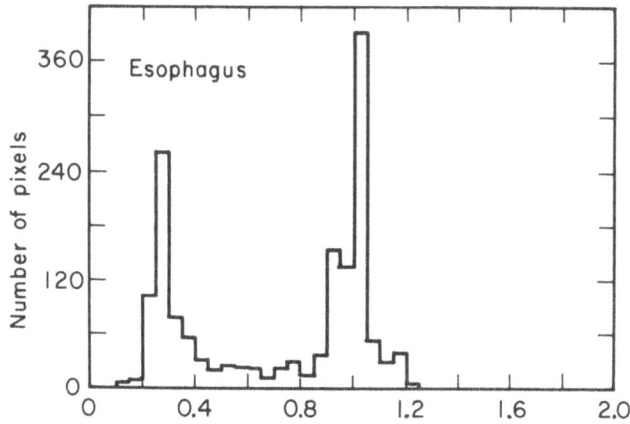

Fig. 9. Histogram of pixels in a thorax irradiation. Originally published in Chen et al. (1979).

were also irradiated to the 50 percent level, thereby substantially increasing the body radiation load. However, without the
use of these laterals, the photon dose distribution would have
resulted in the unacceptable clinical situation of irradiating
the cord to about 102 percent of the target dose (with the use of
only AP-PA fields).

Another site of interest to charged particle therapy groups
is carcinoma of the esophagus. The primary inhomogeneities in
the irradiation of an esophageal lesion are the lungs, trachea,
bronchi, sternum, and ribs. As Fig. 9 shows, the thorax presents
a number of regions with vastly different densities. This
histogram of pixels within the beam path reveals a soft tissue
peak at 1.02 g/cc with a large peak from lung at 0.27 g/cc. The
pixels with intermediate densities between 0.5 and 0.77 g/cc are
located at the edges of the lungs and body, and are due to
partial volume effects.

A neon treatment plan for the esophagus is shown in Fig. 10.
Three equally weighted and compensated fields are used in this
conedown irradiation. The anterior and right portals have wedge
shaped compensators to pull the dose gradient away from the
spinal cord. Again, the target volume is uniformly irradiated to
100 percent, with the mediastinum and lung treated to about
35 percent. The spinal cord is exposed to 10 percent of the
target dose. This plan is presented in the hard copy mode, where
isodose lines are superimposed over the 80 x 80 density matrix.

In the pelvis, beams must penetrate through the iliac crest
before reaching the target volume. A pixel analysis of the
CT scan (Fig. 11a) shows that within the beam paths, about
25 percent of the pixels for this patient are fat, about
17 percent are bone with an average density of 1.17 g/cc, and the
rest is soft tissue. A three-field portal agreement is shown in
Fig. 11b for the irradiation of this lesion. The carbon beam is
perturbed most greatly in the posterior port, which must traverse
the high density bone in the posterior region of the left iliac
crest. From this direction the range shortening may be as great
as 2 cm when compared with a pure soft tissue path. Compensation
would be added to shape the high dose region to the circular
target. With heavy ions, we are able to limit the dose to the
rectum, bladder, and intestinal tract.

FACTORS INFLUENCING PLAN ACCURACY

Thus far, the heavy-ion isodose distributions shown are
static. They do not include the degrading effects of motion,
breathing, compensation misregistration, or multiple Coulomb
scattering. In addition, the particles are assumed to travel in
straight lines, and the water equivalent pathlength is integrated

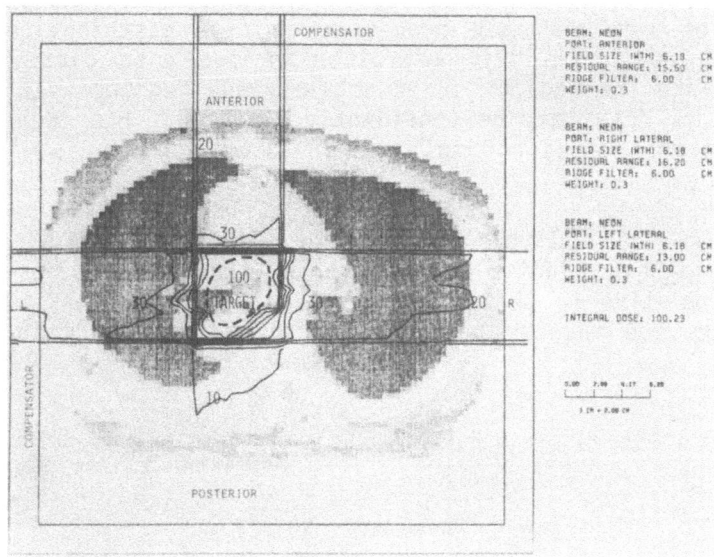

Fig. 10. Treatment plan for an esophagus patient using neon beam. Originally published in Chen et al. (1979).

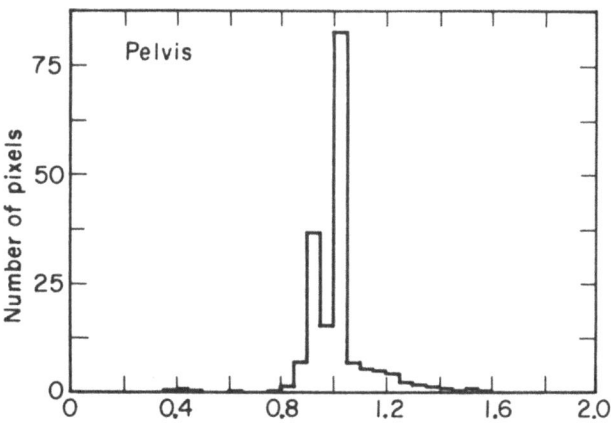

Fig. 11a. Histogram of pixels in pelvic irradiation. Originally published by Chen et al., (1979).

along these lines. In this section, methods to fold some of these effects into the calculation of a treatment plan are now discussed.

Goitein (1978) has developed a technique for analyzing the effects of some of these processes on dose distributions. His approach results in the generation of two dose distributions, which define the regions of uncertainty and regions within which high or low doses may be confidently expected. His technique to calculate the effects of multiple Coulomb scattering will now be described.

Because of multiple scattering, particles do not follow straight paths. Particles may wander through sections of low density relative to the straight path through material along the beam path, and thereby penetrate further than expected. Others

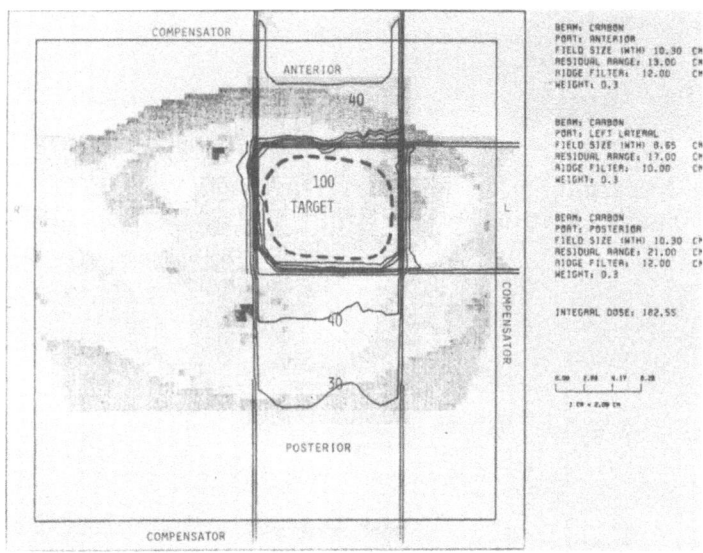

Fig. 11b. Treatment plan for pelvic lesion with carbon beam. Originally published in Chen et al. (1979).

may traverse high density regions and stop upstream of the stopping point as calculated along the line path. Accurate calculations of the effects of multiple scattering on dose distributions could be made with Monte Carlo simulations, but these techniques tend to take large amounts of time, and therefore are not clinically feasible. Goitein suggests a simpler method for estimating the maximum and minimum stopping points (and dose distributions) arising from these effects. First, let us look at Fig. 12, which represents a plan without scatter. The distal straight line is, for illustrative purposes, the line where the

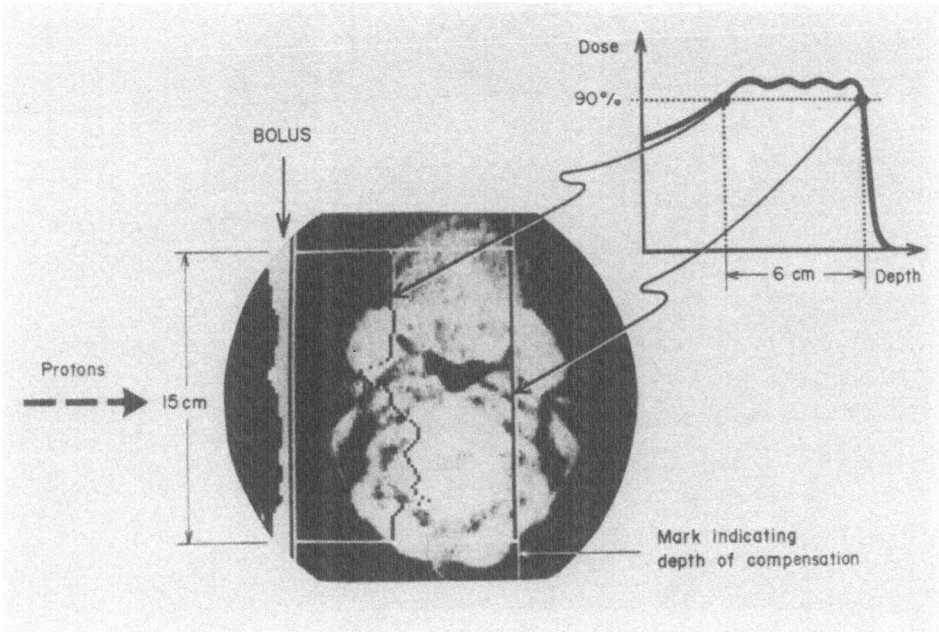

Fig. 12.   Proton treatment plan without multiple scattering.
Reprinted, by permission, from Goitein (1978).

beam should stop. A compensator shown to the left was designed
to stop the beam at the line in the absence of scatter. The
intermediate line represents the 90 percent point of the proximal
peak of the spread Bragg peak. A particle incident along the
dotted line in Fig. 13, would, in the absence of multiple Coulomb
scattering, pass through some well-defined element of the matrix,
as indicated black in the inset of this figure. However, because
of scattering, it might traverse along any of the neighboring
shaded elements instead. To compute the maximum areal density
path possible, which leads to the minimum geometric penetration,
the density of any element is reset to the maximum value of the
density of element or its neighbors within a given number of
pixels. Similarly, one can select the minimum density for
calculation of the maximum geometric penetration. The resulting
distributions will give the limits of maximum and minimum
penetration. If all particles chose the most dense or least

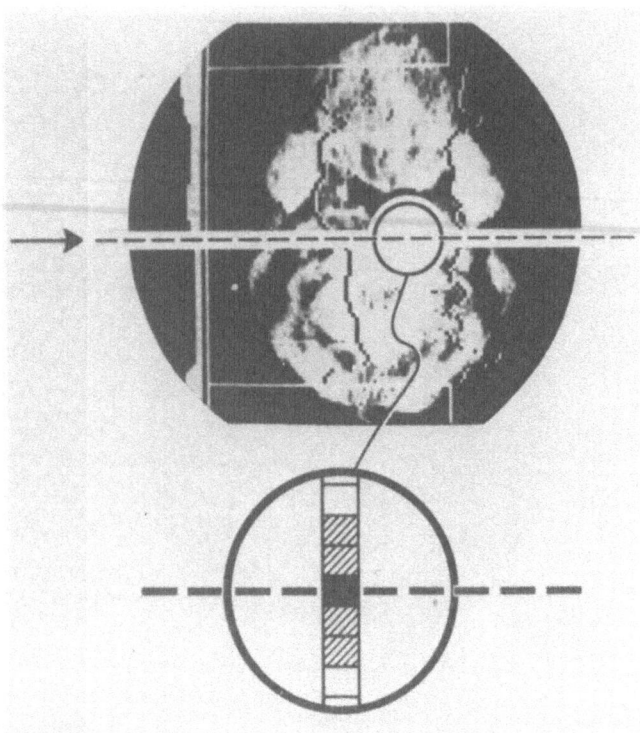

Fig. 13. Particle trajectory through geometrically exact pixel
        and through possible adjacent pixels. Reprinted, by
        permission, from Goitein (1978).

dense path, they would penetrate only as far as the distal
contours in Figs. 14a and 14b.

    Given these results, the strategy calls for the design of a
bolus to assure that the 90 percent isodose reaches the depth of
interest (Figs. 14c and 14d). A similar line of reasoning may
also be used to take into account patient motions or bolus
misregistration. In this approach, one is certain to cover the
tumor adequately, but with the possibility of overshoot, the
extent of which is calculated and presented to the radiotherapist
for evaluation.

    At Berkeley, we have calculated the degradation of CoRE dose
distributions due to breathing. It is assumed that respiration
is sinusoidal in time and has a maximum amplitude of 0.5 cm.
This is corroborated with film stack experiments which measure
the maximum water equivalent pathlength through the body when the

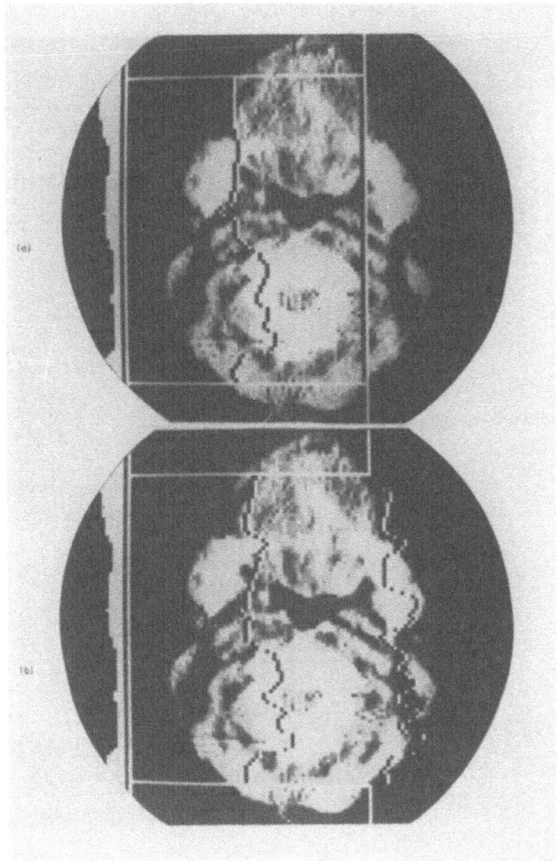

Fig. 14a,b.   Envelope of least and greatest
              possible penetrations. Reprinted,
              by permission, from Goitein (1978).

patient holds his breath at the extrema of inhalation and exhala-
tion. The respiration-modified spread Bragg peak is shown in
Fig. 15, and illustrates that only a small change in the distal
slope of the depth dose curve is produced by respiration. This
new curve may then be used to plan treatments where respiration
is expected to be a factor (primarily in abdominal treatment
plans).

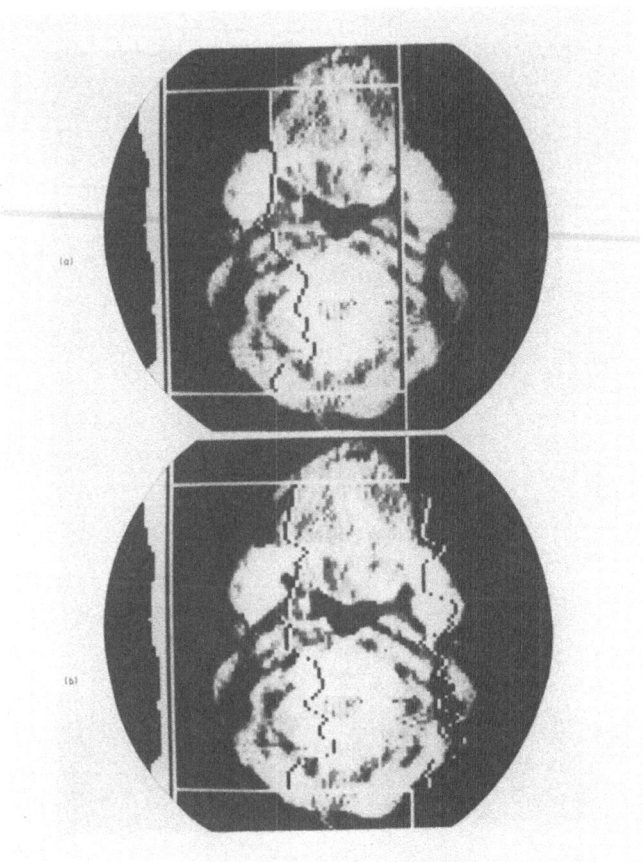

Fig. 14c,d. Bolus designed to compensate for
scattering-envelope of least and
greatest possible penetrations.
Reprinted, by permission, from
Goitein (1978).

Bolus misregistration has also been simulated by the
capability of laterally offsetting the compensator designed from
the static case, and recalculating the dose distribution with the
misaligned compensator. Figure 5 showed an exactly registered
compensator and its dose distribution, while Fig. 16 shows the
resulting dose distribution from a compensator misregistered by

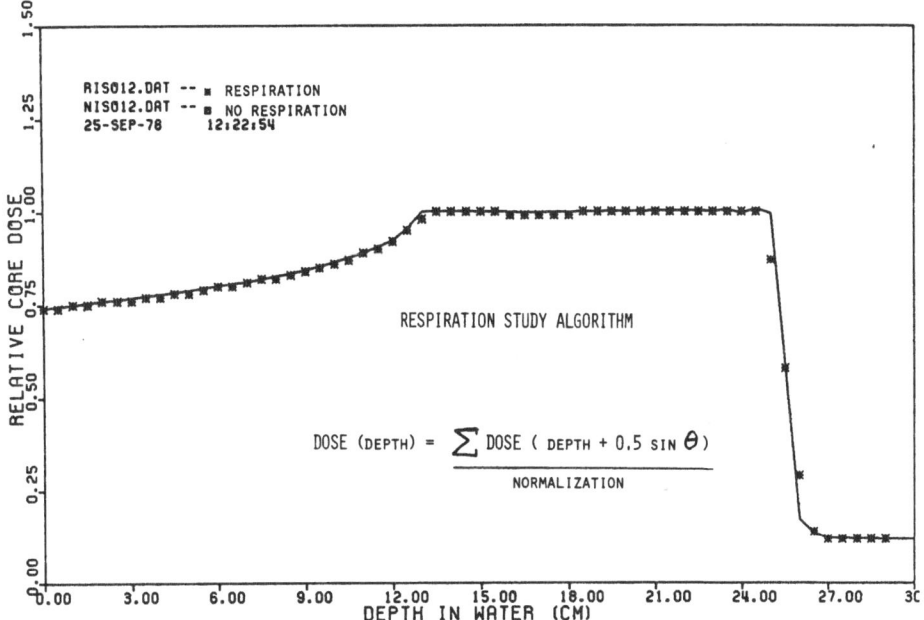

RISO12.DAT -- ▪ RESPIRATION
NISO12.DAT -- ▪ NO RESPIRATION
25-SEP-78   12:22:54

RESPIRATION STUDY ALGORITHM

$$\text{DOSE (DEPTH)} = \frac{\sum \text{DOSE (DEPTH + 0.5 SIN } \theta)}{\text{NORMALIZATION}}$$

Fig. 15.   Spread Bragg peak modified to include
           respiration effects.

1 pixel. In Fig. 16, the 100 percent isodose is seen to extend
into the target contour, indicating inadequate coverage. This
misregistration is most likely to be critical in head and neck
cases, where misregistration can have disastrous effects on the
dose distribution. Calculations of misaligned compensators are
useful in evaluating the allowable limits of misplacement and the
resulting dose distributions.

SUMMARY

    This lecture has described the current status of treatment
planning for charged particle radiotherapy. Rapid evolution of
the treatment planning codes is likely within the next few years
as additional data become available. Dosimetry will provide
information necessary to establish the validity of the algorithms
used. Improvements in the conversion of CT number to water
equivalent range may provide for more accurate range calculations. Radiobiological studies may indicate the need for
different RBEs for different organs, and the validity of the CORE
concept. Accurate treatment planning codes in charged particle
radiotherapy will be an essential tool in the precision delivery

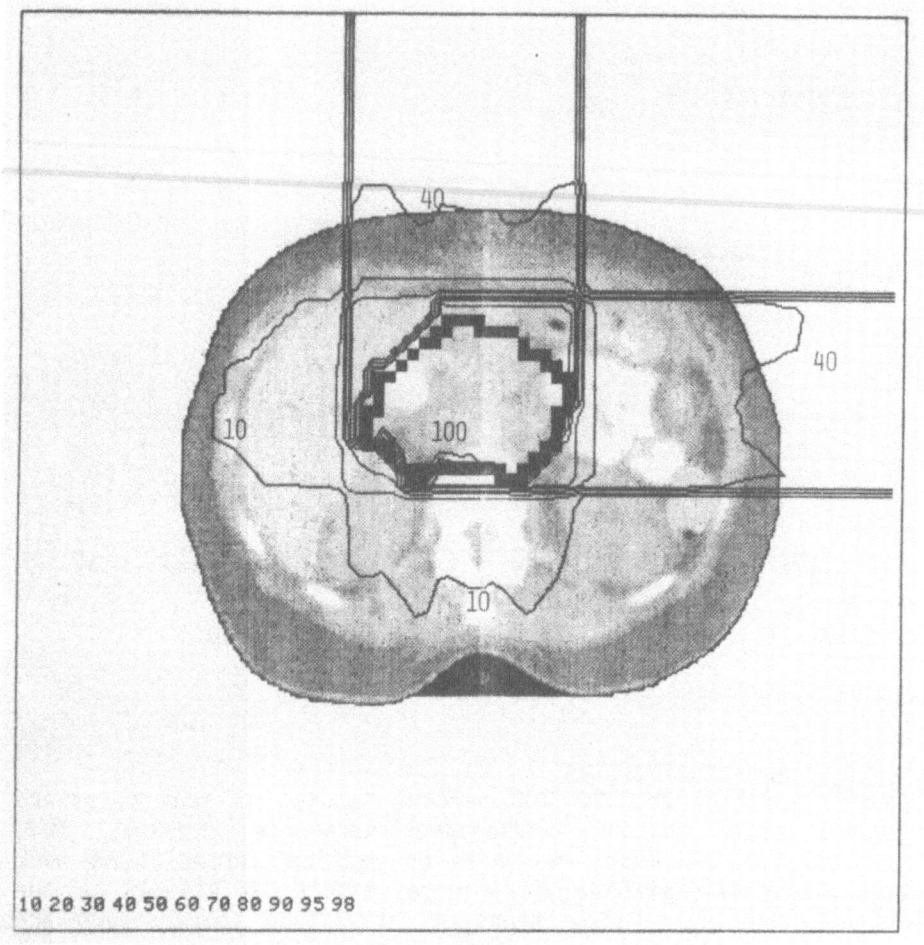

Fig. 16.   Distribution for compensator 1 pixel laterally dis-
           placed as compared with Fig. 5.

of these new particle beams. We hope that with precision dose
localization and their radiobiologic properties, these new
particle beams will have an impact on the local control of cancer.

ACKNOWLEDGEMENTS

     I am pleased to acknowledge extremely useful conversations
with Dr. Michael Goitein of the Massachusetts General Hospital,
and Drs. Peter Berardo, Sondra Zink and Ken Hogstrom of
Los Alamos Scientific Laboratory about treatment planning for

charged particle beams. Some of these studies were supported by grants from the National Cancer Institute and the U. S. Department of Energy under contract No. W-7405-ENG-48.

REFERENCES

Abols, V., Genant, H. K., Rosenfeld, D., Boyd, D. P., Ettinger, B. S., and Gordon, G. S., 1978, Spinal bone mineral determination using CT in patients, controls and phantoms, Proceedings of the 4th International Conference on Bone Measurements, University of Toronto, Canada, June 1978.

Alonso, J. R., Howard, J. and Criswell, T., 1979, The BEVALAC radiotherapy facility, Lawrence Berkeley Laboratory, Report LBL-8961.

Amols, H. I., Liska, D. J., and Halbig, J., 1977, Use of a dynamic range shifter for modifying the depth-dose distributions of negative pions, Med. Phys., 4, 404.

Castro, J. R., Quivey, J. M., Lyman, J. T., Chen, G. T. Y., Kanstein, L. and Walton, R., 1977, Heavy ion radiotherapy, in "Biological and Medical Research with Accelerated Heavy Ions at the BEVALAC, 1974-1977," Lawrence Berkeley Laboratory, LBL-5610.

Chen, G. T. Y., Singh, R. P., Castro, J. R., Lyman, J. T., and Quivey, J. M., 1979, Treatment planning for heavy charged particle radiotherapy. Int. J. Radiat. Oncd. Biol. Phys. Vol. 5 No. 10, p1809-1819.

Crowe, K., Kanstein, L., Lyman, J., and Yeater, F., 1975, A large field medical beam at the 184 in. cyclotron, Lawrence Berkeley Laboratory Report LBL-4235.

Genant, H. K., and Boyd, D., 1977, Quantitative bone mineral analysis using dual energy computerized tomographic scanning, Invest. Radiol., 12, 545 (Abstract).

Goitein, M., 1978, Compensation for inhomogeneities in charged particle radiotherapy using computed tomography, Int. J. Radiat. Oncol. Biol. Phys., 4, 499.

Hogstrom, K. R., Smith, A. R., Kligerman, M. M., vonEssen, C. F., Sala, J. M., Mettler, F. A., Kelsey, C. A., Simon, S. L., Somers, J. W., Lane, R. G., Rosen, I. I., Berardo, P. A., and Zink, S. M., 1979, Static pion beam treatment planning of deep seated tumors using CT scans at LAMPS: Clinical aspects, Int. J. Radiat. Oncol. Biol. Phys., Vol. 5, No. 6, pp. 875-886.

Kijewski, P. K., and Bjarngard, B. E., 1978, The use of computed tomography data for radiotherapy dose calculations, Int. J. Radiat. Oncol. Biol. Phys., 4, 429.

Koehler, A. M., Schneider, R. J., and Sisterson, J. M., 1977, Flattening of proton dose distributions for large field radiotherapy, Med. Phys., 4,, 297.

Raju, M. R., Gnanapurani, M., Martins, B., Howard, J., and Lyman,
    J. T., 1972, Measurement of OER and RBE of the 910 MeV helium
    ion beam using cultured cells (T1), Radiology, 102, 425.
Rutherford, R. A., Pullan, B. R. and Isherwood, I., 1976.
    Measurement of effective atomic number and electron density
    using the EMI scanner, Neuroradiology, 11, 15.
Smith, A. R., Rosen, I. I., Hogstrom, K. R., Lane, R. G., Kelsey,
    C. A., Amols, H. I., Richman, C., Berardo, P. A., Helland,
    J. A., Kittell, R. S., Paciotti, M. A., and Bradbury, J. N.,
    1977, Dosimetry of pion therapy beams, Med. Phys., 4, 408.
Witt, R. M., 1973, Bone standards for the intercomparison of
    photon absorption-metric bone mineral measuring systems,
    Proceedings, International Conference on Bone Mineral
    Measurements, Chicago, 1973, DHEW Pub. No. 75-883.

USE OF COMPUTERIZED TOMOGRAPHY IN PHOTON AND ELECTRON TREATMENT PLANNING.

Andrée Dutreix

Physics Department
Institut Gustave-Roussy
94800 Villejuif - France

INTRODUCTION

In the ICRU report 24 (1976) on the determination of absorbed dose in a patient, the accuracy required in clinical dosimetry is examined (chapter 7.1) and the following statement made : "The conclusion which emerges is that although it is too early to generalize, the available evidence for certain types of tumour points to the need for an accuracy of $\pm$ 5 % in the delivery of an absorbed dose to a target volume, if the eradication of the primary tumour is sought".

In order to ensure such an accuracy in dose delivery, accurate treatment planning is needed. This requires a knowledge of cross sectional anatomy for two main reasons : First, determination of the extent of the tumour and its localization with respect to surrounding tissues and body outline including the determination of the position and size of critical organs which may need to be protected ; second, quantitative information on density distribution throughout the treated volume to enable corrections to be made for the attenuation of X-Ray beams or the absorption of electron beams in different materials.

In the Gustave Roussy Institute, the first determinations of anatomical cross-sections in transverse planes were performed in 1955 with conventional axial tomography for sitting patients (Frain et al., 1955). Since 1974, they have been performed for lying patient with a Toshiba tomograph. The accuracy obtained with these methods was much greater than with conventional orthogonal radiographs. Body contours are usually indistinct but a set of small lead wires placed at the level of the cross-section allows one to achieve a

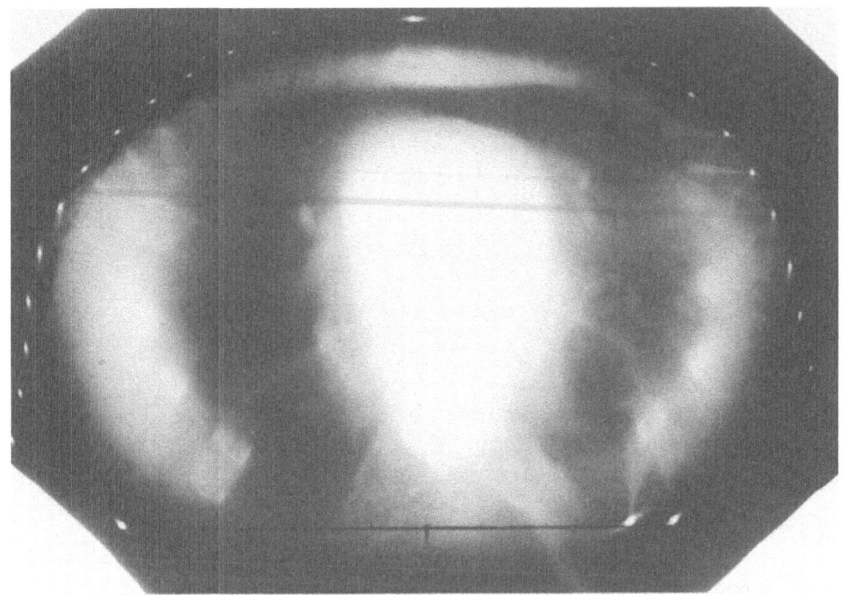

Fig. 1. Transverse section of a thorax. The image was obtained with
        a Toshiba transverse tomograph. Sections of lead wires are
        visible at the skin level. The three white points at the
        bottom of the radiograph are lead markers placed on the
        table top to visualize the couch level.

precision of ± 2 mm on the outline (figure 1) ; contours of organs
have been read out and digitized by mean of a curve follower and
the numerical data used directly in the computer for treatment
planning. Unfortunately, only bones, lungs and a few opacified
organs could be localized with accuracy because of the poor contrast
between the various soft tissues.

The recent development of CT scanners has provided good quality
images of patient cross-sections whose advantages for treatment planning are evident. As the therapist must determine the extent of the
tumour in all directions, as well as the position and nature of
adjacent structures in the various regions of the target volume,
he needs a series of anatomical cross-sections through the entire
volume of the tumour. These data are then used in the various steps
of treatment planning to calculate the three-dimensional dose distribution.

The current generation of CT scanners has spatial and contrast
resolution sufficient to cover the needs of radiotherapy. However
there exists a certain number of causes of error which can be divided

into two groups : factors entailing inaccuracies in target volume
localization and factors leading to errors in dose determination
because of errors in densities.

The first part of this paper will review the various causes
of error and the second part will deal with the improvements in
treatment planning made possible by using CT data.

ERRORS ON PATIENT DATA

Errors on target volume

Whereas the primary problem which a diagnostician addresses
is whether or not a disease is present and what sort of disease it
is, the therapist is supposed to have the knowledge of the tumour
existence, but he must determine its extent in all directions, the
position and size of adjacent structures, the relationships between
the organs and the body outline or eventually external landmarks
which are to be used later on for setting up the patient in the
treatment room. When the position of the patient is slightly modified, variations of a few centimetres can be noted in the position
of organs with respect to the body outline. These variations are
greater than the accuracy reasonably required in the distance
between beam edges and target volume outline (about 0.5 cm). It is
essential then that the patient set up be identical for treatment
and for CT scanning. The scanner couch should be flat, wide and
rigid and as similar as possible to the treatment couch. It should
include a whole set of accessories identical to those used for
treatment, such as arm-rest or head-rest... The reproducibility of
the patient position can be ensured only if the radiotherapist or
at least a well-trained therapy technician assist the diagnosticians
in positioning the patient on the CT couch.

Another cause of error in the determination of organ size is
linked to image characteristics such as image control settings :
window level or window width. As is well known, CT numbers related
to the linear attenuation coefficient of tissues, are stored into
the computer during a scan. These numbers are converted to electrical
signals using a digital analogue converter and fed into a cathode-ray
tube where the brightness modulated image can be viewed and eventually photographed (figure 2). The window width controls the range
of the density scale and the window level sets the midpoint of the
range.

Large differences are evident in the apparent size of organs
depending on the choice of both window level and window width
(figure 3), window level being the most important feature. As an
example, the error in lung thickness in a lateral direction varies
from + 2 mm to - 10 mm when the window level is varied from average

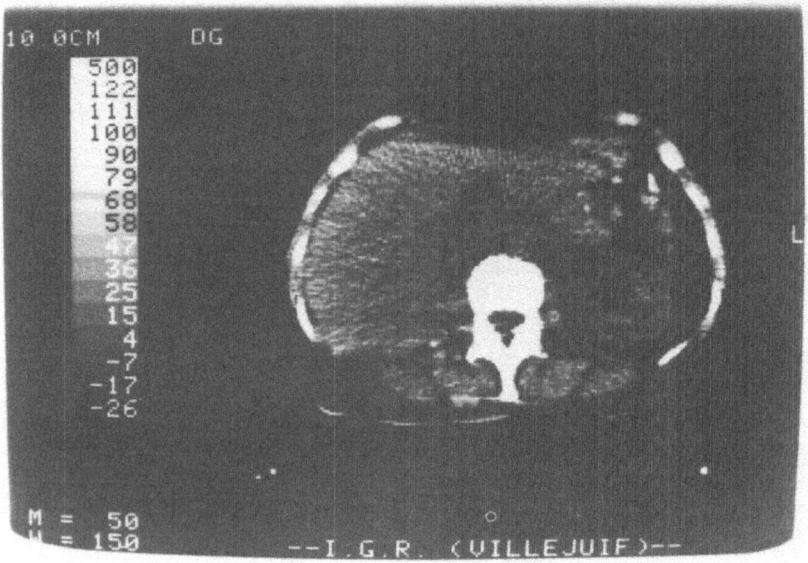

Fig. 2. Polaroid photograph of the image displayed on the cathode-
ray tube. This picture was obtained with a window level
equal to 50 and a window width equal to 150.

lung density to bone density. As a result, the optinum window level
and window width are not identical for the best diagnostic image
and for the determination of organ sizes. The accurate determination
of organ sizes implies the calibration of both window level and
window width for each type of organ and for each CT scanner by
comparison with a body phantom whose organ sizes are known with
accuracy. Such calibrations are especially required in high density
gradient regions such as for patient outlines and lung or bone
contours.

In a few sophisticated radiotherapy departments the computer
used for treatment planning is on-line with the computerized tomograph and the registered numbers are directly worked out and used
for an automatic determination of contours. In other radiotherapy
departments, an intermediate step is necessary to transfer contours
from the CT scanners to a full size image for manual calculations
or to the treatment planning computer. It is a common practice to
magnify to life size the small size photograph of the screen. Such
a procedure can lead to large errors (a few centimetres) because
of the screen curvature. In figure 4, a comparison is made of the
outlines obtained with 4 different devices. It shows that magnification of the polaroid photograph of the screen, using the center

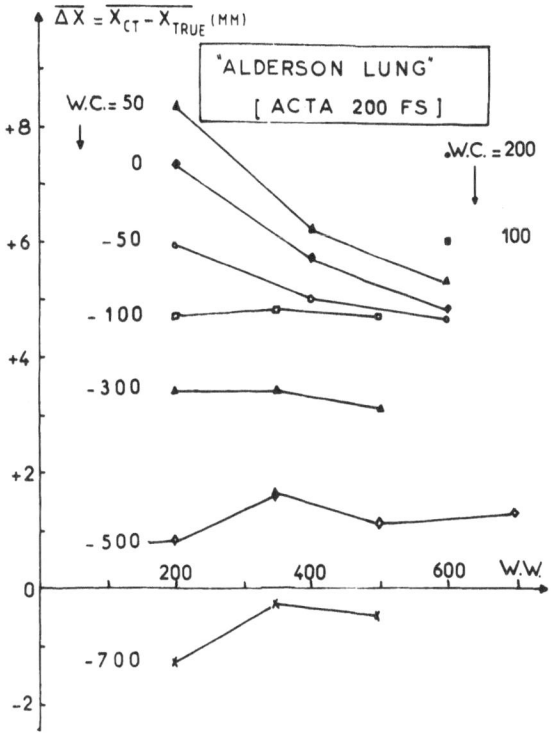

Fig. 3. Error Δx in the width of the lung of the Alderson body
        phantom, versus window width. Each curve corresponds to a
        different value of the window level (W.C.) – The error in-
        creases when the window level is increased. The variation
        with window width (W.W.) is smaller.

square of the grid superimposed on the screen to scale the image,
leads to an error of several centimetres in the overall size of
the body. The best way to overcome this difficulty is by using the
numbers stored in the tomograph computer either directly on line to
the treatment planning computer or on a plotter. Figure 5 shows
the result of image transcription to life size on an electrostatic
plotter used on line with the C.T. at the Gustave Roussy Institute.

    One of the basic postulates of the theory of image reconstruc-
tion in C.T. scanners is that all scans are projections of the same
unchanged object. Unfortunately too often the patient or his organs
move during the scanning cycle leading to a blurring of details
and especially of interfaces between organs whose densities are

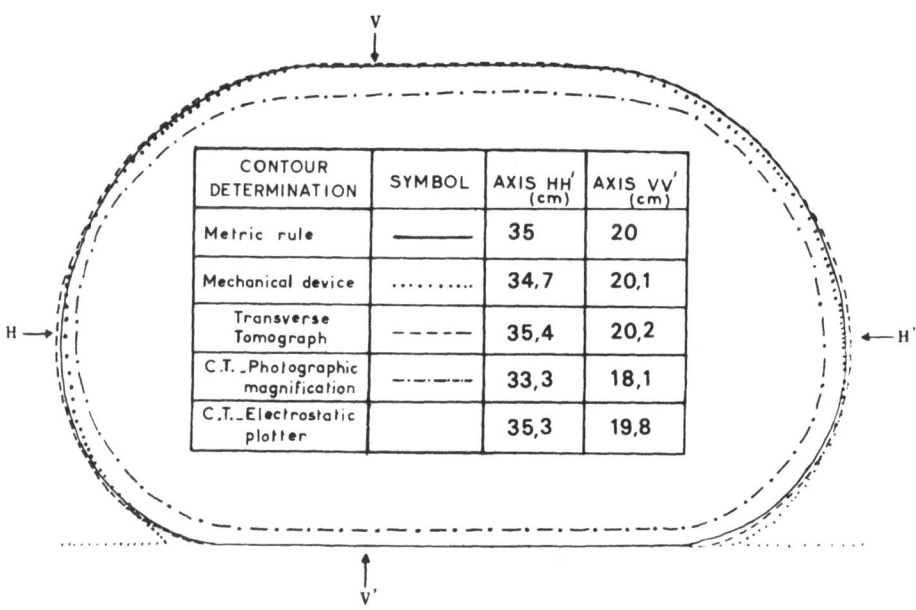

CONTOUR DETERMINATION	SYMBOL	AXIS HH' (cm)	AXIS VV' (cm)
Metric rule	————	35	20
Mechanical device	....	34,7	20,1
Transverse Tomograph	– – – –	35,4	20,2
C.T._Photographic magnification	–·–·–·–	33,3	18,1
C.T._Electrostatic plotter		35,3	19,8

Fig. 4. Comparison of body outlines obtained with 4 different
        devices. The inserted table shows the differences from
        the dimensions as measured with a metric rule (first line).

not very different. Patient motion results also in severe streaking
artifacts even if the amplitude of the movements is only a few
millimetres. These artifacts tend to increase the uncertainty in
target volume delineation. One possibility to reduce these artifacts
would be to decrease the cycle time but this would result, for a
given C.T. scanner, in an increase in noise and therefore in a
decrease in image resolution. The duration of the scanning cycle
was about 20 minutes in the first available C.T. ; it has been progressively reduced to a few minutes a few years later and to a few
seconds in the most recent generation so that the artifacts linked
to patient motion have been progressively decreased.

A special problem appears with patient breathing which blurs
many contours within the thorax and even in the upper part of the
abdomen. When a C.T. of recent generation is used with a cycling
time of a few seconds, it is possible to ask the patient's co-ope-
ration in holding his breath during the exam. In that case, contours
of organs are clearly seen but large differences in size and position
of organs can be noted depending on whether the patient holds his
breath in inspiration or expiration. As the duration of a radiotherapy session is such that breath holding is not possible, one may
want to assess the motion of tissues during treatment and to average

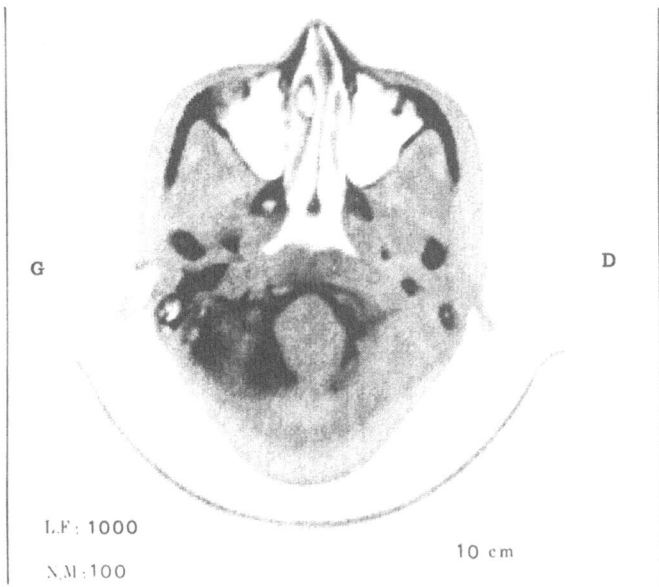

G   D

I.F: 1000

N.M:100

10 cm

Fig. 5. Transcription on paper of a C.T. image of a head to life
size with an electrostatic plotter.

dimensions and densities over several respiratory cycles. An ideal
solution could be to scan every cross-section twice : once, with a
short scan time (3 seconds or so) and the patient breath holding to
get a clear image of the contours of the lesion and a second time
with a rather long scan time (20 seconds or more) in order to get
an average image of the patient in treatment conditions during
normal respiration.

There is a widespread misconception that in treatment planning
the resolution of both spatial and absorption coefficient measurements
need be less accurate than that necessary for diagnostic purposes.
A few papers have appeared in the literature proposing modifications
of radiotherapy simulators to be used instead of C.T. in the radiotherapy department. Some others have seriously suggested that the
first generation C.T. scanners be used for treatment planning in
radiotherapy. However as localization of tumour and normal structures
is an essential step in treatment planning, it is evident that the
requirements for C.T. scanners used for treatment planning are at
least as severe as those for machines used only for diagnostic purposes since the main difficulty is to delineate the tumour tissues
by discriminating them from healthy tissues.

Errors on tissue densities

The relevant parameter for dose calculation in high energy
X-ray therapy is the number of electrons per unit of volume, if one
assumes that interaction of photons takes place primarily by the
Compton process. Dose distribution in the presence of medium h is
usually calculated by applying correction factors to the dose distribution calculated in water. It appears that the correction factor
is a function of the relative electron density ρ_{eh} of the tissues

$$\rho_{eh} = \frac{n_e(h)}{n_e(w)}$$

h under consideration where ρ_{eh} is the ratio of the numbers of
electrons per unit of volume for medium h and water. In a similar
way in high energy electron therapy the relevant parameter is the
relative electron density ρ_{eh} since the slowing-down of electrons
is governed by the number of electrons per unit of volume.

The conventional way to ascertain the relative electron density
of a human tissue is a calculation from measured atomic composition
and physical density $(g.cm^{-3})$.

Table 1. Density in grams per cubic centimetre and number of electrons per gram for several human tissues.

Tissue or material	$\rho(g.cm^{-3})$	Ne el.$g^{-1}.10^{-23}$	ρ_{eh}
Water	1.00	3.349	1.00
Muscle	1.06	3.25 to 3.32	1.03 - 1.05
Bone	1.09 to 1.65	3.10 to 3.25	1.015- 1.60
Lung	0.26 to 1.05	3.25 to 3.33	0.25 - 1.04
Fat	0.92 to 0.94	3.38	0.93 - 0.95
Brain	1.03 to 1.05	3.31 to 3.33	1.02 - 1.045
Liver	1.05 to 1.07	3.32 to 3.34	1.04 - 1.065

References

Densities : Cho, 1975, ICRU, 1963, ICRU, 1976, Jayachandran, 1971,
         Lindskoug et al, 1976, Rao, 1975.
Atomic compositions : ICRU, 1972, Kim, 1974, White, 1974, Woodard,
         1962.

Table 1 shows some of the values recently published. The authors
claim thant the expected variations throughout a given organ or
for the same organ among various people are less than ± 2 %.

Furthermore, apart from lung and bone, variations in ρ_{eh} among
various soft tissues are within the limits 1.03 ± 0.02 that is to
say ± 2 %. Variations of electron densities among various bones
are much larger since ρ_{eh} varies from 1.1 for the centre-part of
spongy bones to 1.85 for cortical bone. The electron density of a
bone varies also with the age and disease of the patient. Large
variations in ρ_{eh} are also encountered for lung (between 0.15 and
1.0) among the various parts of a given patient lung but mainly
among various patients.

Valuable information about ρ_{eh} can be obtained indirectly
from C.T. numbers. As everybody knows, C.T. numbers are a measure
of the linear attenuation coefficient of each element of the matrix
(called a pixel) at the energy of the diagnostic photon beam used
at the depth of the element. The linear attenuation coefficient
should be proportionnal to the electron density if the interactions
of X-rays with tissues were essentially Compton interactions, but
photoelectric interactions are not negligible, especially when high
Z components are present in tissues, as in bone.

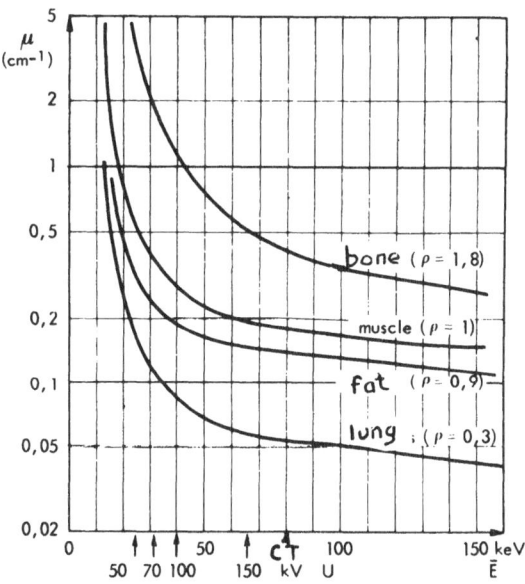

Fig. 6. Variation of the linear attenuation coefficients of human
        tissues versus energy. The vertical arrows at the bottom of
        the figure show the energies equivalent to typical diagnostic
        X-ray beams. The small arrow marked C.T. corresponds to a
        typical C.T. The slopes of the curves decrease when energy
        is increased, that is to say that any filtration effect decrea-
        ses when energy increases.

Figure 6 shows the variation of linear attenuation coefficient versus
energy for various human tissues. If a monoenergetic X-ray beam could
be used the attenuation coefficient of a given material could be
accurately determined and the relationship between C.T. numbers and
electron densities could be accurately measured. Unfortunately the
only way to decrease the width of the emission spectrum of the
X-ray target is to increase the filtration which decreases the
number of photons and increases the "noise". It is thus necessary to
compromise between increasing the filtration and decreasing the
photon fluence. The usual parameters used are between 100 and 140 kV,
30 to 100 mA, 3 to 6 mm of Aluminium filtration, with a HVL equal to
4 to 7 mm of Aluminium. Thus the beam can be approximated by a monoenergetic beam of 60 to 80 keV.

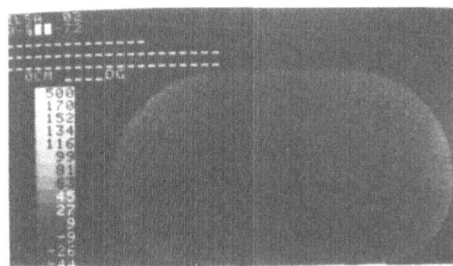

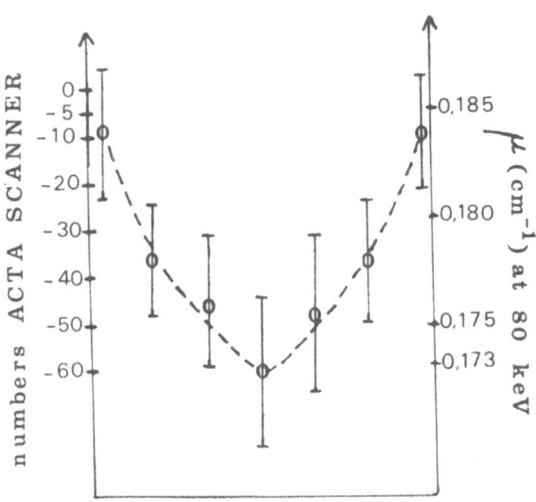

Fig. 7. The upper part of the fig.
shows the display picture
of a homogenous phantom.
The lower part shows the
variation of the C.T. numbers along a profile of the
phantom; C.T. numbers de-
crease rapidly from the
surface to the deepest
points. Linear coefficients
calculated from C.T. num-
bers are indicated on the
scale at the right of the
curve showing a variation
of about 6%.

A supplementary filtration is provided by the tissues themselves and the average energy of the beam increases with depth : the filtration caused by bone being higher than the one caused by soft tissues. As a result the linear attenuation coefficient of a given material that is to say the corresponding C.T. number, decreases with depth (figure 7). Computer programs have been introduced in the recent generation C.T. to correct for such a variation in the average energy of the beam with depth. Results are very good for the head because of the cylindrical symmetry of the skull but more sophisticated programs are necessary for the body.

If the atomic compositions and electron densities of given materials are known, a calibration curve of the C.T. numbers in terms of electron densities can be obtained. Figure 8 shows such a calibration curve. As the beam is not monoenergetic, two materials with the same electron density but different atomic composition may correspond to two different C.T. numbers and the measured points do not lie on a unique line, furthermore the variation of the C.T. numbers versus the electron density is not linear. A computerized tomograph having a good resolution in density allows accurate determination of organ contours but does not allow necessarily an accurate determination of the absolute value of the electron densities. Errors as large as 7 to 10 % can be made on ρ_{eh} when data are not corrected for filtration (figure 7).

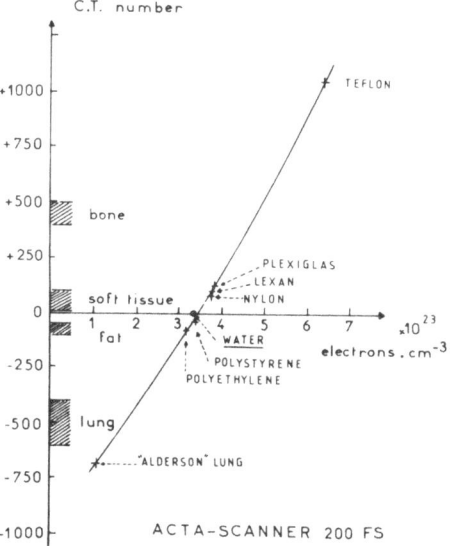

Fig. 8. Calibration curve of C.T. numbers versus electron densities. The average numbers for several human tissues have been shown.

USE OF C.T. DATA IN TREATMENT PLANNING

Localization data

The goal of radiotherapy is to eradicate a tumour without causing a too severe damage to healthy tissues. The probability of producing necrosis of normal tissues rises very rapidly as the dose is increased above the tolerance level whereas the probability of achieving local control of the primary tumour is significantly reduced when the tumour dose is reduced below the curative level. (Wambersie et al., 1969, Herring et al., 1971).

Such a statement implies that one of the best way to improve radiotherapy is to improve localization of tumour and critical structures in order to position more precisely the radiation beams so as to cover the whole tumour with a sufficient dose while reducing as far as possible the dose delivered to the adjacent normal tissues. For the first time because of the good contrast resolution of the C.T. images, many soft tissue structures can be identified and the extent of the tumour can be recognized. One of the many other advantages of computerized tomography is the possible delineation of the tumour in its three dimensions and not only in a single plane. A series of images of transverse planes throughout the whole tumour volume can be easily obtained avoiding the tumour to be partially unirradiated and lesion being not controlled. Several authors and in particular Goitein et al. (1978) have reported on the results of prospective studies to assess the value of C.T. in planning radiation therapy. They have planned first treatment with conventional methods and then they have performed C.T. scans in order to reevaluate the treatment goals. Finally plans were changed when necessary. The C.T. scan was judged to be of major value for 28 of the 77 patients planned and of minor value in a further 12 patients, resulting for these patients (52 % of the total) in changes in treatment plans.

Sophisticated computer programs do exist now for three-dimensional calculation of dose distribution (figure 9). As a first step in the three-dimensional calculation, dose distribution is calculated at least in three different planes for every patient treated with radiations in the Gustave Roussy Institute. It has lead us to increase the number of compensating filters so as to compensate for dose inhomogeneities mainly in the longitudinal direction. With the recent generation of C.T., calculations are feasible in planes other than transverse, for instance sagittal or frontal planes. The combination of these improvements offer to the radiotherapist the possibility to calculate dose distribution in any chosen plane and to superimpose it on the display system with the corresponding C.T. image. Such programs are still in the infancy and much has to be done to make these improvements available in every radiotherapy department.

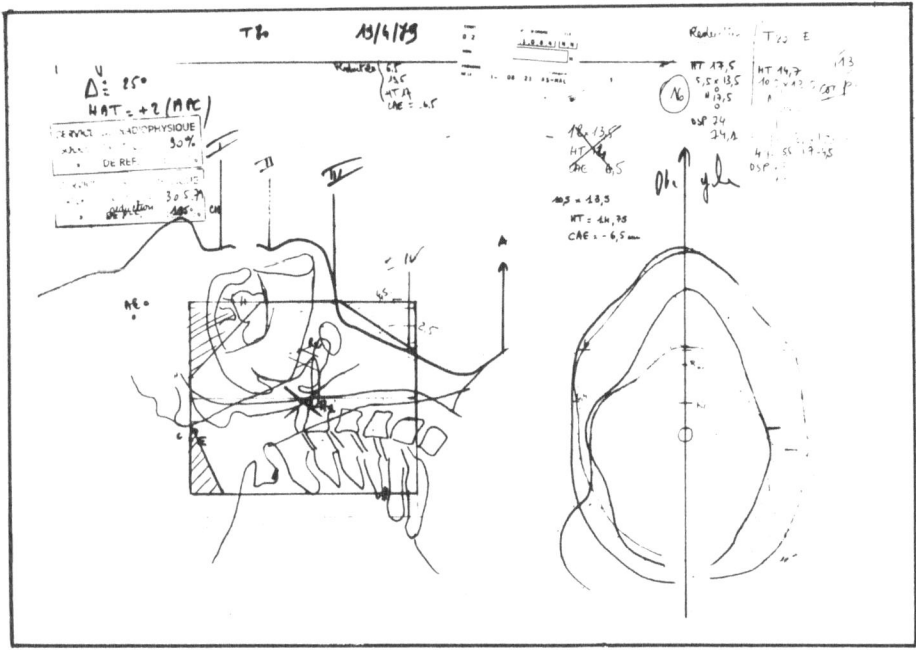

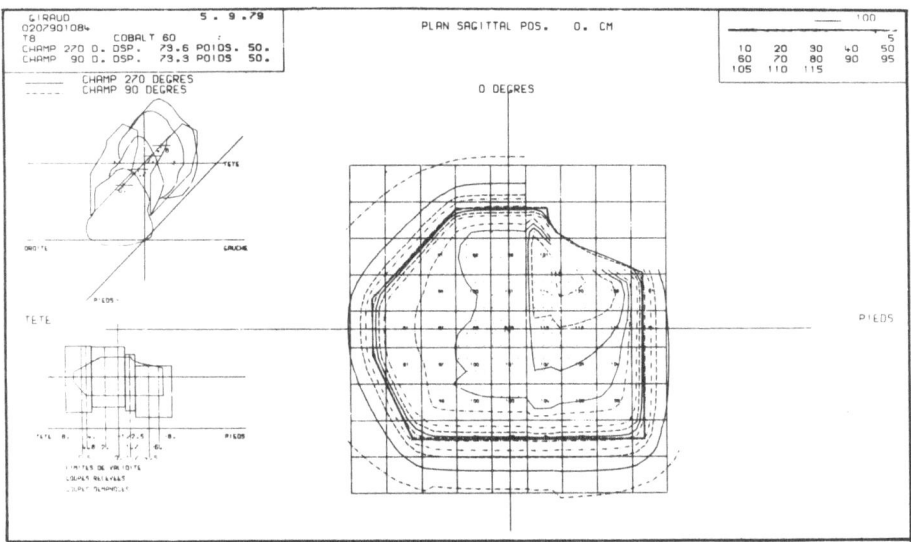

Fig. 9a and b. Several contours of a head are drawn in figure 9(a) together with a sagittal profile of a patient treated for a larynx cancer. In figure 9(b) the dose distribution is computed in the sagittal plane taking account of the shielded parts of the field. The small insert at the upper left shows a perspective view of the various body contours with the two opposed fields.

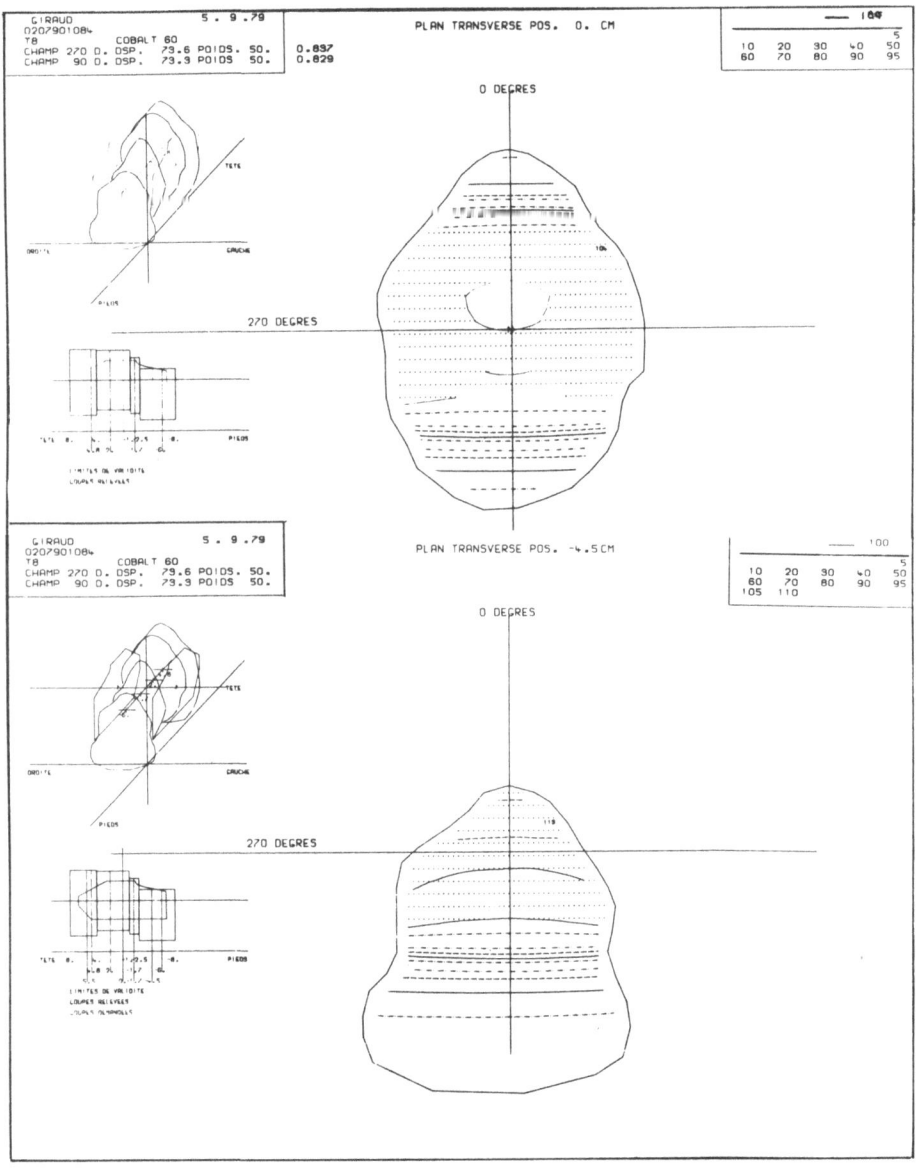

Fig. 9c and b. Dose distributions in two transverse planes at the
beam axis level (c) and 4.5 cm below (d) for the
patient shown in(a). The dotted lines correspond to
the region irradiated with a dose equal to or higher
than the "tumour dose" referred as 100 %. A compensating filter will be built from the calculation in
(b) to homogenize the dose. The dose distributions
can be superimposed on the CT images.

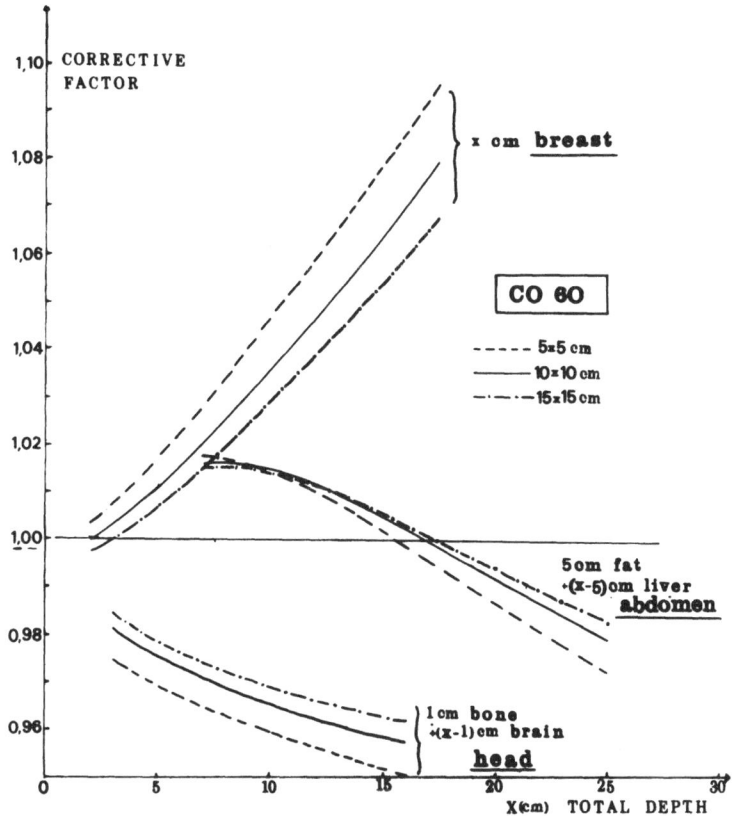

Fig. 10. Correction factor versus depth for three different body
cross-sections irradiated with a Cobalt 60 beam.

Density data

    In a patient cross-section, CT provide a detailed distribution
of numbers which can be converted into electron densities. The array
of electron densities can then be used to correct the in-water dose
distributions. Such corrections are usually known as inhomogeneity
corrections. Strictly speaking, none of the human tissues or organs
is water equivalent and a body-section presents a succession of
layers of tissues of different densities and compositions. Many of
them (table 1) differs in density from water by not more than 3 or
4 % and can be assumed to have an atomic composition similar to
water. Fat and bone differ from water by density and composition.
Lung is in fact the most important inhomogeneity because of its
usual low density. Many manual methods or computer programs have
been described to correct for the presence of lung but in the

author's knowledge, dose distributions until now were not corrected
routinely for fat and bone and certainly never for soft tissue densities. The first question which arises is the magnitude of the

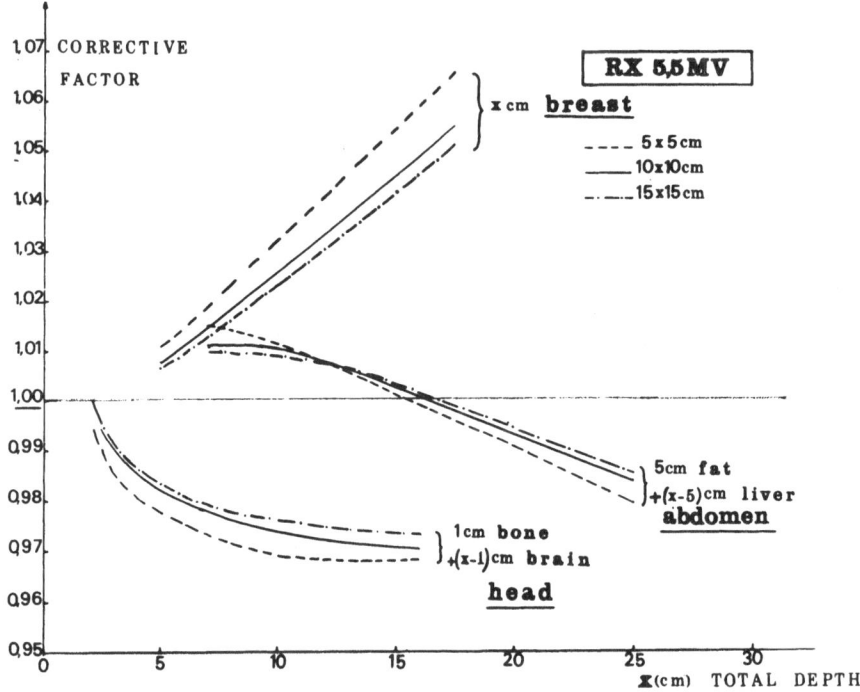

Fig. 11. Correction factor versus depth for three different body
        cross-sections irradiated with 5.5 MV X-rays.

error encountered when soft tissues are assumed to be water-equiva-
lent. We have calculated the correction coefficients (figures 10
and 11) to be applied to in-water dose distributions for three different body sections as a function of depth. Figure 10 refers to
Cobalt 60 γ-rays and figure 11 to 5.5 MV X-rays. For Cobalt 60,
the correction factors vary between 1.02 and 0.98 in the abdomen
because the low density of fat compensate for the higher density
of the liver ; the correction factor is always smaller than 1 for
the head and always larger than 1 for the breast. The magnitude of
these errors decreases when the energy increases because of the
attenuation decrease. In electron beams, the dose variation with
depth is not exponential and it is not recommended to apply correction factors to in-water dose distribution to take into account

Table 2. Variation of the dose delivered at the deepest part of the
tumour for several electron beam energies. It is assumed
that the depth dose is intended to be larger than 80 % of
the maximum dose throughout the target volume for in-water
calculations. Minimum doses at the deepest part of the tum-
pour were calculated when thicknesses were underestimated
by 0.5 or 1 $g.cm^{-2}$ because of the higher density of human
tissues.

$E_e.$ (MeV)	$\%D_T$ at the posterior limit of tumour (Ref : 80 % at post. limit).	
	$- 0,5 \ g.cm^{-2}$	$- 1 \ g.cm^{-2}$
8	50 %	20 %
10	55 %	35 %
19	66 %	58 %
25	74,5 %	67,5 %
32	76,5 %	72 %

inhomogeneities. When a tumour is treated with an electron beam,
the electron energy is usually chosen so that the depth dose at
the deeper limit of the target volume be equal to 80 or 85 % of
the maximum dose. The dose delivered to the healthy tissues deeper
than the tumour is much lower than the tumour dose because of the
high dose gradient in the last part of the electron depth dose
curves.

Table 2 shows the depth doses at the deepest part of a tumour
when the total equivalent thickness is underestimated by 0.5 or
1 $g.cm^{-2}$. It is assumed that the thickness is underestimated because soft tissues were assumed to be water equivalent. To correct
the dose distribution, the radiotherapist has to increase the electron energy which was planned from in-water dose distribution because
of the electron density of the tissues.

The second question to be answered is related to the sensitivity of dose prediction to errors in electron density or organs
thickness (Geise et al, 1977). Table 3 shows errors in dose corresponding to an error of 5 % on density. Table 4 shows errors in
dose for errors in thickness of 0.5 or 1 $g.cm^{-2}$. These two tables
show that the dose is more sensitive to errors in thickness than to
errors in density. From these tables, it appears that accurate CT
data are necessary if accurate dosimetry is needed : uncertainties
as great as 5 % in electron density and 0.5 cm in depth cannot be
tolerated if a final accuracy of $\pm$ 5 % in dose is needed. As we
have seen (figure 7), some computerized tomographs lead to rather
large errors in electron densities (more than 10 %) especially when

Table 3. Errors in dose in percentage of the local dose due to an
error in electron density ρ_e equal to 5 % for three typical
inhomogeneities irradiated either with Cobalt 60 γ-rays
or with 5.5 MV X-rays.

Inhomogeneity nature and thickness.	% dose error from $\frac{\delta \rho_e}{\rho_e} = 0,05$	
	Co-60	RX-5,5 MV
1 cm bone ($\rho_e = 1,575$)	0,45 %	0,3 %
5 cm fat ($\rho_e = 0,92$)	1,3 %	0,9 %
10 cm soft tissue ($\rho_e = 1,03$)	2,7 %	1,9 %
	4,45 %	3,1 %

the C.T. is not correctly calibrated or when no correction is
applied for filtration of the X-ray beam by tissues. In such a
case the direct use of C.T. numbers to correct for inhomogeneities
is debatable.

As the contrast resolution does not depend upon the accuracy
on density determination, the C.T. image can still be used to delineate the organs. Ascribing to an identified organ an average
known density, will not lead to an error in the dose greater than
1 % or so. The only organ for which a density determination is
necessary for individual patients is lung whose density may vary
from 0.25 to 1.0. The calculation of lung electron density from
average C.T. number of the patient (even if it is erroneous) will
lead to better accuracy than assuming a constant electron density
for every patient.

Delineating the organs and ascribing average densities was
proposed first by Geise et al, (1977) and is probably the method
of choice during an intermediate period when many C.T. in use
are not from the last generation and the computer programs available
for treatment planning assume a constant density within a well delineated region. However treatment planning can certainly be improved
by interfacing the C.T. with the treatment planning computer and
computing dose point-by-point using the C.T. numbers adequately
converted into electron densities. New approaches for these calculations have to be found. On the other hand it is necessary to focus
one's attention on accuracy of depth determinations which involves
localization of tumour depth, accuracy of contour delineation and
last, what is now the weakest link, the accurate reproduction of
patient position during each treatment session.

Table 4. Errors on dose in percentage of the local dose due to an error in the equivalent thickness equal to 0.5 or 1 $g.cm^{-2}$. the error in the equivalent thickness can be due to either an error in the thickness or to an error in density. Errors decrease when energy is increased.

Error in x ($g.cm^{-2}$)	x ($g.cm^{-2}$)	Error in the dose ($\Delta D/D_p$)		
		Co-60	RX 5,5 MV	RX 25 MV
a) <u>BREAST</u> : X $g.cm^{-2}$ fat				
- 0,5	5	1,6 %	1 %	-
	> 10	2,7 %	1,7 %	0,9 %
- 1	5	3,3 %	2,6 %	
	> 10	5 %	3,3 %	2,2 %
b) <u>ABDOMEN</u> : 5 $g.cm^{-2}$ fat + X $g.cm^{-2}$ liver				
- 0,5	7	2 %	1,7 %	0,5 %
	> 12	2,7 %	2 %	0,7 %
- 1	7	4,4 %	2,8 %	1 %
	> 12	5,2 %	3,4 %	2,2 %

CT data permit more accurate and complete calculation of the dose distribution but the full benefit of this information will be lost if it cannot be transferred to the therapy unit.

CONCLUSIONS

As it was underlined by several authors (Geise et al, 1977, Goitein et al, 1978, Stewart et al, 1978) CT offers a great potential for improving radiation therapy through accurate localization of tumour and critical normal structures. However the improvement brought by more accurate inhomogeneity corrections is more controversial.

Clinical experience throughout the world has been gained with in-water dose distributions. Tumour doses stated in the literature are tumour doses in error by a few percent because of the non water-equivalence of tissues. Performing inhomogeneity corrections without clear warning for the radiotherapist in charge of the patients could lead to failure to control the patients disease because of systematic over-dosage or under-dosage depending on the tissue densities in the region of interest. In the publication of doses and cure-rates in clinical journals, clear statements are essentials on the inhomogeneity corrections performed. Comparisons with in-water doses should be recommended during a transitional period.

However refusing to perform corrections when the necessary data and methods are available would be against progress. Such corrections vary evidently for the same type of tumour from one patient to the next depending upon the depth of the tumour and the amount of fat tissues. The analysis of clinical results as a function of uncorrected doses may be erroneous because of the blurring of any small variation in local tumour control or normal tissue damage with dose. By improving dose calculations, then, one can expect to learn more accurately the dose-response and dose-time-volume relationships.

The impact of computerized tomography on treatment planning can be summarized as follows :

- Improvement of clinical assessment of tumour extent and tumour localization as well as delineation of adjacent critical organs.

- Improvement of dose calculations within the target volume and within the normal tissues of interest.

- Better evaluation of maximum tolerance dose to healthy tissues and minimum curative dose for the local tumour.

Focusing attention on the necessity to ensure an accurate reproduction of patient position, session after session, and of course between computerized tomograph and treatment unit is an important outcome of the introduction of CT in radiotherapy departments.

REFERENCES

Cho, Z.H., 1975. Study of contrast and modulation mechanism in X-Ray photon transverse axial transmission tomography - Phys. Med. Biol., 20, 879.

Frain, C., Surmont, J., Tubiana, M., Pierquin, B., Dutreix, A., 1955. Intérêt de la tomographie transversale dans le repérage, le centrage et la dosimétrie des tumeurs - J. Radiol. Electrol. 36, 792.

Geise, R.A., McCullough, E.C., 1977. The use of CT Scanners in Megavoltage Photon Beam Therapy Planning - Radiology 124, 133.

Goitein, M., Wittenberg, J., Doucette, J., Friedberg, C., Mendiondo, M., Gunderson, L., Lingood, R., Shipley, W.U., Fineberg, M.V., 1978. The value of CT in radiotherapy treatment planning - J. Comput. Assist. Tomogr. 2, 524.

Herring, D.F., Compton, D.M.J., 1971. The degree of precision required in the radiation dose delivered in cancer radiotherapy - Brit. J. Radiology 55, 51.

ICRU Report 10d, 1963. Clinical Dosimetry - Recommendations of the International Commission on Radiological Units and Measurements - Natl. Bur. Std. Handbook 87.

ICRU Report 21, 1972. Radiation Dosimetry - Electrons with initial energies between 1 and 50 MeV - Natl. Bur. Std.

ICRU Report 24, 1976. Determination of absorbed dose in a patient irradiated by beams of X or Gamma rays in radiotherapy procedures - Natl. Bur. Std.

Jayachandran, C.A., 1971. Calculated effective atomic number and kerma values for tissue-equivalent and dosimetry materials - Phys. Med. Biol. 16, 617.

Kim, Y.S., 1974. Human tissues chemical composition and photon dosimetry data - Radiation Research, 57, 38.

Lindskoug, B., Hultborn, A., 1976. Tissue heterogeneity in the anterior chest wall and its influence on radiation therapy of the internal mammary lymph nodes - Acta Radiologica Therapy, 15, 97.

Rao, P.S., 1975. Attenuation of monoenergetic gamma rays in tissues - Am. J. Roentg. 123, 631.

Stewart, J.R., Hicks, J.A., Boone, M.L.M., Simpson, L.D., 1978. Computed Tomography in Radiation Therapy - Int. J. Radiation Oncology Biol. Phys. 4, 313.

Wambersie, A., Dutreix, J., Dutreix, A., 1969. Précision dosimétrique requise en radiothérapie - J. Belge Radiol. 52, 1.

White, D.R., 1978. Tissue substitutes in experimental radiation physics. Med. Phys. 5. 467.

Woodard, H.Q., 1962. The elementary composition of human cortical bone - Health Physics 8, 513.

SELECTED ADVANCES IN THE USE OF COMPUTERS IN RADIATION THERAPY

George T. Y. Chen

Biology and Medicine Division
Lawrence Berkeley Laboratory
Berkeley, California 94720

INTRODUCTION

Since the 1950s, when they were first used to calculate isodose distributions, computers have played an increasing role in radiotherapy departments. In recent times, their utilization has expanded to include the following broad areas: (1) control of patient flow and records, (2) automatic verification and control of therapy equipment, (3) treatment planning, (4) biological modelling, and (5) tumor registries.

An up-to-date review of the status of these subjects in contained in the Proceedings of the 6th International Conference on the Use of Computers in Radiotherapy (Sternick, 1976). In a previous lecture, Dr. Dutreix covered selected areas in the use of CT in treatment planning. I will limit my presentation to work in the area of computer controlled therapy machines and the potential role of three-dimensional graphics in therapy planning. These topics have relevence to charged particle therapy because (1) computer controlled machines also offer the potential of highly localized dose distributions through conventional radiation beams, and (2) as both photon and charged particle therapy are truly three-dimensional planning problems, methods for the display of results through graphics may play a role in both domains.

DYNAMIC THERAPY

While charged particle therapy offers the potential of highly localized dose distributions, their restricted availability at the present time may be a barrier to widespread use. Therefore, several centers in the United States are also investigating the use of computer controlled photon therapy machines in optimizing complex dose distributions. The objective is identical to that in particle therapy, namely to reduce the dose to sensitive tissues surrounding the tumor while treating the tumor with a uniformly high tumoricidal dose.

About 100,000 deaths occur each year in the U. S. due to failure of local control of the tumor (Bond et al., 1970). Although many of these cases also include distant disease at the time of death, the hypothesis is that with improved control of disseminated disease by adjuvant chemotherapy or immunotherapy, an improvement in local control will have an impact in reducing this large number of treatment failures.

With sophisticated dose delivery techniques, it may be possible to improve local control rates without a corresponding increase in complication rate. It is the dose received by the critical organs adjacent to the target which often limits the actual dose given to the target. The relationship between control success and complication rate is generally depicted by two sigmoidal curves, as shown in Fig. 1 (Paterson, 1948). At dose level A, a certain probability of tumor control is achieved with a low, and therefore acceptable, rate of complications. An increase to dose level B will result in a higher cure rate, but with a corresponding increase in complications. It is interesting to note that a modest increase of 500 rad for T3 lesions of the larynx can increase the local control rate by 35 percent, as described by Stewart and Jackson (1975). The hope is to look for methods which increase the dose without increasing morbidity. An improved dose distribution is one approach toward this goal.

The question addressed by those interested in computer controlled therapy is: can computer controlled therapy machines produce the desired dose distributions which reduce dose to critical tissues and allow higher doses to the tumor, thereby improving clinical results ?

Current irradiation techniques on most machines call for machine parameters to be set and static during therapy. Before treatment, each of the following parameters are set: (1) the field size is defined by rectangular collimators, (2) the gantry angle is set, (3) the table coordinates are defined, and (4) the collimator angle is set. Except for gantry angle, all of the other parameters are fixed during therapy. In rotational therapy, the start and stop angles of the gantry are defined and the gantry rotates during the irradiation.

Several linear accelerators in the United States have been modified to allow collimator and table motions during irradiation. The most advanced hardware implemented for the investigation of dynamic therapy is installed at the Joint Center for

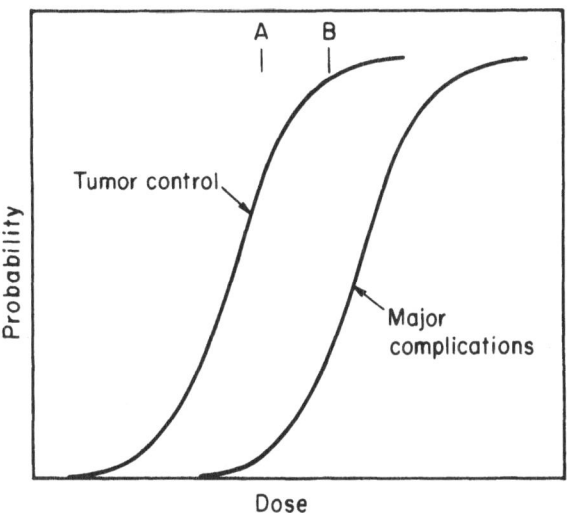

Fig. 1.   Tumor control and complication rate as a function of dose. Reprinted, by permission, from Levine et al. (1978).

Radiation Therapy in Boston (Bjarngard and Kijewski, 1978). The major components of this system are a Siemens Mevatron 12 linear accelerator modified to include all hardware for computer control, and a DEC PDP 11/45 computer with a specially designed interface for therapy machine control. This linear accelerator is identical to the standard factory built model except that it has been modified to enable the following computer controlled operations: (1) the collimators are decoupled and independently controlled; one set of collimator blocks can be moved past the central axis; (2) the collimator may be rotated during therapy; (3) couch translation and rotation is possible during therapy; and (4) the dose rate may be controlled during treatment. An overview of the machine is shown in Fig. 2. Instructions for the computer controlled treatment are considerably more complex

than for a standard irradiation. Here, the motions of all
parameters must be prescribed and properly synchronized. The
approach taken by the Joint Center is to prescribe the position
and velocity for each moving parameter as functions of time
(Levine et al., 1978).

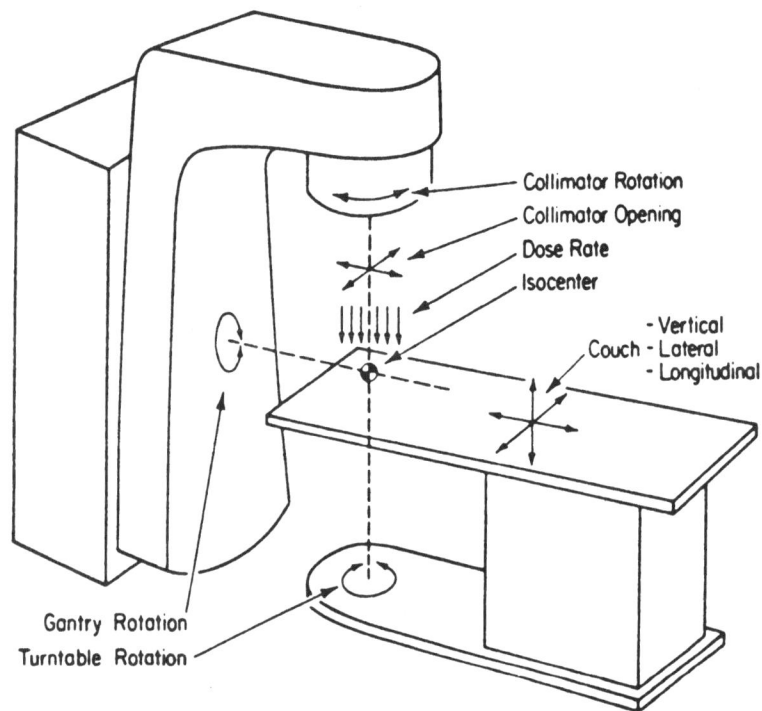

Fig. 2. Variable parameters for the Joint Center for Radiation
Therapy dynamic therapy machine. Reprinted, by
permission, from Bjarngard and Kijewski (1978).

The clinical applications of dynamic therapy are currently
under study. Initial studies of computer controlled therapy
were directed toward the use of moving the collimator jaws to
produce wedged fields (Kijewski et al., 1978). This corresponds
to the use of wedge filters with conventional machines. The

dynamic wedge was selected as a test vehicle because of its simplicity and safety to serve as an initial evaluation of the system operation. In producing a wedged field with collimator motions, the computer directs the movement of one of the independently movable lower collimator jaws across the field during

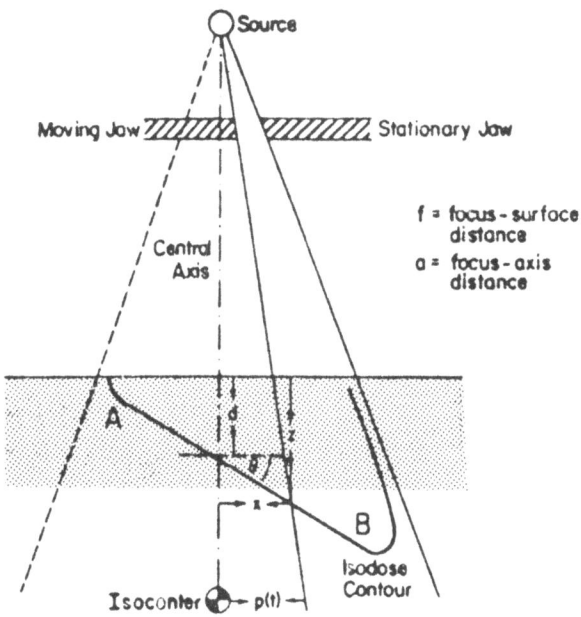

Fig 3(a). Wedge shaped distributions from dynamic therapy machines. Reprinted, by permission, from Kijewski et al. (1978).

the period of irradiation (Fig. 3). The actual position of the jaw as a function of time was calculated by iterative techniques. Conventional photon therapy relies on the use of up to five wedge angles for dose distribution shaping. With a dynamic

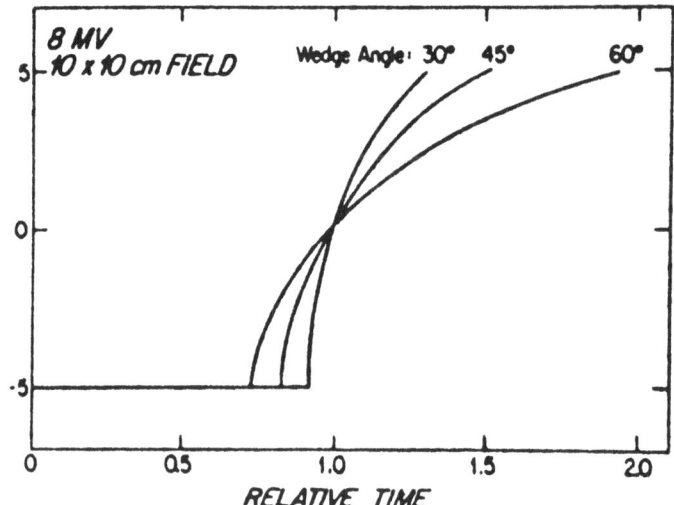

Fig. 3(b). Jaw position as a function of time.   Reprinted, by
            permission, from Kijewski et al. (1978).

wedge, it is possible to have a continuous wedge angle. Initial
studies of the dynamic wedge have shown that it can be implemented repeatedly with high accuracy and without unforseen
difficulties (Levine et al., 1978). The dynamic wedge has also
been used to produce slightly more uniform dose distributions,
since the wedge angle can be tailored to the exact individual
case requirements.

    As of July 1979, eight patients have been treated under
dynamic therapy (Kijewski, private communication). In addition
to the movement of collimator jaws under computer control to
produce a wedge, the collimators have also been controlled
during gantry rotation to conform to the projected target size.
The advantage of this approach is clearly illustrated in Fig. 4.

    One of the most exciting areas of investigation is the
delivery of complex dose distributions of several parameters are
controlled during therapy. For computer controlled radiation
therapy to become useful in the clinic, substantial improvements
of dose distributions should be produced. At the Joint Center,
studies have been performed to evaluate dynamic therapy in the

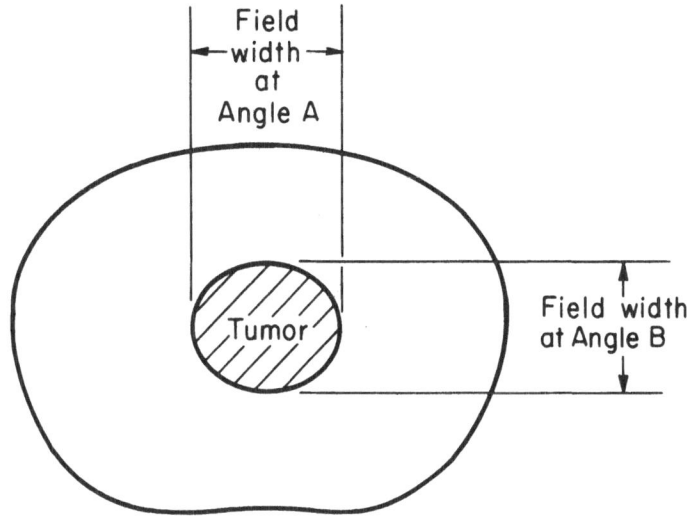

Fig. 4. Field width varies with the irradiation angle.

irradiation of the retroperitoneal lymph nodes. It has been reported that these nodes are involved in Stage III carcinoma of the cervix in about 38 percent of the cases. Irradiation of these nodes by conventional AP-PA fields has resulted in high complication rates of about 50 percent. These complications include small bowel obstruction, perforation, rectal bleeding, and fistulas. The frequency of occurrence of these complications might be decreased with dynamic therapy.

The target volume for adequate irradiation of this structure is complex. Figure 5 shows that the target volume superiorly looks like a curved cylinder, while in the pelvis it appears as two "wings." In the dynamic therapy of this target volume, one solution (Fig. 6) is to irradiate in short longitudinal sections with four fields. The patient is translated longitudinally between each segment of irradiation. The field widths are controlled at all times to agree with the projected dimensions of the target. The dose distribution resulting from such a dynamic treatment is shown in Fig. 7. The dose distribution in the longitudinal plane shows an irregularly shaped high dose region with a relatively steep dose gradient. The transverse dose distributions at two levels are shown in the lower portion of the figure. The machine parameter motions as a function of time

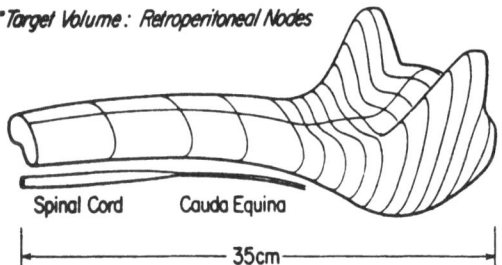

Fig. 5. Para–aortic target volume. Reprinted, by
permission, from Levine et al. (1978).

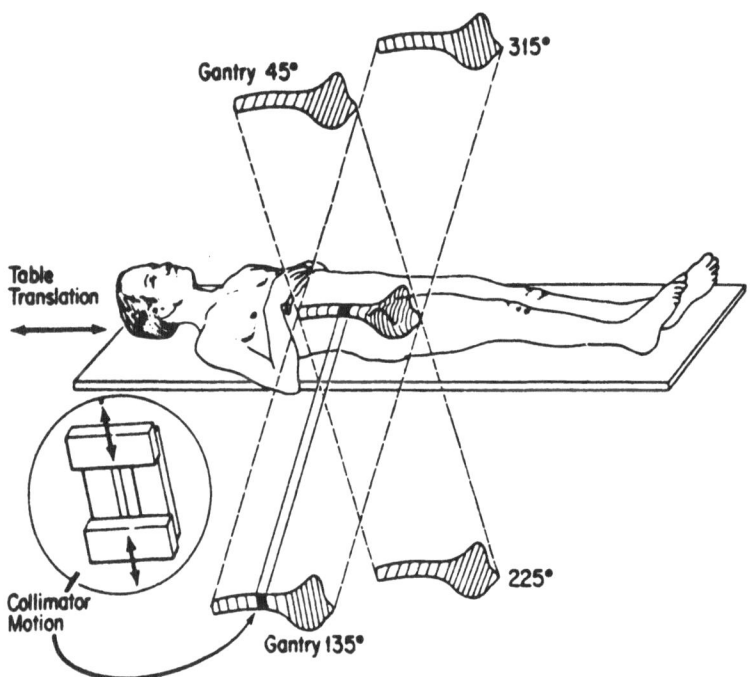

Fig. 6. Irradiation scheme for Fig. 5. Reprinted, by
permission, from Levine et al. (1978).

PLAN R3

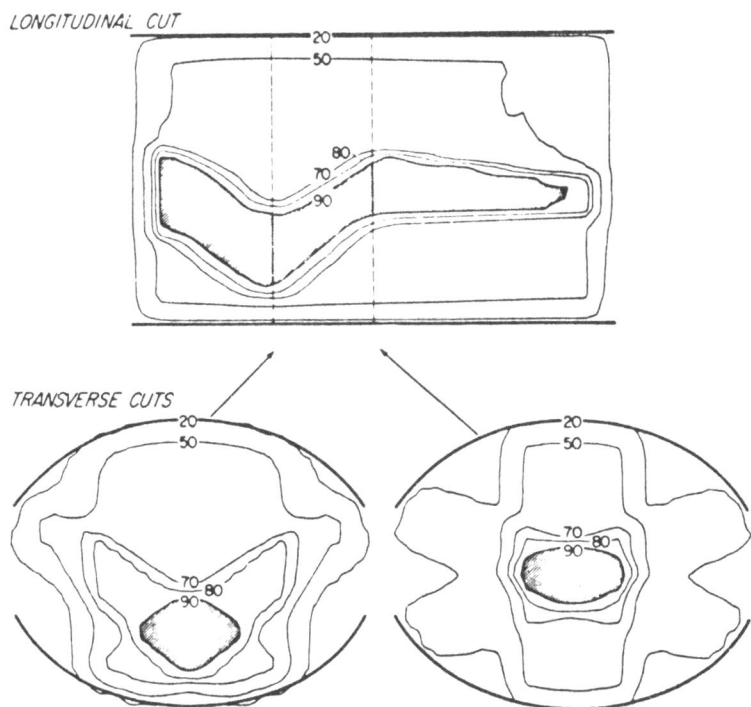

Fig. 7. Dose distribution from a proposed dynamic therapy of
the para-aortic nodes. Reprinted, by permission, from
Levine et al. (1978).

are outlined in Fig. 8. The exact parameters for those complex
irradiations are now in the process of being calculated from
target outlines in sequential CT scans. In early studies such
as this one, some trial-and-error hand planning was required to
produce the detailed control parameters. To date, no patients
have been treated for para-aortic nodes with this technique,
although clinical trials are planned for the near future.

Chung-Bin and his group (1979) have used a computer controlled
Clinac 4 at the Rush Presbyterian Medical Center in Chicago.
They have studied the use of dynamic therapy in the treatment of

MACHINE PARAMETER MOTION (PLAN R3)

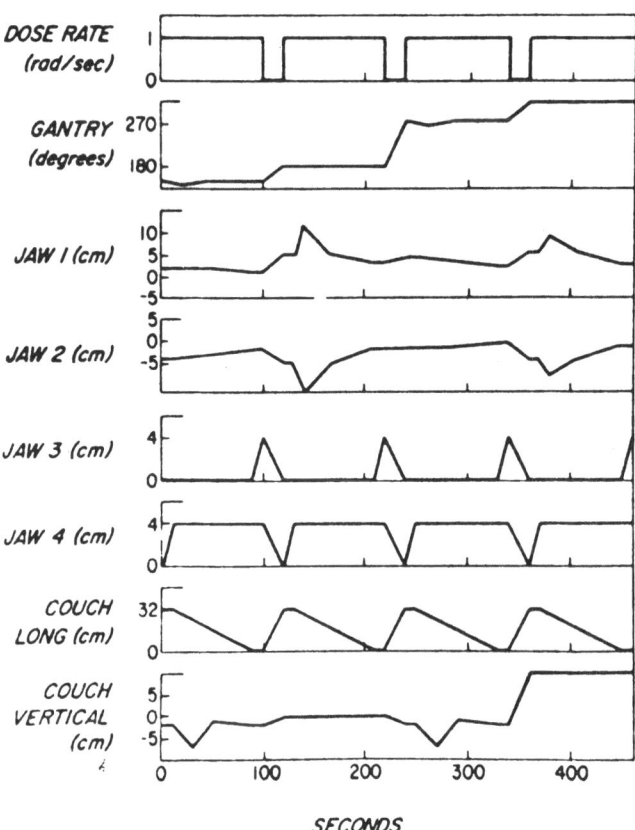

Fig. 8. Parameters as a function of time for irradiation
proposed in Fig. 6. Reprinted, by permission,
from Levine et al. (1978).

mesothelioma of the pleura. In this case, the mediastinum and
chest wall must be treated without overdosing the lung. The
pleura of the lung has an irregular shape, and its location must
be determined by CT. In the treatment of the pleura, the target
volume is an irregularly shaped surface (with finite thickness)
of an irregularly shaped cylinder. The isocenter trajectory is
designed to have a field rotating around the patient such that
the edge of the field is always tangent to the inner part of the

pleura. This treatment involves moving the couch coordinates to keep the isocenter in the proper relationship with the pleura for specific gantry angles. The experimentally measured dose distribution is shown in Fig. 9. The physician had requested that the mediastrinum and pleura be treated to the 70 percent isodose level while sparing the lung as much as possible. In this dose distribution, there is a ±15 percent dose inhomogeneity with the lung receiving about 10 percent of the maximum dose to tumor.

A technique for changing field size and dose rate as a function of gantry angle during rotational therapy has been in use for about six years at the Sinai Hospital in Detroit (Mantel et al., 1977). Their preprocessing program examines the target contour and computes the field sizes by projecting the specified target areas along the beam direction, and then calculates the dose rate in each angle in a manner which attempts to maintain a

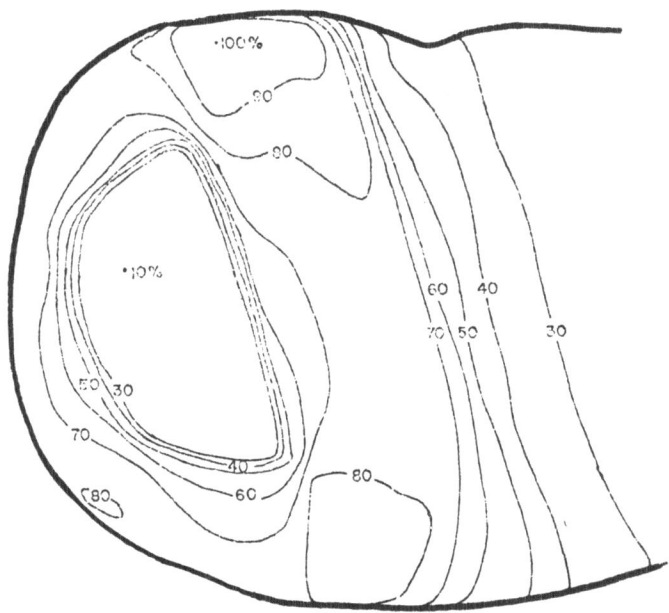

Fig. 9.   Dynamic therapy dose distribution for irradiation of a mesothelioma of the pleura. Reprinted, by permission, A. Chung-bin, T. Wachtor, T. Zusag, and F. Hendrickson (1979).

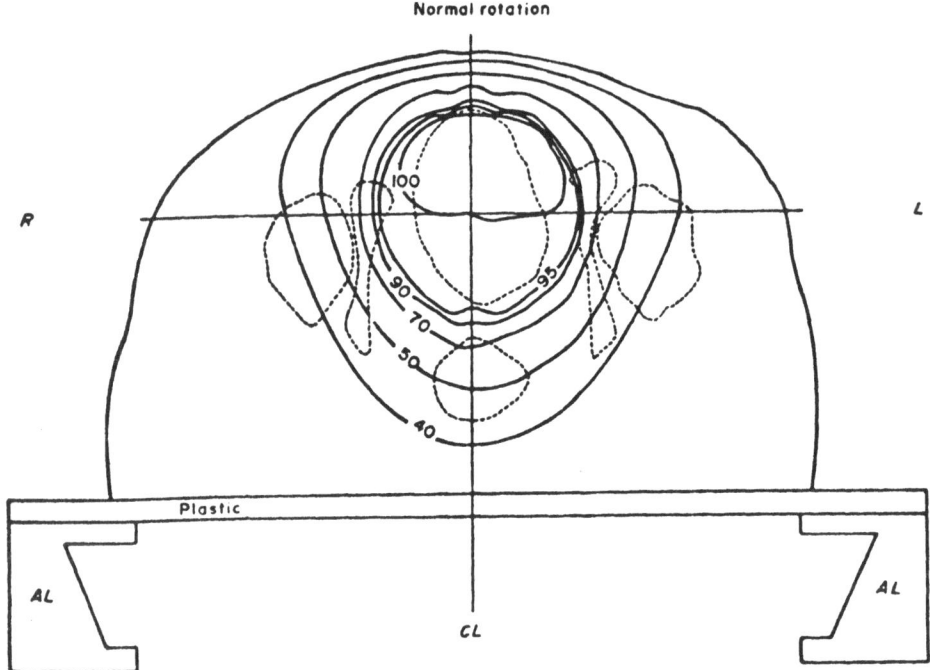

Fig. 10. Normal rotation for the treatment of a bladder
          lesion. Reprinted, by permission, from Mantel
          et al. (1977).

uniform contribution to the target. This program also finds the
isocenter of the target. A conventional rotation for a bladder
irradiation is shown in Fig. 10. This plan has good dose uniformity but has significant areas lateral to the target which
receive a high dose because the shape of the high dose region
does not conform to the target.

The plan (Fig. 11) generated by dynamic therapy has good
dose uniformity within the target and the 95 percent line
follows the shape of the target more closely than in Fig. 10.
Dose to the rectum is less with dynamic therapy than with conventional rotations. Mantel has also used linear programming
techniques to optimize the dose rate for the rotation, and the
optimized plan for this case is shown in Fig. 12. Optimization
has created a 95 percent contour which follows the target
contour very closely. The volume within the rectum where dose
exceeds 50 percent is significantly smaller than in the previous

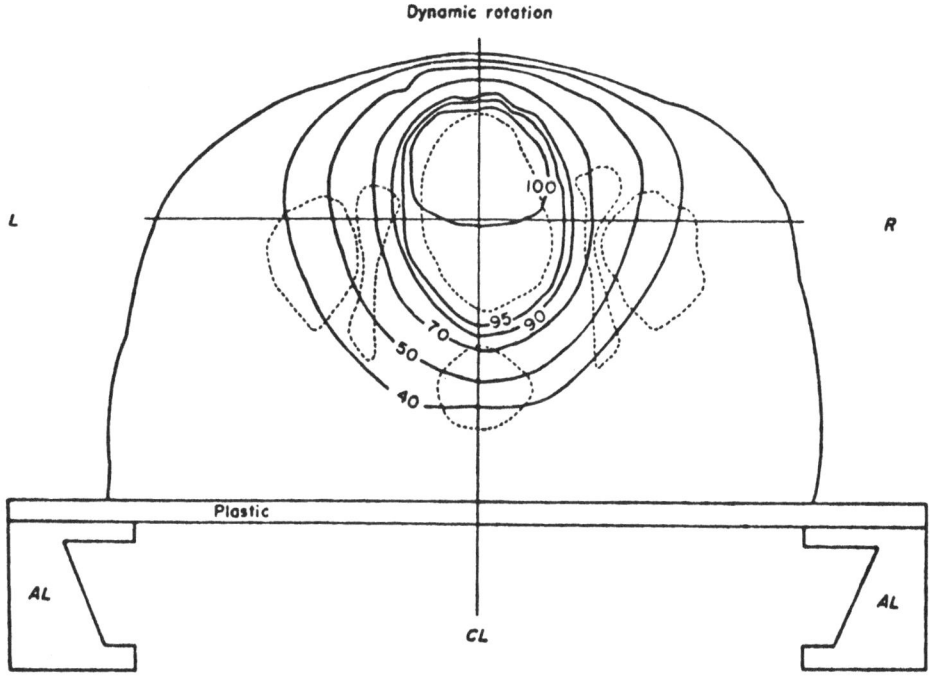

Fig. 11. Dose distribution for a dynamic rotation. Reprinted,
by permission, from Mantel et al. (1977).

two plans. In the optimized plan, the number of actual entry
angles with nonzero dose rates is 16, or about half of the 36
angles used in the nonoptimized dynamic therapy.

Studies in the use of computer controlled therapy machines
have been described at three institutions. Like charged
particle therapy projects, the clinical efficacy of this experimental modality must be proven through randomized trials. The
potential dose distribution advantages of dynamic and charged
particle therapy also impose new and stricter standards in tumor
localization, patient alignment, and immobilization. This area
at present is somewhat underdeveloped and needs much work before
the potential of highly localized dose distributions is
effectively evaluated.

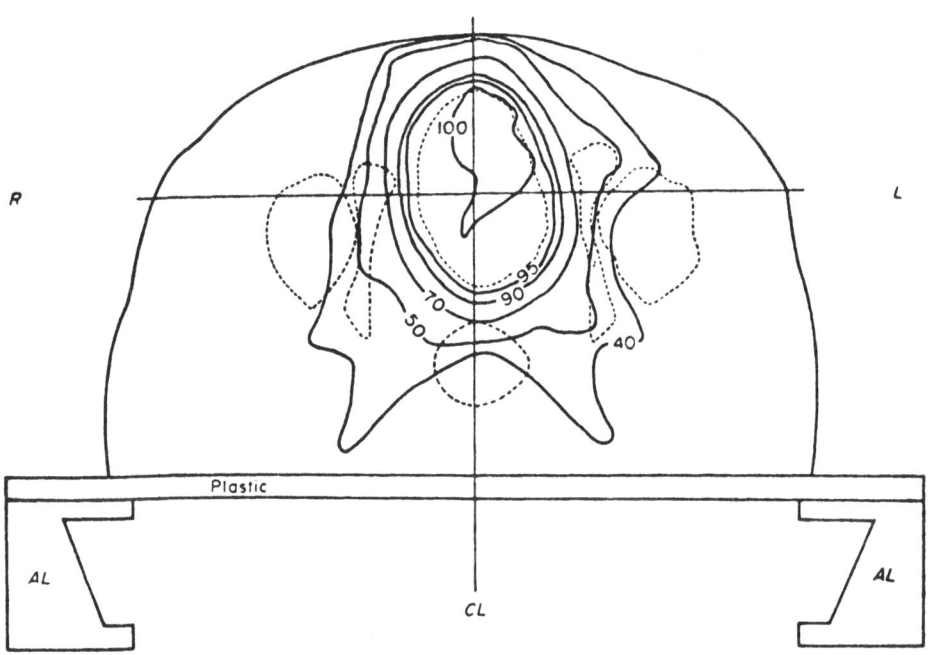

Optimized dynamic rotation

Fig. 12.   Dose distribution for an optimized dynamic rotation.
Reprinted, by permission, from Mantel et al. (1977).

COMPUTER GRAPHICS IN THERAPY

Both particle and dynamic therapy produced complex three-
dimensional dose distributions.   In order to appreciate the
three-dimensional nature involved in these modalities, computer
graphics would be a logical area to investigate.   Relatively
little has been done in this area of graphics as applied specifically to radiotherapy treatment planning; however, some elegant
work (Rhodes, 1979; Herman and Liu, 1977) been done in
diagnostic radiology departments and will now be described.

The basic idea behind three-dimensional utilization of CT
data in therapy and diagnosis involves either (1) creating a
cube of CT information and displaying arbitrary planes of interest (most commonly saggital and coronal views), or (2) defining
contours of interest in sequential scans and displaying these
sets of contours in three dimensions.   The capability of

generating and displaying arbitrary plans of CT data from a cube of CT information is a feature found in most CT display packages. The utility of this display capability in radiotherapy is the ability to visualize tumor volumes and normal structures in the lateral or AP (or oblique) viewpoint without the degrading effects of overlying tissues. These displays allow one to integrate many sequential CT cuts and view them "on edge." An example of an image generated from sequential CT slices, where interpolation is used to smooth from slice to slice, is shown in Fig. 13. Here, 25 CT slices were used to generate a coronal image one pixel thick. From the therapy viewpoint, sequential target contours defined in the axial plane may also be projected onto a coronal plane to indicate the treated area from the conventional A-P (or lateral) perspective. This method may be used to define the collimator shape defined by the sequential target contours, and is useful in collimator positioning.

Contours may also be defined in each slice, stacked, and viewed from a three-dimensional perspective. The contours may be entered manually where the area of interest does not have a CT number distinct from surrounding structure (such as a tumor

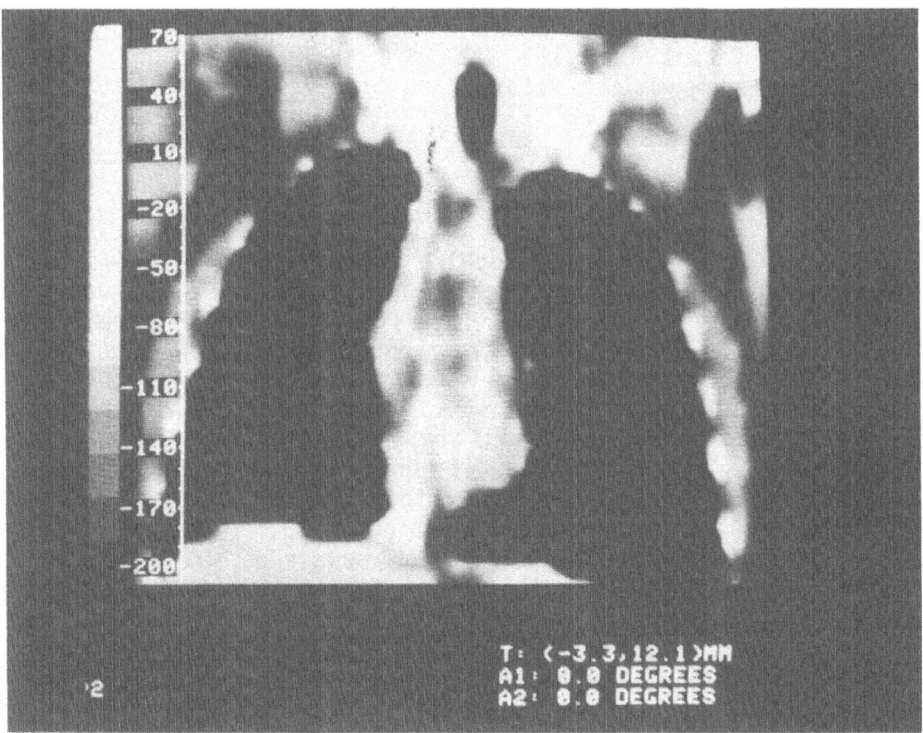

Fig. 13. Data: AP image generated from cube of CT data.

invading the liver), or it may be entered with automatic contouring routines when the CT number is distinct. An elegant automatic contouring algorithm has been developed by Michael Rhodes at Long Beach Memorial Hospital (private communication), which is capable of outlining the entire spine. Using three-dimensional region growing, this algorithm isolates all spatially connected pixels forming a structure's volume. The algorithm also contains a plane-to-plane connection strategy which requires only one seed point for the entire set of CT image planes, and re-enters image planes to find complete structure contours.

This group has also developed software which operates in a PDP 11/34 environment to display the series of contours from a three-dimensional perspective, either on a relatively inexpensive digital graphics unit, as would be found in a CT image display, or on a sophisticated vector graphics unit. Figure 14 demonstrates some examples of their work. To give the appearance of three dimensions to the display, points in the contour in the Evans-Sutherland vector graphics unit are displayed with different intensities. The set of contours may be thought of as being contained in a cube, where the intensity of points in the cube are proportional to their proximity to the front face of the cube. Points closest to the front face are the brightest; points most distant from the face are the least bright. The family of contours may be rotated and translated arbitrarily in real time.

Graphics Specifically in Therapy

The application of these sophisticated display programs has been somewhat limited, and is only now being actively investigated. The need for such systems arises from the desirability of viewing the problem of treatment planning in three dimensions, since one arbitrary CT plane rarely gives correct information on relative positions of tumor and critical organs in other planes. Understanding the relationship of target to surrounding organs would benefit both charged particle and photon therapy. Both external contour and amount of inhomogeneities as well as relative position of tumor and critical organs is likely to change at different levels of the body. The usual method involves assessing these changes at mid, upper, and lower cuts.

Three-dimensional graphics may be of use in assisting the therapist and physicist to integrate information from a series of axial scans. The detailed anatomical information on internal structures from these scans should also be related to standard A-P and lateral radiographs in order to assist the therapist in

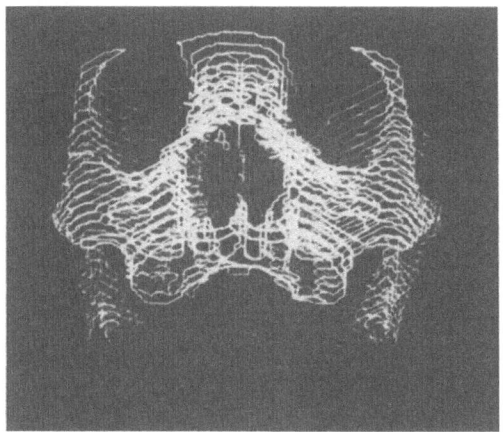

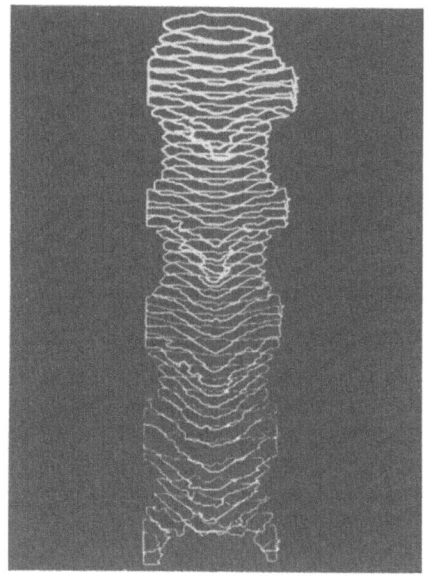

Fig. 14 (a,b). Images of various body
sites generated from auto-
contouring multiple slices.
(a) Spinal canal, (b) skull.
Reprinted, by permission, from
M. L. Rhodes.

field-size selection, placement, and beam entry angle. The
problem is therefore to extract detailed contour information
from CT, and produce three-dimensional graphic representations
on a video monitor capable of real time rotation and translation.

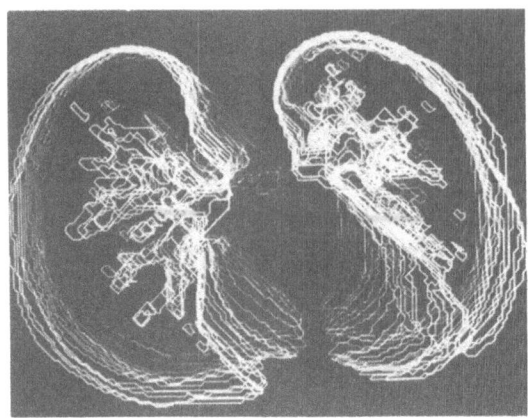

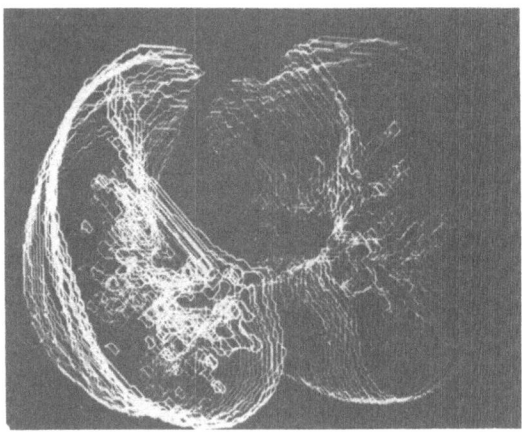

Fig. 14 (c,d). Images of various body
sites generated from auto-
contouring multiple slices.
Oblique views of the lung.
Reprinted, by permission, from
M. L. Rhodes (1979).

Reinstein and co-workers (Reinstein et al., 1978; McShan
et al., 1979) have developed software to solve this problem. In
the particular example cited by them, the problem was to design
a field and select the appropriate angle for irradiation of a
lesion of the upper esophagus. In this region, the relationship
of the esophagus to spinal cord varies with position. The relative positions of the lungs, spinal cord, and esophagus were
obtained from transverse axial tomograms, although they now plan
to use CT data. The tomograms are magnified to life size and

relevant contours entered into the computer by sonic digitizer. From previous experience, a three-field treatment plan utilizing an anterior field and two anterior wedged obliques was used. The beam parameters (gantry and collimator angles, isocenter coordinates, and field sizes) are to be determined such that all three radiation beams irradiate the tumor but the two oblique beams do not irradiate the spinal cord. Figure 15 shows the computer generated display of the patient's esophagus and spinal cord as viewed from an anterior oblique beam incident at a 40 degree angle with respect to direct anterior (0 degrees). The darker areas indicate the locations of the esophagus and spinal cord in each of five planes. A linear interpolation is used to define the position of the organs between tomographic scan planes. In this display, the positions of the trachea, lungs, and external contour were removed for clarity. It can been seen that the tumor volume and the spinal cord are superimposed in both the right and left obliques at various positions, and therefore this plan is unacceptable.

The oblique ports were then angled to 55 degrees and the resulting beam's eye view of the anatomy is presented in

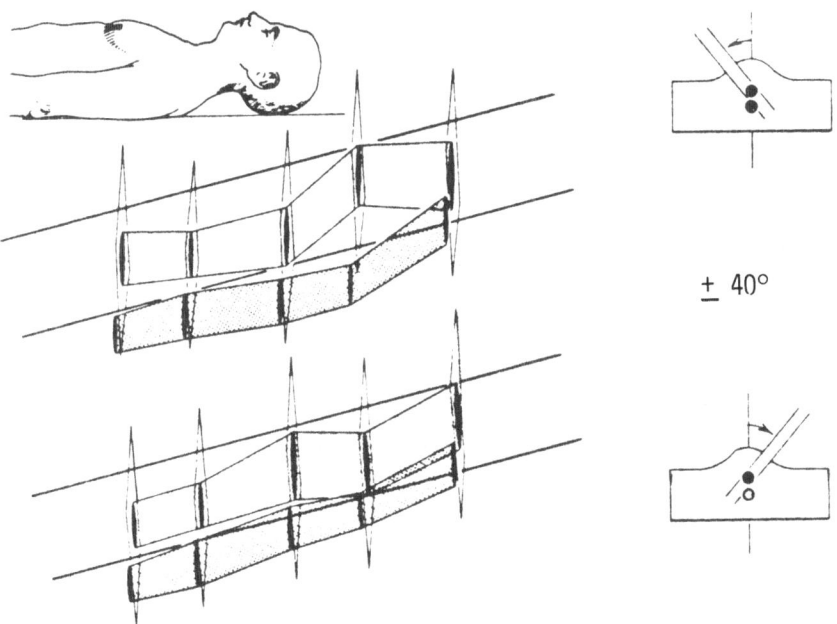

Fig. 15. Esophagus and spinal cord viewed at 40 degrees. Reprinted, by permission, from Reinstein et al. (1978).

Fig. 16. Adequate separation of the cord and esophagus is now
obtained. The outlines of the fields for this irradiation are
shown in the figure as the dark lines above and below the
esophagus. The field lengths, width and collimator angles, and
isocenter positions are outputted. For this patient it was
possible to irradiate the entire esophagus if the appropriate
field size and collimator angles were determined. While this
simulation could have beeen performed on a standard treatment
simulator, the computer graphics technique is an improvement in
two ways. First, more detailed anatomical information is
visible from sequential CT scans than from the ususal fluoro-
scope system used in simulators. Second, this process reduces
the time required to have the patient on the simulator. Simulation may be a difficult and unpleasant experience for a sick
patient, as it may take over one hour.

Goitein (private communication) has used non-CT based three-
dimensional graphics to assist in the treatment planning for
irradiation of ocular melanomas. In this case, the objective is

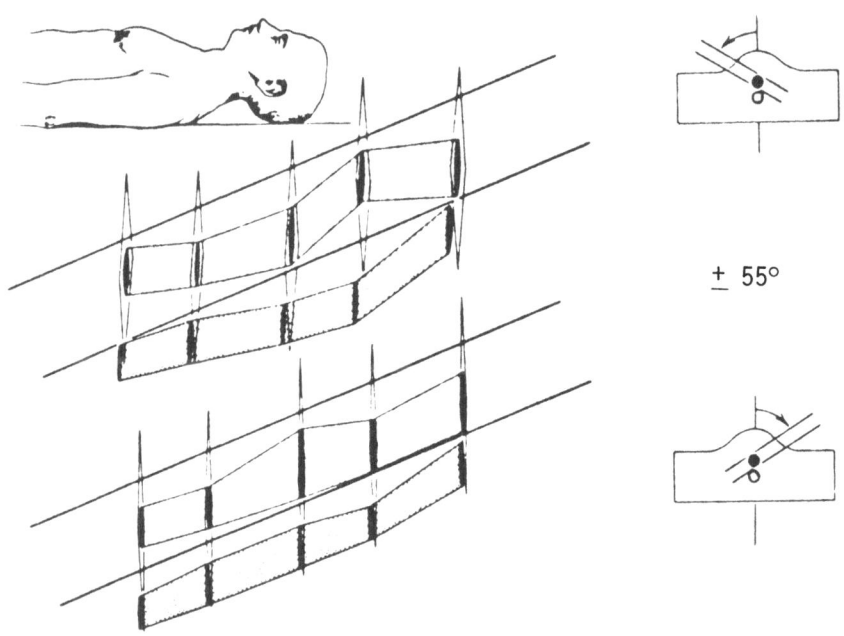

$\pm$ 55°

Fig. 16. Esophagus and spinal cord viewed at 55 degrees.
          Reprinted, by permission, from Reinstein et al.
          (1978).

to use graphics to assist in the design of the irradiation
aperture, to understand the critical structures in the beam
path and behind the lesion, and to choose the best entry angle
for the irradiation of the tumor with protons. An interactive
program is used to define the coordinates of the tumor relative
to a standard eye. Definition of the tumor region involves
defining a contour of the base of the tumor on the retina and
the ultrasound measured height of the lesion. In the treatment
of ocular melanoma by charged particles, the patient is asked to
fix his eye on a light to bring the tumor into the best orientation of a fixed horizontal beam. An appropriate fixation point
is one in which the lens, macula, and optic nerve are not at
risk during the irradiation. The program allows one to view the
eye from several viewpoints, including the lateral, anterior,
and beam viewpoint (Fig. 17). The fixation point may then be
interactively moved, and by observing the relationship of the
projected target to the lens, optic nerve, and macula, a suitable fixation point may be chosen. The projected target volume
as viewed from the beam viewpoint defines the appropriate

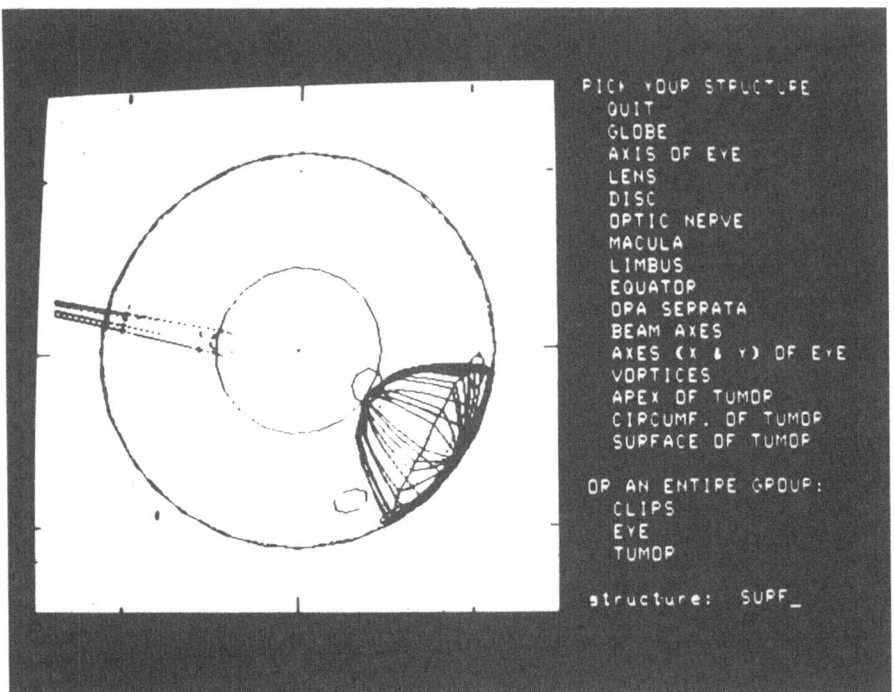

Fig. 17. Ocular melanoma treatment planning with three-
dimensional graphics. Reprinted, by permission,
from M. Goitein.

aperture for irradiation. These exciting examples of computer graphics in treatment planning illustrate the power of computer simulation and its usefulness to radiotherapy.

ACKNOWLEDGEMENTS

I am pleased to acknowledge the work and generous use of materials from Drs. Bjarngard, Kijewski and Chin of the Joint Center of Radiation Therapy, Dr. Chung-Bin of Rush Presbyterian Medical Center, Dr. Mantel of Mt. Sinai Hospital, Detroit, Dr. Michael Rhodes of Long Beach Memorial Hospital, Dr. Reinstein of Rhode Island Hospital, and Dr. Michael Goitein of the Massachusetts General Hospital. These studies were supported by the National Cancer Institute, and the U. S. Department of Energy under contract No. W-7405-ENG-48.

REFERENCES

Bjarngard, B. E., and Kijewski, P. K., 1978, Computer controlled radiation therapy, IEEE, CH1413-4/78/0000-0086.
Bond, V. P., Suit, H. D., and Marcial, V., eds., 1970, "Time and Dose Relationship in Radiation Biology as Applied to Radiotherapy," Proceedings, NCI-AEC Conference, Sept. 15-18, 1969, Carmel, CA, Upton, NY, Brookhaven National Laboratory, BNL50203 (C-57).
Levine, M. B., Kijewiski, P. K., Chin, L. M., Bjarngard, B. E., and Helman, S. H., 1978, Computer controlled radiation therapy, Radiology, 129, 769.
Herman, G. T., and Liu, H. K., 1977, Display of three-dimensional information in computed tomography, J. Comput. Assisted Tomogr., 1, 155.
Kijewski, P. K., Chen, L. M., and Bjarngard, B. E., 1978, Wedge-shaped dose distributions by computer-controlled collimator motion, Med. Phys., 5, 426.
Chung-bin, A., Wachtor, T., Zusag, T., and Hendrickson, F., 1979, Paper No. 92, Proceedings, Combined Meeting XII Intern. Conf. on Medical and Biological Engineering and the V Intern. Conf. on Medical Physics, Jerusalem, Israel.
Mantel, J., Perry, H., and Weinkam, J. J., 1977, Automatic variation of field size and dose rate in rotational therapy, Int. J. Radiat. Oncol. Biol. Phys., 2, 697.
McShan, D. L., Silverman, A., Lanza, D. M., Reinstein, L. E., and Glicksman, A. S., 1979, A computerized three-dimensional treatment planning system utilizing interactive colour graphics, Brit. J. Radiol., 52, 478.
Paterson, R., 1948, "The treatment of Malignant Disease by Radium and X-rays," Arnold Press, London.

Reinstein, L. E., McShan, D., Webber, B. M., and Glicksman, A., 1978, A computer assisted three-dimensional treatment planning system, Radiology, 127, 259.

Rhodes, m. 1979, An algorithmic approach to controlling search in three dimensional image data, SIGGRAPH 1979 Proceedings, Vol. 13, No. 2, pp. 134-142.

Sternick, E., ed., 1976, "Computer Applications in Radiation Oncology," University Press of New England, Hanover, NH.

Stewart, J. G., and Jackson, A. W., 1975, The steepness of the dose response curve both for tumor cure and normal tissue injury, Laryngoscope, 85, 1107.

RADIATION MONITORING

Ralph H. Thomas

Lawrence Berkeley Laboratory
University of California
Berkeley, California

INTRODUCTION

In the strict sense, this paper will not discuss any real "advances" in radiation monitoring. What is presented here has been known and applied at high-energy accelerator laboratories for several years. However, the increasing application of a variety of high-LET radiations, produced by accelerators, to radiodiagnosis and radiotherapy which have been described in this course has led to the need to more widely disseminate this knowledge.

At the present time the number of people exposed to man-made high-LET radiations is rather small,[1] but it may well be that the increasing application of accelerators to the problems of medicine will substantially increase the fraction of the general exposure resulting from high-LET radiations. This is perhaps particular interest because of the recently voiced concerns over the incidence of leukemia induced by fission neutrons.[2]

The radiation phenomena at high energy accelerators have been rather thoroughly investigated because it was necessary to be able to construct these instruments so that their radiation environments were safe for human occupation and that their use for research was not inhibited by high radiation backgrounds. This latter consideration was in many instances the over-riding factor.[3]

Most of our basic knowledge of accelerator health physics has been obtained at high-energy laboratories because these institutions have available the resources needed for the required fundamental investigations.[4] Experience has shown that the radiation environments of high-energy accelerators are in many respects similar to those produced by lower-energy accelerators. Many of the techniques of radiation monitoring developed at high-energy laboratories may be applied equally to low or high-energy accelerators. Unfortunately there is some evidence that the available information is not widely known.[5]

TYPES OF MONITORING

A Comprehensive radiation monitoring program has three components:

(1) Area monitoring
(2) Environmental monitoring
(3) Personal monitoring

There is a strong argument for performing all three types of monitoring at accelerator installations. Each component may be subject to occasional difficulties in interpretation, but all components are mutually supportive.

Two examples will suffice. In the first, if a member of the staff inadvertantly contaminated his personal dosimeter (film badge, thermoluminescent dosimeter) with radioactive material, the individual's dosimeter reading would then be suspect. Investigation of the accelerator operations log, personal dosimeter readings of colleagues, area monitoring records--and even environmental monitoring records--then shows no unusual exposure to radiation had occurred. These facts would warrant entering the normally expected radiation exposure into the individual's record.

The second example actually occurred at the research laboratories of a well-known electrical engineering company. Physicists working with an electron accelerator were puzzled when they noted that their area radiation monitors continued to indicate radiation for some minutes after the accelerator had been turned off. Investigation revealed that increasing the electron beam energy had led to significant photo-neutron production, with the consequent induction of radioactivity in beam collimators and the accelerator structure. This rediscovery of neutron induced activity, some forty years after the original work of Fermi and his colleagues in Rome,[6] led to a reevaluation of the personal dosimetry and area monitoring program. The inadvertent neutron production would, however, have gone unnoticed if only a β-γ personal dosimetry program had been pursued.

Little will be said in this lecture on personal dosimetry because the techniques of personal dosimetry are well understood, and because few advances have occurred in the past five years and, finally, because of the constraints of space and time.

Area Monitoring

"The most important parts of a program of monitoring for external radiation in workplaces is the conduct of a comprehensive survey when any new installation is put into service or when any substantial changes have been made, or may have been made, in an existing installation. An example of this type of monitoring is the surveying of the area round a research reactor immediately on restarting after a shutdown".[7]

The terms "area monitoring" and "environmental monitoring" are often used equivalently. For our purposes here area monitoring is taken to mean radiation measurements made in the work place by either fixed or portable radiation detectors, not worn by individual members of the staff. "It is largely of a confirmatory nature but may include the use of fixed detectors to identify the onset of abnormal or emergency conditions, such as criticality accidents."[8]

The ICRP has suggested that routine area monitoring, referred to as "environmental monitoring", is only necessary under certain conditions:

"If the radiation situation in the workplace is not liable to change, except as a result of substantial alterations to the protective equipment or the processes carried out in the workplace (which should be followed by comprehensive surveys), then routine environmental monitoring is not needed. If, however, the radiation fields in the workplace are liable to change, but the changes are not likely to be rapid or severe, then occasional checks, mainly at fixed points, will usually give sufficient warning of deteriorating conditions. Alternatively, the results of individual monitoring for external radiation may be used for this purpose.

"If the radiation fields are liable to increase rapidly to serious levels, then a system of warning

instruments will be required, either in the environment or worn individually by the workers.[10] It is
particularly important to identify situations calling
for this type of warning monitoring because, if
carried out effectively, a program of warning monitoring may prevent the receipt of high doses at· high
dose-rate and thus eliminate a genuinely dangerous
situation. Other types of monitoring, while contributing to the overall safety of the operation, rarely
fulfil such a positive function."[9]

The interpretation of environmental monitoring data has been
discussed in paragraphs 45-47 of ICRP Publication 12:[11]

"(45) The general problem of interpreting radiation
dose-rate measurements in the workplace in terms of
the dose to organs and tissues of the workers is
extremely complex. The dose rate and the quality of
the radiation will vary in space and with time, while
the workers move through their environment in a way
which is neither predictable nor accurately known or
recorded. It is therefore essential to introduce
major simplifying assumptions. One simplification is
to assume that measured γ and neutron dose rates at a
point accurately reflect the dose rates in the gonads
and red bone marrow of a man at the same point, while
the total dose rate due to β- and γ-radiation and
neutrons reflects the dose rate to skin. The errors
caused by ignoring (or, in the case of neutrons,
standardizing) back-scatter and depth-dose effects are
trivial compared with the difficulty of relating the
dose rate at a number of points in space and time to
the integrated dose to a worker.

"(46) For some applications, it is convenient to
assume that someone will be present for the whole of
his working time at the point of highest dose rate in
the workplace. This method, which establishes an
upper limit to the dose which can be received, has the
advantage of needing no restriction of movement within
the workplace. In practice, actual doses received in
such an environment are likely to be well below this
maximum and, if the design of protection of the workplace and process has been adequate, well below the
Maximum Permissible Doses recommended by the
Commission.

"(47) If it is not practicable to keep dose rates in
the workplace low enough for this simple system of
interpretation to be of value, it becomes necessary to

assess, and sometimes restrict, the access time in areas of high dose rate. This is achieved qualitatively by the use of warning signs or by the planning of operations. Any quantitative interpretation is provided by individual monitoring."

The techniques that may be used for area monitoring have been discussed in preceding lectures of this course[12,13] and a further description is not necessary here. Rather an example of area monitoring at particle accelerators used in radiotherapy will be given.

Comprehensive Radiation Survey made at the LBL 184" Synchrocyciotron Medical Facility

The 184 inch synchocyclotron can accelerate helium ions (alpha particles) to an energy of 920 MeV. Pilot studies are underway using the helium-ion beam for the treatment of brain-, eye-, pancreas- and other tumours.[14] The production of the large uniform irradiation fields necessary for this application utilizes beam scatterers, collimators and energy-degraders to obtain the desired depth-dose characteristics. These beam line elements are sources of secondary particles which produced an unwanted radiation field which is a source of patient exposure outside the area of treatment and might also result in exposure to the medical and ancilliary staff. It was therefore necessary to determine the characteristics of this secondary radiation field and determine potential patient and/or staff exposure.

Schimmerling and his colleagues[15-17] have described the radiation surveys made at this facility in some detail. Figure 1 shows a plan view of the accelerator facility and the path of the transported 920 MeV alpha-particle into the medical treatment area. The medical treatment area is shown in more detail in Fig. 2. Secondary particles are produced in the first and second scatterers, at the edge of the iron pipe collimators, in the water column and, finally, in the patient collimater. The radiation field inside the medical treatment area is entirely due to beam interactions with the beam transport system, there being no significant leakage through the cyclotron shielding wall. Neutrons are the dominant component of the radiation field at the patient table.

Measurements were made to:

1. Identify the major sources of secondary particles in the beam transport system.
2. Measure the source strengths of these components.

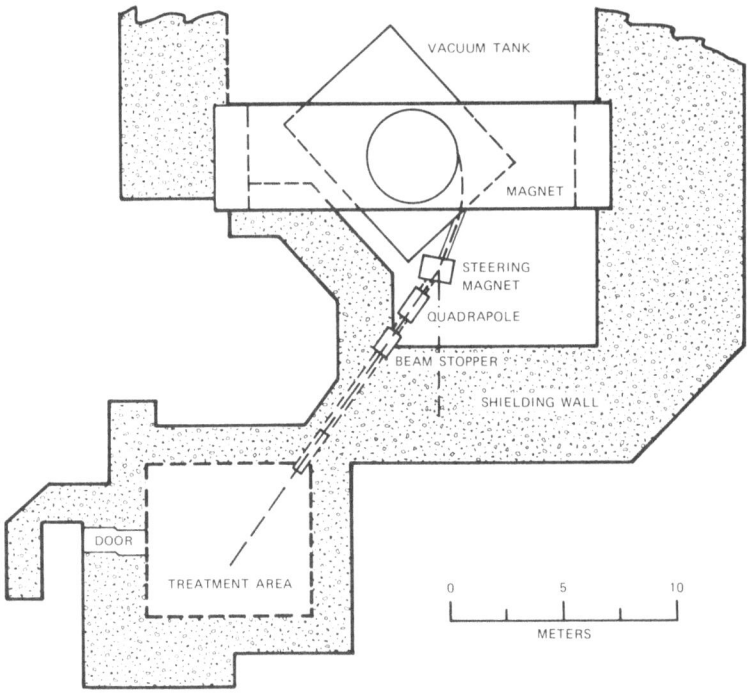

Fig. 1. Plan view of the 184-in. synchrocyclotron.

3. Determine the neutron intensity and spectrum at locations on the patient table.
4. Calculate the absorbed dose to the patient resulting from the secondary particle.

The neutron detectors selected for this series of measurements included moderated BF_3 counters, thorium and bismuth fission chambers, and aluminum, carbon and indium activation detectors. Figure 3 shows the neutron fluence profiles measured along the line parallel to and 20 cm. distance from the beam line. The curves correspond to different beam line configuration listed in Table 1. Curve B approximately corresponds to the configuation used in patient therapy with a phantom placed in the beam to simulate the patient. The measured profiles are consistent with the hypothesis that each secondary source is dominated by an isotropic component but with a smaller forward, directed contribution. Figure 4 shows an analysis based upon this assumption while Table 2 gives the estimated source strengths.

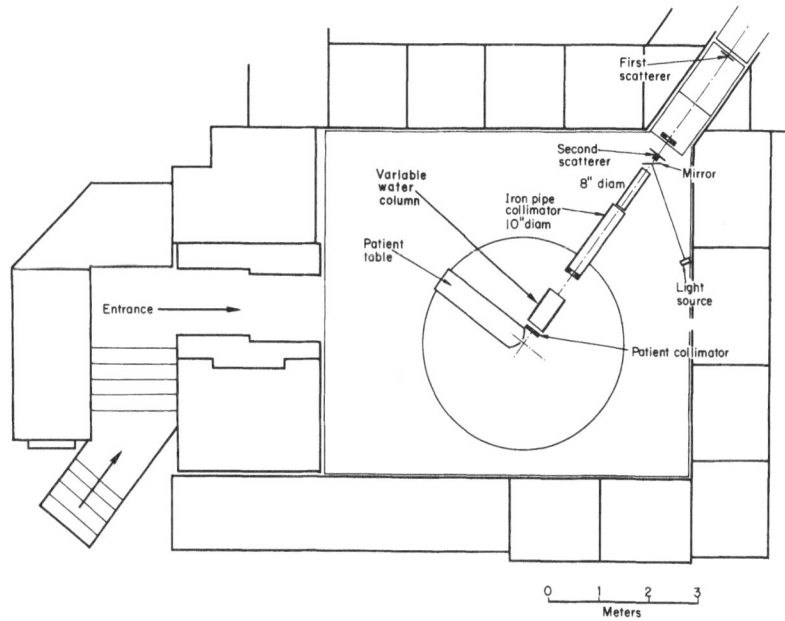

Fig. 2. Plan view of the biomedical cave of the 184-in. synchrocyclotron.

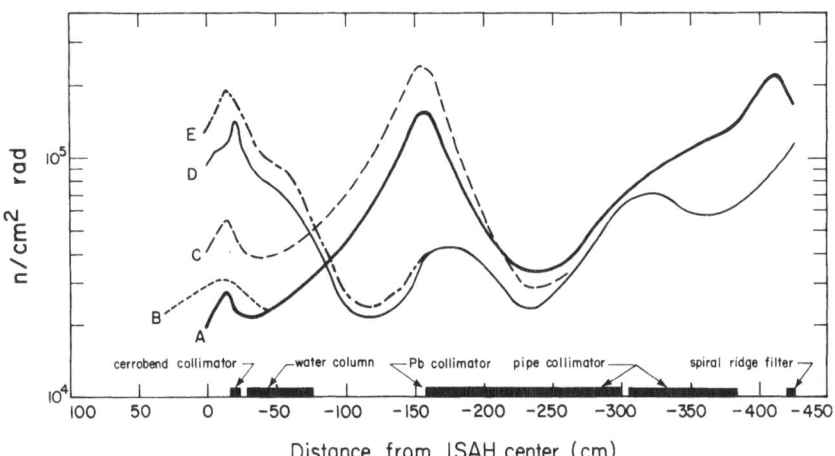

Fig. 3. Neutron fluence profiles along the beam line, measured by the ^{27}Al $\rightarrow$ ^{24}Na reaction, for the beam line configurations given in Table 1. The equivalent 14-MeV flux density is shown.

Table 1. Beam Line Configurations

Exposure (See Fig. 3)	Brass Collimator	Spiral Ridge Filter	Lead Collimator	Water Column (cm, water)	Cerrobend Collimator	Head Phantom
A	In	In	In	0.0	In	Out
B	In	In	In	16.0	In	In
B'	In	In	In	16.0	In	Out
C	In	Out	Out	0.0	In	Out
D	In	Out	Out	0.0	In	Out
E	In	Out	Out	0.0	Out	Out

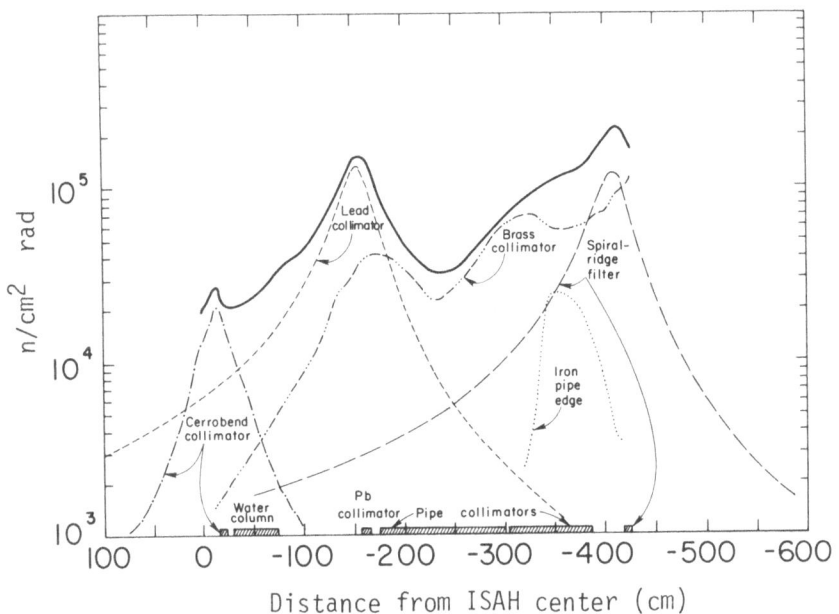

Fig. 4. Analysis of neutron fluence profiles. The heavy solid
        curve shows the experimentally determined profile,
        and the dashed curves show the profiles for the three
        major sources of neutrons.

Table 2. Point Source Strengths

Z Location Along Beam Line (cm)	Beam Element	Source Strength per Incident Rad (n/sec)
-25	Water column/Corrobend collimator	1.5×10^8
-165	Lead collimator	7.0×10^8
-350	Iron pipe	1.7×10^8
-420	Spiral ridge filter	6.0×10^8

From the data of Table 2 the relative contributions to the neutron fluence along the patient table from the sources of secondary radiations may be calculated (Table 3).

At distances out to 50 cm. from the beam axis, the water column and patient collimator are the dominate sources of neutron exposure to the patient outside the treatment region. At larger distances from the beam axis, the lead collimator dominates.

Table 3. Relative Fluence Contribution

X Location on Patient Table (cm)	Percent Fluence Contribution From:			
	Water Column and Cerrobend Collimator	Lead Collimator	Iron Pipe	Spiral Ridge Filter
0	89	9.2	0.5	1.2
50	64	30	1.7	4.3
100	39	49	3.4	8.6
150	27	56	4.6	12
200	21	57	5.7	15

Figures 5 and 6 show neutron fluence measurements along the patient table, both with and without a phantom in the primary beam. From these measurements the neutron spectrum may be estimated by solving the set of Fredholm equations obtained.[18] When the neutron spectrum is known the absorbed dose in soft tissue, D, may be calculated from:

$$D = \int_{E_{min}}^{E_{max}} f(E) \; \phi(E) dE$$

where $f(E)$ are fluence to absorbed dose conversion factors. Values of $f(E)$ given by Rindi[19] were used in the work reported here. The absorbed dose is not a sensitive function of the neutron spectrum and consequently, although the spectrum is not well determined from the solutions of the degenerate Fredholm equations of the first kind,[18] the absorbed dose can be quite well determined. Figure 7 shows the distribution of absorbed dose along the patient table.

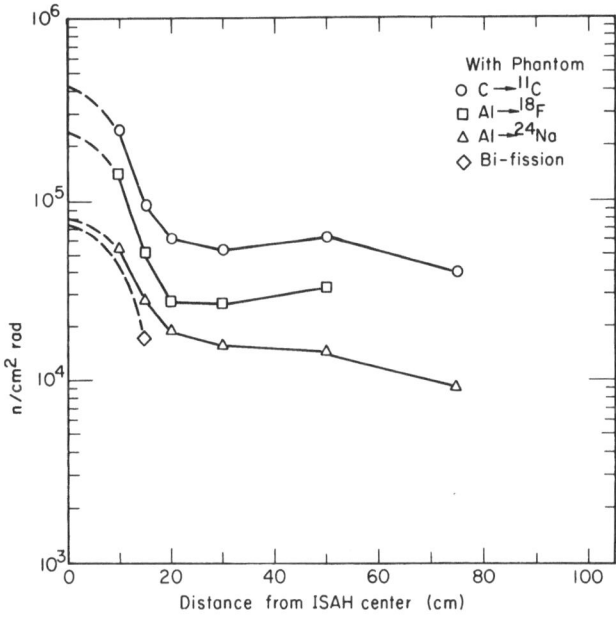

Fig. 5. Neutron fluence, as a function of distance from ISAH center, measured along the patient table with four detectors. (Head phantom in place; 16-cm water energy degrader), The equivalent 14-MeV flux density is given for the $^{27}Al \rightarrow ^{24}Na$ reaction.

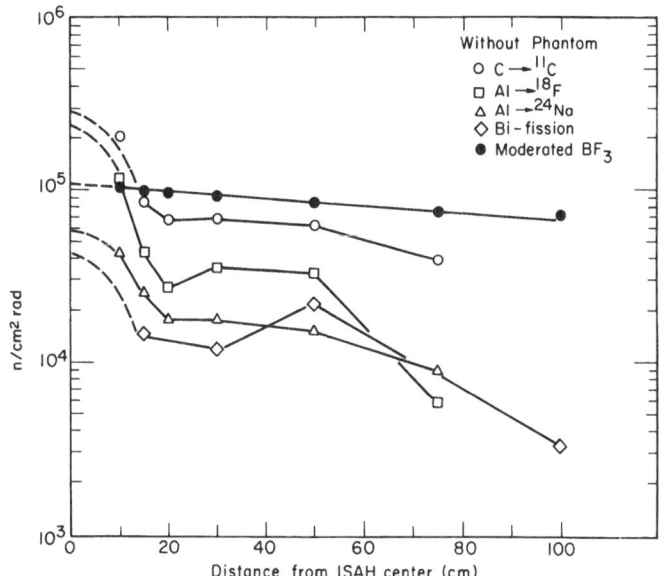

Fig. 6. Neutron fluence, as a function of distance
from ISAH centr, measured along the patient
table with five detectors. (No head phantom
in place; 16-cm water energy degrader.) The
equivalent 14-MeV flux density is given for the
^{27}Al $\rightarrow$ ^{24}Na reaction.

Table 4 summarizes the maximum absorbed dose to the patient
at various locations along the patient expressed as a ratio to
the primary beam dose in the plateau region. The dose to the
irradiated tumor would be a factor of about 1.4 higher.

Environmental Monitoring

Function. The function of environmental monitoring may be
defined as follows:

"Environmental monitoring is intended to show that the
working environment is satisfactory for continued
operations and that no change has taken place calling
for a reassessment of operating procedures. It is
largely of a confirmatory nature but may include the
use of fixed detectors to identify the onset of

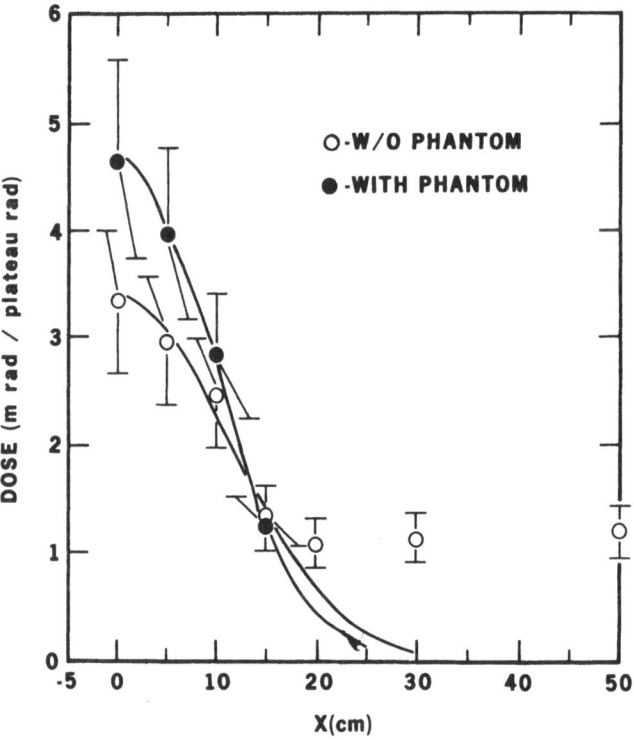

Fig. 7. Estimated absorbed dose distribution
along patient table.

Table 4. Ratio of Absorbed Dose from Neutrons to that from
            Primary Beam

Distance from ISAH Center (cm)	No Phantom ($\times 10^{-3}$)	Phantom ($\times 10^{-3}$)
0	3.33	4.64
5	2.97	3.96
10	2.49	2.83
15	1.35	1.27
20	1.10	
30	1.15	
50	1.21	

abnormal or emergency conditions, such as criticality
accidents. The term is also used for monitoring outside the boundaries of installations handling radioactive materials or radiation sources, but the context
is usually sufficient to avoid confusion between the
two meanings."[8]

Objectives: The objectives of a program of environmental
monitoring are principally:

1. To estimate the actual and potential exposure of man (and
   in some cases other organisms) to radioactive materials
   and radiation. Insofar as possible individual exposures
   should be determined, and estimates made of probable
   upper limits.
2. To check the effectiveness of control measures within the
   facility.
3. To carry on scientific investigations which are related
   to the overall program.
4. To predict trends in radiation levels.
5. To identify sources of specific contaminants.
6. To ensure compliance with regulations.
7. To establish good public relations.

Techniques. About five years ago a significant change in
attitude by government agencies towards man-made radiation
became evident.

"The recent trend toward the quantitative definition
of "as low as practicable" guidelines pertaining to
the release of radionuclides to the environment from
nuclear facilities and the resulting dose places a
significantly increased burden on environment
surveillance programs. It was previously believed
that adherence to the admonitions of expert bodies
such as the ICRP to limit unnecessary radiation
exposure was recommended "maximum permissible" annual
levels of 500 mrem to individuals or 170 mrem to a
"suitable sample."[20] The U. S. Regulatory agencies are
now preparing numerical limitations on environmental
radiation dose to man from lightwater power reactors
and the nuclear power fuel cycle.[21]

"The net effect of these limitations is to lower the
"maximum permissible" dose to off-site individuals by
two orders of magnitude. While the merits of such a
reduction in terms of public health and realistic
benefit-risk assessment are arguable, the rationale
for this change has been that practical, though

costly, techniques for the treatment of nuclear facilities effluents will permit plant operations within the limits.

"Questions immediately arise relating to how well the actual doses can be assessed and documented, given the fact that most existing environmental surveillance programs were designed to assure that critical populations groups do not receive doses that are much higher than the proposed limits. If the public and regulatory agencies are to be assured that nuclear facilities are operating within their design specifications, both experimental and calculational methods are required to allow accurate dose assessment at the very low exposure levels that are expected to exist."[22]

The assessment of the population dose equivalent around nuclear facilities requires, first, the development of reliable techniques in determining the man-made contributions to radiation exposure. Population dose inference must then be based on the incorporation of these experimental data into suitable models which take into account population distribution, living habits, meteorological conditions, and other significant factors which influence population dose.

Radiation Measurements. We have seen that we may need to make measurements of man-made radiation of the order of 1 millirem/yr in a natural background of about 100 millirem/yr. Furthermore, the natural background may show seasonal fluctuations of as much as 20 millirem/yr.

At first sight such a task seems impossible! The separation of various components of an observed fluctuating radiation level and the derivation of the component due to the operation of a nuclear installation at levels of ~1 millirem year is indeed a formidable task. The normal strategy employed is to place fixed radiation monitors around the installation that can record either the integrated dose equivalent or dose equivalent rates as a function of time. The selection of detector locations will depend upon several factors, such as population density, wind direction and local rainfall patterns.[23]

In many cases the dominant component of the radiation field will be γ-rays. In such cases Geiger counters,[24,25] ionization chambers,[26,27] scintillation detectors,[28,29] or thermo-luminescent dosimeters[30,31] have proved to be convenient detectors. Under some circumstances, for example around high energy accelerators, neutrons may be the dominant component of the radiation field.[32] Stephens[33] and his colleagues have described environmental monitoring techniques used at the Lawrence Berkeley Laboratory under such conditions. Here moderated BF_3 counters are used to detect the neutrons while Geiger counters are used to measure the γ-rays.[34]

Ionization Chambers. Beck et al[35] have described the use of a pressurized ionization chamber that can detect changes in ambient radiation levels of about 0.1 $\mu r/hr$ (~1 mr/yr). The chamber consists of an 8 liter stainless steel sphere filled with argon at a pressure of 25 atmospheres. Chamber current is measured with MOSFET electrometer. At exposure rates of 0.1 $\mu r/hr$ the chamber current is ~2 x 10^{-5} amps. Sensitivity of the instrument is essentially limited by leakage current which could be lower than 10^{-16} amps but in field use may be several times higher.

In routine use the output of the chamber is sampled every ten seconds and the data recorded on magnetic tape. A commercial version of this instrument is produced by Reuter-Stokes, Cleveland, Ohio (see Fig. 8). The output may be recorded on strip-chart.

Fig. 8. Ionization chamber and electrometer.
The chamber is filled with argon at a
pressure of 40 atmospheres and is
capable of measuring exposure rates
of about 1 mr/yr. (Reuter-Stokes,
Cleveland, Ohio).

Readings of the exposure rates as a function of time from instruments at several locations may be used to separate the various components of a radiation field. Several factors contribute to the variations in the reading of an instrument continuously monitoring background. Fluctuations in natural background occur over periods of a few hours to a few days.

Burke[36] has studied the temporal variations in natural background at several residential locations in New York and New Jersey. These locations were selected to assure that there were no radiation contributions from man-made sources (except for residual nuclear weapons in test global fallout). Radiation exposures were measured with LiF thermoluminescent dosimeters (Harshaw TLD-700), placed for monthly periods at the selected sites. Figure 9 shows typical results for the period of September 1972-September 1973. The accuracy of each measurement is estimated to be ±3.5 percent. These data clearly show a

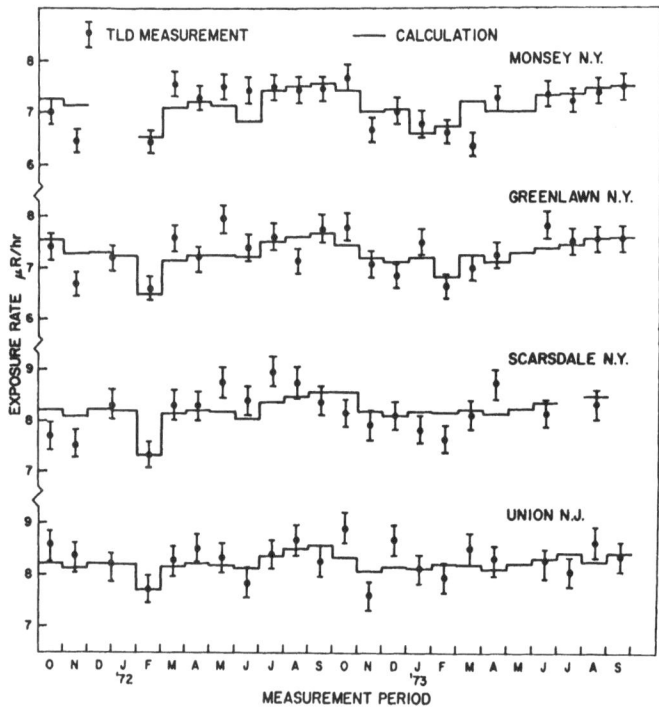

Fig. 9. Comparison of exposure rates measures at residences near New York City with calculated values using a climatic exposure model (from de Planque Burke, 1974).

significant variation in background level with time even when averaged over periods as long as a month. Such temporal variations in exposure rates can primarily be due to three causes:

1. Variations in climate conditions.
2. Variations in the rate of fallout deposition.
3. Variations in cosmic ray intensity.

Burke and O'Brian[37] have shown the variations of the type shown in Fig. 9 exhibit a seasonal pattern. An analysis of fallout deposition pattern during the period of measurement and estimates of the possible fluctuations in cosmic ray intensity showed that these components could account for only a small fraction of the observed variations. Variations in climatic conditions therefore remain as the probable cause of the observed fluctuations in background exposure rate.

The moisture content of soil, or the presence of standing water or snow cover can influence ambient radiation levels in two ways. First, the increased soil density resulting from high moisture content (or the water or snow cover itself) leads to increased attenuation of radiation emitted in decay of the radionuclides contained in the soil. Secondly, the presence of water in the interstices of the soil may inhibit the diffusion of radon and thoron into the atmosphere.

^{226}Ra decays by α-emission to its daughter ^{222}Rn which has a half-life of 3.8 days. Similarly, ^{224}Ra (a descendant of 232 Th) decays to "thoron" (^{220}Rn) which has a half life of 54 sec. Both these gases diffuse with the atmosphere. Eisenbud[38] quotes Pearson and Jones as estimating the average diffusion of ^{222}Rn as 1.4 pCi/m^2/sec.

The atmospheric concentration of the noble gases, thoron and radon in the atmosphere, depend upon many geological and meteorological factors. The complex diffusion processes from the soil into the atmosphere and the subsequent dispersion in the air are still largely not understood. In general, radon concentrations are 50-100 times greater than those of thoron found at a given location, largely because of the difference in radioactive half lives [3.8 days for ^{222}Rn compared to 54 sec for ^{220}Rn (thoron)]. The concentrations of radon at a given location show great variation from day-to-day. For example, Lockhart measured variations of more than two orders of magnitude in Washington, D.C. during 1957.[38]

"It is likely that these variations are dependent on meteorological factors that influence the rate of emanation of the gases from the earth. Thus, the rate of emanation from soil may increase during periods of diminishing atmospheric pressure and decrease during periods of high soil moisture, owing to the solubility of radon. It is also likely that the history of an air mass for several days prior to observation influences its radon and thoron concentration".[39]

Figure 10 shows the correlation between radon daughter washout peaks and precipitation reported by the USAEC Health and Safety Laboratory, New York. Measurements were made near the site of the Harkness Park Nuclear Power Stations during reactor shutdown. Exposure rates recorded by an ionization chamber are shown on the upper graph. Rainfall (in cm) are indicated directly below. Rain washes down dust to which radon daughters have become attached and the radiation level rises. Following washout a significant reduction of ~ several tenths μr/h in the background level is observed. This is due to the addition of water to the soil. As the water evaporates the radiation level slowly increases back to the original level.

Although the daily variations in radon concentration may be quite large the variations in averaged monthly exposure attributable to these fluctuations is extremely small compared to

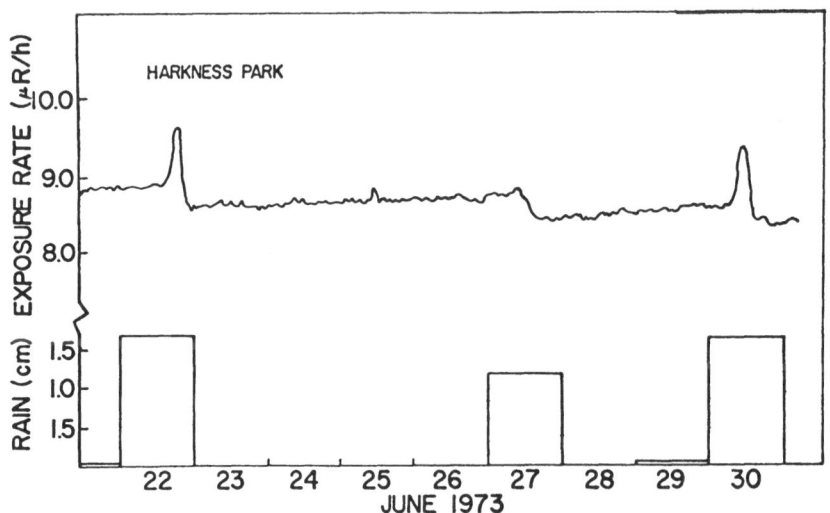

Fig. 10. Correlation between radon daughter washout peaks and precipitation data during reactor shutdown. Addition of rain water to soil lowers background levels but evaporation in the days that follow results in a slow rise. (From USAEC Health and Safety Laboratory.)

observed   month–to–month   variations.   Burke[36]   therefore
suggests that the observed fluctuations in terrestrial gamma
radiation is primarily due to changes in the density of the
soil–water medium and in standing water or snow cover on the
ground surface.[40–42]

By comparison with the time periods of variations in natural
background variations due to instrument noise and random fluctuations occur in time intervals comparable to the instrument
response time (~5 secs). Both the time characteristics and
magnitude of fluctuations in exposure rates measured by an
instrument such as the Reuter–Stokes ionization chamber may be
used to discriminate between natural and man–made radiation.
For example, substantial variations in the exposure rate due to
plumes of gaseous effluents from stacks occur due to changes in
wind speed, wind direction and atmospheric turbulence can occur
over periods of a minute or less.

An increased exposure rate from a man–made source, such as
the plume from the stack of nuclear power reactor, may have
strong directional properties. For example, we would expect to
find elevations in exposure rates up wind of the stack of
nuclear reactor but not down wind. On the other hand, if rainfall is widespread, radon daughter washout may produce an elevation in radiation level observable at all monitoring stations
around the reactor. Such criteria may prove important in
discriminating between natural and manmade radiation levels.
Another important parameter to be considered is the magnitude of
deviations from the average radiation level produced by man–made
and natural radiation sources. Gogolok and Miller[43] have
described how chambers that continuously read exposure rate may
be used to estimate exposures resulting from gaseous effluents
from nuclear installations. Figure 11 shows the exposure rates,
averaged over an interval of one hour at two locations near a
boiling water power reactor. The large fluctuations are due
either to radon daughter washout or the presence of gaseous
radioactive effluent. Figure 11 also shows the standard deviation of exposure rate averaged over a 5 min. period from the
hourly average. The standard deviation of natural background is
found to be less than 0.2 μr/hr. Analysis using calculation of
"the standard deviation clearly shows that the peak occurring at
the beginning of the 6th day of monitoring is due to radon
daughter washout, while the peaks occurring on the 5th, 8th and
9th at location one and on the 3rd, 5th and 11th days at location two are due to the noble gas plume. On the 3rd day, radon
daughter washout occurred at both locations, while the gaseous
plume was detectable only at location two. The standard deviation variations distinguish these contributions even though they
occurred simultaneously."[22]

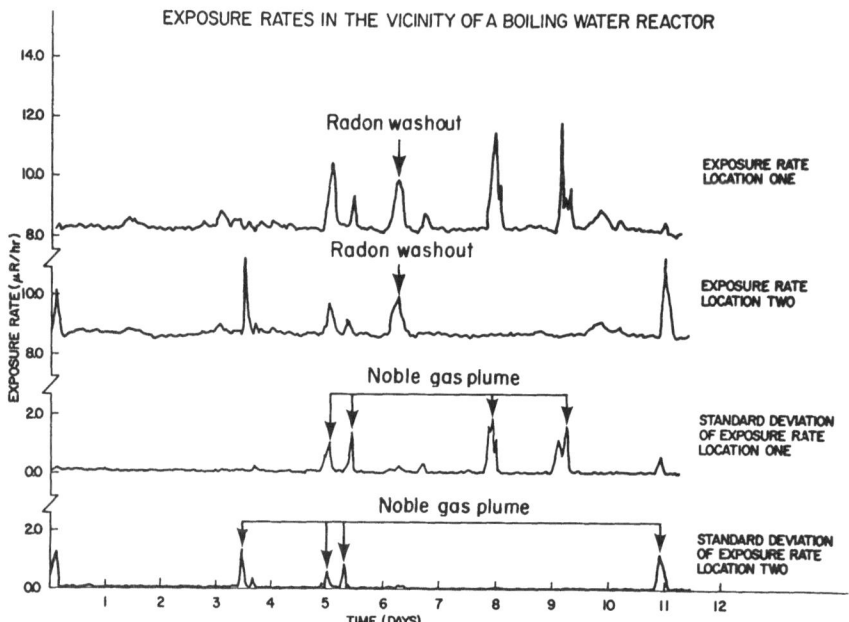

EXPOSURE RATES IN THE VICINITY OF A BOILING WATER REACTOR

Fig. 11. Exposure rates measured at two monitoring
stations in the vicinity of a boiling water
reactor.

Scintillation Counters. A NaI scintillation counter has been
found extremely valuable in environmental radiation studies
carried out in the Bay Area by the Health Physics Department of
the Lawrence Berkeley Laboratory.[28]

     The detector is a thallium-activated sodium iodide crystal
3 in. in diameter and 3 in. thick. The crystal is optically
coupled to a Dumont 6363 multiplier phototube, also 3 in. in
diameter. The crystal and phototube assembly is shock-mounted
in 1/2 in. foamed rubber, and is contained in a stainless steel
canister 1/16 in. thick, 5 in. in diameter, and 12 in. high.
(see Fig. 12) The casing excludes β-particles below about 3 MeV
and this makes it possible to correlate field readings with
laboratory pulse-height analysis of field samples, without being
influenced by a changing β-γ ratio caused by changes in the
field environment.

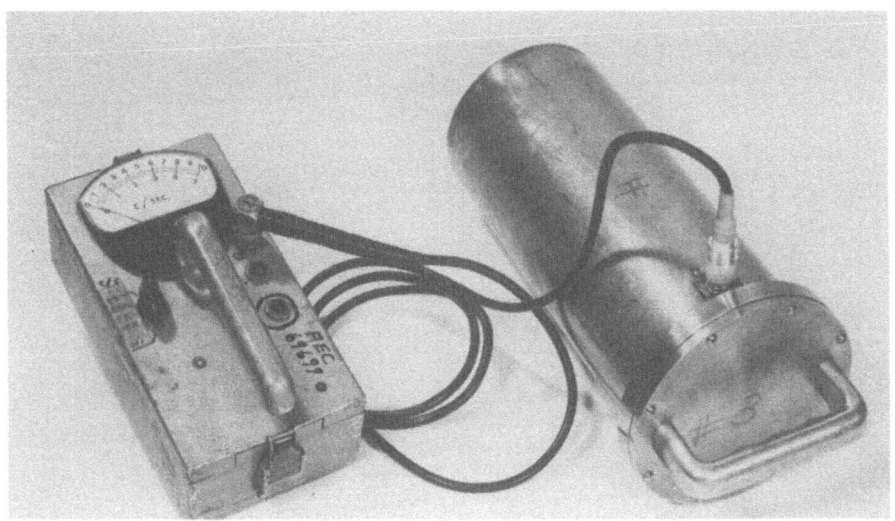

Fig. 12. Portable NaI scintillation counter
used for radiation surveys at LBL.

The transistorized count-rate meter was designed by
Goldsworthy.[44] His design has been slightly modified for our
purpose but is basically the same. The instrument contains a
Cockcroft-Walton high-voltage supply, a four-transistor linear
pulse amplifier, an integral pulse-height selector circuit, and
a rate-meter circuit. The electrical power for the instrument
is supplied by a 10.8V mercury battery, which provides 300 hr or
more of operation. The count-rate meter has four linear ranges
spanning an interval from 0-50,000 counts/sec, or from
0-1.25 mr/hr. Experience at LBL has shown the instruments to be
specific detectors of terrestrial gamma radiation and to be
reliable and rugged in field service.[45-47] Figure 13 shows
the instrument in use.

With the threshold set at about a photon energy of 100 keV
the sensitivity of the instrument is such that 400 counts per
sec corresponds to an exposure rate of 10 μr/hr. The instrument
is calibrated using a radium source. Because the spectra
produced by terrestrial gamma radiation and by fallout in a
field environment are closely similar to that produced by our
standard radium sources, this conversion factor can be used in
any field situation.

Fig. 13. The portable NaI scintillation counter in use surveying
exposure rates around a granite rock.

    In the field measurements are made within areas roughly 30
by 30 ft. over which several readings are taken during a visit.
Readings for a location are then averaged and the value entered
into a log book. Readings are made with the detector ≈3 ft.
above the ground surface. A simple geometric analysis shows
that above a uniformly radioactive surface ≈90 percent of the
detectable radiation comes from within a circle of ≈20 ft.
radius. The effective area increases markedly as local relief
increases.

    One feature of the NaI scintillation counter that should be
noted is its different response characteristics to environmental
radiation compared to an ionization chamber. The ionization
chamber measures ionization produced in a gas and, therefore,
responds proportionally to the exposure rate produced by terrestrial radionuclides and cosmic radiation. For example at
Berkeley where the environmental levels due to terrestrial
radionuclides are ~30 millirem/yr, the relative response of an
ionization chamber will be about 2/3 due to terrestial radionuclides and 1/3 due to cosmic radiation. A scintillation
counter, on the other hand, responds to ionizing event rates
(not exposure rate). Thus, for example, a 3" X 3" NaI has a

count rate of 400 cps in a γ-radiation field of 10 μr/hr while
the corresponding counting rate due to μ mesons it is about
1 count per sec.

<u>γ-Spectrometers</u>. Although detectors which measure gross γ-ray
exposure may provide a great deal of information, it is sometimes desirable that particular radionuclides be identified.
Such an example might be in studies of the relative contribution
to background due to terrestrial and fallout radionuclides.[28]

The portable scintillation counter described by Wollenberg
et al.[28] measures the gross γ exposures rates. If soil
samples are removed from the area in which gross measurements
are made they may be assayed in the laboratory for uranium,
thorium potassium as well as fallout radionuclides by
γ-spectrometry.

Wollenberg et al. have described the use of a 4 in. dia.,
2 in. thick NaI (Tl) γ-spectrometer located in low background
facility.[48]   Figure 14 shows a typical spectrum obtained from

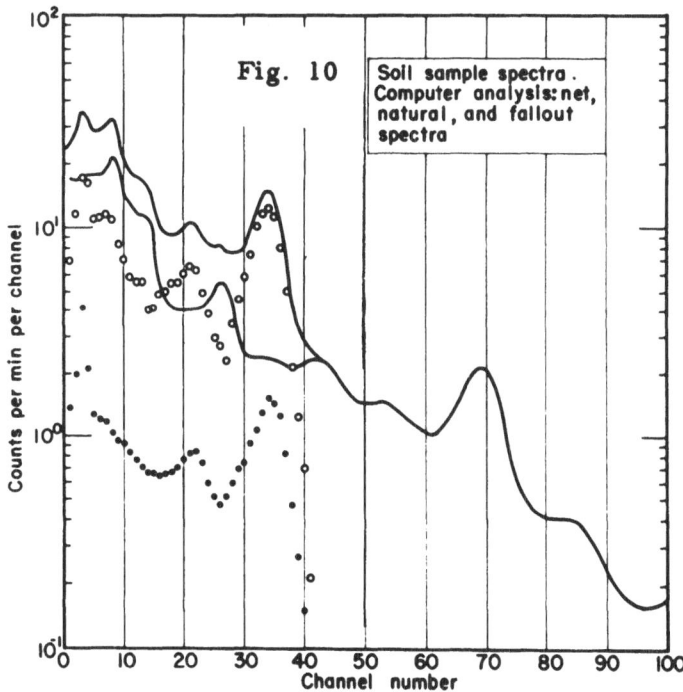

Fig. 14. Comparison of measured exposure rates.

a surface-soil sample taken in the San Francisco Bay Area. The upper curve shows the total spectrum resulting from natural and fallout radionuclides, after coorection for instrumental back-ground. The lower curve shows the spectrum calculated from the uranium, thorium and potassium concentrations derived from the upper curve. (Prominent peaks in the natural spectrum are listed in Table 5.)

The two curves are seen to coincide at an energy of about 0.9 MeV. Subtraction of the two curves gives the spectrum due to fallout nuclides, shown as open circles in Fig. 14. Recognizable peaks are attributed to ^{144}Ce at 0.14 MeV, Rh and Ru isotopes at 0.45-0.52 MeV, and ^{95}Nb-^{95}Zr at ≈0.76 MeV. Recounting of this sample about 6 months later would show a ^{137}Cs peak at 0.66 MeV, presently masked by the Nb-Zr peak because of the shorter effective half-life (≈50 days) of ^{95}Nb-^{95}Zr compared with the 28 yr. half-life of ^{137}Cs.

An interesting comparison may be made between the calculated fission product spectrum derived from the soil sample with that obtained from dried weeds collected from a ditch in the Santa Cruz Mountains. (Black dots in Figure 14) The only activities in the weed sample are from fission products.

Studies such as this with a γ-spectrometer then make it possible to distinguish between fallout and natural sources of natural background. Thus, if it is assumed that fallout was evenly deposited in time Wollenberg et al.[28] show that in the years from 1958 to 1968 an integrated γ-ray dose of 282 mr was superimposed on a natural terrestrial of 482 mr and a cosmic-ray dose of 350 mr. Roughly three times as much γ-emitting radio-activity was deposited in the San Francisco Bay Area between March 1958 and the end of 1960 as has been deposited since 1960.

Table 5. Prominent Peaks in the Natural Spectrum

Energy (MeV)	Decay series	Nuclide
0.24	Th	^{224}Ra
≈0.60	U and Th	^{214}Bi and ^{208}Tl
0.90	Primarily Th	^{208}Tl
1.12	U	^{214}Bi
1.46	^{40}K	^{40}K
1.76	U	^{214}Bi

Intercomparison between estimates using these techniques and measurements made by the U.S. Health and Safety Laboratory using an ionization chamber and a portable γ-ray spectrometer showed good agreement. It is also of interest to note that Wallenberg et al.[28] have reported excellent agreement between exposure rates measured directly in the field and those calculated from the measured radionuclide composition of rock samples taken to the laboratory. If the composition of the soil or rock is known, the γ-ray level may be calculated quite accurately. Table 6 gives the exposure rate in μr/hr for measured contributions in ppm.

Figure 15 shows the agreement obtained. It is of interest also to note that Fig. 15 illustrates the influence of moisture content on the ambient radiation level for the San Francisco Bay Area, where the soil.moisture measurements indicate an average of 30 percent difference in ambient radiation level between dry (summer) and wet (winter) conditions. When it is necessary to measure man-made radiation levels as low as 5 millirem/yr it is necessary that these fluctuations be well understood in the locality where observations are made.

Beck et al.[49] have discussed the relative merits of Ge(Li) and NaI detectors. Table 7 summarizes this comparison. For many applications the lower cost, greater sensitivity, and facility of a large NaI crystal make it very convenient.

Figure 16 shows a typical preoperational γ-ray spectrum measured at a reactor site. This spectrum together with a measurement of total exposure rate may be used to identify the five principal components of the background radiation exposure rate.

Table 6. Exposure Rate in μr/hr for Measured Contributions in ppm.

Element	Exposure rate in μr/hr per ppm of element. (detector 3 ft above ground surface)
U	0.76
Th	0.36
K	1.71×10^{-4}

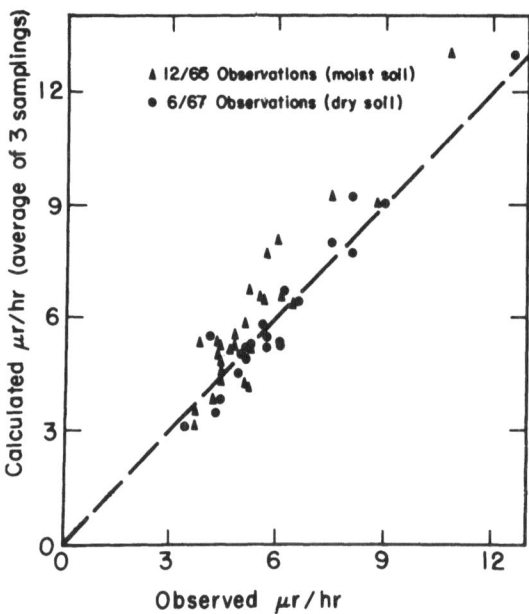

Fig. 15. Comparison of calculated exposure rates
due to terrestrial radioactivity with
measurements made in the field. Measurements made in the field are depressed when
the soil moisture content in significant.

Figures 17 and 18 show data taken simultaneously at the same
site in the Northeastern United States. Figure 17 shows data
obtained with a 10 cm X 10 cm NaI(Tl) crystal in a counting time
of 20 minutes. γ-rays from the fallout nuclides ^{137}Cs and
[95]Zr-[95]Nb are clearly resolved from γ-rays produced by
naturally occurring radionuclides. Figure 18 shows the dramatic
improvement in resolution that may be obtained using a Ge(Li)
detector.[49]

For special problems which may require this good resolution,
Ge(Li) detectors may be invaluable, but NaI detectors are
generally adequate. Thus, for example: Nakamura has reported
the use of a NaI detector to investigate the energy spectrum of
"Skyshine" photons around accelerators at the Institute for
Nuclear Study.[50]

Table 7. Characteristics of HASL Field Spectrometer Systems

	NaI (Tl)	Ge(Li)
Detector size	4 x 4 in. (820 cm^3)	60 cm^3
Detector geometry	rgt. circ. cyl.	closed coax. cyl.
Detector resolution (0.662 MeV)	53 keV (FWHM)	2.3 KeV (FWHM)
Detector efficiency (0.662 MeV)	36.5 cts/unit flux	2.2 cts/unit flux
Analyzer capacity	400 ch.	4000 ch.
Data accumulation times	10–20 min.	30–90 min.
Data recording	(1) magnetic tape (2) decimal printout	(1) magnetic tape
Data output	(1) peak areas (2) energy bands (4)	(1) peak areas

Phelps et al.[51] have used the environmental monitoring system developed at the Health and Safety Laboratory to make extensive in situ measurements of radionuclides in soil. This has been done at the Nevada Test Site and at several nuclear reactor sites. They have mounted their analytical equipment in a small van making field measurements very convenient.

Thermoluminescent dosimeters. Thermoluminescent dosimeters are preferred over film in the measurement of photon exposures above a few millirem. They are intrinsically more accurate and less subject to the problems of latent image fading found in film.

Several thermoluminescent materials are capable of the measurement of exposures below 1 mr with good accuracy. Thus, in 1974 de Planque Burke[30] reported the use of LiF and CaF_2:Mn to measure integrated environmental gamma exposures. Over exposure periods of 4 weeks and 2 weeks respectively accuracies of ±3 percent at exposures of 1 mR were reported. Lindeken et al.[31] reported measurements of the γ-exposure in houses in the Livermore Valley using CaF_2:Dy dosimeters in 1973.

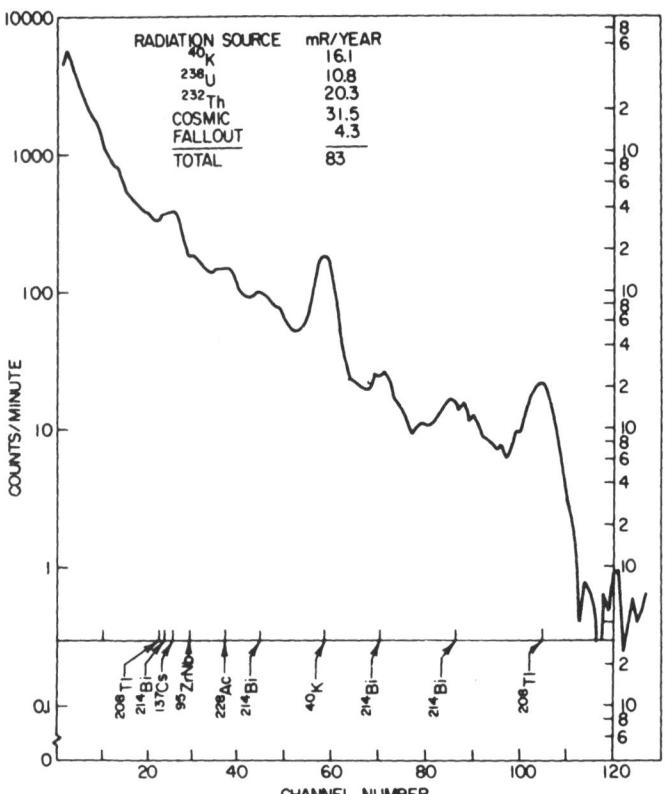

Fig. 16.   Typical preoperational γ spectrum
            at reactor site. This spectrum,
            a conjunction with a pressurized
            ion chamber measurement of the
            total γ dose rate, can be reduced
            to provide the portion of the total
            dose contributed by the five sources
            of background radiation (Environmental
            Analysis Inc.).

    The use of thermoluminescent dosimeters for environmental
radiation measurement is now widespread and the results of an
intercomparison of dosimeters under both field and laboratory
conditions with 85 participants from 26 countries has recently
been published.[52]   Tuyn et al.[53] have used LiF thermoluminescent dosimeters to determine contours of equal dose equivalent from γ-rays and neutrons around the complex of accelerators
at CERN.

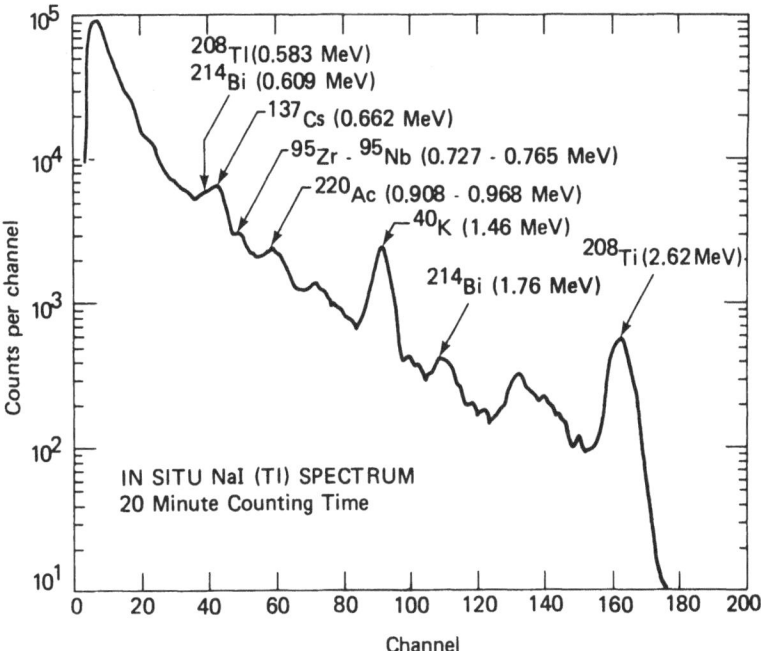

Fig. 17. *In situ* spectrum, northeastern U.S.A. location, taken in 1971 with a 10 cm by 10 cm NaI (Ti) crystal, 20 min counting time.

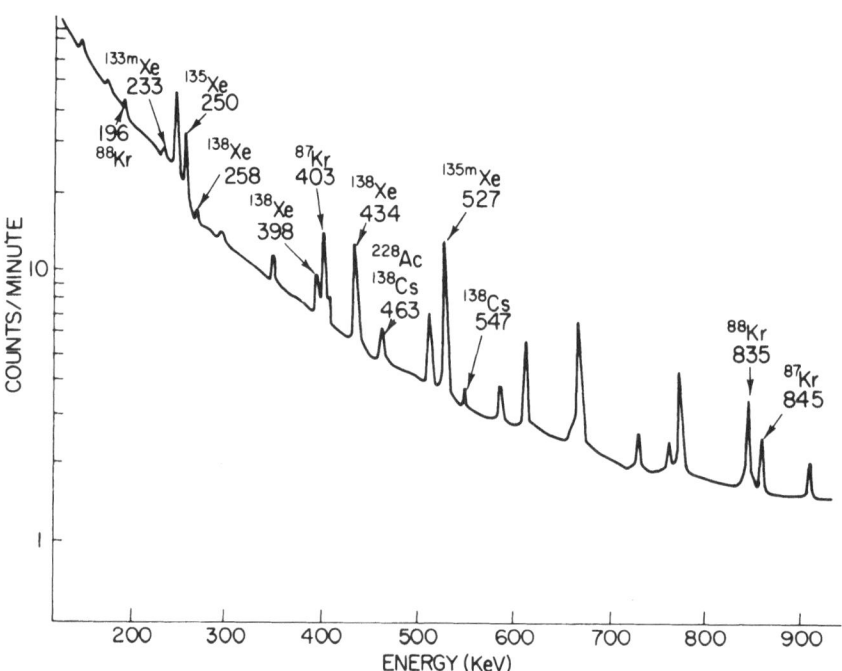

Fig. 18. In situ Ge(Li) spectrum taken near stack
          of BWR power plant with relatively short
          noble gas holdup (<1 hr). Peaks not
          identified are due to natural and fall-
          out emitters (Harold Beck, U. S. Atomic
          Energy commission, Health and Safety
          Laboratory).

REFERENCES

1. International Commission on Radiological Protection. "The RBE for High-LET Radiations with Respect to Mutagenesis" in ICRP Publication No. 18, Pergamon Press. Oxford (1972).
2. H. H. Rossi and C. W. Mays, Leukemia Risk from Neutrons. Health Physics 34: 333 (1978).
3. A. Rindi, R. H. Thomas, 1973, The Radiation Environment of High-Energy Accelerators, Annu. Rev. of Nucl. Sci. 23: 315, (1973).
4. R. H. Thomas, and A. Rindi, (Eds), "Proceedings of the First Course on High-Energy Radiation Dosimetry and Protection" Erice, Italy. October 1975, in I.E.E.E. Trans Nuclear Science Ns-23, No. 4. (1976).
5. R. H. Thomas, Book Review, NCRP Report No. 51 Health Physics 36: 92 (1979).
6. E. Fermi, E. Amaldi, O. D'Agostino, F. Rasetti, and E. Segre, Proc. Roy Soc. (London) A: 146, 483 (1934).
7. International Commission on Radiological Protection, "General Principles of Monitoring for Radiation Protection of Workers" in ICRP Publication No. 12, Pergamon Press, Oxford, Par. 4 (1969).
8. Op. Cit. Ref. 7. Par. 28.
9. Op. Cit. Ref. 7. Par. 42.
10. Op. Cit. Ref. 7. Par. 86.
11. Op. Cit. Ref. 7. Par. 45-47.
12. V. Perez-Mendez, Instrumentation--Active Detectors. Lecture No. 9, in "Advances in Radiation Protection and Dosimetry in Medicine" Course Proceedings, International School of Radiation Damage and Protection, Ettore Majorana Centre for Scientific Culture, Erice, Italy (Sept. 1979).
13. R. H. Thomas, Instrumentation--Passive Detectors. Lecture No. 10, in "Advances in Radiation Protection and Dosimetry in Medicine" Course Proceedings, International School of Radiation Damage and Protection, Ettore Majorana Centre for Scientific Culture, Erice, Italy (Sept. 1979).
14. J. R. Castro, J. M. Quivey, J. T. Lyman, G. T. Y. Chen, C. A. Tobias, L. L. Kanstein, and R. E. Walton, Heavy-ion Therapy in Biological and Medical Research with Accelerated Heavy Ions at the Bevalac 1974-1977, Lawrence Berkeley Laboratory Report LBL-5610, pp. 182-218 (1977).
15. A. R. Smith, et al., Neutron Flux Density and Secondary-Particle Energy Spectra at the 184 Inch Synchrocyclotron Medical Facility, Lawrence Berkeley Laboratory Report LBL-6721, (1978).
16. J. B. McCaslin, W. R. Schimmerling, A. R. Smith and R. H. Thomas, Neutron Fluence Rates and Energy Spectra at the 184 Inch Synchrocyclotron Medical Facility, Paper read at Health Physics Society Meeting, Philadelphia, July 8-13, 1979.

17. W. S. Schimmerling, A. R. Smith, and R. H. Thomas, Neutron Flux Density and Secondary Particle Energy Spectra at the 184 Inch Synchrocyclotron Medical Facility, XII International Conference on Medical and Biological Engineering, Jerusalem, August 19-24, 1979.

18. J. T. Routti, High-Energy Neutron Spectroscopy with Activation Detectors, Incorporating New Methods for the Analysis of Ge(Li) Gamma-Ray Spectra and the Solution of Fredholm Integral Equations, Ph.D. Thesis—University of California at Berkeley, Lawrence Berkeley Laboratory Report UCRL-18514 (1969).

19. A. Rindi, An Analytical Expression for the Neutron Flux to Absorbed Dose Conversion Factor. Health Physics 33: 264 (1979).

20. Title 10, Part 50, Appendix I, Code of Federal Regulations, Federal Register 36: 111 (1971).

21. Environmental Radiation Protection Requirements for Normal Operations of Activities in the Uranium Fuel Cycle, Notice of Proposed Rulemaking, U. S. Environmental Protection Agency (1973).

22. W. M. Lowder and C. V. Gogolak, Experimental and Analytical Radiation Dosimetry Near a Large BWR IEEE Trans., Nucl. Sci. NS-21, No. 1: 423 (1974).

23. W. M. Lowder and C. V. Gogolak, Experimental and Analytical Radiation Dosimetry Near a Large BWR IEEE Trans., Nucl. Sci. NS-21, No. 1: 429 (1974).

24. A. R. Jones, A Gamma Monitor for Measuring Environmental Gamma Doses and Dose Rates, Atomic Energy of Canada Limited Report AECL-3989 (1974).

25. A. R. Jones, Measurement of Low Level Environmental Gamma Dose with TLD's and Geiger Counters, IEEE Trans., Nucl. Sci. NS-21, No. 1: 456 (1974).

26. H. L. Beck, J. A. DeCampo, et al., New Perspective on Low Level Environmental Radiation Monitoring Around Nuclear Facilities, Nuclear Technology 14: 232 (1972).

27. M. E. Cassidy, S. Watnick, et al., A Computer-Compatible field Monitoring System, IEEE Trans, Nucl. Sci. 21, No. 1: 461 (1974).

28. H. W. Wollenberg, H. W. Patterson, A. R. Smith, and L. D. Stephens, Natural and Fallout Radioactivity in the San Francisco Area, Health Physics 17, No. 2: 313, (1969).

29. J. D. Chester, R. L. Chase, and S. Wood, A Digital Environmental Monitor, Brookhaven National Laboratory Report BNL-16922 (1972).

30. G. de Planque-Burke, Thermoluminescent Dosimeter Measurements of Perturbations of the Natural Radiation Environment, in "Proc. of Second Intl. Symp. on the Natural Radiation Environment," U.S. Atomic Energy Commission Symposium Series (1974).

31. C. L. Lindeken, D. E. Jones, and R. E. McMillen, Environmental Radiation Background Variations Between Residences, Health Physics 24: 81 (1973).

32. H. W. Patterson and R. H. Thomas, Accelerator Health Physics, Academic Press, New York (1973).

33. L. D. Stephens and H. S. Dakin, A High Reliability Environmental Radiation Monitoring and Evaluation System, Proc. of the VIth International Congress of the Société Francaise de Radioprotection, Bordeaux, France, March 27-31, 1972.

34. R. H. Thomas (Ed), The Environmental Surveillance Program of the Lawrence Berkeley Laboratory, Lawrence Berkeley Laboratory Report LBL-4827 (1976).

35. H. L. Beck, W. M. Lowder and J. C. McLaughlin, In Situ External Environmental Gamma Ray Measurements Utilizing Ge(Li) and NaI(Tl) Spectrometry and Pressurized Ionization Chambers, IAEA SM/148-2, IAEA, Vienna.

36. G. de Planque-Burke, Variations in Natural Environmental Gamma Radiation and its Effect on the Interpretability of TLD Measurements made Near Nuclear Facilities. USAEC, Health and Safety Laboratory, (1974).

37. G. de Planque-Burke and K. O'Brien, USAEC Report, HASL-283 (1974).

38. M. Eisenbud, Environmental Radioactivity, chapter 7, in "Natural Radioactivity," Academic Press, New York (1973).

39. L. B. Lockhart, Atmospheric Radioactivity Studies at U. S. Naval Research Laboratory, U. S. Naval Research Laboratory, Rep. 5249 (1958).

40. R. M. Sievert and B. Hulquist, Acta Radiologica, 37: 388 (1952).

41. P. R. J. Burch, J. C. Duggleby, B. Oldroyd, and F. W. Spiers, in "The Natural Radiation Environment", p. 767, The University of Chicago Press, Chicago.

42. H. L. Beck and G. de Planque-Burke, USAEC Report HASL-195 (1968).

43. C. V. Gogolak and K. M. Miller, Method for Obtaining Radiation Exposure due to a Boiling Water Reactor Plume from Continuously Monitoring Ionization Chambers, Health Physics 27: 132 (1974).

44. W. W. Goldworthy, Transistorized portable counting rate meter, Nucleonics 18: 92 (1960).

45. L. D. Stephens, H. W. Patterson and A. R. Smith, Fallout and Natural Background in the San Francisco Bay Area, Health Physics 4: 267 (1961).

46. H. W. Patterson, and R. W. Wallace, Report on a Radiation Survey Made in Egypt, India and Ceylon in January 1963, Health Physics 12: 935 (1966).

47. H. A. Wollenberg and A. R. Smith, Studies in terrestrial gamma radiation in "The Natural Radiation Environment", University of Chicago Press, Chicago (1964).

48. H. A. Wollenberg and A. R. Smith, A concrete low-background counting enclosure. Health Physics 12: 53 (1966).

49. H. L. Beck, J. DeCampo and C. Gogolak, In Situ Ge(Li) and NaI(Tl) Gamma-Ray Spectrometry, USAEC Health and Safety Laboratory Report NASL-258 (1972).

50. T. Nakamura, et al., Skyshine of Neutrons and Photons from the INS F M Cyclotron, p. 43, Institute for Nuclear Study, Tokyo Univ. Ann. Report (1975).

51. P. L. Phelps, et al., Ge(Li) Low Level in-situ Gamma Ray Spectrometer Application, IEEE Trans, Nucl. Sci. NS-21, No. 1: 543 (1974).

52. G. de Planque-Burke and T. F. Gesell, Second International Intercomparison of Environmental Dosimeters, Health Physics 36: 221 (1979).

53. A. Bonifas, et al., On the Use of Thermoluminescent Dosimetry for Stray Radiation Monitoring on the CERN Site, CERN Health Physics Internal Report HA-74-138 (1974).

ACKNOWLEDGMENT

This work was supported by the U.S. Department of Energy under contract No. W07405-ENG-48.

PARTICIPANTS

- Angela ALBERICI, Istituto Nazionale Tumori, Via Venezian 1, 20133, MILANO, Italy

- Antonio BARBIERI, Escola Paulista de Medicina, Setor Radio-terapia, R. Botucatu 715, Vila Clementino, SAO PAULO, Brasil

- Luciano BENINI, Servizio di Fisica Sanitaria, Ospedale Civile, 35100 PADOVA, Italy

- Daniela BETTEGA, Istituto di Fisica, Università di Milano, Via Celoria 16, 20133 MILANO, Italy

*† - Claudio BIRATTARI, Istituto di Fisica "A. Pontremoli" Università di Milano, via Celoria 16, 20133 MILANO, Italy

 * - Johan J. BROERSE, REP-Institutes Organisation Health Research TNO, 151 Lange Kleiweg, RIJSWIJK (ZH), The Netherlands

- Gianfranco CALICCHIO, Istituto Tumori, Via M. Semmola 1, 80100 NAPOLI, Italy

- Ugo CERCHIARI, Istituto Nazionale Tumori, Via Venezian 1, 20133 MILANO, Italy

 * - George T. Y. CHEN, University of California, Lawrence Berkeley Laboratory Bldg. 55/106, BERKELEY, CA 94720, USA

- Leopoldo CONTE, Servizio Fisica Sanitaria Ospedale di Circolo, Viale Borri 57, 21100 VARESE, Italy

- Tiziana COSTI, Servizio Fisica Sanitaria Arcispedale S. Maria Nuova, Viale Risorgimento 80, 42100 REGGIO EMILIA, Italy

 * - Stanley B. CURTIS, University of California, Lawrence Berkeley Laboratory, BERKELEY, CA 94720, USA

- Alberto DEL GUERRA, Università di Pisa, Istituto di Fisica, Piazza Torricelli 2, 56100 PISA, Italy

* - John F. DICELLO, Los Alamos Scientific Laboratory M.S. 844,
  P.O. Box 1663, LOS ALAMOS, New Mexico 87545, USA

* - Mme Andrée DUTREIX, Institut Gustave Roussy, 16 Ave.
  Vaillant Coutourier, 94800 VILLEJUIF (Val de Marne), France

  - Andreas FORMICONI, Via Pier La Torre 1, 50065 PONTASSIEVE
  (Firenze), Italy

  - Rosina GALLINI, Ente Universitario Lombardia Orientale,
  Via Valsabbina 19, 25100 BRESCIA, Italy

  - Diego GOMEZ VELA, Facultad de Medicina de Cadiz, Catedra de
  Terapeutica Fisica, CADIZ, Spain

  - Sigmund GULDBAKKE, Physikalisch-Technische Bundesanstalt,
  Bundesallee 100, D-3300 BRAUNSCHWEIG, Germany

  - Søren HOLM, Physics Laboratory II, H. C. Oersted Institute,
  University of Copenhagen, Universitetsparken 5, 2100
  COPENHAGEN, Denmark

* - Börje LARSSON, University of Uppsala, Dept. of Physical
  Biology "Gustaf Werner Institut", Box 531, S-75121,
  UPPSALA, Sweden

  - Heinrich LESIECKI, Physikalisch-Technische Bunesanstalt,
  Bundesallee 100, D-3300 BRAUNSCHWEIG, Germany

  - Maria Concetta MACCARONE, Gruppo Fisica Applicata a Problemi
  Biomedici, Istituto di Fisica, Università di Palermo, Via
  Archirafi 36, 90123 PALERMO, Italy

  - Alberto MACHI, Gruppo Fisica Applicata a Problemi Biomedici,
  Istituto di Fisica, Università di Palermo, Via Archirafi 36,
  90123 PALERMO, Italy

  - Francesco MALGIERA, Istituto Tumori, Via M. Semmola 1, 80100
  NAPOLI, Italy

  - Ehrenfried MORODER, Ospedale Regionale di Bolzano, Servizio
  di Fisica Sanitaria, Via Sernesi 1, 39100 BOLZANO, Italy

* - Herwig G. PARETZKE, GSF-Institut für Strahlenschutz, D-8042
  NEUHERBERG, Germany

  - Guido PEDROLI, Servizio Fisica Sanitaria Ospedale di Circolo,
  Viale Borri 57, 21100 VARESE, Italy

** - Victor PEREZ-MENDEZ, University of California, Lawrence
Berkeley Laboratory, BERKELEY, CA 94720, USA

- Rita PETIT, Université Catholique de Louvain, Faculté de
Medicine, Avenue Hippocrate 54, 1200 BRUXELLES, Belgium

- Luigi RAFFAELE, Studio Tecnico di Radioprotezione, Via
Battisti 75, MESSINA, Italy

***- Alessandro RINDI, INFN, LNF, C.P. 13, 00044 FRASCATI (Roma),
Italy

* - Harald H. ROSSI, Columbia University Dept. of Radiology 630
West 168th Street, NEW YORK, N.Y. 10032, USA

- Leif SARHOLT-KRISTENSEN, University of Copenhagen H. C.
Oersted Institute Physics Lab II, Universitetsparken 5,
DK-2100 COPENHAGEN, Denmark

- Baruch ROSNER, Physics Department Technion Israel Institute
of Technology, 32000 HAIFA, Israel

* - Walter SCHIMMERLING, University of California, Lawrence
Berkeley Laboratory Bldg. 29, BERKELEY, CA 94720, USA

* - Giovanni SILINI, CNEN, Division of Radiation Protection CNS
La Casaccia, C.P. 2400, 00100 ROMA, Italy

- Stig STEENSTRUP, University of Copenhagen, H. C. Oersted
Institute Physics Lab. II, Universitetsparken 5, DK-2100,
COPENHAGEN, Denmark

- Luis Garcia TEJERA, Private Address, c/Zaragoza 9, CADIZ,
Spain

** - Ralph H. THOMAS, University of California, Lawrence Berkeley
Laboratory Bldg. 50A, BERKELEY, CA. 94720, USA

- Guido TONELLI, Istituto di Fisica, Università di Pisa, Piazza
Torricelli 2, 56100 PISA, Italy

- Ilkka UOTILA, Institute of Radiation Protection, P.O. Box 268,
00101 HELSINKI 10, Finland

* Lecturer.
** Director of the Course.
*** Director of the School.
† Unable to attend because of illness.

LECTURERS

J. J. Broerse	Radiobiological Institute REP-Institutes Organisation Health Research TNO, Rijswijk, The Netherlands.
G. T. Y. Chen	Lawrence Berkeley Laboratory, University of California, Berkeley, California, U.S.A.
S. B. Curtis	Lawrence Berkeley Laboratory, University of California, Berkeley, California, U.S.A.
J. F. DiCello	Los Alamos Scientific Laboratory, University of California, Los Alamos, New Mexico, U.S.A.
A. Dutreix	Institut Gustave Roussy, Villejuif (Val de Marne), France.
B. Larsson	University of Uppsala, Uppsala, Sweden.
H. G. Paretzke	Gesellschaft Für Strahlen-und Umweltforschung MBH, München, West Germany
V. Perez Mendez	University of California at San Francisco, San Francisco, California, U.S.A.
H. H. Rossi	Columbia University, New York, New York, U.S.A.
W. Schimmerling	Lawrence Berkeley Laboratory, University of California, Berkeley, California, U.S.A.
G. Silini	CNEN, Division of Radiation Protection CNS La Cassaccia, Roma, Italy
R. H. Thomas	Lawrence Berkeley Laboratory, University of California, Berkeley, California, U.S.A.